Physical Properties of Foods

Novel Measurement Techniques and Applications

Contemporary Food Engineering

Series Editor

Professor Da-Wen Sun, Director

Food Refrigeration & Computerized Food Technology
National University of Ireland, Dublin
(University College Dublin)
Dublin, Ireland
http://www.ucd.ie/sun/

Physical Properties of Foods: Novel Measurement Techniques and Applications, *edited by Ignacio Arana* (2012)

Handbook of Frozen Food Processing and Packaging, Second Edition, *edited by Da-Wen Sun* (2011)

Advances in Food Extrusion Technology, *edited by Medeni Maskan and Aylin Altan* (2011)

Enhancing Extraction Processes in the Food Industry, *edited by Nikolai Lebovka, Eugene Vorobiev, and Farid Chemat* (2011)

Emerging Technologies for Food Quality and Food Safety Evaluation, *edited by Yong-Jin Cho and Sukwon Kang* (2011)

Food Process Engineering Operations, *edited by George D. Saravacos and Zacharias B. Maroulis* (2011)

Biosensors in Food Processing, Safety, and Quality Control, *edited by Mehmet Mutlu* (2011)

Physicochemical Aspects of Food Engineering and Processing, *edited by Sakamon Devahastin* (2010)

Infrared Heating for Food and Agricultural Processing, *edited by Zhongli Pan and Griffiths Gregory Atungulu* (2010)

Mathematical Modeling of Food Processing, *edited by Mohammed M. Farid* (2009)

Engineering Aspects of Milk and Dairy Products, *edited by Jane Sélia dos Reis Coimbra and José A. Teixeira* (2009)

Innovation in Food Engineering: New Techniques and Products, *edited by Maria Laura Passos and Claudio P. Ribeiro* (2009)

Processing Effects on Safety and Quality of Foods, *edited by Enrique Ortega-Rivas* (2009)

Engineering Aspects of Thermal Food Processing, *edited by Ricardo Simpson* (2009)

Ultraviolet Light in Food Technology: Principles and Applications, *Tatiana N. Koutchma, Larry J. Forney, and Carmen I. Moraru* (2009)

Advances in Deep-Fat Frying of Foods, *edited by Serpil Sahin and Servet Gülüm Sumnu* (2009)

Extracting Bioactive Compounds for Food Products: Theory and Applications, *edited by M. Angela A. Meireles* (2009)

Advances in Food Dehydration, *edited by Cristina Ratti* (2009)

Optimization in Food Engineering, *edited by Ferruh Erdoğdu* (2009)

Optical Monitoring of Fresh and Processed Agricultural Crops, *edited by Manuela Zude* (2009)

Food Engineering Aspects of Baking Sweet Goods, *edited by Servet Gülüm Sumnu and Serpil Sahin* (2008)

Computational Fluid Dynamics in Food Processing, *edited by Da-Wen Sun* (2007)

Physical Properties of Foods

Novel Measurement Techniques and Applications

Edited by
Ignacio Arana

CRC Press is an imprint of the
Taylor & Francis Group, an **informa** business

CRC Press
Taylor & Francis Group
6000 Broken Sound Parkway NW, Suite 300
Boca Raton, FL 33487-2742

First issued in paperback 2016

ISBN 13: 978-1-138-19848-7 (pbk)
ISBN 13: 978-1-4398-3536-4 (hbk)

Visit the Taylor & Francis Web site at
http://www.taylorandfrancis.com

and the CRC Press Web site at
http://www.crcpress.com

Contents

Series Preface

CONTEMPORARY FOOD ENGINEERING

Food engineering is the multidisciplinary field of applied physical sciences combined with the knowledge of product properties. Food engineers provide the technological knowledge transfer essential to the cost-effective production and commercialization of food products and services. In particular, food engineers develop and design processes and equipment in order to convert raw agricultural materials and ingredients into safe, convenient, and nutritious consumer food products. However, food engineering topics are continuously undergoing changes to meet diverse consumer demands, and the subject is being rapidly developed to reflect market needs.

In the development of food engineering, one of the many challenges is to employ modern tools and knowledge, such as computational materials science and nanotechnology, to develop new products and processes. Simultaneously, improving food quality, safety, and security continue to be critical issues in food engineering study. New packaging materials and techniques are being developed to provide more protection to foods, and novel preservation technologies are emerging to enhance food security and defense. Additionally, process control and automation regularly appear among the top priorities identified in food engineering. Advanced monitoring and control systems are developed to facilitate automation and flexible food manufacturing. Furthermore, energy saving and minimization of environmental problems continue to be important food engineering issues, and significant progress is being made in waste management, the efficient utilization of energy, and the reduction of effluents and emissions in food production.

The *Contemporary Food Engineering Series*, consisting of edited books, attempts to address some of the recent developments in food engineering. Advances in classical unit operations in engineering applied to food manufacturing are covered as well as such topics as progress in the transport and storage of liquid and solid foods; heating, chilling, and freezing of foods; mass transfer in foods; chemical and biochemical aspects of food engineering and the use of kinetic analysis; dehydration, thermal processing, nonthermal processing, extrusion, liquid food concentration, membrane processes, and applications of membranes in food processing; shelf life, electronic indicators in inventory management; sustainable technologies in food processing; and packaging, cleaning, and sanitation. The books are aimed at professional food scientists, academics researching food engineering problems, and graduate-level students.

The books' editors are leading engineers and scientists from many parts of the world. All the editors were asked to present their books to address the market's need and pinpoint the cutting-edge technologies in food engineering.

All contributions are written by internationally renowned experts who have both academic and professional credentials. All authors have attempted to provide critical, comprehensive, and readily accessible information on the art and science of a relevant topic in each chapter, with reference lists for further information. Therefore, each book can serve as an essential reference source to students and researchers in universities and research institutions.

Da-Wen Sun
Series Editor

Preface

Foods are characterized by their physical properties. These properties intensely affect the quality of foods and can be used to classify or identify them. Formerly, the quality of a food was given by its geometric characteristics, but is now evaluated as *total quality*, and takes into account the entire spectrum of physical properties of foods. In addition, in a globalized market, foods must be differentiated to better compete and the differentiation has to be based on their physical properties. Thus, it is necessary to characterize the properties of foods and to evaluate them by means of physical parameters. These parameters should be able to be measured objectively, quickly, individually, at a low cost, and in a manner that will not destroy the food; the measurement methods should be applicable on-line.

Many technical publications about the physical properties of foods have appeared for years, but the techniques of measurement are continuously evolving and have not been sufficiently described in the literature. Thus, it is appropriate to publish *Physical Properties of Foods: Novel Measurement Techniques and Applications*. The book is divided into two parts. Part 1 deals with principles and measurement techniques of the main physical parameters, highlighting the newest techniques and their ability to replace the traditional ones. Part 2 covers their application to the measurement of these parameters and to classify, identify, and differentiate various foods, for example, fruits, vegetables, cereals, meat and meat products, and dairy products.

Physical Properties of Foods: Novel Measurement Techniques and Applications is written by international peers who have both academic and professional credentials, highlighting the truly international nature of work. It aims to provide the engineer, the researcher, the food market, and the food industry with critical and readily accessible information on the modern techniques for measuring physical parameters affecting food quality and food characterization. The book should also serve as an essential reference source to undergraduate and postgraduate students and researchers in universities and research institutions.

Series Editor

Professor Da-Wen Sun, PhD, was born in Southern China and is a world authority on food engineering research and education; he is a Member of Royal Irish Academy, which is the highest academic honor in Ireland. His main research activities include cooling, drying, and refrigeration processes and systems; quality and safety of food products; bioprocess simulation and optimization; and computer vision technology. His innovative studies on vacuum cooling of cooked meats, pizza quality inspection by computer vision, and edible films for shelf-life extension of fruits and vegetables have been widely reported in the national and international media. Results of his work have been published in nearly 600 papers including over 250 peer-reviewed journal papers (h-index = 35). He has also edited 12 authoritative books. According to Thomson Scientific's Essential Science Indicators^{SM} updated as of 1 July 2010, based on data derived over a period of ten years plus four months (1 January 2000 – 30 April 2010) from ISI Web of Science, a total of 2,554 scientists are among the top one percent of the most cited scientists in the category of Agriculture Sciences, and Professor Sun tops the list with his ranking of 31.

Dr. Sun received first-class BSc honors and an MSc in mechanical engineering, and a PhD in chemical engineering in China before working at various universities in Europe. He became the first Chinese national to be permanently employed in an Irish university when he was appointed college lecturer at National University of Ireland, Dublin (University College Dublin) in 1995, and was then continuously promoted in the shortest possible time to senior lecturer, associate professor, and full professor. Dr. Sun is now professor of food and biosystems engineering and director of the Food Refrigeration and Computerized Food Technology Research Group at University College Dublin (UCD).

As a leading educator in food engineering, Dr. Sun has contributed significantly to the field of food engineering. He has trained many PhD students who have made their own contributions to the industry and academia. He has also, on a regular basis, given lectures on the advances in food engineering at academic institutions internationally and delivered keynote speeches at international conferences. As a recognized authority in food engineering, Dr. Sun has been conferred adjunct/visiting/consulting professorships from 10 top universities in China including Zhejiang University, Shanghai Jiaotong University, Harbin Institute of Technology, China Agricultural University, South China University of Technology, and Jiangnan University. In recognition of his significant contribution to food engineering worldwide and for his outstanding leadership in the field, the International Commission of Agricultural and

Biosystems Engineering (CIGR) awarded him the CIGR Meri Award in 2000 and again in 2006; the Institution of Mechanical Engineers based in the United Kingdom named him Food Engineer of the Year 2004; in 2008 he was awarded the CIGR Recognition Award in recognition of his distinguished achievements as the top 1% of agricultural engineering scientists around the world; in 2007, Dr. Sun was presented with the only AFST(I) Fellow Award in that year by the Association of Food Scientists and Technologists (India); and in 2010, he was presented with the CIGR Fellow Award; the title of "Fellow" is the highest honor in CIGR, and is conferred to individuals who have made sustained, outstanding contributions worldwide.

Dr. Sun is a fellow of the Institution of Agricultural Engineers and a fellow of Engineers Ireland (the Institution of Engineers of Ireland). He has also received numerous awards for teaching and research excellence, including the President's Research Fellowship, and has received the President's Research Award from the University College Dublin on two occasions. He is editor-in-chief of *Food and Bioprocess Technology—An International Journal* (Springer) (2010 Impact Factor = 3.576, ranked at the 4th position among 126 ISI-listed food science and technology journals); series editor of *Contemporary Food Engineering Series* (CRC Press/Taylor & Francis); former editor of *Journal of Food Engineering* (Elsevier); and an editorial board member for *Journal of Food Engineering* (Elsevier), *Journal of Food Process Engineering* (Blackwell), *Sensing and Instrumentation for Food Quality and Safety* (Springer), and *Czech Journal of Food Sciences*. Dr. Sun is also a chartered engineer.

On May 28, 2010, Dr. Sun was awarded membership to the Royal Irish Academy (RIA), which is the highest honor that can be attained by scholars and scientists working in Ireland. At the 51st CIGR General Assembly held during the CIGR World Congress in Quebec City, Canada, in June 2010, he was elected as incoming president of CIGR, and will become CIGR president in 2013–2014; the term of his presidency is six years, two years each for serving as incoming president, president, and past president. On 20 September 2011, he was elected to Academia Europaea (The Academy of Europe), which is functioning as European Academy of Humanities, Letters and Sciences and is one of the most prestigious academies in the world, election to the Academia Europaea represents the highest academic distinction.

Contributors

Ana Cristina Agulheiro Santos
Universidade de Évora
Évora, Portugal

Nuria Aleixos
Instituto Interuniversitario de Investigación en Bioingeniería y Tecnología Orientada al Ser Humano
Universidad Politécnica de Valencia
Valencia, Spain

Alejandro Arana
Risk MR Pharmacovigilance Services
Zaragoza, Spain

J. Ignacio Arana
Universidad Pública de Navarra
Navarra, Spain

Silvia Arazuri
Universidad Pública de Navarra
Navarra, Spain

Iñigo Arozarena
Universidad Pública de Navarra
Navarra, Spain

Maria José Beriain
Escuela Técnica Superior de Ingenieros Agrónomos
Universidad Pública de Navarra
Navarra, Spain

José Blasco
IVIA- Instituto Valenciano de Investigaciones Agrarias
Centro de Agroingeniería
Valencia, Spain

Gloria Bobo
Universidad Pública de Navarra
Navarra, Spain

Paulo Cesar Corrêa
Department of Agricultural Engineering
Federal University of Viçosa
Viçosa – MG, Brazil

Emílio de Souza Santos
Department of Agricultural Engineering
Federal University of Viçosa
Viçosa – MG, Brazil

Satyanarayan R. S. Dev
Department of Bioresource Engineering
McGill University
Quebec, Canada

Belén Diezma
Universidad Politécnica de Madrid
Madrid, Spain

Abelardo Gutierrez
Fundación AgroAlimed
Unidad Investigación IVIA-Fundación AgroAlimed Centro de Agroingeniería
Valencia, Spain

Begoña Hernández Salueña
Universidad Pública de Navarra
Navarra, Spain

Gabriel Henrique Horta de Oliveira
Department of Agricultural Engineering
Federal University of Viçosa
Viçosa – MG, Brazil

Asunción Iguaz
Universidad Pública de Navarra
Navarra, Spain

Kizkitza Insausti
Escuela Técnica Superior de Ingenieros Agrónomos
Universidad Pública de Navarra
Navarra, Spain

Enrique Moltó
IVIA- Instituto Valenciano de Investigaciones Agrarias
Centro de Agroingeniería
Valencia, Spain

Maria José Noriega
Universidad Pública de Navarra
Navarra, Spain

Maria Dolores Pérez
Tecnología de los Alimentos
Facultad de Veterinaria
Universidad de Zaragoza
Zaragoza, Spain

G. S. Vijaya Raghavan
Department of Bioresource Engineering
McGill University
Quebec, Canada

Cristina Roseiro
Universidade de Évora
Évora, Portugal

Margarita Ruiz-Altisent
Universidad Politécnica de Madrid
Madrid, Spain

Carlos Sáenz Gamasa
Universidad Pública de Navarra
Navarra, Spain

Lourdes Sánchez
Tecnología de los Alimentos
Facultad de Veterinaria
Universidad de Zaragoza
Zaragoza, Spain

Maria Victoria Sarriés
Escuela Técnica Superior de Ingenieros Agrónomos
Universidad Pública de Navarra
Navarra, Spain

Pedro Casanova Treto
Department of Agricultural Engineering
Federal University of Viçosa
Viçosa – MG, Brazil

Paloma Virseda
Universidad Pública de Navarra
Navarra, Spain

1 Basics of Electronic, Nondestructive Technologies for the Assessment of Physical Characteristics of Foods

Abelardo Gutierrez, José Blasco, and Enrique Moltó

CONTENTS

1.1 ELECTRONIC DEVICES FOR MEASURING PHYSICAL PROPERTIES OF FOOD

Increasing awareness of quality and enhanced perspicacity of consumers are leading a strong drive for improved quality of fruits and vegetables, in both the fresh market and the food industry. In this sense, food technologists as well as agricultural engineers are interested in physical properties of food materials in order to determine how foods or fruits will handle during processing, to get an indication of the product quality, and to understand why consumers prefer certain foods or fruits. Quality of produce encompasses sensory properties, nutritive values, chemical constituents, mechanical properties, functional properties, and lack of defects, each of which has been the subject of many studies.

The need to measure the physical properties of food also arises from the increased regulatory action and heightened consumer concerns about food and fruit safety. It was estimated that this industry spends on average 1.5–2.0% of the value of its total sales on quality control and appraisal (Luong et al., 1997). Moreover, the trend toward continuous automated production in place of human-assisted operation necessitates the measurement of food properties, particularly in the area related to on-line or rapid at-line process control applications. Automation not only optimizes quality assurance, but more importantly, it helps to remove human subjectivity and inconsistency. It usually increases the productivity and changes the character of factory or farmworkers, making it less arduous and more attractive. The fact that the productivity of man working in mechanized and automated environments is approximately ten times that of manual workers has stimulated progress in the development of many novel sensors and instruments for the food industry, often by technology transfer from other industrial sectors, including medical and nonclinical sectors.

Agricultural products have low unit values and economic viability, and industrialized agriculture generally implies high-speed automation. One example was provided by Gall et al. (1998) for the UK potato industry. The market standards were still increasing and tolerance values reducing. To be viable, automated sorting systems should be able to handle 100,000 or more potatoes in 1 hour.

Thus, there is an increased need for better quality monitoring, and this has to be based on electronic sensors. Over the years, many electronic devices have been developed, most of them trying to mimic human sensory characteristics, in order to automatically measure quality and quality-related attributes and often oriented toward real-time and nondestructive testing.

Electronics has been widely adopted in fruit grading machines since the 1980s, initially to support the use of load cells, for more accurate and faster weighing than the mechanical predecessors. Widespread use of photocells and microprocessors enabled the development of color-sorting systems in the 1980s. They were followed quickly by machine vision–based systems that used charge-coupled device (CCD) cameras and PC frame grabbers for sorting produce based on shape and color. Implementation of morphometric algorithms, allowed estimation of fruit volume to be implemented.

Although machines were initially marketed with vision or weighing capabilities, the combination of the two capabilities proved to be useful for the assessment of produce density. In the mid-1990s, automated blemish detection systems, employing more sophisticated algorithms (e.g., neural networks, fuzzy logics, etc.) provided new utilities to machines based on image processing. At the same time, the fact that computer images are 2D projections of a 3D scene, made researchers envisage different methods to obtain information on the whole surface of the produce, that is, imaging the fruits when they rotate on rollers or the use of mirrors.

Furthermore, use of wavelengths outside the visible spectral range, or combinations of visible and invisible information (ultraviolet, near-infrared) from images is becoming commercially available. A wide range of noninvasive imaging technologies (x-ray, ultrasonics, magnetic resonance, etc.) are currently used in medicine, and soon they will be available in the agri-food sector. Widespread adoption of soft x-ray inspection systems by the security industry is driving prices down for this technology. It is likely that this technology will migrate to the food sorting industry at some stage.

Spectrophotometry is also capable of extracting information related to the chemical composition of fruit and vegetables, thus providing data to estimate the quality of these products. Other technologies, such as electrical impedance, chlorophyll fluorescence, gas sensors, and biosensors are under development in order to generate more complete and accurate assessment of quality.

This chapter reviews the electronics basis of these devices in order to facilitate the understanding of the following chapters.

1.2 SPECTROPHOTOMETERS

Spectrophotometers are employed to measure the amount of light that a sample absorbs or reflects. A beam of monochrome light is directed toward a sample and the intensity of light reaching a detector situated on the opposite part (absorption measurement) or at a certain angle (normally 45º, reflection measurement) is quantified. Thus, a spectrophotometer consists of two parts—a light source whose spectrum is decomposed in narrow bands of certain wavelengths, and a **photometer** for measuring the intensity of the monochromatic light coming from the object.

Historically, spectrophotometers use a monochromator, containing diffraction grating to produce the analytical spectrum composed of narrow bands of specific and close wavelengths by changing the position of the grating with respect to the incident light. The single photometer is a photomultiplier tube.

Photomultiplier tubes consist of an evacuated envelope with a photocathode that emits electrons when exposed to light. There are two main types of photomultiplier tubes: *side-on photomultipliers* and *end-on photomultipliers*. Side-on detectors are more economical than end-on models, and have faster rise times. Their vertical configuration takes up less space than the end-on versions and they mount in standard or pulsed housings. The main disadvantage of these photomultiplier tubes is their nonuniform sensitivity. End-on photomultiplier tubes, sometimes known as *head-on photomultipliers*, offer better spatial uniformity and photosensitive areas from tens of square millimeters to hundreds of square centimeters (Hamamatsu Photonics, 2006).

Some modern ultraviolet-visible (UV-Vis) spectrometers use solid-state detector arrays instead of a single detector. These solid-state detectors (photodiodes) are not as sensitive as the conventional detector (photomultiplier tubes), but they are much smaller and can be made into an array and can detect a large number of wavelength elements at the same time. By measuring all the wavelength elements within the whole spectrum simultaneously, spectral data are acquired much faster.

Figure 1.1 shows a schematic of an absorbance spectrophotometer with diffraction grating and a single photometer. The instruments are arranged so that liquid in a cuvette can be placed between the light source, whose wavelengths have been separated by a monochromator, and the photometer. The changes in wavelengths are produced by changing the orientation of the monochromator with respect to the source. The intensity of light passing through the cuvette is measured by the photometer. The

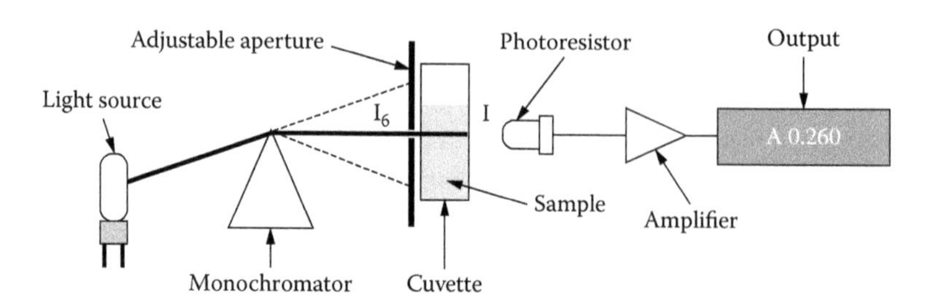

FIGURE 1.1 Basics of a spectrophotometer: spectrometer and photometer. (From http://commons.wikimedia.org/wiki/User:YassineMrabet/Gallery.)

photometer delivers a voltage signal to a display. The signal changes as the amount of light absorbed by the liquid changes depending on the incident wavelength.

When monochromatic light (light of a specific wavelength) passes through a solution there is usually a quantitative relationship (Beer's law) between the solute concentration and the intensity of the transmitted light, which is given by Equation (1.1),

$$I = I_0 * 10^{-kel} \tag{1.1}$$

where I_0 is the intensity of transmitted light using the pure solvent, I is the intensity of the transmitted light when the studied compound is added, c is concentration of the studied compound, l is the distance the light passes through the solution, and k is a constant. If the light path l is a constant, as is the case with a spectrophotometer, Beer's law may be written as Equation (1.2),

$$I \div I_0 = 10^{-kc} = T \tag{1.2}$$

where k is a new constant and T is the transmittance of the solution. There is a logarithmic relationship between transmittance and the concentration of the studied compound (Equation [1.3]). Thus,

$$-\log T = \log 1/T = kc = \text{optical density (O.D.)} \tag{1.3}$$

O.D. is directly proportional to the concentration of the studied compound. Most spectrophotometers have a scale that reads both in O.D. (absorbance) units, which is a logarithmic scale, and in percentage of transmittance, which is an arithmetic scale.

In addition to the familiar absorbance measurements described above, there are other types of measurements that can be carried out using UV-Vis spectrophotometers. Among such additional measurements are those that are based on the ability of a spectrophotometer to measure the reflectance of materials. Reflectance measurements are of great value in providing a reference standard for the comparison of the color of different samples.

A reflectance spectrophotometer is similar to a standard UV-Vis spectrophotometer. It should have a bandwidth narrow enough to provide well-resolved visible spectra yet wide enough to provide a good energy level for diffuse reflectance

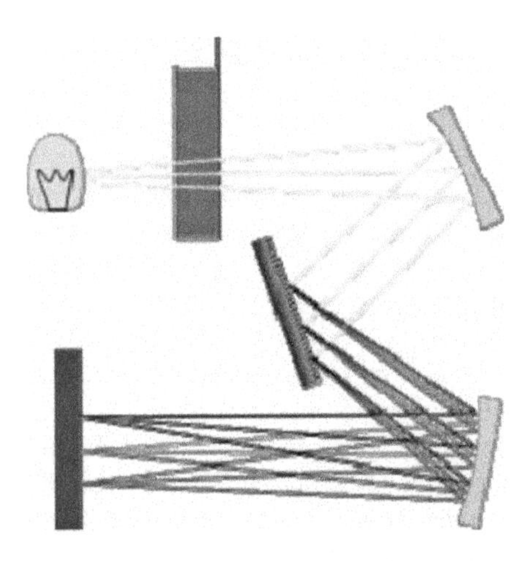

FIGURE 1.2 Diagram of a single-beam UV/Vis spectrophotometer. (From Freeman V S, (1995). Spectrophotometry. *In Laboratory instrumentation*, ed. Haven, M.C., Tetrault G A, and J.A. Schenken, 72–96. New York: John Wiley & Sons.)

measurements. The reflectance spectrophotometer must also have optics and electronics systems of high sensitivity, and should be able to physically accommodate reflectance and transmission accessories. The adapted spectrophotometer must be able to make measurements both at selected fixed wavelengths or perform scans over the complete wavelength range.

There are two major classes of devices: single-beam and double-beam spectrophotometers. A double-beam spectrophotometer compares the light intensity between two light paths, one path containing a reference sample and the other the test sample. A single-beam spectrophotometer (Figure 1.2) measures the relative light intensity of the beam before and after a test sample is inserted. Although comparison measurements from double-beam instruments are easier and more stable, single-beam instruments can have a larger dynamic range, are optically simpler, and are more compact (Nielsen, 2010).

According to electromagnetic spectra, spectrophotometers can be divided in the two groups discussed in the next two sections.

1.2.1 Ultraviolet-Visible (UV-Vis) Spectrophotometers

The most common spectrophotometers are used in the UV and visible regions. Most of these instruments also operate into the near-infrared region. Traditional visual region spectrophotometers cannot detect if a colorant or the base material has fluorescence. In this case, a bispectral fluorescent spectrophotometer is used (Nielsen, 2010).

This device measures the intensity of light passing through a sample (I), and compares it to the intensity of light before the sample (I_0). The ratio I / I_0 is called the *transmittance*, and is usually expressed as a percentage (%T). The absorbance, A, is based on the transmittance (Equation [1.4]):

$$A = -\log(\%T / 100\%) \tag{1.4}$$

The radiation source is often a Tungsten filament (typical radiation between 300 and 2500 nm wavelengths). For studies in the UV region, deuterium arc lamps (190–400 nm) are used. More recently, light-emitting diodes (LEDs) and xenon arc lamps are used for the visible wavelengths. Typically, photosensors are either single photodiodes or CCD.

1.2.2 Infrared (IR) Spectrophotometers

Spectrophotometers designed for the mid- and far-infrared regions are quite different because of the technical requirements of measurement in those regions, especially on longer wavelengths (more than 5 μm), when the thermal radiation is reached. Many materials, such as glass and plastic, absorb infrared light, making them incompatible as optical media. Ideal optical materials are salts, which do not effectively absorb IR radiation (Tetrault & Schenken, 1995).

In Fourier transform infrared (FTIR) spectroscopy, infrared light is guided through an interferometer and then through the sample (or vice versa). A moving mirror inside the apparatus alters the distribution of infrared light wavelengths that pass through the interferometer. The signal directly recorded, called an interferogram, represents light output as a function of mirror position (horizontal axis). A data-processing technique called Fourier transforms this raw data (vertical axis) into the desired result—light output as a function of infrared wave number.

1.3 VIDEO CAMERAS

Cameras are devices that convert UV, visible, or near-infrared (NIR) radiation coming from a scene in an electronic signal that can be input into a computer. The scene is projected by a set of lenses that refract the incident radiation and concentrate it in on a photosensitive device.

Most popular industrial cameras are based on a charge-coupled device (CCD). Basically, a CCD is composed of multiple capacitors that receive light from a particular location of a projected scene as input, and accumulate an electric charge that is proportional to the light intensity at that location. Each of these capacitors is called a *pixel* (abbreviation of picture element). Their electric charge is subsequently converted to a standard electronic signal.

Some cameras are based on a linear CCD, composed of a one-dimensional array of capacitors that acquire a narrow strip of the scene. These cameras, known as *line-scan cameras*, are suitable for use in applications where the object is moved below the camera or the camera moves above the object, so that the complete image of its surface is gradually acquired line by line.

Matrix cameras use a two-dimensional CCD and are currently the most commonly used in industry and research. They acquire images by using a bidimensional CCD (Figure 1.3), which consists of an array of sensors (pixels), each of which is a photocell and a condenser (Janesick, 2001). The load acquired by the condenser depends on the amount of light received by the photocell. In the CCD, these charges are converted into voltage and subsequently converted into a video signal.

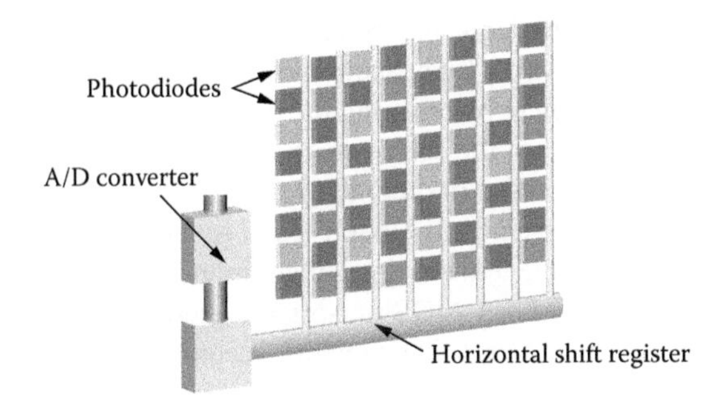

FIGURE 1.3 Schematic of a visible CCD camera.

The most advanced cameras have three different monochromatic CCD sensors. Light enters through the lens and is divided into three equal parts by a prism, then directed to each CCD. Before reaching it, the light is conveniently filtered, so that each CCD represents the red, green, and blue bands (Figure 1.4). Greater color fidelity is obtained with these cameras than with those based on a single CCD, but they are more expensive. Three CCD cameras are currently the most widely used in research and many examples are found in the scientific literature (Cubero et al., 2010).

CCDs in matrix cameras can be interlaced or noninterlaced (progressive scan cameras). Interlaced cameras first scan odd lines of pixels and then even ones, building what are known as the odd and even fields. This is done to preserve the sensation of movement in the human eye while watching video on a screen. However, inspected food produce travels at a high speed under the camera, so both fields are displaced in relation to each other, which deforms the shape of the objects and complicates the analysis of the image.

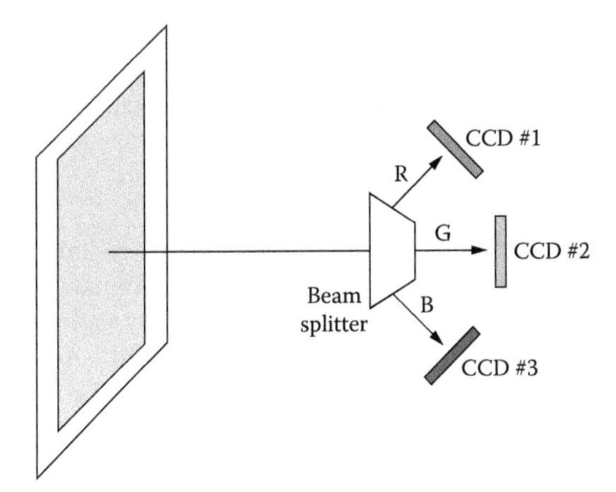

FIGURE 1.4 Schematics of a three CCD camera. (From http://www.kodak.com/global/plugins/acrobat/en/business/ISS/supportdocs/chargeCoupledDevice.pdf)

In order to avoid these problems, progressive scan cameras produce noninterlaced images. This is combined with a high electronic shutter speed, which decreases the effect of the object's movement by reducing the time that the CCD is exposed to the influence of light coming from the scene. As the shutter speed increases, the intensity of lighting must be increased in order to avoid underexposure.

Newer cameras use complementary metal-oxide semiconductor (CMOS) sensors in a similar way, but consume less energy and are more economical to manufacture (Wang and Nguang, 2007). For CMOS sensors the production process is the same as for all microprocessors. With CCD cameras, incoming photons create negative charge carriers, which have to be integrated over a certain time, which is determined by shutter speed. By contrast, CMOS sensors are, in principle, photosensitive diodes that are in series with a resistor. While in CCD cameras electrons have to be transported via shift registers, the principle of the CMOS camera allows a continuous transformation of the incoming photons into a resulting voltage. A CMOS sensor is nothing but an array of exposure meters.

The enlargement of the size of images due to the great increase of computational power since 2000 has brought about an improvement in the performance of machine vision systems, since it is now possible to detect small defects of a few mm^2. Another important advancement is related to the spread of high-speed protocols for data transfer between external devices and computers, like the universal serial bus (USB) (Axelson, 2005) or FireWire IEEE (Institute of Electrical and Electronics Engineers) (Shelly and Vermaat, 2010). These have modified the traditional configuration of a video camera to include an acquisition board (which initially was basically an analog-to-digital converter) to direct communication between the camera and the computer.

Intelligent cameras are also reaching the market. They incorporate a microprocessor that allows analysis of the image inside the camera. The spread of the Internet has increased the availability of so-called IP cameras, which transfer the images using a network standard like TCP/IP, and allow the control of the camera via the Internet.

Although the camera-based technology is the most widespread in machine vision for agricultural applications, there are other technologies, such as flatbed scanners (Evening, 2005), that allow the acquisition of images of small objects like nuts or leaves.

1.4 X-RAY COMPUTED TOMOGRAPHY

X-ray computed tomography (CT) is a nondestructive technique for visualizing interior features within solid objects, and for obtaining digital information on their 3D geometries and properties. A CT image is typically called a *slice*, as it corresponds to what the object being scanned would look like if it were sliced open along a plane. A CT slice corresponds to a certain thickness of the object being scanned. So, while a typical digital image is composed of pixels, a CT slice image is composed of *voxels* (volume elements). A complete volumetric representation of an object is obtained by acquiring a contiguous set of CT slices.

The gray levels in a CT slice image correspond to x-ray attenuation, which reflects the proportion of x-rays scattered or absorbed as they pass through each voxel. X-ray attenuation is primarily a function of x-ray energy and the density and composition of the material being imaged. Tomography imaging consists of directing x-rays at an object from multiple orientations and measuring the decrease in intensity along a series of linear paths. This decrease is characterized by Beer's law, which describes intensity reduction as a function of x-ray energy, path length, and material linear attenuation coefficient. A specialized algorithm is then used to reconstruct the distribution of x-ray attenuation in the volume being imaged (ASTM, 1992).

The elements of x-ray tomography are an x-ray source, a series of detectors that measure x-ray intensity attenuation along multiple beam paths and a rotational geometry with respect to the object being imaged. Different configurations of these components can be used to create CT scanners optimized for imaging objects of various sizes and compositions.

The great majority of CT systems use x-ray tubes as sources, although tomography can also be done using a synchrotron or gamma-ray emitter as a monochromatic x-ray source. Important tube characteristics are the target material and peak x-ray energy, which determine the x-ray spectrum that is generated; current, which determines x-ray intensity; and the focal spot size, which is related to the spatial resolution (ASTM, 1992).

Most CT x-ray detectors utilize scintillators (Ketchman & Carlson, 2001). Important parameters of the scintillator are material, size, geometry, and the means by which scintillation events (a flash of light produced in a transparent material by one ionization event) are detected and counted. In general, smaller detectors provide better image resolution, but reduced count rates because of their reduced area compared to larger ones. To compensate, longer acquisition times are used to reduce noise levels (Ketchman & Carlson, 2001).

Figure 1.5 shows some of the most common configurations for CT scanners. In planar-beam scanning, x-rays are collimated and measured using a linear detector array. Typically, slice thickness is determined by the aperture of the linear array. Collimation is necessary to reduce the influence of x-ray scatter, which results in spurious additional x-rays reaching the detector from locations not along the source–detector path. Linear arrays can generally be configured to be more efficient than planar ones, but have the drawback that they only acquire data for one slice image at a time (ASTM, 1992).

In cone-beam scanning, the linear array is replaced by a planar detector, and the beam is no longer collimated. Data for an entire object, or a considerable thickness of it, can be acquired in a single rotation. The data are reconstructed into images using a cone-beam algorithm. In general, cone-beam data are subject to some blurring and distortion as the distance to the central plane increases. They are also more subject to artifacts stemming from scattering if high-energy x-rays are utilized. However, the advantage of obtaining data for hundreds or thousands of slices at a time is considerable, as more acquisition time can be spent at each turntable position, decreasing image noise (Scannavino & Cruvinel, 2009).

Parallel-beam scanning is performed using a specially configured synchrotron beam line as the x-ray source. In this case, volumetric data are acquired without distortion. However, the object size is limited by the width of the x-ray beam.

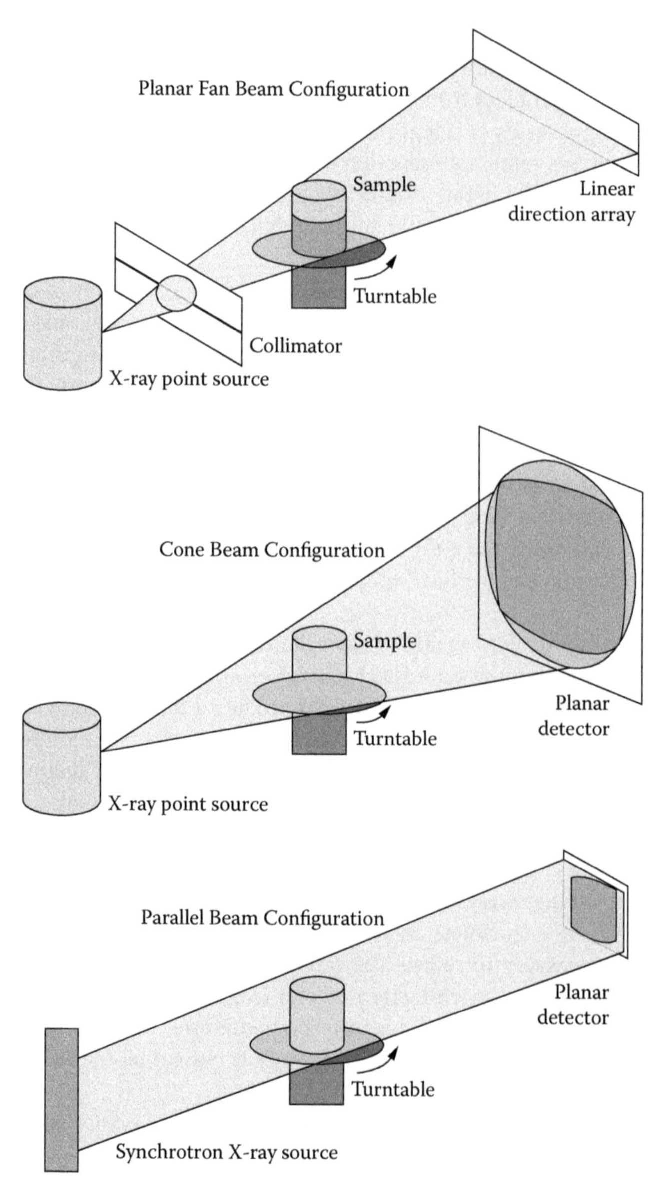

FIGURE 1.5 Configurations for CT x-ray scanners. (From ASTM (American Society for Testing and Materials). 1992. Standard guide for computed tomography (CT) imaging, ASTM Designation E 1441-92a. In *1992 Annual Book of ASTM Standards*, Section 3 Metals Test Methods and Analytical Procedures, 690–713. Philadelphia: ASTM.)

Depending on beam line configuration, objects up to 6 cm in diameter may be imaged. Synchrotron radiation generally has very high intensity, allowing data to be acquired quickly, but the x-rays have generally low energy (< 35 keV), which can preclude imaging samples with thick materials (Barcelon et al., 1999).

Other variants are multiple-slice acquisition, in which a planar detector is used but data are processed with a fan-beam reconstruction algorithm and spiral scanning in which sample elevation is changed during data acquisition, potentially reducing cone-beam artifacts (Ketchman & Carlson, 2001).

1.5 MAGNETIC RESONANCE IMAGING (MRI)

MRI is actually a map of very weak magnetization, originated from some of atomic nuclei in the body tissue, in the presence of an external magnetic field. Since this magnetization is proportional to the density of those nuclei, MRI shows the distribution of the selected atoms. Hydrogen atoms are the most studied in MRI, since many materials contain large amounts of water molecules. Softer tissues are generally seen easily in proton MR images that show even small differences in chemical composition.

Protons are subatomic particles having a single positive electrical charge that constitute the nucleus of the ordinary hydrogen atom. They have a rotation movement due to a quantum mechanical property, called *spin*, around its own axis. Thus, a single atomic nucleus can be thought of as a spinning charged body, which acts as a tiny magnet. An external magnetic field can exert torque on the nucleus, which forces alignment of the nuclear magnetic field with the external field (Figure 1.6). Since the nucleus is spinning, it will precess about the magnetic field instead of aligning with it. The angle of precession of the nucleus's magnetic field is quantized (due to the quantization of angular momentum), which is often referred to as being *up* or *down*.

The external magnetic field modifies the natural equilibrium of the energies of the spin states of the nuclei of the material. Nuclei whose spin is aligned with the applied field drop in energy; the lower energy state spin is more frequent than the higher energy state. Electromagnetic radiation of appropriate frequency (with energy that can cover the gap between the up and down states) applied to the nuclei make them resonate, thus producing rapid switches of nuclei states. This change of state may be detected with accurate apparatus since the required frequency is in the radio waves range (McRobbie et al., 2003). The strength of the transmitted energy is proportional to the number of protons in the tissue. This strength is also modified by properties

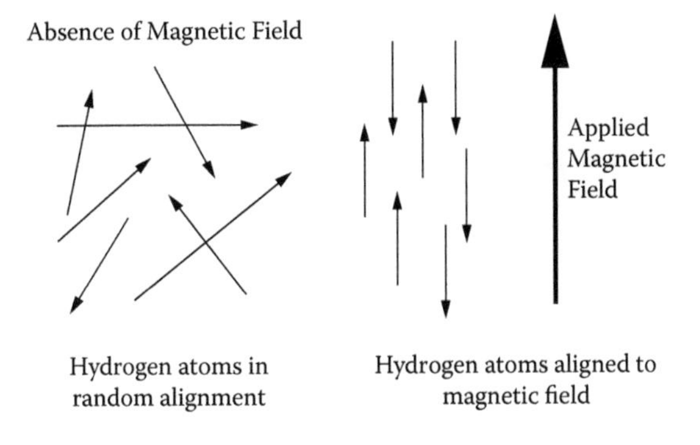

FIGURE 1.6 Principle of operation of magnetic resonance imaging. (From Neuroimaging Primer, Keith A. Johnson, M.D., Harvard Medical School.)

of each proton's microenvironment, such as its mobility and the local homogeneity of the magnetic field. MR signal can be *weighted* to accentuate some properties and not others.

The frequency of radiation at which resonance occurs is distinct for each nucleus and depends on its gyromagnetic ratio and the local environment. It is possible, therefore, to distinguish between (for example) the carbonyl and the methyl carbon atoms in ethanal (acetaldehyde). In this manner, nuclear magnetic resonance (NMR) is an extremely useful analytical tool for chemists, as it allows the chemical structure of an unknown compound to be probed to the degree that the structure may be completely deduced (Hill, 1998).

When an additional, convenient spatially variable magnetic field is superimposed, each voxel of the material has a unique radio frequency at which the electromagnetic signal is received and transmitted. This makes it possible to generate a tomographic image of the sample (Figure 1.7).

As has been said, protons in a magnetic field oscillate and are capable of absorbing electromagnetic energy of the same frequency of oscillation. After they absorb energy, they release it to return to their initial equilibrium state. This reradiation is what is observed as the MRI signal. The return of the nuclei to their equilibrium state does not take place instantaneously, but rather takes place over some time and is governed by two physical processes:

- One that is due to the component of the nuclear magnetization parallel to the external magnetic field that requires T1 time
- Another one that is due to the component of the nuclear magnetization perpendicular to the magnetic field that requires T2 time

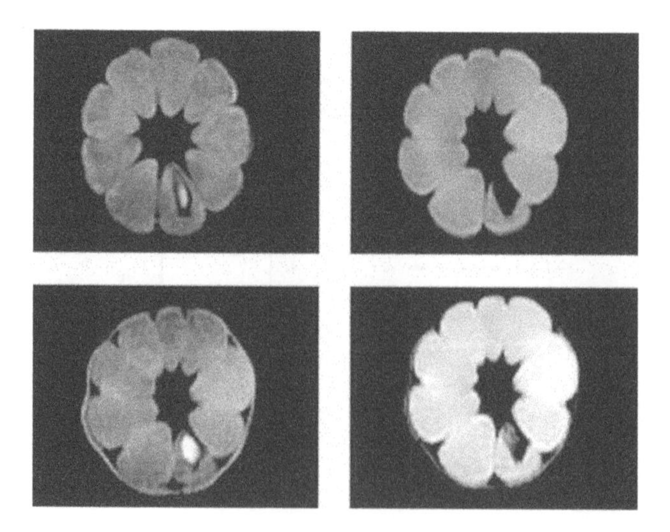

FIGURE 1.7 MR images of a mandarin with one seed inside, acquired using different MR sequences.

T1 and T2 can be dramatically different for different soft tissues and for this reason may be used to increase the contrast between soft tissues in an image. Moreover, T1 and T2 are strongly influenced by the viscosity or rigidity of tissues. The greater the viscosity and rigidity, the smaller the value will be for T1 and T2 (Hill, 1998).

Summarizing, the strength and characteristics of the MRI signal from a voxel depend primarily on the proton density and on T1 and T2, so it is possible to manipulate the MR signal by changing the way in which the nuclei are initially subjected to electromagnetic energy. Hence, one has a number of different MR imaging techniques (weightings) to choose from, which accentuate some properties and not others.

As has been said, MRI machines use a stable magnetic field and a variable magnetic field. Stable fields are produced by three types of magnets: resistive, permanent, and superconducting (McRobbie et al., 2003). Resistive magnets are based on wired coils that conduct an electric current that produces the magnetic field. They require electric power, usually up to 50 kW. Permanent magnets use magnetized materials and do not require electricity. However, they are extremely heavy. Superconducting magnets differ from the traditional resistive magnet in that their coils are submerged in liquid helium and kept at an extremely low temperature, close to absolute zero, thus increasing the conductivity of the wires and requiring much less electric power to produce high-intensity magnetic fields.

In all three cases, variable fields are produced by three gradient magnets that have lower power than the main magnet and are much smaller.

1.6 ULTRASONIC SCANNING SYSTEMS

Ultrasonic waves are mechanical or acoustical waves of frequency ≥ 20 kHz, above the human hearing capabilities. With respect to sensing and measurements, high frequency avoids interference from many audible, low-frequency noises due to wind, machinery, pumps, and vibration of large bodies. High frequency allows resolution of the *small* in both the temporal and special senses. Attenuation often imposes the upper limit on the maximum usable frequency (Lynnworth, 1989). In most solids and liquid industrial materials, as well as in most biological tissues, such as in the human body, ultrasound energy is easily propagated, which facilitates diagnostic or detection procedures (Walls, 1969).

Ultrasound technology has been used for many years, its main application areas being medical diagnostics and industrial processes and inspections. At high frequencies and low power it can be used as an analytical and diagnostic tool, and at a very high power it can assist processing. Throughout the scope of its applications, ultrasounds are generated in the same way. A transducer contains a ceramic crystal that is excited by a short electrical pulse that has a typical form of several sine cycles. Due to the piezoelectric effect, this electrical energy is converted to a mechanical wave that is propagated as a short sonic pulse at the fundamental frequency of the transducer. This energy is transferred into the material or body under analysis and propagated through it (Krautkramer and Krautkramer, 1990).

Either the same or another piezoelectric element acts as a receiver, converting ultrasound waves to electrical energy. When the system operates in *pulse–echo mode*,

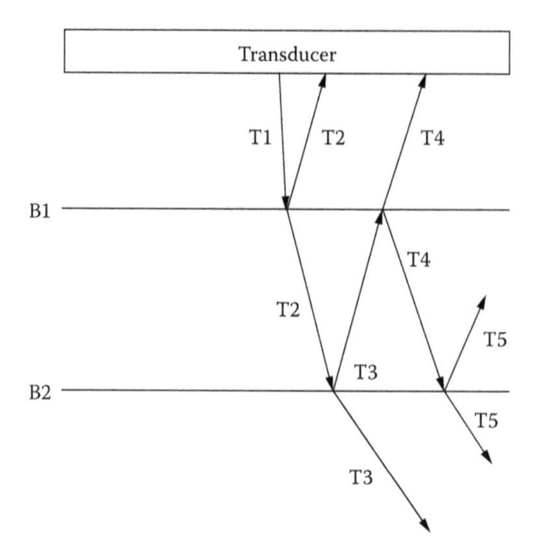

FIGURE 1.8 The reflection and refraction of sound waves from an ultrasonic scanner. (From Muzzolini, R. E. 1996. A volumetric approach to segmentation and texture characterisation of ultrasound images. PhD thesis, University of Saskatchewan.)

the same piezoelectric element acts as a transmitter and a receiver alternately; when a *through transmission mode* is used, a second piezoelectric element acts as a receiver.

Ultrasonic energy propagates through a material until the sound waves encounter an impedance change due to changes in the material density or the velocity of the sound wave (Kuttruff, 1991). This can occur inside the material, due to tissue changes, the presence of hollow areas, or reflective materials.

The amount of reflected energy depends on the impedance change or the size of the reflector. If there are no internal reflectors, waves continue until the energy is totally attenuated. The energy attenuation of the ultrasound beam and the speed of wave propagation depend on the nature of the material and its structure (Kuttruff, 1991). Most physical or chemical changes in the materials cause changes in the attenuation and velocity of the propagated beam.

Ultrasound image processing has gained support in both the medical and agricultural fields. The most common ultrasound imaging technique is known as *pulse–echo ultrasound*. A simplified example of this ultrasound technique is presented in Figure 1.8. A single sound pulse, whose primary direction is shown as T1, leaves the ultrasound transducer and proceeds to enter the subject being scanned. As the sound pulse continues outward, the density of the medium through which the pulse travels will change. This occurs most significantly at boundaries between two tissues of different densities. When the pulse reaches these types of acoustical boundaries, a portion of the pulse reflects back toward the transducer while a now weakened pulse refracts further into the subject. Once the reflected portion of the sound wave, referred to as an *echo*, returns to the transducer, its amplitude is recorded and using the formula *distance* = 2 **velocity* **time*, the depth at which the echo was produced can be determined (Muzzolini, 1996).

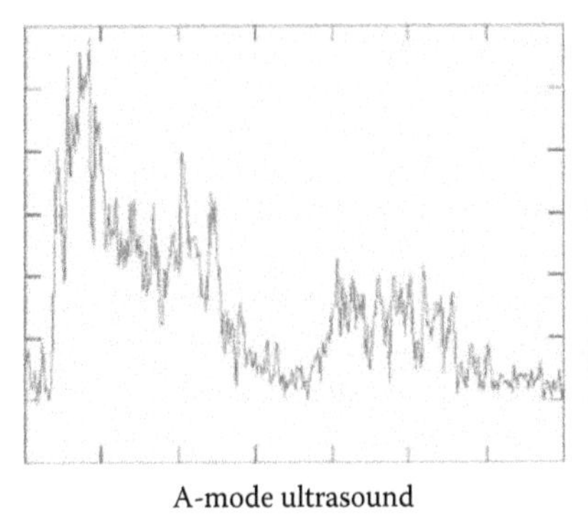

A-mode ultrasound

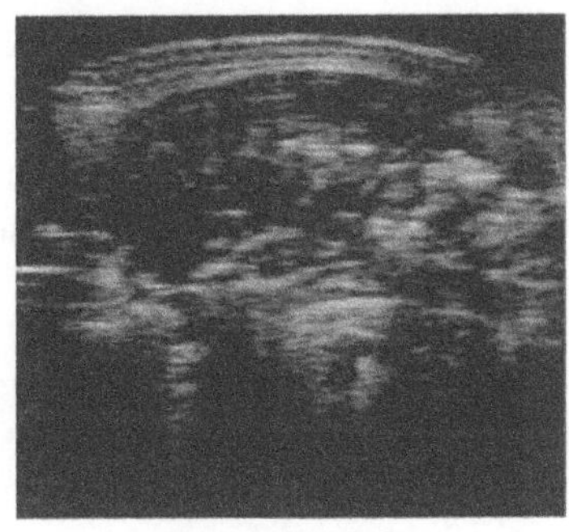

B-mode ultrasound

FIGURE 1.9 Ultrasound image types.

Using a single transducer, only one-dimensional images can be obtained, using the method known as an amplitude mode (A-Mode) scan. An example of such a scan is presented in Figure 1.9 (a). The brightness mode (B-Mode) scan is used to create two-dimensional images and requires an array of transducers. As a result, a 2D ultrasound image is an array of A-Mode ultrasound scans. An example of such is shown in Figure 1.9 (b). Pixel intensity in an ultrasound image corresponds directly to the amplitude of the echoes received by the associated transducer.

This image acquisition technique is susceptible to noise in fruit and vegetable characterization. Most biological tissues do not have a homogeneous density, resulting in echoes being recorded that are not near tissue boundaries (Mizrach, 2008). The superposition of echoes originated from different transducers also adds a particular type of high-intensity noise known as *speckle noise*. However, the most problematic issue with ultrasound in terms of noise is movement of the object.

Other imaging artifacts are due to the fact that echoes recorded by the receiver depend on the angle at which the sound pulse meets a tissue boundary. As a result, the amplitude of an echo, and in turn the pixel intensity in an ultrasound image, will be lower the less perpendicular to a tissue boundary the incident sound pulse is. Moreover, as the echo returns, it is common that a portion of the echo reflects on the transducer surface and reverberates between the transducer and the skin of the object. This produces a smooth decay in amplitude after a strong echo is received, instead of a sharp drop-off (Muzzolini, 1996). These two artifacts may prevent to detect region boundaries.

Current image segmentation algorithms applied in this field use assumptions about the appearance of a region or its boundary that are common to general image processing. Unfortunately, these segmentation approaches are very limited, due to the nature of ultrasound images. Assumptions that hold for the segmentation of other images, particularly those whose region boundaries produce a high gradient of intensity, generally do not hold for ultrasound images. For this reason, current research is directed at proposing new approaches for detecting edge points in ultrasound images (Booth et al., 2006).

1.7 ELECTRONIC ESTIMATION OF WEIGHT

Electronic weighing systems based on load cells (Figure 1.10) have made a significant difference in the reliability of food sorting processes. Thanks to electronics, the accuracy of measurement is greatly improved, and this means that weight tolerance levels can be reduced.

A strain-gauge load cell essentially converts weight into electronic signals. As weight increases, it deforms a structural component of the load cell. As this component changes shape, the electrical resistance of the gauge changes (Figure 1.11).

The strain gauges, usually four or a multiple of four, are connected into a Wheatstone bridge configuration in order to convert the very small change in resistance into a usable electrical signal. Passive components such as resistors and temperature-dependent wires are used to compensate and calibrate the bridge output signal.

In other load cells, the force applied to the device is transferred to a piezoelectric crystal and voltage across the crystal is measured. In both cases, the electronic signal produced is so small that subsequent amplification is needed.

Nowadays, two types of strain-gauge cells are in widespread use in the food sector. *Bending load cells* are widely used, in many configurations, for commercial transducers. Bending beams offer high strain levels at relatively low forces, which make them ideal for load cells integrated in weighing systems for fruit and vegetables. Furthermore, in case of a beam with a symmetrical cross section about the bending axis, there are always two surfaces subjected to equal strains of opposite sign. This offers a convenient means for implementing a full bridge circuit, while temperature compensation is relatively easy. Most products using the bending principle are of the parallelogram or double bending type.

Shear (beam) load cells have become increasingly popular for all types of applications. Shear as a measuring principle offers a standard profile for a given capacity, good resistance against side loads, and a relatively small sensitivity to the point of loading.

A recess is machined in each side at section A-A (Figure 1.12), leaving a relatively thin section in the center. Just as in a structural beam, most of the shear force imposed by the load is carried by this section, while the bending moment is resisted primarily by the flanges. At the neutral axis, where the bending stress is negligible, the state of stress on the web is one of pure shear, acting in the vertical and horizontal directions. As a result, the principle axes there are at 45° to the longitudinal axis of the beam, and the corresponding principal strains are of equal magnitude and opposite sign. Pairs of strain gauges are installed on both sides of the web and connected in a full-bridge circuit for load measurement. Although it is more difficult to install the strain gauges in some form of recess, they can readily be sealed and protected against environmental effects. Shear beam load cells are relatively insensitive to the point of loading and offer a good resistance to side loads. This simplifies its use in many weighing applications.

1.8 MEASUREMENT OF RHEOLOGICAL PARAMETERS

Rheology is the branch of science that deals with the flow and deformation of materials. At the most basic level, rheology can be described as the study of how materials

FIGURE 1.10 Typical load cells to be integrated in a weighing system.

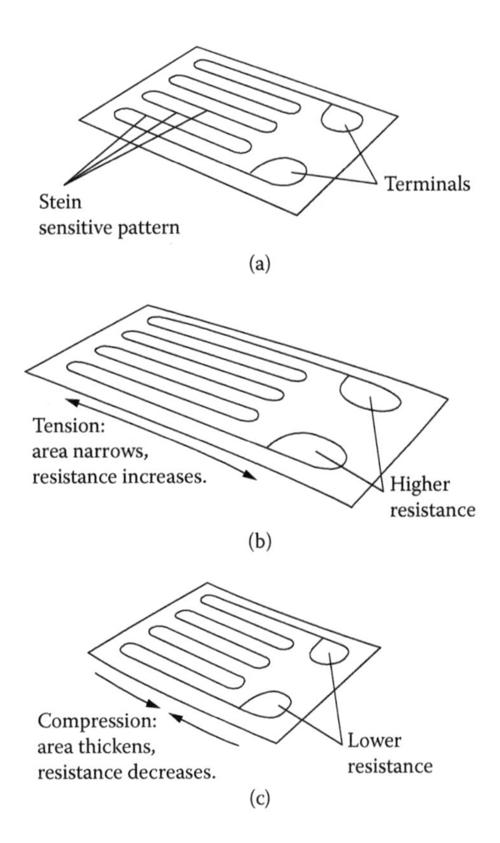

FIGURE 1.11 Working concept behind the strain gauge on a beam under exaggerated bending. (From Omega Engineering Technical Reference, http://*www.omega.com/techref/*.)

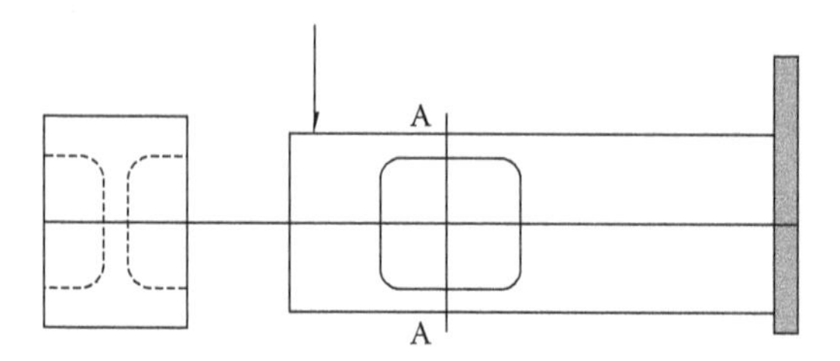

FIGURE 1.12 Principle of shear-web sensing element. (From Revere transducers. Application Note 07/06–13/01.)

flow, or in a broader sense, how materials respond to the application of deformational energy. This definition could obviously encompass the entire field of solid and fluid mechanics. In practice, both the field of rheology and the development of rheometric instrumentation have been linked to the study and development of viscoelastic material systems, which may be simply defined as material systems whose properties are time and/or temperature dependent (Macosko, 1994). Some material systems, such as food products usually display viscoelastic behavior. In these, texture is a key quality factor. Rheological behavior is associated directly with textural qualities such as mouth feel, taste, and shelf-life stability.

Rheological instrumentation and rheological measurements have become essential tools in the analytical laboratories of food companies for characterizing ingredients and final products, as well as for predicting product performance and consumer acceptance. The materials under investigation can range from low-viscosity fluids to semisolids, and gels to hard, solid-like food products. Based on their measuring principles, basic rheometric instruments are either viscometers or rheometers (Gerth, 1980; Muschelknautz and Heckenbach, 1980). Viscometers record the torque and the shear rate in a geometrically defined gap during steady-state shear flow. A typical rheometer incorporates the following features: wide torque, shear stress, temperature, shear rate, and frequency range. The output of a rheometer is typically a set of curves that depicts these forces as the pressure on the flowing material changes (Figure 1.13).

The most basic and widely used form of rheometric instrumentation is the *simple steady-shear viscometer*. A wide variety of existing devices have been developed for the measurement of steady-shear viscosity, many of which are specific to a particular industry or material. In principle, however, all of these devices share a common goal: to measure the bulk viscosity of a material as it flows in a steady or continuous fashion. In fact, the definition of the term *viscosity*, which is the ratio of the shear stress applied to a material and the resulting shear rate, suggests the design of most of these simple viscometers (Weipert, 1990).

As a comprehensive materials characterization tool, however, the traditional viscometer is of limited use. The most powerful and versatile form of rheometric instrumentation currently in use may be described in general terms as a *dynamic shear rheometer* or simply a *dynamic rheometer*. The modern dynamic rheometer shares a basic common concept with the simpler viscometer in that it uses well-defined

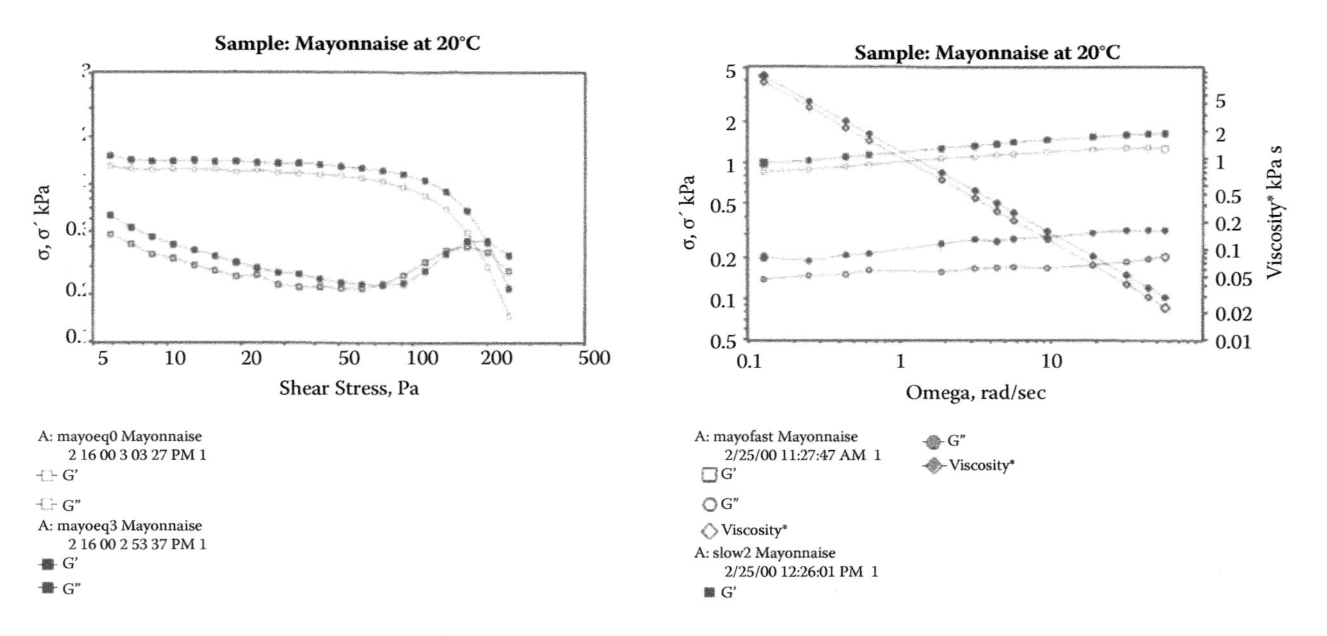

FIGURE 1.13 Strain and frequency sweep on mayonnaise. (From Herh, K. W., Colo, S. M., Roye, N., and Hedman, K. 2000. *Rheology of foods: New techniques, capabilities and instruments.* Application Note June 2000. Bordentown, NJ: ATS RheoSystems.)

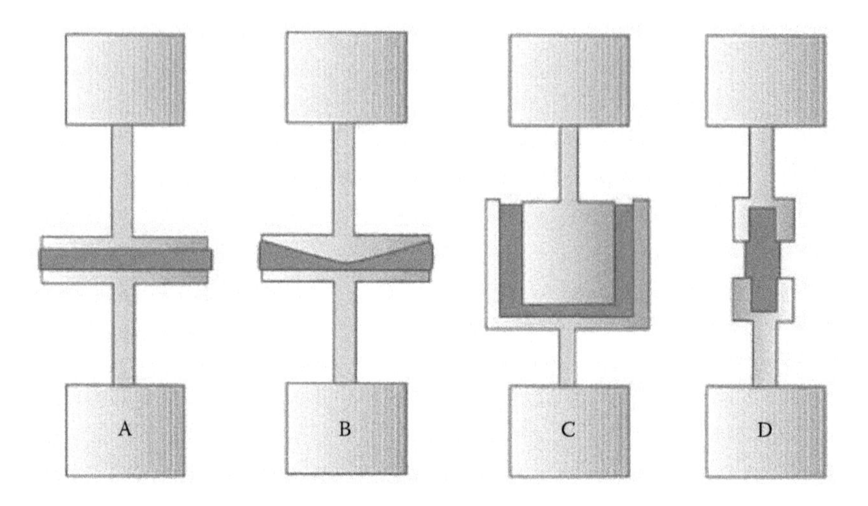

FIGURE 1.14 Typical sample testing geometries for dynamic rheometers: (A) parallel plates, (B) cone and plate, (C) concentric cylinder (cuvette), and (D) solid or torsion rectangular. (From Herh, K. W., Colo, S. M., Roye, N., and Hedman, K. 2000. *Rheology of foods: New techniques, capabilities and instruments.* Application Note June 2000. Bordentown, NJ: ATS RheoSystems.)

geometries, such as cone plates, parallel plates, or concentric cylinders, to isolate and deform the material in a controlled fashion (see Figure 1.14).

A dynamic rheometer applies very small amounts of rotation or deformation in a dynamic or oscillatory fashion. It is often useful to visualize this type of dynamic shear testing as if the sample were being *vibrated* between parallel plates or concentric cylinders while controlling the sample temperature. The resulting measurement is delivered in terms of discrete components of the material's viscosity or shear modulus, as opposed to the simple bulk viscosity reported by traditional viscometers.

As mentioned previously, viscoelastic materials display time- and temperature-dependent properties. When analyzed using a dynamic rheometer, the viscosity or shear modulus of a viscoelastic material may be resolved into component parts referred to as the *elastic* and *viscous* components (Equation [1.5]):

$$G^* = [(G')^2 + (G'')^2]^{1/2} \tag{1.5}$$

where G^* is the dynamic shear modulus, G' is the elastic or storage modulus, and G'' is the viscous or loss modulus.

These component parts of the bulk viscosity or modulus have specific meaning in the context of the bulk properties of the material and are individually very sensitive to specific events occurring in the morphology or microstructure, or even the nanostructure, of the material system. These same structural effects or phenomena are often invisible to traditional, steady-shear viscometry.

REFERENCES

ASTM (American Society for Testing and Materials). 1992. Standard guide for computed tomography (CT) imaging, ASTM Designation E 1441-92a. In *1992 Annual Book of ASTM Standards*, Section 3 Metals Test Methods and Analytical Procedures, 690–713. Philadelphia: ASTM.

Axelson, J. 2005. *USB complete: Everything you need to develop custom USB peripheral.* Madison, WI: Lakeview Research.

Barcelon, E. G., Tojo, S., and Watanabe, W. 1999. X-ray computed tomography for internal quality evaluation of peaches. *Journal of Agricultural Engineering Research* 73 (4):323–330.

Booth, B., Neighbour, R., and Li, X. 2006. On agricultural ultrasound image segmentation. In *Proceedings of IEEE international conference on signal processing*. Washington, DC: IEEE.

Cubero, S., Aleixos, N., Moltó, E., Gómez-Sanchis, J., and Blasco, J. 2010. Advances in machine vision applications for automatic inspection and quality evaluation of fruits and vegetables. *Food and Bioprocess Technology*. DOI 10.1007/s11947-010-0411-8:

Evening, M. 2005. *Adobe Photoshop CS2 for photographers*. Oxford, UK: Focal Press.

Freeman V. S. (1995). Spectrophotometry. In *Laboratory instrumentation*, ed. Haven, M.C., Tetrault G A , and J. A. Schenken, 72–96. John Wiley & Sons Inc, New York (USA).

Gall, H., Muir, A., Fleming, J., Pohlmann, R., Göcke, L., and Hossack, W. 1998. A ring sensor system for the determination of volume and axis measurements or irregular objects. *Measurement Science and Technology* 9:1809–1820.

Gerth, C. H. 1980. Rheometry. In *Analysis and measurements. Uhlmanns encyclopedia of technical chemistry*, Vol. 5. Beerfield Beach, FL; Basel, Switzerland: Chemie Weinheim.

Hamamatsu Photonics. 2006. *Photomultipliers tubes. Basics and applications*. London, UK: Hammamatsu Photonics.

Herh, K. W., Colo, S. M., Roye, N., and Hedman, K. 2000. *Rheology of foods: New techniques, capabilities and instruments*. Application Note June 2000. Bordentown, NJ: ATS RheoSystems.

Hill, B. 1998. *Magnetic resonance imaging in food science*. New York: John Wiley & Sons.

Janesick, J. R. 2001. *Scientific charge-coupled devices*. Bellingham, WA: SPIE Press.

Ketcham, R. A., and Carlson, W. D. 2001. Acquisition, optimization, and interpretation of x-ray computed tomographic imagery: Applications to the geosciences. *Computers and Geosciences* 27:381–400.

Krautkramer, J., and Krautkramer, H. 1990. *Ultrasonic testing of materials*. Heidelberg, Germany: Springer-Verlag.

Kuttruff, H. 1991. *Ultrasonics: Fundamentals and applications*. New York: Elsevier.

Luong, J. H. T., Bouvrette, P., and Male, K. B. 1997. Developments and application of biosensors in food analysis. *Tibtech* 15:369– 377.

Lynnworth, L. C. 1989. *Ultrasonic measurements for process control: Theory, techniques, applications*. San Diego: Academic Press.

Macosko, C. W. 1994. *Rheology: Principles, measurements and applications*. New York: John Wiley & Sons.

McRobbie, D. W., Moore, E. A., Graves, M. J., and Prince, M. R. 2003. *MRI: From picture to proton*. New York: Cambridge University Press.

Mizrach, A. 2008. Ultrasonic technology for quality evaluation of fresh fruit and vegetables in pre- and postharvest processes. *Postharvest Biology and Technology* 48:315–330.

Muschelknautz, E., and Heckenbach, M. 1980. *Rheological measurements in chemical practice*. Berlin: Springer.

Muzzolini, R. E. 1996. A volumetric approach to segmentation and texture characterisation of ultrasound images. PhD thesis, University of Saskatchewan.

Nielsen, S. S. 2010. *Food analysis*. New York: Springer.

Scannavino, F. J., and Cruvinel, P. 2009. Instrumental opportunities of x-ray computed tomography on soil compaction characterization to sustainability in agricultural systems. Paper presented at CIGR Proceedings, Technology and Management to Increase the Efficiency in Sustainable Agricultural Systems, Rosario (Argentina), September 1–4.

Shelly, G. R., and Vermaat, M. E. 2010. *Discovering computers: Fundamentals*. Boston: Thompson Course Technology.

Tetrault, G. A., and Schenken, J. A. 1995. Laboratory instrumentation. Edited by M. C. Haven. New York: John Wiley & Sons.

Walls, P. N. T. 1969. *Physical principles of ultrasonic diagnosis*. New York: Academic Press.

Wang, T. Y., and Nguang, S. K. 2007. Low cost sensor for volume and surface area computation of axi-symmetric agricultural products. *Journal of Food Engineering* 79(3):870–877.

Weipert, D. 1990. The benefits of basic rheometry in studying dough rheology. *Cereal Chemistry* 67 (4):311–317.

2 Rheological Properties of Foods

Ana Cristina Agulheiro Santos and Cristina Roseiro

CONTENTS

2.1 INTRODUCTION

For profitable and effective communication methods and procedures, standardization is absolutely necessary, and it is even more important for biological materials like food. Despite many efforts and effective work done by prominent researchers over the years (Malkin, 1994; Steffe, 1996; Rao, 2007; Bourne, 1978; Szczesniak et al., 1963) there is still little consensus about methods or measurements of rheological properties of food.

Nowadays, another challenge is to develop adequate methods for new food materials and also to develop new preservation methods.

Published literature on methods about this topic is based on old methods that sometimes are not adequate for today's needs and exigencies; however, some new approaches have also appeared. Results are only valid with precisely defined conditions of measurement. A variety of testing methods makes comparison between results difficult, if not impossible, and often results in inconsistencies.

To relate these modern methods and the use of classic methods to new foods is the aim of this chapter. However, we offer a brief overview of classic concepts, dedicated to those who are just beginning their understanding of this area.

It should be always remembered that biological materials are biomechanical systems with complex structures, and their mechanical behaviors cannot be explained from a physical point of view because of this complexity.

2.2 BACKGROUND INFORMATION FOR SOLID VISCOELASTIC FOOD

Food rheology studies the relationships among stress, strain, and time scale of foods in order to understand the effects of processing on products, probe the system structure, reveal critical aspects of food texture, and correlate their texture characteristics with sensory evaluation.

Three regimes should be considered in understanding rheological behavior during stress of viscoelastic solid food material. The first region, in which Hooke's law is obeyed, is the linear region where the relationship between stress and strain is proportionate. The second region is characterized by nonlinearity and reflects a more complex relationship between stress and strain. The last point occurs at sample fracture.

The second part is very important in order to describe sensory evaluation of food products and its contribution to understanding of texture analysis.

2.2.1 Large Strain Methodologies: Compression, Torsion, and Tension

The study of behavior related to regions two and three is done through large strain methodologies like compression, torsion, or tension.

Compression tests are really simple to perform, and because of that, they are very often used. Compression tests use a cylindrical sample that is squeezed between two flat plates at a constant velocity, and the force required to deform the material to failure is measured to compute a fracture stress, and the degree of deformation determines fracture strain. Young's modulus can be calculated using the force deformation curve that is obtained, if the material probe has parallel and plane surfaces. The total energy of fracture represented by the area behind the curve until the failure point is reached is another useful measurement.

During tension testing, the sample is extended until rupture, and similar parameters are calculated. These tests require adequate equipment for securing the probes and are often rejected because of this added difficulty.

Torsion testing involves twisting the material to fracture. The degree of rotation and magnitude of torque required to break the sample is monitored to calculate the strain and stress at fracture. Torsion places a condition known as *pure shear* on the material, creating an equivalent distribution of shear and normal stresses throughout the sample. Each of these fracture approaches can create a combination of shear and normal stress distributions throughout the material and care must be taken to report the appropriate failure conditions (Foegeding et al., 2003).

Nowadays, enterprises dedicated to food production and evaluation require practical methods that allow reproducible results, are easy to perform, and provide a fast and adequate answer. So, empirical methods are often used in food, such as texture profile analysis, whose parameters can be easily correlated with sensory evaluation of several foods. Compression and penetration tests with different probes and apparatuses are the most popular among food technicians; both are based on uniaxial compression. Instrumental tests such as creep/recovery tests are performed in order to find rheological characteristics like elasticity and the ability of the products to recover their original shape after the removal of stress (Drake, et al., 1999a; Drake et al., 1999b). According to Breuil and Muellenet (2001), tests of compression, puncture, and penetration are easily done with a universal testing machine, and they give more parameters than any other rheological test.

2.2.2 Young's Modulus and Stiffness

When stress and strain result from axial loading tests, the force acting normal to the surface causes stress in an uniaxial direction, making it possible to calculate Young's modulus (E). It gives a useful indication of how easily the studied material can be contracted and stretched, sometimes referred to as the *stiffness* of the material, and it corresponds to the slope of the stress–strain curve (Equation [2.1]).

$$E = d\rho/d\varepsilon \qquad (2.1)$$

where ρ = stress in Nm^{-2} and ε strain, and E = Young's modulus in Nm^2 or Pa.

Known values of Young's modulus are, according to Figura and Teixeira (2007), $9{,}9 \times 10^9$ for ice, 3×10 for ice, 3×10^9 for dry spaghetti, 0,6 until $1{,}4 \times 10^7$ for raw apple, 2×10^3 for gelatin, and 0,8 until 3×10^6 for banana. Some Young's moduli have been known for a many years. Fridley et al. (1968) obtained values of 1.03 and 5.3 MPa for peaches and pears, respectively. Misra and Young (1981) observed that the Young's modulus of soybeans varied between 125 and 126 MPa, at the loading rate of 5 mm/min and moisture content of 13% (w.b.). For biological materials, Poisson's ratio depends on moisture content, stress magnitude, and loading rate.

2.2.3 Bulk Modulus and Compressibility

On the other hand, the *bulk* compression loading modulus (K) is used to characterize material under pressure from all directions, causing changes in the volume of the studied material. This reflects how a material withstands an elastic compression, sometimes referred to as the *firmness* of the material (Equation [2.2]).

$$K = dP/(dV/V) \tag{2.2}$$

where

K = Bulk modulus in Nm^2 or Pa
dP = differential change in pressure on the object
dV = differential change in volume of the object
V = initial volume of the object

The behavior of food materials subject to high pressure treatment is quite different for liquids and gases, which are isotropic, or for solids, which are mainly anisotropic.

Both stress–strain curves reflecting the application of uniaxial and omnidirectional stresses, exhibit a linear elastic behavior during the first part of the stress–strain curve and their slope allows calculation of the previously referenced modulus of elasticity and K. Those moduli can be mathematically related trough Poisson's ratio.

2.2.4 Poisson's Ratio

Apparent elastic properties, such as Poisson's ratio and Young's modulus, can be used to understand load-deformation behavior of food materials and to compare the relative strengths of different food materials. Some Poisson's ratio values of food material have been known for a long time: it varies from 0 to 0.5 for most materials (Mohsenin, 1986; Peleg, 1987); for gels, it is 0.3 until 0.5; for apple it is 0.23, for potato 0.49, for peaches 0.2, for pears 0.5, and for water it is 0.50 (Fridley et al., 1968; Mohsenin, 1986). The technique described by Sitkei (1986) is often used to determine Poisson's ratio.

2.3 NOVEL MEASURING TECHNIQUES

2.3.1 Measuring Techniques for Poisson's Ratio and Young's Modulus

The Poisson's ratio and Young's modulus of red bean grains as a function of moisture content (5, 7.5, 10, 12.5, and 15% w.b.) and loading rate (3, 6, 9, 12, and 15 mm/min) for Goli and Akhtar varieties were studied and experimentally measured with the Sitkey technique (Kiani et al., 2009). Before initializing the tests, the original length and diameter of the studied specimens should be recorded.

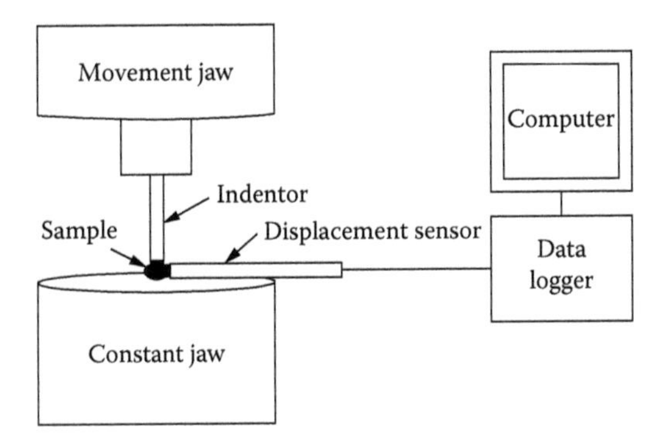

FIGURE 2.1 Schematic diagram of the setup for measuring Poisson's ratio. (From Kiani, M.K., H. Maghsoudi, and S. Minaei. 2009. Determination of Poisson's ratio and Young's modulus of red bean grains. *Journal of Food Process Engineering*, doi: 10.1111/j.1745-4530.2009.00391.x. Reprinted with permission from John Wiley and Sons.)

In this case, a digital caliper (CD-6″CS, Mitutoyo, Kanagawa, Japan) was used. The specimens were loaded in a material testing machine (H50 k-s, Hounsfield Ltd, Surrey, UK) and quasi-statically compressed until the material failed. Axial and lateral deflections of the sample were recorded at the cracking limit of the axial load. Strain (axial displacement) was measured using the material testing machine (Figure 2.1). Lateral deflection was measured using an instrumented bending beam (CE-10, Tokyo Sokki Kenkyujo, Tokyo, Japan), which contacted the sample.

The ratio of transverse strain to axial strain produces a proportionality factor called Poisson's ratio.

For determining Young's modulus, tests were conducted at moisture levels of 5, 7.5, 10, 12.5, and 15% (w.b.). Also, the effects of the loading rate on Young's modulus were studied at 3, 6, 9, 12, and 15 mm/min, as recommended by the ASAE (American Society of Agricultural Engineering) Standards (2004). Ten samples were tested in the material testing machine with a 500 N compression load cell, at each of the loading rates and moisture levels. Grains were placed between two parallel plates (Figure 2.2) and compression force exerted along the thickness of the samples; data on the strength properties were automatically obtained from an integrator.

As expected, Poisson's ratio of these grains exhibits a negative relationship with moisture content. On the other hand, a positive correlation was observed between Poisson's ratio values and loading rate, for increasing loading rates from 3 to 15 mm/min. Young's modulus, in both varieties, significantly decreased with increasing moisture content and with increasing loading rates from 3 to 15 mm/min. Other complementary information from this work concerns the different behavior of varieties under different loading rates and their different elastic properties, which is useful information for designing adequate machinery.

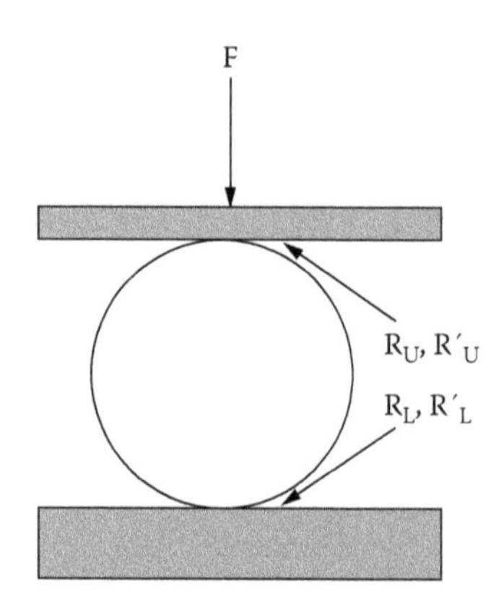

FIGURE 2.2 Sample loading using parallel plate contact (ASAE Standards 2004). (From Kiani, M.K., H. Maghsoudi, and S. Minaei. 2009. Determination of Poisson's ratio and Young's modulus of red bean grains. *Journal of Food Process Engineering*, doi: 10.1111/j.1745-4530.2009.00391.x. Reprinted with permission from John Wiley and Sons.)

2.3.1.1 Measuring Techniques for Compressibility in High-Pressure Processing of Foods

Nowadays, high-pressure processing (HPP) to preserve foods is increasingly used for many different foods, from fruit juice to meat. In theory, the practice of HPP of food should provide uniform temperature and pressure distribution during process duration. It corresponds to a situation of bulk compression loading on the material (Denys et al., 2000; Hartmann and Delgado, 2005). All compressible substances experience changes in their temperature during physical compression; this is an unavoidable thermodynamic effect. At pressures of 400–1000 MPa during HPP and under adiabatic conditions, water changes 3°C for every 100 MPa of pressure change. Fats and oils show the highest compression heating values (6 to 8.7°C per 100 MPa) (Ting et al., 2002; Rastogi et al., 2007). It is generally assumed that food systems have thermodynamic properties similar to water, mainly for high-moisture liquid foods.

Therefore, food pressurization induces changes in thermodynamic properties, in rheological properties, and induces compression heating. These changes could have consequences in the safety and quality of food (Otero and Sanz, 2003; Balasubramaniam and Farkas 2008). Physical properties of foods with high moisture content remained similar after high-pressure (HP) treatment; gas-containing foods can show changes after HP treatment due to gas displacement and liquid infiltration. Finally, foods without air voids don't exhibit changes. Even HP treatment can cause desirable texture and sensory changes, for example in surimi and cheese (Ohsima et al., 1993; O'Reilly et al., 2002).

According to Otero and Sanz (2003) and Barbosa-Cánovas and Rodriguez (2005), it is absolutely necessary to measure food properties in situ under conditions of pressure to obtain accurate data and to go on with high-pressure food processing.

Min et al. (2010) worked with 16 different types of food with the aim of measuring compressibility and density of selected liquid and solid foods at 25°C and pressures between 0.1 and 700 MPa.

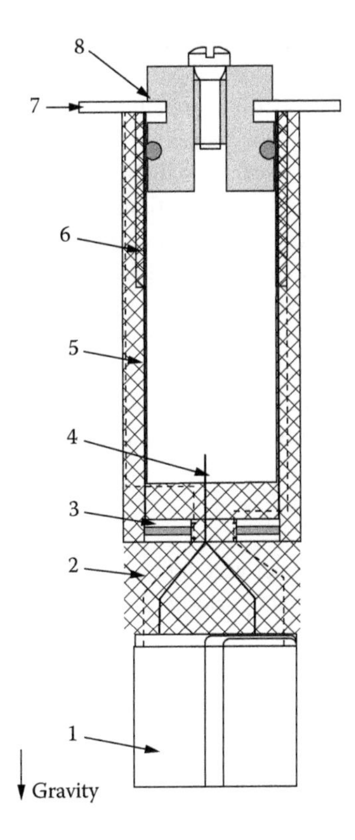

FIGURE 2.3 Cross-section drawing of sensor in the sample loading position. The sensor was inverted to a piston-downward position when installed in the pressure vessel. (1) Pressure vessel closure; (2) epoxy coated junction of copper and thermocouple wires; (3) sensor end cap; (4) thermocouple; (5) epoxy coated polycarbonate tube; (6) magnet wire coil; (7) tool used for piston positioning, removed during operation; (8) piston pressure transmitting fluid required to compress the pressure, measured with a strain gage (C-6211, Harwood Engineering, Walpole, MA, ±0.4% accuracy) and thermocouple temperature, accurate to ±1°C, were recorded with a data logger (34970, Agilent, Santa Clara, CA). Pressurization rates were approximately 20 MPa/s, and depressurization rates were approximately 2–5 MPa/s. (From Min S., S.K. Sastry, and V.M. Balasubramaniam. 2010. Compressibility and density of select liquid and solid foods under pressures up to 700 MPa. *Journal of Food Engineering* 96: 568–574. Reprinted with permission from Elsevier.)

A hydrostatic pressure system (26190, Harwood Engineering, Walpole, MA), rated to 1000 MPa was used to generate pressure on food samples. To control sample temperature, propylene glycol at controlled temperature is recirculated through the vessel's jacket. Sample volumes were measured using a piezometer that was specially designed and constructed for this purpose.

Min et al. (2010) provided a detailed sensor description and its operating principle and procedure. Figure 2.3 shows a cross-sectional drawing of the sensor.

Compressibility and density were measured at 25°C and pressures up to 700 MPa with a variable volume piezometer on sucrose solutions (2.5–50%), soy protein solutions (2.5–10%), soybean oil, chicken fat, clarified butter, chicken breast, ham, Cheddar cheese, carrot, guacamole, apple juice, and honey. This methodology allowed the authors (Min et al., 2010) to conclude that compressibility of tested foods changed with pressure at a rate that decreased with increasing pressure. Variability in compressibility of different materials is likely due to differences in concentration, chemical composition, and complex interactions between components within a food system. Compressibility of sucrose and soy protein solutions decreased as a function of concentration. Protein solutions were less compressible than sucrose solutions at pressures greater than 200 MPa for equal mass concentrations. Compared with water, fats showed high compressibility to 100 MPa, similar compressibility from 100 to 300 MPa, and less compressibility from 300 to 700 MPa. Chicken breast, ham, Cheddar cheese, carrot and guacamole showed relatively large compressibility from 0.1 to 100 MPa. Honey showed the smallest volume decrease over 700 MPa.

2.3.2 Mechanical Impact

The concept of Mechanical Impact was defined by Goldsmith (1960) as an instantaneous load, non continuous and not periodic. Later, Manor (1978) corroborated those fundamental theory aspects.

To develop this experimental method, Hertz's theory should be accepted. The contact between two convex bodies, one of them moving at high velocity, causes an impact with two different effects—the contact between them in the impact area and the wave propagation (Moshenin, 1970; Timoshenko and Goodier, 1970). Works carried out by several researchers, some of them in food engineering, contributed the definition of *time of impact* as the time when the impact load acts. According to Ruiz-Altisent (1986), impact force should act during a brief time of 10 ms, which is defined as the *impact time*. The following are the fundamental equations for the impact study.

Maximum deformation (D_{max}):

$$D_{max} = \left[\frac{15}{16}\pi v^2 A\right]^{\frac{2}{5}} \left[\frac{R_1 + R_2}{R_1 R_2}\right]^{\frac{1}{5}} \left[\frac{m_1 m_2}{m_1 + m_2}\right]^{\frac{2}{5}} \tag{2.3}$$

Impact time (t):

$$t = 2.9 \frac{D_{max}}{v} \tag{2.4}$$

Maximum normal stress:

$$\sigma_{max} = 0,2515 \left[\left[\pi^4 \frac{v^2}{A^4} \frac{m_1 m_2}{m_1 + m_2}\right]^{\frac{1}{5}} \left[\frac{R_1 R_2}{R_1 + R_2}\right]^3\right]^{\frac{2}{5}} \tag{2.5}$$

Radius of the tension surface:

$$a = \left[\frac{15}{16} v^2 A \frac{m_1 m_2}{m_1 + m_2}\right]^{\frac{1}{5}} \left[\frac{R_1 R_2}{R_1 + R_2}\right]^{\frac{2}{5}} \tag{2.6}$$

$A = (1 - \eta_1^2)/E_1 + (1 - \eta_2^2)/E_2$; V represents the relative velocity between the bodies; m_1 and m_2 represent the mass of bodies 1 and 2, respectively; R_1 and R_2 represent the radius of bodies 1 and 2, respectively; η is Poisson's ratio of the studied body, a fruit per example value of 0.49; and E represents the elasticity modulus of the fruit.

2.3.2.1 Hertz Contact and Impact Measurement Parameters

Some parameters should be defined to obtain credible information about mechanical impact test performance. Impact energy, absorbed energy, rebound energy, and load velocity are absolutely indispensable according to Ruiz-Altisent et al. (1987). Chen et al. (1985) referred also to maximum force, velocity after impact, and maximum deformation not recoverable.

These parameters can be organized in three categories in a practical approach to fruit study (Garcia, 1988).

The first group includes: maximum deformation; permanent deformation; critical deep, bruise dimensions; extension; size and depth of the bruise; and maximum impulse. These parameters are dependent on the drop height. The impulse can be calculated by integration of the force–time curve, which is a good measure of impact energy that is linearly correlated with drop height and independent of fruit ripeness stage.

The second group of parameters is also correlated with drop height and consists of maximum impact force (*FI*), slope of the force–time curve (*FT*), ratio between the two referred parameters (*FI*/*FT*), and rebound velocity.

The third group consists of time of impact (*TT*), final duration of impact (*TF*), the difference between them (*TT* − *TF*), time necessary to reach maximum force, optimum slope of the force–deformation curve, modulus of apparent elasticity, and maximum tangential force. All of these parameters are linearly correlated with the ripeness stage of the fruits. This fact is very important because it allows the use of the parameters in building new quality and maturity indexes for fruits using nondestructive methods, if mechanical impact is done behind the bioyield point.

The elasticity modulus in pears revealed a very strong correlation with their maturity stage, and in 1979, Chen and Friedley used it to develop a new nondestructive method for classifying pears according to their ripeness stage. They used a flat plate with an infinitum radius ($R_2 = \infty$), and several impacting spheres with different diameters. The use of the flat probe with $R_2 = \infty$ caused a strong influence of the radius of the fruit on the elasticity modulus. On the other hand, if the impacting sphere is too small, the tension applied on the surface of the fruits increases and causes damage to the fruit tissue (Jarén, 1994).

Delwiche et al. (1987), and Lichtensteiger et al. (1988, quoted by Bellon et al., 1994), concluded that time characteristics of impact were sensitive to elasticity but independent of the shape of the fruit.

Experiments done by Delwiche et al. (1988) revealed that the ratio of maximum force/(time for maximum force)^2 among other parameters were strongly correlated with the elasticity modulus of fruits and with penetrometer measurements of firmness. Those measures didn't reveal a strong connection with mass and radius of fruits.

Horsfield et al. (1972) adapted the impact theory basis to Hertz's theory. The maximum compression stress should be inferior to the tissue bioyeld resistance, and when the impact occurs, this value can be calculated by using Equation (2.7):

$$T\max = K_1 \times mgh^{1/5} \times \left(K_2 \frac{1}{\frac{1}{E_1} + \frac{1}{E_2}} \right)^{\frac{4}{5}} \times (1/R_1 + 1/R_2)^{3/5} \tag{2.7}$$

R_1 and R_2 are the curvature radii of the bodies that are in contact, K is a constant, m is the mass of the body, g is the force of gravity, and h is the drop height of the impacting sphere.

2.3.2.1.1 *Load Velocity*

According to Hellebrand (1985) 0.1 mm/s is the minimum velocity necessary for the occurrence of an impact. The definition refers an impact time of 10 milliseconds (Ruiz-Altisent, 1986) or less than 15 milliseconds (Garcia, 1988).

The impact is mainly distinguished from the quasi-static force by the duration of the contact of the bodies (Fluck and Ahmed, 1973).

The damages observed in fruits are quite different, even with the same amount of energy, due to different velocities (Mohsenin, 1972; Holt and Schoorl, 1977; Sitkei, 1986). Chen et al. (1985) reported that the parameters that depend on velocity could be analyzed and that their relation to bruise susceptibility in fruits could be studied.

Fruits behave like viscoelastic bodies; the elastic component is more evident with the application of an impact. The rupture in this kind of rheological material depends mainly on the load velocity (Sinobas, 1988).

The significant effect of the deformation velocity on the fracture velocity was studied by Chen and Sun (1984) in cylindrical apple samples, varying the deformation velocity between 0.05 to 0.15 mm/s. The importance of viscosity behavior of vegetal material, such as fruits, to mechanical impact was deducted.

During the application of a mechanical impact, Chen et al. (1985) recommended the study of several parameters: impact energy, impact velocity taking into account the velocity before the impact occurs, absorbed energy, maximum force, velocity after the impact, maximum deformation, and deformation not recoverable. The relation between those parameters could also be very useful in studying the impact.

The measurement of both energy and velocity is quite easy, but the other parameters require very fast equipment that is able to read parameters like acceleration that change very quickly.

The velocity of impacting mass $v(t_1)$ is given by Equation (2.8):

$$v(t_1) = \sqrt{2gh} \tag{2.8}$$

where t_1 is the time at the beginning of the impact, g is the acceleration due to gravity, and h is the drop height.

The velocity, $v(t)$, the acceleration, $a(t)$, and the displacement of the sphere, $x(t)$, are related by Equations (2.9) and (2.10):

$$v(t) = v(t_1) + \int_{t_1}^{t} a(t)dt \tag{2.9}$$

$$x(t) = x(t_1) + \int_{t_1}^{t} v(t)dt \tag{2.10}$$

The deformation of the fruit during the impact is given by Equation (2.11):

$$D(t) = x(t) - x(t_1) = \int_{t_1}^{t} v(t) \tag{2.11}$$

with $t_1 \leq t \leq t_2$.

The force of the impacting sphere on the fruit is given by Equation (2.12):

$$F(t) = ma(t) \tag{2.12}$$

where m is the value of the impacting mass.

Finally, the energy transferred from the sphere to the fruit can be calculated by using Equation (2.13):

$$E(t) = m \int_{0}^{(D)t} a(t)dD(t) \tag{2.13}$$

where m represents the mass of the impacting body, D the correspondent deformation, and $dD(t)$ the derivative of the deformation as a time function.

2.3.2.2 Methods to Study Impacts

There are some adequate methods to study impact that allow to obtain characterizing parameters.

According to Manor (1978) and Ruiz-Altisent et al. (1987), there are different experimental approaches involving impact methods for studying fruits:

- Free fall of fruits on a hard or a soft surface
- Free fall of a mass on a fruit or part of a fruit
- Impact produced by a simple or complex pendulum upon a material, such as fruits
- Impact produced by a pneumatic embolus

- Impact produced by an impulse projectile attached to a string or by another mechanism

The more common methods are those of free fall of fruits upon surfaces, free fall of a mass on fruits, or even impacts produced by pendulums. All of them allow mechanical study of the material.

It is necessary to considerer the movement of the fruit. If the fruit is static, there is transference of the moment at the impact point and also at the support point, and two damages are produced. If the fruit is moving while the impact occurs, there is transference of the moment between the fruit and the surface that suffered the impact and the force is only applied at a single point called the *contact point*.

According to Fluck (1975), much information can be obtained from a single impact procedure and its corresponding curve: the maximum force, which corresponds to the highest point of the curve, the time of impact, the time necessary until the end of the contact, and the movement. The area under the curve corresponds to the change in velocity, according to Wright and Splinter (1968), and can be used to calculate the energy of impact.

If the fruit doesn't recover its initial shape, the energy should be calculated by the Equation (2.14):

$$Eimp = \frac{1}{2} m \left(v_i^2 - v_f^2 \right) + mg \, \Delta dt \tag{2.14}$$

where v_i is the initial velocity, v_f is the final velocity, m is the impacting mass, g is the acceleration of gravidity, and $\Delta\ dt$ is the nonrecoverable deformation.

Finney and Massie (1975) developed a computational system using a pendulum to produce an impact that allows force, deformation, and their relation during impact to be registered.

In order to simplify the data registration and their analysis, a system was developed in which the signal through a force transducer and a photoelectric sensor is sent to a computer through an amplifier, a multiplexor, and an analogical digital structure converser. Jarimopas (1984) continued these studies on impacts using that system (Figure 2.4).

Chen et al. (1985) developed a new mechanism to measure firmness. They used a predefined mass to impact the sample in free fall conditions. This device can register the impacting mass deceleration when it contacts the fruit, and the impact time. Thus, the parameters velocity of impact and deformation of the fruit can be established, using the equations of a uniformly decelerated movement.

Today, this method allows very different and useful studies of fruit texture, such as those at Davis University, directed by Paul Chen, and at Universidad Politécnica de Madrid, directed by Margarita Ruiz-Altisent, both of whom are authors of the LPF 1.0 vertical impacting system.

However, these research teams worked on different approaches. At Davis University, the goal was to obtain a single parameter to describe firmness, as can be seen in the published works of Delwiche (1986) and Delwiche et al. (1988). In these

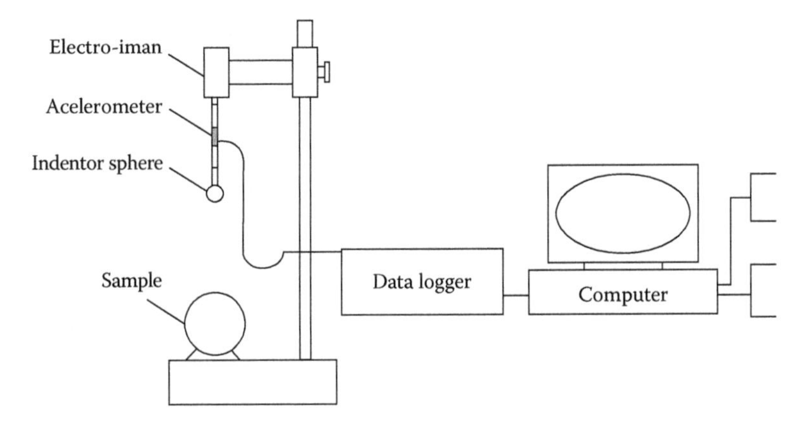

FIGURE 2.4 Representation of the device to study mechanical impact developed by Paul Chen's team at the University of California in 1985.

works, the authors propose the use of the force–time slope and their relations. Zhang and Brusewitz (1991) recovered the idea of studying the force–time slope.

The aim of the team working at the Universidad Politécnica de Madrid was to study the impact and its parameters searching for a relationship based on the ripeness stage of several fruits (apple, pear, peach, and avocado) and, on the other hand, look for their relation with standardized methods, such as penetration (Magness and Taylor, 1925).

A lot of research has been done concerning this last goal, like those of Jarén (1994) with pear and apple; Jarén et al. (1992) and Ruiz-Altisent (1993), achieving a practical goal of automatically selecting fruits using several impact parameters to distinguish firmness; Santos (1991), with a study of melon during cold storage; Correa (1992), with determination of avocado firmness; and Barreiro (1994) studying apple, pear, peach and apricot.

Several research works done with pear and apple by Barreiro and Ruiz-Altisent (1994), comparing different methods for firmness measurements with impact, concluded that impact measurements can be stricter and more powerful than some standardized measurements like compression.

According to Ruiz-Altisent et al. (1986), the performance of an impact using the previously referenced device with appropriate software, allows a large amount of data to be recorded: 100 values of acceleration are immediately visible at the monitor and include the curves of velocity, acceleration, force, maximum values of acceleration and deformation, as well as time of impact.

The existing programs allow easy calculation of values for modulus of elasticity (Fekete and Felföldi, 1994), *bioyield point*, force change, time to reach maximum force, time necessary to reach maximum deformation, and force and deformation during impact based on the elastic model.

Accuracy of the impact data is absolutely necessary for correct study results and allows comparison of data obtained in different studies. It should be noted that with the described device, Chen et al. (1985) obtained measures of acceleration from 0 until 1000 m/s^2 perfectly lineal and with accuracy above 99.9%.

Chen and Tjan (1998) tried to develop a method to calibrate the device, which is essential for its correct performance. They used a small sphere and took into account the fact that the acceleration is quite independent of the fruit mass and relatively insensitive to the radium of curvature. Values of correlation of 0.85 between parameters of impact and firmness of Magness-Taylor were obtained for peaches.

2.3.2.3 New Approaches Using Other Different Methods Based on Mechanical Impact and Study of Vibrations

Other perspective of the impact applications is the study of vibrational characteristics, often used for evaluation of fruit firmness through elasticity modulus. The work of the following researchers should be noted: the work done with low frequency by Hamann and Carroll (1971), and with sonic vibration by Finney (1970, 1971, 1972); by Sarker and Wolfe (1983) that studied ultrasounds and the study of acoustic response to impact by Yamamoto et al. (1980, 1981), Yamamoto and Haginuma (1984a, 1984b), Armstrong et al. (1990), and Chen and Huarng (1992). The acoustic signs produced by impact were studied by Hayashi et al. (1995) to obtain a reliable quality evaluation of different fruits.

The results through waves transmitted on the surface of the fruit and the velocity of transmission allow prediction of the firmness of fruit flesh because they discovered that the velocity of propagation of the waves change dramatically based on ripeness stage.

Sugiyama et al. (1994) verified that the impact waves of acoustic signals produced by impacts were transmitted at the equatorial zone of the fruit at uniform velocity. Considering that information, they proposed firmness indices, and a correlation with a value of 0.823 between propagation velocity of the acoustic signal produced by impact and flesh firmness was found. They also concluded that transmission velocity decreases with increases in maturity stage.

Harker et al. (2002) studied various instrumental tests in order to determine which were more intensely correlated with sensory measurements, and then to identify the minimum instrumental difference that was required before a trained sensory panelist could detect a difference in apple texture. Instrumental tests included a nondestructive test based on impact responses of fruit (SoftSense) as well as the recording of chewing sounds, sensory panelists, and other instrumental tests such as puncture, tensile, twist, and Kramer shear tests. SoftSense is an instrument developed to nondestructively measure firmness using impact response analysis (McGlone and Schaare, 1993). The general principle of impact response is well established, because fruit bounces differently depending on firmness. In this study, fruits were dropped 10 mm onto the SoftSense load cell before making any measurement of destruction. The resulting impact response was characterized by the contact with the load cell. The three instrumental tests that more accurately predicted a sensory response were puncture, twist, and chewing sounds.

Subsequent work was done by Grotte et al, (2002) using three different measurement methods, falling impact, the resonance impulse method, and puncture test, in order to determine some mechanical properties of Golden Delicious apples. The assumption of Hertz's theory modified by Kosma and Cunningham (1962) for impact between two elastic bodies permitted the calculation of Young's modulus E, Poisson's ratio v, and Lame's coefficients μ and λ from elasticity coefficients measured by puncture test and acoustic impulse method was assumed by that research team. Young's modulus and Lame's coefficients decreased with storage time and with the degree of fruit ripeness. Poisson's ratio increased differently during ripening. At the beginning of storage, increases in Poisson's ratio are linked to the decrease of percent contribution of the flesh to the overall firmness. The relationship between Young's modulus and Lame's coefficient m is linear.

Another interesting practical approach is to develop nondestructive methods to increase the sample number. Valero et al. (2003) tried nondestructive methods on peaches, nectarines, and plums. A commercial nondestructive impact sensor (the Sinclair Internal Quality Firmness Tester; SIQ-FT, Sinclair Systems International, LLC, Fresno, CA) was used to measure firmness of the samples, and their relation with the standard penetrometer was studied. The SIQ-FT pneumatically operated sensor has a head equipped with a piezoceramic generator, which is pushed out of the bellow's end each time the device hits a fruit sample. The electronic sensor is capable of converting force to voltage. The resultant voltage signal depends on fruit firmness. The voltage signal passed through an analog-to-digital converter interfaced to a personal computer and was processed by software (Sinclair IQ version PIQ01-v2.18.01) to return a measure of fruit firmness as a number indexed from 0–100 (arbitrary units)—the Sinclair firmness index (SFI). Softer fruits are assigned lower index values than firmer fruit. The University of California, Davis fruit firmness tester (UCF) is a hand-driven press (Western Industrial Supply Co., San Francisco, CA) equipped with an Ametek penetrometer (Ametek, Hatfield, PA) and a 7.9-mm diameter tip.

Although scientific studies are necessary, results obtained with nectarine, peach, and plums allow the conclusion that SIQ-FT is less sensitive to very soft fruits and more adequate to evaluate firmer fruits. For UCF values below 5 N, a range of SFI measurements between 5 and 15 was obtained for nectarines and peaches. The range for plums was between 0 and 7. It was also noticeable that correlation coefficients were high for the relationship between penetrometer and Sinclair in all cases studied. The creation of good estimative models was possible.

2.3.3 Texture Profile Analysis

It is an enormous challenge to perform practical tests that depend on the rheological behavior of food materials while correlating mechanical parameters of texture with sensory methods of texture evaluation. Despite this, the use of texturometers or Inston Universal Testing Machines allows the performance of a large range of tests related to the rheological properties of foods. A possible definition of food texture, proposed by Lawless and Heyman (1998), cited by Foedingen et al. (2011), is "all the rheological and structural (geometric and surface) attributes of the product perceptible by means of mechanical, tactile, and, where appropriate, visual and auditory receptors."

In the 1960s, some researchers began to publish scientific documents about food texture. It was an innovative approach to apply methods and definitions used in classical laboratories, factories, orchards, and markets, and make them accessible to a large number of professionals and technicians. However, that popularization, as well as the use of texture as a sensory concept, caused a a number of miscellaneous sensory and rheological/mechanical definitions.

In 1963, Friedman et al. published some important studies on the use of the General Foods Texturometer that approached the simulated mechanical studies of textural characteristics of foods. Bourne (1978) used texture profile analysis (TPA) by uniaxial compression of food.

The main goal of these textural and rheological tests, called *imitative tests*, is not to improve knowledge of fundamental rheological properties, but to measure physical properties as well as to understand the relations of these properties with a dynamic perception of texture. Nowadays, the most commonly used instrumental method is probably the compression method of TPA. This analysis mimics the conditions of the mastication process (Foegedinga et al., 2003).

Today, the need for standardized methods that permit measurement of textural parameters and correlation of them with sensory classification is still necessary. Many fundamental publications should be reread in order to improve adequate and correct use of texture profiling methods.

The performance of TPA should be still considered a new approach to studying food texture. This instrumental method, texture profiling, consists of compressing the test material twice and quantifying the mechanical parameters using force–deformation curves.

One of the advantages of using this method is the possibility of getting intense correlations between some instrumental parameters and sensory ratings, as can be noted in several works subsequently described for cheeses and meat (Szczesniak, 2002).

2.3.3.1 TPA Performance

Proof of the real need for this test is the large amount of work done using TPA tests through the years, with diverse food materials, from cheese to meat.

Jack et al. (1993) concluded that firmness and springiness were highly correlated with sensorial evaluation. Studying the effect of fat reduction on the texture of Cheddar cheese, Bryant et al. (1995) concluded that hardness, springiness, and adhesiveness in instrumental texture analysis were intensely correlated with sensory texture data for the same attributes.

Meullenet et al. (1998) also studied the correlation between sensory texture data and instrumental texture using twenty-one different food samples, including cheese. They found high linear correlations of sensory data with hardness and springiness.

Drake et al. (1999a) studied nine commercial and six processed experimental cheeses and concluded that the sensory term *firmness* was highly correlated to the instrumental term *hardness*.

Breuil and Muellenet (2001) worked with twenty-nine different cheese samples and performed three different instrumental analyses: uniaxial compression (as a control), needle puncture, and cone penetration. They found satisfactory correlations between instrumental and sensory attributes, mainly hardness, springiness, and cohesiveness of mass. They concluded that in order to optimize the prediction of sensory attributes, a combination of instrumental methods should be used.

Adhikari et al. (2003) studied the textural differences between low-fat, full-fat, and smoked Cheddar, Gouda, and Swiss cheeses by sensory and instrumental texture analysis. They also tried to correlate the sensory texture attributes with the instrumental texture parameters. They found a correlation between the sensory and the instrumental texture data sets. Hardness1 and hardness2 (instrumental parameters) were positively correlated to sensory texture attributes such as *dry*, *hardness*, and *crumbly*. These sensory attributes can be used to describe cheese varieties that are perceived as hard during mastication.

Foegedinga et al. (2003) affirmed that to understand the molecular basis of texture, fundamental tests are required. They reported that resilience is highly correlated with rheological properties, while adhesiveness and cohesiveness of cheese are weakly correlated with rheological parameters. In spite of all these works done in the 1990s, recently Joshi et al. (2004) presented a work that allows comparison among different rheological approaches. Differently processed Cheddar cheeses were evaluated using instrumental TPA, stress relaxation characteristics, and viscoelastic characteristics (elastic modulus-G' and viscous modulus-G"). Results indicated that the methods of instrumental texture and rheology evaluation had common outcomes for hardness and viscoelasticity of cheeses. Results of stress relaxation tests were more useful in differentiating the cheese characteristics as compared to the results of instrumental TPA and dynamic rheology tests.

It is important to understand and measure meat texture, from a practical point of view, in order to control quality. This goal improved the research on rheological methods with good results. In spite of the inexistence of standardized methods, a difficult task to describe and compare meat texture has been developed. At 1996 a group supported by the Organization for Economic Cooperation and Development (OECD), was asked to study and standardize methods, based on rheological knowledge, to measure texture. Honikel presented different rheological methods: the penetrometer measurement that resembles the process of mastication, the first bite between the teeth, and which measurements can be related to sensorial evaluation. However, it is not clear which structural properties of the meat are evaluated. Honikel (1997) proposed a procedure using a cylindrical flat-ended plunger (diameter 1.13 cm, area = 1 cm*) that is driven vertically 80% of the way through a 1-cm-thick meat sample cut so that the fiber axis is perpendicular to the direction of the penetration. The plunger is driven (100 mm per min) twice into the meat at each location and the work and force deformation curves are recorded. The parameters that should be recorded are: hardness as the maximal force for first deformation (*N*); cohesiveness as the ratio of work done during the second penetration, relative to the first; gumminess as hardness × cohesiveness; other parameters can also be defined. However, the author strongly recommended that the methods for determining tenderness should be validated against sensory panels. The ideal of a single measurement to accurately predict consumer perceptions under all conditions may not be achievable.

The compression parameters obtained with TPA have been used by many authors in their evaluations of meat products to determine the quality of the product, to select adequate functional ingredients, and to study new techniques. Herrero et al. (2007) performed TPA, physicochemical measurements (pH, aw, dry matter, fat content), and breaking strength by tensile test to study fermented sausages. The results obtained allowed these meat products to be grouped according to four different textural profiles. These profiles were characterized ($p < 0.05$) by the values of breaking strength (BS), hardness, adhesiveness, cohesiveness, and springiness. Multivariate analysis confirmed that BS and TPA parameters were correlated significantly ($p < 0.00005$). Based on these results, TPA parameters could be used to construct regression models to predict BS, and therefore to obtain a more complete textural property description of dry fermented sausages.

Rubio et al. (2007) studied the microbiological, physicochemical and sensory properties of salchichon with high unsaturated fat content, packed under vacuum and 20/80% CO_2/N_2 modified atmosphere, in order to evaluate its quality changes during storage under refrigeration. Instrumental TPA was performed; the parameters determined from the force–time curves were hardness, springiness, cohesiveness, and chewiness.

Yang et al. (2007) investigated new prepared formulas of low-fat sausages with added hydrated oatmeal or tofu as texture-modifying agents at three different levels of 10%, 15%, and 25% (w/w). They tried to determine the effects of the type and level of texture-modifying agents on the physical and sensory properties of low-fat sausages. TPA was one of the tests performed, and from the resulting force–deformation curves, the textural parameters of hardness, cohesiveness, springiness, gumminess, and chewiness were calculated. The influence of texture-modifying agents on hardness associated with the water binding property of the agents is complicated and remains in dispute.

Wang et al. (2009) also performed TPA to study the effects of phosphate level on water-holding capacity and texture of emulsion-type sausage during storage.

Spaziani at al. (2009) investigated low-acid sausages in order to characterize their physicochemical, microbiological, and textural properties during ripening. Texture profile analysis was performed and once more the determined parameters were: hardness (kg; the peak force during the first compression cycle), cohesiveness (dimensionless; ratio of the positive force area during the second compression to that during the first compression excluding the areas under the decompression portion of each cycle), adhesiveness (J; the negative force area after the first compression). These traditional low-acid sausages could be considered more like dried sausages than fermented sausages and they are easily distinguishable by sensory analysis and characterized by textural parameters showing low hardness and cohesiveness.

2.4 VISCOSITY

2.4.1 Definition

The viscosity of fluid food is an important property, which has many applications in food technology, such as developing food processes, the control of products, quality evaluation, and an understanding of the structure of food (Alvarado and Romero, 1989, Walker and Prescott, 2000).

2.4.2 Fluid Behavior in Steady Shear Flow: Newtonian and Non-Newtonian Fluids

Viscosity of a fluid is the internal resistance to flow when a shear force is applied. According to flow behavior, fluids can be classified into two main groups: the Newtonian and non-Newtonian fluids. When the viscosity of a liquid remains

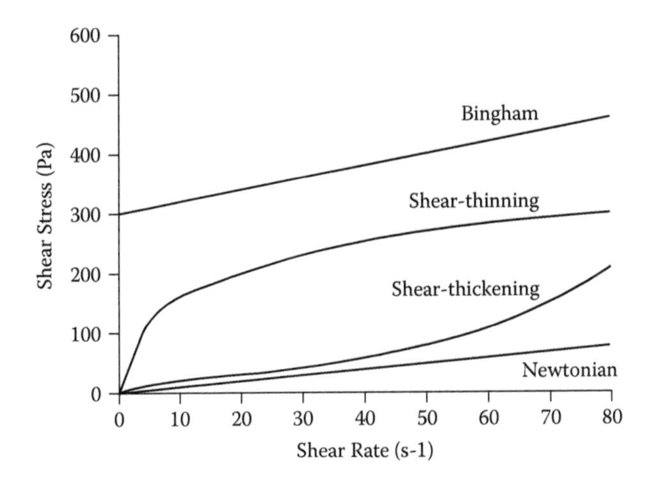

FIGURE 2.5 Flow curves for time-independent fluid foods.

constant and is independent of the applied shear stress, the liquid is termed a *Newtonian liquid*. In this type of fluid, the relationship between shear stress and shear rate is a straight line (Figure 2.5).

A *non-Newtonian fluid* is defined as one for which the relationship between shear stress and shear rate is not a constant. The viscosity of such fluids depends on the applied shear force and time. In this case, the experimental parameters used (viscometer model, spindle, and speed) have an effect on the measured viscosity of a non-Newtonian fluid. This viscosity is called the *apparent viscosity* of the fluid and is accurate only when experimental parameters are known.

Non-Newtonian flow can be explained by thinking of any fluid as a mixture of molecules with different shapes and sizes. As they pass by each other, during flow, their size, shape, and cohesiveness will determine how much force is required to move them. At each specific rate of shear, the alignment may be different and more or less force may be required to maintain motion. There are several types of non-Newtonian flow behavior, characterized by changes in the fluid's viscosity in response to variations in shear rate. The most common types of non-Newtonian flow behaviors include: *shear-thinning*, *shear thickening*, and *plastic*.

Fluids with a *shear-thinning* behavior will display a decreasing viscosity with an increasing shear rate, as shown in the Figure 2.5. The breakdown of structural units in a food due to the hydrodynamic forces generated during shear is the main reason for this flow behavior (Rao, 2007). The most common of the non-Newtonian fluids, shear-thinning behavior is exhibited for many foods, including salad dressings and some concentrated fruit juices. This type of flow behavior is sometimes called *pseudoplastic*. Increasing viscosity with an increase in shear rate characterizes the *shear-thickening*, also called *dilatant* fluid. The shear-thickening behavior presents a flow curve concave downward, representing that an increasing shear stress gives a less than proportional increase in shear rate. This behavior should be due to an increase in the size of the structural units as a result of shear (Rao, 2007). Although rarer than shear-thinning, shear-thickening behavior is frequently observed in fluids containing high levels of deflocculated solids, such as corn starch in water, and partially gelatinized starch dispersions.

The *plastic* fluid will behave as a solid under static conditions. A certain amount of force must be applied to the fluid before any flow is induced; this force is called the *yield stress*. Once the yield value is exceeded and flow begins, plastic fluids may display Newtonian (called Bingham plastic model), shear-thinning, or shear-thickening flow characteristics. Shear-thinning behavior with yield stress is exhibited by foods such as tomato concentrates, tomato ketchup, mustard, and mayonnaise.

Some fluids will display a change in viscosity with time under conditions of constant shear rate. There are two categories to consider: *thixotropic* and *rheopectic*. A thixotropic fluid undergoes a decrease in viscosity with time, while it is subjected to constant shearing. The time element is extremely variable; under conditions of constant shear, some fluids will reach their final viscosity value in a few seconds, while others may take up to several days.

When subjected to varying rates of shear, a thixotropic fluid will react as illustrated in Figure 2.6. A plot of shear stress versus shear rate was made as the shear rate was increased to a certain value, then immediately decreased to the starting point. This *hysteresis loop* is caused by the decrease in the fluid's viscosity with increasing time of shearing. Such effects may or may not be reversible; some thixotropic fluids, if allowed to stand undisturbed for a while, will regain their initial viscosity, while others never will.

Most of the foods with *thixotropic* behavior are heterogeneous systems containing a dispersed phase. At rest, the particles or molecules in the food are linked together by weak forces, but when the hydrodynamic forces during shear are sufficiently high, the interparticle linkages are broken, resulting in a reduction in the size of the structural particles that offer lower resistance to flow during shear (Mewis, 1979). Thixotropy is frequently observed in foods such as salad dressings and soft cheeses. *Rheopexy* is essentially the opposite of thixotropic behavior, in that the fluid's viscosity increases with time as it is sheared at a constant rate. Rheopectic fluids are rarely encountered.

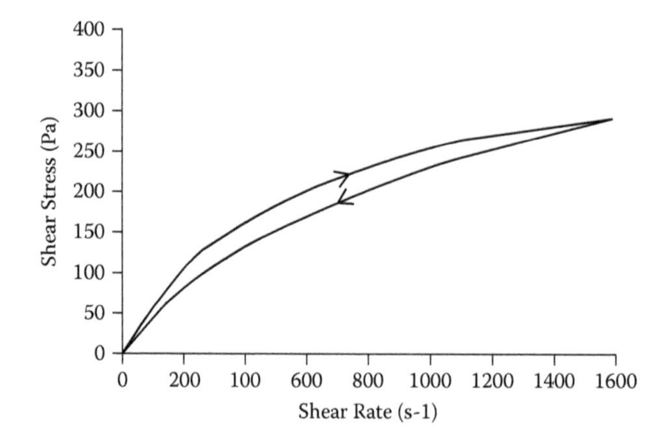

FIGURE 2.6 Tixotropic behavior.

Viscosity measurement has emerged as an integral and necessary component of many quality control procedures in food processing. The study of the Newtonian and non-Newtonian flow behavior of foods requires considerable care and adequate instrumentation. Viscosity of a fluid can be measured by a number of approaches and methods. Rotational, vibrational, and tube viscometers are the more common types of traditional measuring techniques based upon controlled deformation of the sample.

2.4.3 Current Measuring Techniques

2.4.3.1 Rotational Viscometers

Rotational viscometers may be operated in the steady or oscillatory mode. A spindle (cylinder, cone, or plate) is rotated continuously in a liquid and the torque required to rotate the spindle at this rate is measured. Therefore, only two parameters are measured, the torque and angular velocity on the rotating spindle. Shear stress is calculated from the torque, and shear rate from the angular velocity. Viscosity is the relation between shear stress on the surface of the turning cylinder and shear rate.

In rotational methods, the test fluid is continuously sheared between two surfaces, one or both of which are rotating. These devices have the advantage of being able to shear the sample for an unlimited period of time, under controlled rheometric conditions. Rotational methods can also incorporate oscillatory and normal stress tests for viscoelastic properties characterization. In general, rotational methods are better suited for the measurement of concentrated suspensions, gels, and pastes, but are generally less precise as compared to capillary methods.

Rotational measurements fall into one of two categories: stress or rate controlled. In stress-controlled measurements, a constant torque is applied to the measuring tool in order to generate rotation, and the resulting rotation speed is then determined. In rate-controlled measurements, a constant rotation speed is maintained and the resulting torque generated by the sample is determined using a suitable stress-sensing device. Some commercial instruments have the capability of operating in both modes: stress or rate controlled. Rotational systems are generally used to investigate time-dependent behavior.

2.4.3.2 Vibrating Viscometers

Vibrational viscometers are the most common instruments for process control systems (Steffe, 1996). Simplicity of probe design makes the technique attractive for process engineers; they are robust, easy to clean in place, and have the added advantage of no moving parts (Cullen et al., 2000). Vibrational viscometers are considered surface-loaded systems, responding to a thin layer of fluid that surrounds the oscillating probe. Measurement depends on the surrounding fluid dampening probe vibration, in proportion to its viscosity and density (Cullen et al., 2000). An advantage of vibration viscometry over rotational is that it may be employed in high-pressure applications, such as spray drying. Shear rate, however, is undefined as it is not only a function of instrument parameters (i.e., probe geometry, frequency and amplitude of oscillation), but also fluid parameters (i.e., density and viscosity) (Cullen et al., 2000).

2.4.3.3 Tube Viscometers

A tube viscometer consists in a cylinder where a fluid flows, based in pressure difference established. This method is based on measuring the time that a determinate amount of fluid takes to flow through a capillary of known diameter and length. The length-to-diameter ratio may range from 2 to 400, but is typically on the order of 100. Raw data for tube or capillary viscometers are pressure drop and volumetric flow rate. Their construction is simple, flow is continuous, and sample-holding time can be very short (Cullen et al., 2000).

2.4.4 Novel Measuring Techniques

2.4.4.1 Tomographic Velocity Profiling

Numerous tomographic techniques are available or emerging, based upon determining the velocity profile at a specified cross section of a pipe and the corresponding pressure drop over a known length. Shear rate information may be determined by differentiating velocity values as a function of position, ranging from zero at the tube center to a maximum at the tube wall (Cullen et al., 2000).

Shear stress data is determined from auxiliary pressure drop and conservation of linear momentum (tube viscometry). In contrast to conventional tube viscometry, which depends on volumetric formulas to determine one viscosity data point from a single pressure drop, tomographic viscometry provides multiple shear viscosity/shear rate data points from one combined viscosity profile/pressure drop measurement (Arola et al., 1998). Advantages of these techniques include their noninvasive and nondestructive mode of operation along with determination of rheological model parameters (shear viscosity, yield stress, slip velocity) on-line or in-line (Cullen et al., 2000).

2.4.4.2 Ultrasonic Doppler Velocimetry (UDV)

UDV has proven to be an effective tool for on-line viscosity determination of Newtonian, non-Newtonian, and multiphase opaque foods (Cullen et al., 2000). Rheological analysis of flow processes in real time on an industrial scale has been evaluated for crystallization processes: concentrated chocolate (Windhab et al., 1996) and fat suspensions (Ouriev et al., 2000). This technique, teamed with a slit rheometer, was reported by Ouriev (2000) to be employed in an industrial dough extrusion process. The technique may also be extended to determine flow behavior, which affects rheological measurements (slip velocity) and more complex rheological behavior such as viscoelasticity, extensional viscosity (McCarthy, 2000) and yield stress (Ouriev, 2000). The response time for UDV is fast, on the order one second or less.

2.4.4.3 Ultrasonic Reflectance Sensor

Ultrasonic sensors have been used for process monitoring of viscosity measurement in real time. Ultrasonic sensors are based upon reflectance at the liquid interface rather than transmittance through the liquid. A density and viscosity sensor employing longitudinal and shear ultrasonic reflectance was applied to measure the viscosity of water-based solutions and slurries displaying potential for many food products (Greenwood et al., 1999). Reported advantages of the technique include robustness and nonsensitivity to flow rate, entrained air, or vibrations. Using a similar in-line technique, Saggin (2000) found shear reflectance as a sensitive measurement of oil viscosity and both shear and longitudinal reflectance as a measure of solid content of lipids.

2.4.4.4 Mass-Detecting Capillary Viscometer

Shin and Keum (2003) studied the application of a newly designed mass-detecting capillary viscometer (MDCV) to measure the rheological characteristics of dairy food continuously over a range of shear rates by single measurement of liquid–mass variation with time. The instrument consists of a pipette pump, falling tube, glass capillary tube, glass adapter, receptacle, load cell, and computer data acquisition system. The essential feature in an MDCV is the use of a precision mass balance to measure the fluid collected in the receptacle, $m(t)$, every 0.05 s with a resolution of 0.01 g. According to the authors, this new viscometer presents important advantages when compared with established viscosity measurement techniques with a RotoVisco, such as simplicity (i.e., ease of operation and no moving parts) and the ability to make accurate measurements over a relatively broad shear rate range (Shin and Keum, 2003).

2.4.5 Viscosity Measuring Applied to Fluid Foods Submitted to High Pressure

High-pressure processing (HPP) has been steadily gaining popularity over the past fifteen years since it can inactivate pathogenic microorganisms with minimal heat treatment and offers good retention of nutritional and sensory (color and flavor) characteristics (Mertens and Deplace, 1993; Mozhaev et al., 1994; Tedford et al., 1998). HPP is well known to affect the conformation of macromolecules within foods such as proteins (Cheftel and Culioli, 1997). Covalent bonds within food constituents are less affected than weaker interactions such as hydrogen bonds, van der Waals forces, and hydrophobic bonds. This situation results in the retention of food quality attributes including vitamins, pigments, and flavor components (Indrawati and Hendrickx, 2002). HPP influences the viscosity of fluids, and in particular, oils. Oils and lipids are related to the organoleptic properties of many foods in which mouthfeel is an important attribute that relates to viscosity. It is important to have an understanding of the nature of viscosity at high pressure since it has an expressive influence in terms of plant design, quality control of both raw material and product at different stages of the process, evaluation of sensory attributes, and assessment of food structure and molecular conformation (McKenna and Lyng, 2001).

Schaschke et al. (2006) reported the measurement of the viscosity of olive oil at high pressure using a falling sinker–type viscometer. The design of the viscometer relies on a close-fitting self-centering sinker enclosed within a vertical tube maintained at a controlled temperature in which gravity was used as the applied force. The viscometer consists of a vertical tube through which the sinker falls. Samples of oils are sealed into the tube using a shrinkable polytetrafluoroethylene (PTFE) expansion sheath.

Kiełczyński et al. (2008) studied a new method for measuring the viscosity of liquids at high pressure. The authors used an ultrasonic method involving the Bleustein-Gulyaev (BG) surface acoustic wave. According to the authors, the change in the complex propagation constant of the BG wave produced by the layer of liquid loading the waveguide surface is proportional to the shear mechanical impedance of the liquid. With this new method it is also possible to measure the viscosity of liquids during the phase transition and during the decompression process.

REFERENCES

Adhikari, K., H. Heymann, and H.E. Huff. 2003. Textural characteristics of low-fat, full-fat, and smoked cheeses: Sensory and instrumental approaches. *Food Quality and Preference* 14:211–218.

Alvarado, J.D., and C.H. Romero. 1989. Physical properties of fruits: density and viscosity of juices as functions of soluble solids and content and temperature. *Latin American Applied Research* 19(15):15–21.

Armstrong, P.R., H.R. Zapp, and G.K. Brown. 1990. Impulsive excitation of acoustic vibration in apples for firmness determination. *Transactions of the ASAE* 33(4): 1353–1359.

Arola, D.F., G.A. Barrall, R.L. Powell, K.L. McCarthy, and M.J. McCarthy. 1998. A simplified method for accuracy estimation of nuclear magnetic resonant imaging. *Review of Scientific Instruments* 69:3300–3307.

ASAE (American Society of Agricultural Engineering) Standards. 2004. Compression test of food materials of convex shape. *American Society of Agricultural Engineering* S368.4:585–592.

Balasubramaniam, V.M., and D. Farkas. 2008. High pressure food processing. *Food Science and Technology International* 14(5):413–418.

Barbosa-Cánovas, G.V., and J. Rodriguez. 2005. Thermodynamic aspects of high hydrostatic pressure food processing. In *Novel food processing technologies*, ed. Barbosa-Cánovas, G.V., Tapia, M., Pilar Cano, M.. Boca Raton, FL: CRC Press.

Barreiro, P. 1994. Modelos para la Simulación de Daños Mecánicos, y Desarrollo de un Algoritmo de Evaluación de Maquinaria para los Principales Cultivares de Albaricoque, Manzana, Melocotón y Pera, 256. PhD Thesis. Universidad Politécnica de Madrid, Escuela Técnica Superior de Ingenieros Agrónomos. Madrid, Spain.

Barreiro, P., and M. Ruiz-Altisent. 1994. Bruise susceptibility in pome fruits under different loading and storage conditions. *Fruit, Nut and Vegetable Production Engineering* 2:185–192.

Bellon, V., J.L. Vigneau, and M. Crochon. 1994. Non destructive sensing of peach firmness. *Fruit, Nut, and Vegetable Production Engineering* 2:291–297.

Bourne, M.C. 1978. Texture profile analysis. *Food Technology* 32(7):62–66.

Breuil, P., and J.F. Meullenet. 2001. A comparison of three instrumental tests for predicting sensory texture profiles of cheese. *Journal of Texture Studies* 32:41–55.

Bryant, A., Z. Ustonol, and J. Steffe. 1995. Texture of cheddar cheese as influenced by fat reduction. *Journal of Food Science* 60:1216–1219.

Cheftel, J.C., and J. Culioli. 1997. Effects of high pressure on meat: A review. *Meat Science* 46(3):211–236.

Chen, P., and R.B. Fridley. 1979. Quality evaluation of agricultural products based on their mechanical properties. *Proceedings workshop on design applications of mechanical properties*. Pennsylvania University, Philadelphia, USA.

Chen, P., and L. Huarng. 1992. Factors affecting acoustic responses of apples. *Transactions of the ASAE* 35(6):1915–1920.
Chen, P., and Z. Sun. 1984. Critical strain failure criterion: Pros and cons. *Transactions of the ASAE* 27(1):278–281.
Chen, P., S. Tang, and S. Chen. 1985. Instrument for testing the response of fruit to impact force. *Transactions of the ASAE* 30(1):249–254.
Chen, P., and Y. Tjan. 1998. A real-time impact sensing system for online firmness sensing of fruits. International Conference on Agricultural Engineering, AgEng 98, Part I: 314–315. Oslo.
Correa, P.C. 1992. Estudio de los Índices de Madurez Y Calidad del Aguacate Tratado por Frio Mediante Ensayos no Destrutivos, 226. PhD Thesis. Universidad Politécnica de Madrid. Escuela Técnica Superior de Ingenieros Agrónomos, Madrid.
Cullen, P.J., A. Duffy, C.P. O'Donnell, and D. O'Callaghan. 2000. Process viscometry for the food. *Trends in Food Science and Technology* 11:451–457.
Delwiche, M.J. 1986. Theory of fruit firmness sorting by impact forces. *Transactions of the ASAE* 30 (4):1160–1166.
Delwiche, M.J., T. McDonald, and S.V. Bowers. 1987. Determination of peach firmness by analysis of impact force. *Transactions of the ASAE* 30(1):249–254.
Delwiche, M.J., S. Tang, and J.J. Mehlschan. 1988. A fruit firmness sorting system. Paper presented at the A.E.I. Conference, Paris.
Denys, S., L. Ludhikhuyze, A. Van Loey, and M. Hendrickx. 2000. Modeling conductive heat transfer and process uniformity during batch high pressure processing of foods. *Biotechnology Progress* 16:92–101.
Drake, M.A., P.D. Gerard, V.D. Truong, and C.R. Daubert. 1999a. Relationship between instrumental and sensory measurements of cheese texture. *Journal of Texture Studies* 30:451–476.
Drake, M.A., V.D. Truong, and C.R. Daubert. 1999b. Rheological and sensory properties of reduced fat processed cheeses containing lecithin. *Journal of Food Science* 64:744–747.
Fekete, A., and J. Felföldi .1994. Fruit Firmness Tester. International Conference on Agricultural Engineering, AgEng 94: 885–886. Milano, Itália.
Figura, L.O., and A.A. Teixeira. 2007. *Food physics: Physical properties-measurement and application*. Berlin: Springer-Verlag.
Finney, E.E. 1970. Mechanical resonance within Red Delicious apples and its relation to fruit texture. *Transactions of the ASAE* 13(2):177–180.
Finney, E.E. 1971. Random vibration techniques for non destructive evaluation of peach firmness. *Journal of Agricultural Engineering Research* 16(1):81–87.
Finney, E.E. 1972. Vibration techniques for fruit firmness. *Journal of Texture Studies* 3:263–283.
Finney, E.E., and D.R. Massie. 1975. Instrumentation for testing the response of fruits to mechanical impact. *Transactions of the ASAE* 18(6):1184–1192.
Fluck, R.C. 1975. Use of impact parameters to control handling damage to fruit and vegetables. Design application of mechanical properties of solid foods materials. *Proceedings of workshop, Department of Agricultural Engineering*. Pennsylvania University, USA
Fluck, R.C., and E.M. Ahmed. 1973. Impact testing of fruit and vegetables. *Transactions of the ASAE* 16(4):660–666.
Foedingen, E.A., Daubet, C.R., Drake, M.A., Essick, G., Trulsson, M., Vinyard, C.J., and F. Van de Velde. 2011. A comprehensive approach to understanding textural properties of semi and soft-solid foods. *Journal of Textural Studies* 42(2):103–129.
Foegeding, E.A., J. Brown, M.A. Drakea, and C.R. Daubert. 2003. Review: Sensory and mechanical aspects of cheese texture. *International Dairy Journal* 13:585–591.

Fridley, R., R. Bradley, and P. Adrian. 1968. Some aspects of elastic behaviour of selected fruits. *Transactions of the ASAE* 11:46–49.

Friedman, H.H., J.E. Whitney, and A.S. Szczesniak. 1963. The Texturometer—A new instrument for measurement. *Journal of Food Science* 28:390–396.

Garcia, C.A. 1988. Impacto Mecanico en Frutos: Técnicas de Ensayo y Aplicación en Variedades de Pera e Manzana. PhD Thesis. Universidad Politécnica de Madrid, Escuela Técnica Superior de Ingenieros Agrónomos, Spain.

Goldsmith, W. 1960. *Impact: The theory and physical behaviour of cooling solids*. London: Arnold Publishers LTD.

Greenwood, M.S., J.R. Skorpik, and J.A. Bamberger. 1999. Online sensor for density and viscosity measurement of a liquid or slurry for process control in the food industry. *Proceedings of AIChE Sixth Conference on Food Engineering*, Dallas, TX, October 31–November 5, ed. G.V. Barbosa-Canovas and S.P. Lombardo, 691–696.

Grotte, M., F. Duprat, E. Piétri, and D. Loonis. 2002. Young's modulus, Poisson's ratio, and Lame's coefficients of Golden Delicious apple. *International Journal of Food Properties* 5(2):333–349.

Hamann, D.D., and D.E. Carroll. 1971. Ripeness sorting of muscadine grapes by use of low-frequency vibration energy. *Journal of Food Science* 36:1049–1051.

Harker F.R., J. Maindonald, S.H. Murray, F.A. Gunson, I.C. Hallett, and S.B. Walker. 2002. Sensory interpretation of instrumental measurements 1: Texture of apple fruit. *Postharvest Biology and Technology* 24:225–239.

Hartmann, C., and A. Delgado. 2005. Numerical simulation of thermal and fluid dynamical transport effects on a high pressure induced inactivation. *Simulation Modelling Practice and Theory* 13:109–118.

Hayashi, S., J. Sugiyama, and K. Otobe. 1995. Nondestructive quality evaluation of fruits and vegetables by acoustic transmission waves. *Proceedings of International Symposium on Automation and Robotics in Bioproduction and Processing*. Kobe University, Japan.

Hellebrand, J. 1985. Mechanical properties of materials under dynamic loading. III Congress of Physical Properties of Agricultural Materials, Praga 1985: 89–95.

Herrero, A.M., J.A. Ordoñez, R. de Avila, B. Herranz, L. de la Hoz, and M.I. Cambero. 2007. Breaking strength of dry fermented sausages and their correlation with texture profile analysis (TPA) and physico-chemical characteristics. *Meat Science* 77:331–338.

Holt, J.E., and D. Schoorl. 1977. Bruising and energy dissipation in apples. *Journal of Texture Studies* 7(4):421–432.

Honikel, K.O. 1997. Reference methods supported by OECD and their use in Mediterranean meat products. *Food Chemistry* 59(4):573–582.

Horsfield, B.D., R.B. Fridley, and L.L. Clay. 1972. Application of elasticity to the design of fruit harvesting and handling equipment for minimum bruising. *Transactions of the ASAE* 15(4):746–750.

Indrawati, A.V.L., and M. Hendrickx. 2002. High pressure processing. In *The nutrition handbook for food processors*, ed. C.J.K. Henry and C. Chapman, 433–461. Cambridge, UK: Woolhead Publishing.

Jack, F.R., A. Paterson, and J.R. Piggott. 1993. Relationships between composition and rheology of cheddar cheeses and texture as perceived by consumers. *International Journal of Food Science and Technology* 28:293–302.

Jarén, C., M. Ruiz-Altisent, and R. Pérez de Rueda. 1992. Sensing physical stage of fruits by their response to nondestructive impacts. AgEng 92. International Conference on Agricultural Engineering. Paper no. 9211-13. Uppsala, Sweden.

Jarén, M.C. 1994. Detección de la Textura de Frutos por Medio de Impactos No Destructivos: Desarrollo y Aplicaciones del Procedimiento de Clasificación. PhD Thesis. Universidad Politécnica de Madrid, Escuela Técnica Superior de Ingenieros Agronomos.

Jarimopas, B. 1984. Failure of apples under dynamic loading. Research Thesis. Israel, Faculty of Agricultural Engineering, Institute of Technology.

Joshi N.S., R.P. Jhala, K. Muthukumarappan, M.R. Acharya, and V. Mistry. 2004. Textural and rheological properties of processed cheese. *International Journal of Food Properties* 7(3):519–530.

Kiani, M.K., H. Maghsoudi, and S. Minaei. 2009. Determination of Poisson's ratio and Young's modulus of red bean grains. *Journal of Food Process Engineering* (doi: 10.1111/j.1745-4530.2009.00391.x).

Kiełczyński, P., M. Szalewski, R.M. Siegoczyński, and A.J. Rostocki. 2008. New ultrasonic Bleustein-Gulyaev wave method for measuring the viscosity of liquids at high pressure. *Review of Scientific Instruments* 79:026109–026109-3.

Kosma, A., and H. Cunningham. 1962. Tables for calculating the compressive surface stresses and deflections in the contact of two solid elastic bodies whose principle planes of curvature do not coincide. *Journal of Industrial Mathematics* 12:31–40.

Lawless, H.T., and H. Heymann. 1998. *Sensory evaluation of food: Practices and principles.* New York: Chapman and Hall.

Lichtensteiger, M.J., R.G. Holmes, M.Y. Hamdy, and J.L. Blaisdell. 1988. Impact parameters of spherical viscoelastic objects and tomatoes. *Transactions of the ASAE* 31(2): 595–602.

Magness, J.R., and G.F. Taylor. 1925. *Na improved type of pressure tester for the determination of fruit maturity.* US Department of Agriculture Circular No. 350.

Malkin, A.Y.1994. *Rheology: Fundamentals.* Toronto-Scarborough, Canada: ChemTec Publishing.

Manor, A.N. 1978. Critical analysis of the mechanics of fruit damage under impact conditions. PhD Thesis, Pennsylvania State University.

McCarthy, K. 2000. Tomographic techniques to measure fluid viscosity: MRI and UDV. In *IFT Annual Meeting: Book of Abstracts*, Chicago, IL: Institute of Food Technologists.

McGlone, V.A., and P.N. Schaare. 1993. The application of impact response analysis in the New Zealand fruit industry. American Society of Agricultural Engineers Paper No. 93-6537, 9.

McKenna, B.M., and J.G. Lyng. 2001. Rheological measurements of foods. In *Instrumentation and sensors for the food industry*, 2nd ed., ed. E. Kress-Rogers and J.B. Brimelow, 425–452. Cambridge, UK: Woodhead Publishing.

Mertens, B., and G. Deplace. 1993. Engineering aspects of high-pressure technology in the food industry. *Food Technology* 47:164–169.

Meullenet, J.F., B.G. Lyon, J.A. Carpenter, and C.E. Lyon. 1998. Relationship between sensory and instrumental texture profile attributes. *Journal of Sensory Studies* 13:77– 93.

Mewis, J. 1979. Thixotropy: A general review. *Journal of NonNewtonian Fluid Mechanics* 6:1–20.

Min S., S.K. Sastry, and V.M. Balasubramaniam. 2010. Compressibility and density of select liquid and solid foods under pressures up to 700 MPa. *Journal of Food Engineering* 96: 568–574.

Misra, R.N., and H. Young. 1981. A model for predicting the effect of moisture content on the modulus of elasticity of soybeans. *Transactions of the ASAE* 24(5):1338–1341.

Mohsenin, N.N. 1972. Mechanical properties of fruits and vegetables. Review of a decade of research. Applications and future needs. *Transactions of the ASAE* 15:1064–1070.

Mohsenin N.N. 1986. *Physical properties of plant and animal materials*, 2nd ed. New York: Gordon and Breach Science Publishing.

Mozhaev, V.V., K. Heremans, J. Frank, P. Masson, and C. Balny. 1994. Exploiting the effects of high hydrostatic pressure in biotechnological applications. *TIBTECH* 12:493–501.

Ohsima, T., H. Ushio, and C. Koizumi. 1993. High-pressure processing of fish and fish products. *Trends in Food Science Technology* 4:370–375.

O'Reilly, C.E., P.M. Murphy, A.L. Kelly, T.P. Guinee, M.A.E. Auty, and T.P. Beresford. 2002. The effect of high pressure treatment on the functional and rheological properties of Mozzarella cheese. *Innovative Food Science and Emerging Technologies* 3(1):3–9.

Otero, L., and P.D. Sanz. 2003. Modeling heat transfer in high pressure food processing: A review. *Innovative Food Science and Emerging Technologies* 4:121–134.

Ouriev, B. 2000. Ultrasound doppler based in-line rheometry of highly concentrated suspensions. *Applied Rheology* 10:148–150.

Ouriev, B., B. Breitschuh, and E.J. Windhab. 2000. Rheological investigation of concentrated suspensions using a novel inline doppler ultrasound method. *Colloid Journal* 62:234–237.

Peleg, M. 1987. The basics of solid food rheology. In *Food Texture Instrumental and Sensory Measurement*, ed. P. Moskowitz, 391–394. New York: Marcel Dekker.

Rao, M.A. 2007. *Rheology of fluids and semisolid foods*, 2nd ed. Pullman, WA: Springer Science and Business Media, LLC.

Rastogi, N.K., K.S. Raghavarao, V.M. Balasubramaniam, K. Niranjan, and D. Knorr. 2007. Opportunities and challenges in high pressure processing of foods: Critical reviews. *Food Science and Nutrition* 47(1):69–112.

Rubio, B., B. Martínez, M.J. Sánchez, M.D. García-Cachán, J. Rovira, and I. Jaime. 2007. Study of the shelf life of a dry fermented sausage "salchichon" made from raw material enriched in monounsaturated and polyunsaturated fatty acids and stored under modified atmospheres. *Meat Science* 76:128–137.

Ruiz-Altisent, M. 1986. Las Propriedades Físicas de los Produtos Hortícolas en Relación con su Recolección y Manipulación Mecanica. 18ª Conferencia International de Mecanización Agrária: 113–123. Zaragoza, Spain.

Ruiz-Altisent, M. 1993. Nondestructive quality measurement and modelling in fruits. In *Proceedings of the 5th International Conference on Physical Properties of Agricultural Materials*. Bonn, Germany.

Ruiz-Altisent, M., J.G. Sierra, C.G. Alonso, and L.R. Sinobas. 1986. Daños por Impacto en Frutos: Parâmetros e Métodos Experimentales. II Congreso Nacional de la Sociedad Española de Ciencias Hortícolas, Cordova, Spain.

Ruiz-Altisent, M., J.G. Sierra, P. Chen, and F.M. Lu. 1987. Methods for studying resistance to impact and compression in fruits: Application to four varieties of Asian pears. Congreso Mundial de Tecnologia de Alimentos, Barcelona, Spain.

Saggin, R. 2000. Ultrasonic characterization of oil viscosity and solids content. In *IFT Annual Meeting: Book of Abstracts*, session 49(5).

Santos, A.C.A. 1991. Determinação de Características Físicas em Melões 'Branco da Lezíria' e 'Piel de Sapo'. Masters Thesis, Instituto Superior de Agronomia, Universidade Técnica de Lisboa. Portugal.

Sarker, N., and R.R. Wolfe. 1983. Computer vision based system for quality separation of fresh market tomatoes. *Transactions of the ASAE* 28(5):1714–1718.

Schaschke, C.J., S. Allio, and E. Holmberg. 2006. Viscosity measurement of vegetable oil at high pressure. *Food and Bioproducts Processing* 84(C3):173–178.

Shin, S., and D.Y. Keum. 2003. Viscosity measurement of non-Newtonian fluid foods with a mass-detecting capillary viscometer. *Journal of Food Engineering* 58:5–10.

Sinobas, L.R. 1988. Reacción a la Magulladura Producida por Impacto en Frutos: Estudio Fsiológico, Histológico, Mecanico e Técnicas para sua Evaluación. PhD Thesis. Universidad Politécnica de Madrid, Escuela Técnica Superior de Ingenieros Agronomos. Spain.

Spaziani, M., M. del Torre, and M.L. Stecchini. 2009. Changes of physicochemical, microbiological, and textural properties during ripening of Italian low-acid sausages. Proteolysis, sensory and volatile profiles. *Meat Science* 89:77–85.

Steffe, J.F. 1996. *Rheological methods in food process engineering*, 2nd ed. East Lansing, MI: Freeman Press.
Stikey, G. 1986. *Mechanics of agricultural materials*. Amsterdam: Elsevier Science.
Sugiyama, J., K. Otobe, S. Hayashi, and S. Usui. 1994. Firmness measurement of muskmelons by acoustic impulse transmission. *Transactions of the ASAE* 37(4):1235–1241.
Szczesniak, A.S. 2002. Texture is a sensory property. *Food Quality and Preference* 13:215–225.
Szczesniak, A.S., M.A. Brandt, and H.H. Friedman. 1963. Development of standard rating scales for mechanical parameters of texture and correlation between the objective and sensory methods of texture evaluation. *Journal of Food Science* 28:397–403.
Tedford, L.A., S. M. Kelly, N.C. Price, and C.J. Schaschke. 1998. Combined effects of thermal and pressure processing on food protein structure. *Food and Bioproducts Processing* 76(2):80–86.
Timoshenko, S.P., and J.N. Goodier. 1970. *Theory of elasticity*. London: McGraw Hill.
Ting, E., V.M. Balasubramaniam, and E. Raghubeer. 2002. Determining thermal effects in high-pressure processing. *Food Technology 56*(2):31–34.
Valero, C., C. H. Crisosto, and D. Garner. 2003. Introducing nondestructive flesh color and firmness sensors to the tree fruit industry. *Acta Horticulturae* 604:597–603.
Walker, S., and J. Prescot. 2000. The influence of solution viscosity and different viscosifying agents on apple juice flavor. *Journal of Sensory Studies* 15(3):285–307.
Wang, P., X.L. Xu, and G.H. Zhou. 2009. Effects of meat and phosphate level on water-holding capacity and texture of emulsion-type sausage during storage. *Agricultural Sciences in China* 8(12):1475–1481.
Windhab, E., B. Ouriev, T. Wagner, and M. Drost. 1996. Rheological study of non-Newtonian fluids. In *Proceedings of 1st International Symposium on Ultrasonic Doppler Methods for Fluid Mechanics and Fluid Engineering*, Villigen, Switzerland.
Wright, F.S., and W.E. Splinter. 1968. Mechanical behaviour of sweet potatoes under slow loading and impact loading. *Transactions of the ASAE* 11(6):765–770.
Yamamoto, H., and S. Haginuma. 1984a. Dynamic viscoelastic properties and acoustic properties of watermelons. *Report of National Food Research Institute* 44:30–35.
Yamamoto, H., and S. Haginuma. 1984b. Dynamic viscoelastic properties and acoustic properties of Japanese radish (Shogoin) roots. *Report of National Food Research Institute* 44:36–44.
Yamamoto, H., M. Iwamoto, and S. Haginuma. 1980. Acoustic impulse response method for measuring natural frequency of intact fruits and preliminary applications to internal quality evaluation of apples and watermelons. *Journal of Texture Studies* 1:117–136.
Yamamoto, H., M. Iwamoto, and S. Haginuma. 1981. Nondestructive acoustic impulse method for measuring internal quality of apples and watermelons. *Journal of Japanese Society of Horticultural Science* 50(2):247–261.
Yang, H.S., S.G. Choi, J.T. Jeon, G.B. Park, and S.T. Joo. 2007. Textural and sensory properties of low fat pork sausages with added hydrated oatmeal and tofu as texture-modifying agents. *Meat Science* 75:283–289.
Zhang, X., and G.H. Brusewitz. 1991. Impact force model related to peach firmness. *Transactions of the ASAE* 34(5):2094–2098.

3 Textural Properties of Foods

J. Ignacio Arana

CONTENTS

3.1 INTRODUCTION

Acceptance of foods is far more dependent on sensory quality than on nutritional quality. The texture of an edible material is defined by the British Standards Institution as the attribute of a substance resulting from a combination of physical properties perceived by the senses of touch (including kinesthesia and mouthfeel), sight, and hearing. Texture was defined by Szczesniak (1963) as "the sensory manifestation of the structure of the food and the manner in which the structure reacts to the applied forces, the specific sense involved being vision, kinaesthesia, and hearing." It is mainly a physical attribute and for many vegetables, texture seems to be a primary quality attribute (Lapsley, 1989). Texture perception is a highly dynamic process because the physical properties of food change continuously when they are manipulated in the mouth.

The need to characterize the global quality of foods, including not only flavor, size, shape, and color, but also the whole of the characteristics perceived by consumers has made it necessary to recognize texture as an important quality attribute. Texture can be defined by subjective terms such as: firmness, crispness, crunchiness, mealiness, woolliness, and so on.

The texture of foods has been traditionally assessed using sensory analysis, but it is subjective, costly, time consuming, and destructive. During recent years, much progress has been made in developing nondestructive techniques for the assessment of the texture of foods. A large number of devices and techniques have been developed to determine the firmness of foods; most of the tests are based on puncture, compression, or response to shear stress, creep, or impact, but sonic or ultrasonic vibration, spectroscopy, and other techniques are also used to evaluate the textural characteristics of foods.

3.2 NOVEL MEASURING TECHNIQUES FOR TEXTURAL FEATURES OF FOODS

3.2.1 Novel Techniques for Firmness Measurement

Firmness is one of the most important features characterizing fruit quality. It can be used to determine fruit and vegetable quality and freshness, to choose the moment to collect horticultural products, and to evaluate their ability to be preserved. In postharvest research, one of the main concerns is how to preserve the quality of the product so that it fulfills the expectation of consumers. For most foods, firmness is the key factor in deciding whether the product is accepted by the consumer, in view of the fact that it is related to the maturity of the product and it can be an indicator of its shelf life.

The firmness of food is commonly measured using sensory or mechanical tests. The firmness of vegetables is measured using compression or puncture tests. Cantwell (2004) reported that firm enough tomatoes offered a resistance of at least 18 N during compression tests and Taherian et al. (2009) took the force-to-deformation ratio, from the compression test, as a measure of the firmness of pieces of turnip (*Brassica napobrassica*) and beet root (*B. vulgaris L*). The firmness of a fruit has been traditionally estimated in a destructive manner by means of the Magness Taylor test.

In this test a cylindrical plunger with a standardized diameter is driven into the fruit over a well-defined distance, and the maximum force is expressed as Magness-Taylor (MT) firmness (N). The penetrometer test simulates the mastication of fruit tissue in the mouth, and the rupture properties of the fruit tissue. It is a destructive method that can be only used for sampling batches of fruit, and does not allow for firmness-based grading.

The influence of firmness on food quality has made it necessary to develop novel nondestructive techniques to evaluate food firmness that can be incorporated into the grading chains. These techniques can be based on nondestructive impact response, microdeformation, vibration measurements, near-infrared (NIR) spectroscopy, ultrasonic wave propagation, imaging analysis, and nuclear magnetic resonance (NMR) spectroscopy, time-resolved spectroscopy (TRS), and magnetic resonance imaging (MRI).

3.2.1.1 Nondestructive Impact-Based Measurements

There are two techniques based on impact measurements: dropping the fruit on a force transducer (Finney and Massic, 1975; De Baerdemaeker et al., 1982) and impacting the fruit with a sensing element (Jarén and García-Pardo, 2001).

Schaare and McGlone (1997) detected kiwifruit firmness by dropping kiwifruits on an aluminum and rubber pad with four sensor units made of piezo film located under the pad. They achieved a good sorting efficiency compared to hand grading results. Ragni et al. (2010) used a conveyor belt that throws fruit onto a flat horizontal plate connected to a load cell for measuring the flesh firmness of kiwifruits. Using the Magness-Taylor firmness test as a reference, regression models were developed having a coefficient of determination of 0.823. The impact device did not cause mechanical damage to kiwifruits.

A sensor based on this technique has been patented by IVIA (the Valencian Institute for Agricultural Research) and FOMENSA Company. It has been used to detect puffed clementines on-line at a speed of 5 fruits per second with over 90% accuracy. This device has been mounted in an experimental production line together with a NIR sensor to sort apples, peaches, and nectarines (Gutierrez et al., 2003).

Mohsenin (1970) described a technique for measuring firmness that involved impacting fruits with a pendulum. This system is still used to measure damage in tomatoes (Desmet et al., 2002). There are many ways of using impact sensors: hitting the fruit with elements that include the sensor, putting the fruit over a load cell and letting a weight fall on it, placing the fruit on a flat plate with a load cell located beneath it (García-Ramos et al., 2005).

Chen et al. (1985) developed an instrument to measure fruit response to impacts. The device includes a small, semispherical mass with an accelerometer that can be dropped from different heights onto the fruit. García et al. (1988) developed a vertical impact sensor based on this system. This was used to sort apples and pears according to their firmness (Jarén and Carcía-Pardo, 2002) and to test and classify different padding materials in fruit packing lines (García-Ramos et al., 2002). Figure 3.1 shows this device.

FIGURE 3.1 Impact sensor.

Delwiche and Sarig (1991) developed a sensor operated by an air cylinder that released and returned an impact mass in the probe body. The acceleration of the mass gave a measurement of the impact force, which was analyzed. The correlation obtained between penetration firmness and the peak acceleration was 0.92 for peaches and 0.8 for pears. The sensor acted horizontally and was installed successfully in an experimental packing line (Delwiche et al., 1996).

Hung and Prussia (1995) presented a nondestructive laser-puff detector to measure the firmness of various foods. The excitation of the food was performed by a short puff of pressurized air, which could be regulated to a certain degree according to the firmness scale of each product. A laser displacement sensor measured the deflected surface of the fruit. They found that the laser-puff values correlated well ($R^2 = 0.78$) with destructive measurements in peaches.

Chen and Ruiz Altisent (1996) developed a lateral low-mass impact sensor. It consists of a small swing-arm sensor with lateral movement that impacts the fruit laterally with a semispherical head. The fruit firmness is estimated by a piezoelectric accelerometer located on the head. They found that the variables of impact force and impact duration were directly related to fruit firmness and reported a good performance of the system when testing rubber falls, kiwifruits, and peaches. The device was able to sense fruit firmness at a speed of 5 fruits per second. García Ramos et al. (2003) adapted a modified version of the low-mass impact method for an experimental packing line having an operation speed of 5 to 7 fruits per second. The correlation coefficient between the impact parameter and the force deformation slope during a destructive compression test was 0.93 for peaches. A manual impact sensor shaped like a pistol was developed to be used in orchards to determine optimum harvest date (Chen et al., 2000). Figure 3.2 shows a lateral impact sensor similar to those used by the cited authors.

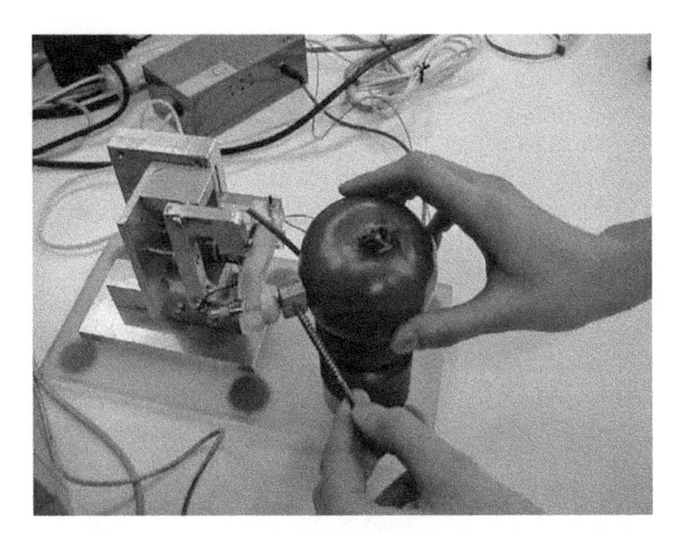

FIGURE 3.2 Lateral impact sensor used in the laboratory of the Public University of Navarre.

Greefa Ltd. introduced an on-line nondestructive firmness detection system (Armstrong, 2001), the intelligent Firmness Detection (iFD) system. Firmness measurements are taken by means of a large wheel equipped with multiple sensor heads that rotates over the packing line impacting the top of the fruits. During the sorting process, a firmness sensor takes 9 to 20 measurements around the fruit and the system operates at speeds of 5–7 fruits per second. This device has been successfully applied for apples, avocados, mangoes, peaches, and kiwifruits.

Sinclair International Ltd. has developed the Sinclair IQ™ – Firmness Tester (SIQ-FT), which is based on a low-mass impact sensor (Horwarth and Ioannides, 2002). This on-line system measures firmness using a piezoelectric sensor surrounded by a rubber bellow activated by compressed air. The system is able to take readings at a speed of 10 fruits per second and has been tested with avocados, citrus, kiwis, plums, nectarines, and peaches. Shmulevich et al. (2003) found medium to high correlation coefficients between SIQ-FT values and variables from penetration tests performed on nectarines (0.83–0.92), plums (0.80), avocados (0.81–0.84), and kiwifruits (0.83–0.92).

García Ramos et al. (2003) modified the impact system for firmness sorting of fruits (TOUCHLINE) developed by Chen and Ruiz Altisent (1996) and installed it in an experimental fruit packing line. The system consists of a lateral impact sensor with a control electronic circuit, control software, and an ejection system with three outlets regulated by a microcontroller. The lateral impact sensor detects the presence of fruit by means of an optical sensor and impacts it. The impact signal is obtained by an accelerometer and is sent to a PC where it is processed by specific software to yield a firmness index. According to this index the software gives an order to the microcontroller of the ejection system and the fruit is sent to its corresponding outlet. Diezma et al. (2006) used nondestructive acoustic and impact tests to monitor firmness evolution of peaches during storage. They found that the maximum force in ball compression correlated well with the maximum acceleration from the impact test (R^2 = 0.75) and with a band magnitude parameter from an acoustic test (R^2 = 0.71). They developed classification models able to classify correctly more than 90% of the peaches. Slaughter et al. (2009) designed a handheld impact sensor, based on low-mass impact, for nondestructive measurement of fruit firmness while the fruit remains attached to the tree. The device worked correctly because validation tests met (r^2 = 0.92 and 0.96, respectively) the ASAE Standard method S368.2 for determining the apparent modulus of intact fruit and the impact firmness scores from a commercial bench-top impact firmness instrument.

Homer et al. (2010) evaluated a nondestructive impact sensor that is another adaptation of the cited laboratory model, installed in an experimental fruit packing line with a commercial sizer chain. They found that sensors work correctly at a speed of 7 fruits per second and allow fruit classification at three levels of firmness.

3.2.1.2 Nondestructive Microdeformation-Based Sensors

Nondestructive sensors have been developed to measure very small deformations of fruits during compression. The principle of measurement consists of indenting the surface of the fruit, usually with a spherical plunger in such a way that it causes no damage. The penetration of the sphere into the skin is measured with high precision using an analogue (spring) or a piezoelectric sensor positioned at the back of the compression plunger. The force–deformation curve, which determines fruit firmness, can be recorded by applying a small load for a fixed period of time (Macnish et al., 1997) or by calculating the force necessary to reach a pre-et deformation (Fekete and Felfoldi, 2000).

Prussia et al. (1994) developed a sensor that uses an air blast to deform the fruit surface instead of the spherical plunger. The device comprises the following components: a means to generate an impulsive jet of a fluid (such as air) aimed at the surface of the object under test, a laser to generate a beam of coherent light aimed at the area on the surface under test impacted by the fluid jet, a detector to sense the light reflected off of the surface of the object from the laser beam, an analyzer to determine the amount of deformation of the surface caused by the fluid jet based on the input to the detector, and a controller to coordinate the release of the impulsive jet with the analyzer. The pressure of the air reservoir is very accurately controlled. The fruit surface deformation is measured using the displacement sensor. Fruit firmness is then related to this local deformation. Lesage and Destain (1996) designed a nondestructive mechanical sensor (Cantifruit) to measure the firmness of tomatoes. It consists of a small plunger constrained to penetrate slightly into the fruits, by using an accurate lever mechanism. They found a significant correlation between firmness measurements performed with this device and the Stable Micro System (SMS), fitted with the same plunger diameter. Lu et al. (2005) evaluated a newly developed bioyield tester for measuring fruit firmness and its variability within individual fruits. They compared the results obtained using this device to the Magness-Taylor firmness method and concluded that the device will be useful for measuring and monitoring fruit firmness during growth, harvest, and postharvest operations. The bioyield probe was a steel cylinder of 6.4-mm diameter with a rubber tip. The rubber tip was 3.2 mm thick and had a secant modulus of 3.24 MPa at 1.5 mm deformation. This probe design was chosen for detecting the bioyield force of apple fruit based on finite element simulations and experimental studies of different probe designs (including probe diameter and rubber thickness and elasticity) performed by Ababneh (2002). The bioyield probe was mounted onto a handheld digital force gauge (Model DPS-44R, Imada, Inc., Northbrook, Ill.), which had a loading capacity of 200 N with a force resolution of 0.01 N. The digital force gauge was connected to a computer via a RS-232 port, allowing data to be recorded in the computer at a rate up to 20 data points per second. The force gauge was mounted on a tabletop motor-driven stand, which provided a constant loading rate during the bioyield test.

A microdeformation device has been developed by CEMAGREF (French acronym for Centre National du Machinisme Agricole, du Génie Rural, des Eaux et Forêts) (Steinmez et al., 1996). A flexible positioning cup (a soft articulation) with a contact plunger (a probe with a sphere at the end) in the center helps the operator to slightly deform the fruit surface (maximum 2 mm approximately). Then, a spring shows the firmness indexed on a scale.

Arazuri et al. (2007) used a Durofell of Copa Rechnologie S.A.: (France), shown in Figure 3.3, to measure tomato hardness nondestructively. It is a nondestructive dynamometer equipped with a metallic, flat-ended probe with three possible contact areas (10, 25, and 50 cm^2). The hardness value is determined by the penetration of the probe into the fruit. The measurement is not based on S.I. units because the measurement scale was 0–100 in Durofel units. Values higher than 70 units indicated hard tomatoes and less than 60 units indicating soft tomatoes (Durofel, 1996).

Macnish et al. (1997) described two other nondestructive devices to measure fruit firmness: the Analogue CSIRO Firmness Meter (AFM) and the Digital CSIRO Firmness Meter (DFM).

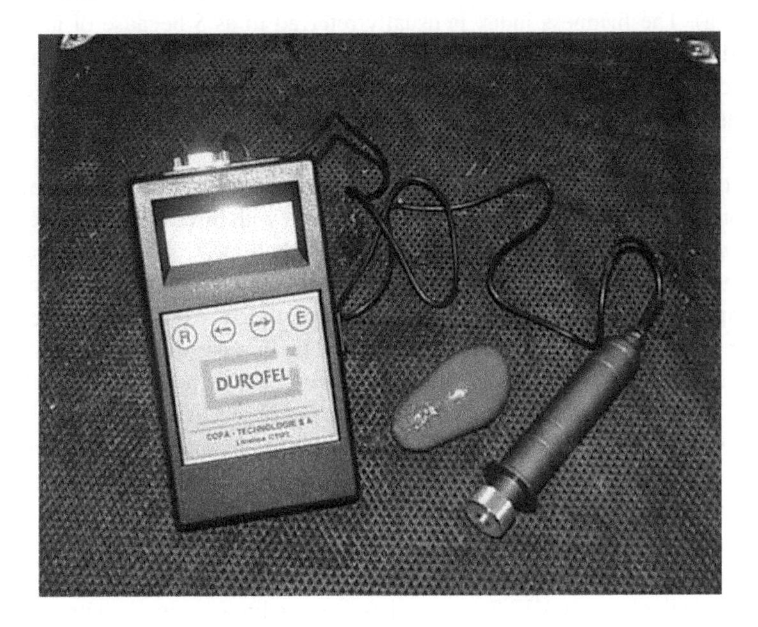

FIGURE 3.3 Durocell of Copa Rechnologie S.A. (France).

3.2.1.3 Vibration-Based Firmness Measurements

In these techniques, the fruit is impacted with a small hammer. The resulting mechanical vibration is then directly measured using accelerometers or laser vibrometers or indirectly using a microphone to capture the corresponding sound wave. A computer, which is hooked up to the measurement device, calculates the frequency response spectrum from the time domain signal by means of a fast Fourier transform (Nicolaï et al., 2006). The resonant frequencies depend on the mechanical properties of the fruit and therefore can be used to characterize fruit firmness. The higher the resonance frequency, the firmer the fruit.

Farabee and Stone (1991) developed a sonic impulse firmness tester to measure the maturity of watermelons. The probe is a closed-end Plexiglas cylinder approximately 5 cm in diameter and 15 cm long. A thin disk-shaped ceramic piezoelectric element, bonded to a similar sized thin brass disk, was mounted at the end of the cylinder in contact with the watermelon. A solenoid, inside the cylinder, was used to deliver a mechanical impulse to the flat face of the piezo ceramic. The impulse is transferred through the ceramic to the watermelon. The resulting vibration of the fruit drives the piezo element and the signal is amplified and filtered before digitalization by the data acquisition unit. Abbott et al. (1992) used a sonic transmission technique to compare the relationship between sensory ripeness of stored Delicious apples with soluble solids content, titratable acidity, Magness-Taylor firmness, and sonic transmission spectra. They found that sonic transmission functions correlated significantly with sensory ripeness and Magness-Taylor firmness. Cooke and Rand (1973) calculated a firmness index $S = f^2 m^{2/3}$ where f is the first resonance frequency (Hz) and m is the mass of the fruit (kg). Different equations based on the modulus of elasticity have been developed to obtain a firmness index as a function of f (Hz) and m (kg). The firmness index is usually referred to as S because of its relationship with the stiffness of the fruit tissue. In these equations, different authors have proposed different exponents affecting m, depending on the importance given to the fruit mass. Abbott and Massie (1998) and Fekete and Felföldi (2000) proposed the equation $S = f^2 m$.

Eshet Eilon Ltd. constructed a piezoelectric-based measurement system, named Firmalon, which includes an instrumented fruit bed with three equally spaced piezoelectric sensors, three electromechanical impulse hammers, and a force transducer that measures the fruit mass and compensates the signal. The piezoelectric sensors are composed of a polyvinylidene fluoride film bonded to soft polyethylene foam padding, to enable free vibrations of the fruit. The impulse devices, consisting of a push-type solenoid and a pendulum, are located opposite to the piezoelectric film sensors. A data acquisition computer program was used to control the test operations, to select the resonance frequencies, and to calculate the acoustic parameters of the fruit: the natural frequency of the fruit, damping ratio, and the centroid of the frequency response. An automated search algorithm was used to identify the first frequency of fruit response to impulse excitation and firmness. It was calculated using the expression set up by Cooke and Rand (1973). Commercial laptop devices are currently available from AWETA as well as grading lines equipped with high-speed sensors. The acoustic firmness sensor (AFS™) is a benchtop device that enables the nondestructive measurement of fruits and vegetables. This equipment detects the vibration modules of the acoustic wave traveling across the fruit.

Sugiyama et al. (1997) developed a portable firmness tester for melons. They found a correlation of about $R = 0.942$ between the transmission velocity measured by two microphones and the apparent elasticity of the melons, measured destructively. De Belie et al. (2000a) developed a monitoring system to estimate the loss of firmness of apples during storage using a nondestructive acoustic impulse response technique to measure firmness. They demonstrated that the sensitivity of this technique to firmness changes of apples was greater than the penetrometer technique. The system is composed of a rotating disc on which a representative fruit sample is located, an electromagnetic excitation mechanism, and an optical sensor to detect the position of the apples. A microphone records the apple vibrations at impact and is linked to a computer with a data acquisition and analysis program. This computer is placed outside the cool storage area. A first-order degradation model, fitted to the measured firmness data, is used to estimate the time when cool stores should be opened to guarantee an average firmness after storage. De Bellie et al. (2000b) used this methodology to measure pear firmness with the fruits remaining on the trees. Terasaki et al. (2001) used a laser Doppler vibrometer to evaluate the softening of kiwifruit during ripening. They defined a stiffness coefficient (E) as $E = f_n^2 m^{2/3}$, where f_n was the frequency of the second resonance peak and m was the fruit mass, and a loss coefficient (I) defined as $I = (f_2 - f_1)/f_n$, where frequencies f_1 and f_2 were determined at 3 dB below the second resonance peak ($f_2 > f_1$). They found a highly significant relationship between the stiffness coefficient and the firmness of the kiwi core determined by measuring the force required to insert a conical probe of 5 mm into the surface of a fruit slice ($R^2 = 0.967$). Their results indicate that early stages of fruit softening are reflected by the stiffness coefficient, and late stages are reflected by the loss coefficient. Therefore, the two coefficients clearly distinguish between ripe and unripe kiwifruit. They also used this technique to monitor ripening behavior of La France pears and to develop a model to simulate the ripening process of pear stored for various periods at low temperature. The E value declined steadily as a quasi-exponential function only for short intervals following storage at low temperature. This time-dependent pattern in decline of the elasticity index, as they designed E, was not observed after long-term storage. They reported that the most acceptable texture quality for the La France pears is within the initial elasticity index range of $(8.6 - 11.4) \times 10^6$ Hz^2 $g^{2/3}$. De Katelaere et al. (2006) compared two commercial nondestructive firmness sensors, one based on acoustic impulse response (AFS) and one based on low mass impact (SIQ-FT), for the quality assessment of apples and tomatoes during storage. They used a commercial desktop unit (AFS, AWETA, Nootdorp, The Netherlands) to measure the acoustic firmness of apples and tomatoes. The fruits were placed in a weight cell to be weighed and were excited by means of an electromagnetic-driven probe and a small microphone captured the vibration. The only two parameters needed to obtain the firmness index S are the resonant frequency of the elliptical mode and the mass of the fruit. In the case of the low-mass impact sensor, a benchtop version (Sinclair iQ™ Firmness Tester) was used for the nondestructive impact test. The sensor hits the fruit by air pressure and captures the impact signal. A special data acquisition and signal analysis program was employed to determine the internal quality index of the tested sample. The firmness was expressed according to the equation set up by Schmulevich et al. (2003). They concluded that both sensors have good repeatability, mainly the AFS sensor on apples, and the correlation between the measurements of both devices was rather high and taking a single measurement using AFS provides sufficient information for grading apples.

Other novel techniques to measure food firmness are based on the response of the fruit to mechanical vibration. The food is placed on a surface that vibrates at certain frequencies and the response is measured by means of an accelerometer. Peleg (1993) developed a nondestructive method for measuring apple firmness based on vibrational excitation. The fruit is located on a support, affected by an electromechanical vibrator, and the response signal of the fruit is measured by an accelerometer attached to a finger that comes into contact with the top of the fruit. The softer the fruit, the larger the attenuation of the signal. This system has been installed on-line and was able to measure firmness at a speed of 4 to 6 fruits per second with an estimated error for testing peaches and apples around 10% and 20%, respectively.

3.2.1.4 Ultrasonic Wave Propagation–Based Sensors

Low-intensity ultrasonics has become an established nondestructive technique in food science. Both velocity and attenuation measurements may provide texture information. Ultrasound devices generally consist of four basic components: a wave generator (pulser/receiver), a transmitter and receiver, a microprocessor (or computer) equipped with signal-processing software, and a display.

Single-touch systems have been developed and used for collecting and processing the acoustic information emerging from tissue segments of whole fruit specimens and continuous-touch systems for whole-fruit measurements. The single-touch systems were usually modified from their industrial configuration, intended for the determination of physical properties of attenuated materials such as cement and composite material, and were adapted for the measurement of the highly attenuated fruit and vegetable tissues (Povey, 1998). This technique has been successfully used to assess firmness on avocados, mangoes, tomatoes, plums, and pears (Mizrach, 2008).

A typical continuous-touch system consists of a high-power, low-frequency ultrasound pulser-receiver, two identical ultrasound transducers, and a mounting structure that provides three-axis movement for the transducers (Mizrach et al., 2000). In the through-transmission mode, with a transducer acting as a transmitter and the other as a receiver, the mounting structure allowed the transducers to move relative to one another. Emitted waves penetrated the peel and propagated through the adjacent tissue along the gap between the probe tips. The output pulse amplitude and the paths of the transmitted waves could be observed visually on a monitor screen. In parallel, a built-in peak detector and microprocessor-controlled serial interface captured the attenuated waves that crossed the known gap and sent digitalized data to an external microcomputer. These data were then used to calculate the wave velocity and the attenuation coefficient of the signal passing through the tissue. The wave propagation velocities C_p were calculated by measuring the time required for the pulse to traverse the gap between the two probes using the Equation (3.1)

$$C_p = l/t \tag{3.1}$$

where l was the distance between probes and t was the transit time between the input and collection probe tips. The attenuation coefficient, α, was calculated by means of the exponential expression, shown in Equation (3.2):

$$A = A_0 e^{-\alpha l} \tag{3.2}$$

where A and A_0, respectively, are the ultrasonic signal amplitudes at the beginning and at the end of the propagation path of the ultrasonic wave (Krautkramer and Krautkramer, 1990). The system has been used to study the changes in physiochemical parameters of avocado fruit during maturation, storage, and shelf life, and can be used for the estimation of the maturity of fruits, a precise determination of the appropriate harvest time and the nondestructive assessment of their firmness.

The single- and continuous-touch systems were built and used as benchtop devices. Portable devices were developed later for rapid measurements of attenuation and for field applications on monitoring pre- and postharvest processes. Three different types of portable systems were designed to collect data by means of two different mounting methods and two modular signal processing routines. A single-touch, permanent-load system operating in the time domain has been developed for field use (Mizrach, 2000). It consisted of three 50-kHz ultrasonic transducers; one of them functioned as a transmitter and the other two as receivers. A single-touch, permanent-load system operating in the frequency domain was used to evaluate the physiochemical properties of mango fruit (Mizrach et al. 1999). The system is based on an ultrasonic pulser-receiver device similar to the system that operated in the time domain, but only one transmitter and one receiver are used. A single-touch, variable-load system operating in the time domain was developed for nondestructive measurements of firmness in whole apple fruit (Mizrach et al. 2003; Bechar et al. 2005).

3.2.1.5 Image Analysis–Based Methods

Texture and appearance are among the most important characteristics of food. Considerable research work has been done on applying machine vision systems to inspect foods. Light scattering is related to the structural characteristics of fruit and hence is potentially useful for estimating fruit firmness.

Duprat et al. (1995) developed an optical system to measure apple firmness. The apples were illuminated with a 670-nm laser and their scattering areas were measured with a stereomicroscope and a camera. They found a correlation coefficient of 0.84 between the area and Young's modulus. Tu et al. (1995) used a 3-mW He-Ne laser beam as the light source to determine apple and tomato properties related to ripeness in a nondestructive way. The fruits were monitored at different maturity stages using the laser source and a red, green, blue (RGB) camera. The images were captured and the total number of pixels exceeding a threshold was taken as a texture indicator. The fruit firmness, measured by nondestructive acoustic response, showed a strong negative correlation with the laser image analysis. Lu (2004) used multispectral imaging to quantify light backscattering profiles from apple fruit for predicting firmness. Spectral images of the backscattering of light at the fruit surface, which were generated from a focused broadband beam, were obtained from Red Delicious apples for five selected spectral bands (10 nm bandpass) between 680 and 1060 nm. Ratios of scattering profiles for different spectral bands were used as inputs to a back-propagation neural network with one hidden layer to predict fruit firmness. Three ratio combinations with four wavelengths (680, 880, 905, and 940 nm) gave the best predictions of fruit firmness, with $r = 0.87$ and the standard error of prediction (SEP) = 5.8 N. Multispectral imaging technique is promising for predicting firmness of apples. Lu and Peng (2006) used a hyperspectral imaging system to simultaneously acquire 153 spectral scattering profiles, generated by a broadband light beam, from Red Haven and Coral Star peaches between 500 and 1000 nm. They developed firmness prediction models using multilinear regression coupled with cross validation. The wavelength of 677 nm, corresponding to chlorophyll absorption, had the highest correlation with fruit firmness among all single wavelengths. They concluded that hyperspectral scattering is potentially useful for rapid and nondestructive estimation of peach firmness.

3.2.1.6 NIR Spectroscopy–Based Methods

In near-infrared (NIR) spectroscopy, the food is irradiated with NIR radiation, which covers the range of the electromagnetic spectrum between 780 and 2500 nm. The reflected or transmitted radiation is measured using a spectrophotometer. When radiation hits a sample, the incident radiation may be reflected, absorbed, or transmitted, and the relative contribution of each phenomenon depends on the chemical composition and physical parameters of the sample. Most absorption bands in the NIR region are overtones or combination of fundamental C–H, O–H, and N–H vibrations. In large molecules, as well as in complex mixtures such as foods, the multiple bands and the effect of peak-broadening result in NIR spectra that have a broad envelope with few sharp peaks (Nicolaï et al., 2007). The spectrum of a sample contains information on absorption as well as scattering. While absorption is related to the presence of chemical components, scattering is related to the microstructure of the tissue. Therefore, the spectra can be used to predict features depending on the chemical composition or the structure of foods. In Figure 3.4 a typical spectrum is shown.

A spectrophotometer consists of a light source, sample presentation accessories, monochromator, detector, and optical components. There are three different methods to obtain NIR spectra: reflectance, transmittance, and interactance. In reflectance mode, the reflected radiation is measured by the detector, which is mounted under a specific angle with respect to the light source. In transmittance mode, the transmitted radiation is measured by the detector, which is positioned opposite to the detector. In interactance mode, the light source and the detector are positioned parallel to each other in such a way that the light due to specular reflection cannot directly enter the detector. In selecting the measurement method, it is important to know that the penetration of NIR radiation into fruit tissue decreases exponentially with depth (Lammertyn et al., 2000). There are portable spectrophotometers that can be used by farmers to make measurements in the field.

Depending on the uniformity of the quality attribute within the fruit, it might be necessary to repeat the spectral acquisition at several positions of the fruit. Kuriyati

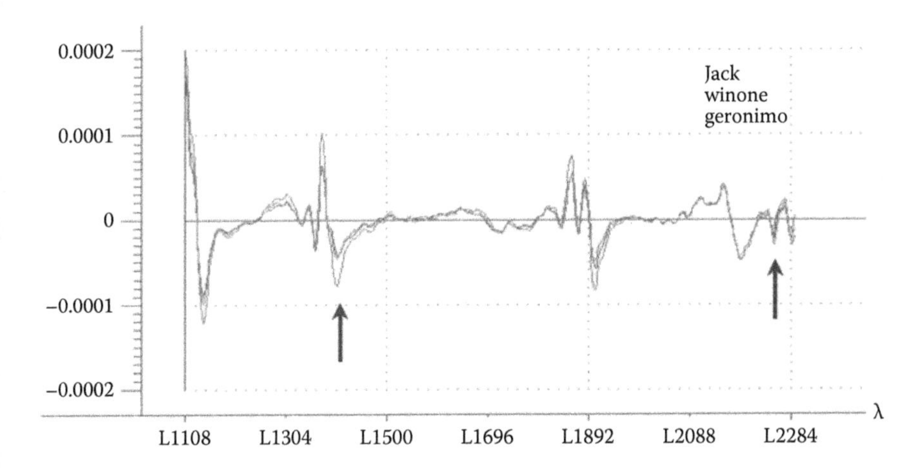

FIGURE 3.4 NIR spectrum acquired from three tomatoes belonging to three different cultivars.

et al. (2004) calculated four types of spectra, which were averaged from 1, 2, 3, and 6 positions, to be used as input vectors for dry matter calibration of tomatoes. They concluded that to obtain a highly efficient and stable calibration equation, the 3- or 6-position spectra should be used.

NIR spectra often contain broad bands, which are the result of many individual overlapped peaks, and the presence of Fermi resonances can also increase the complexity of NIR spectra. Multivariate statistical techniques are therefore required to extract the information about quality attributes that are buried in the NIR spectrum. These techniques include spectral processing techniques used to remove irrelevant information as well as regression techniques. The most important spectral processing techniques are averaging, centering, smoothing, standardization, normalization, and transformation. The regression techniques can be linear and nonlinear. Linear multivariate regression techniques attempt to establish a relationship between the value of reference and the values from the spectral matrix $n \times N$, with n the number of spectra and N the number of wavelengths. Multiple linear regression (MLR), stepwise multiple linear regression (SMLR), principal component regression (PCR), and mainly partial least squares (PLS) regression were used to develop calibration models for the spectrophotometers. These models make it possible to predict the values of certain reference values of the food, measured by means of a reference method. Some nonlinear regression techniques, such as artificial neural networks (ANN), kernel-based techniques, and least squares support vector machines (LS-SVM) have been used to construct NIR calibration models.

Once the calibration model has been developed, it is necessary to select the model that is able to predict the values of the reference parameter with the highest accuracy. The parameters characterizing the ability of prediction of the model are root mean square error for cross validation (RMSECV) when the method of cross validation is used and the root mean square error of prediction (RMSEP) when a validation using independent samples is performed. The coincidence or little difference between errors of calibration and validation is another index of the accuracy of the model.

The reported parameters are independent of the variability of the sample. The ability of a model to accurately predict the reference values depends on the variability of the samples. It is necessary to use a parameter depending on this variability when comparing the ability of prediction of various models obtained from samples having different variability. The residual predictive deviation (RPD) is defined as the ratio of the standard deviation (SD) of reference data for the validation samples and the standard error of prediction (SEP) and provides a standardization of the SEP (Williams, 2001). An RPD between 1.5 and 2 means that the model is able to discriminate against high and low values of the response variable; a value between 2 and 2.5 indicates that coarse quantitative predictions are possible; and 3 or above corresponds to a good or an excellent prediction accuracy, respectively.

Another useful statistic is the coefficient of determination (R^2), which represents the proportion of explained variance of the response variable in the calibration or validation. Calibration models are called *robust* when the prediction accuracy is relatively intense toward unknown changes of external factors.

Many calibration models for spectrophotometers have been developed to predict food firmness. Calibration models have been developed to predict firmness of apples (Lovász et al., 1994; Moons et al., 1997; Peirs et al., 2002; Park et al., 2003), Satsuma mandarin (Hernández Gómez et al., 2006), kiwifruit (McGlone and Kawano, 1997), peaches (Fu et al., 2008), tomato (Shao et al., 2007), carrots (De Belie et al., 2003), cucumbers (Kavdira et al. 2007), wheat (Manley and McGill 1996), rice (Windham et al. 1997), bread (Xie et al., 2003), and cheese (Karoui et al., 2007).

Zude et al. (2006) used a portable desktop device including an acoustic impulse resonance frequency sensor and a miniaturized visible/near infrared (VIS/NIR) spectrometer to predict apple flesh firmness directly on the tree and its shelf life. Partial least-squares calibration models on acoustic data and VIS spectra of Golden Delicious/Idared apple tree fruits were built for predicting fruit flesh firmness. They obtained coefficients of determination ($R^2 = 0.93/0.81$) and standard errors of cross validation (SECV = 7.73/10.50 N/cm^2) for firmness measured on tree fruits and during their shelf life, respectively.

3.2.1.7 Time-Resolved Diffuse Spectroscopy (TRS)

Time-resolved reflectance spectroscopy (TRS) is a nondestructive method for optical characterization of highly diffusive media. In TRS, a short laser light pulse is injected into the medium to be analyzed. Due to photon absorption and scattering events, the diffusely reflected pulse is attenuated, broadened, and delayed. The absorption coefficient μa and the transport scattering coefficient $\mu's$ are simultaneously and independently estimated by fitting the time distribution of the diffusely reflected light pulse. This is detected by time-correlated single photon counting techniques with a theoretical model of light propagation. In TRS, light penetration into a diffusive medium depends on the optical properties of the medium as well as on the source–detector distance. In most biological tissues such as fruits and vegetables, the depth of the probed volume is of the same order as the source–detector distance, which is 1–2 cm (Cubeddu et al., 1999). Consequently, the measurements probe the bulk properties, not the superficial ones, and may provide useful information on internal quality. The scattering coefficient $\mu's$ is associated with the fruit structure while the absorption coefficient μa is associated with the chemical composition of the fruit.

Valero (2001) classified apples, peaches, tomatoes, and kiwis into three categories according to their firmness and used TRS coefficients to make a parallel classification, classifying correctly 76% of apples, 77% of peaches, 81% of tomatoes, and 75% of kiwis. Nicolaï et al. (2008) observed a highly nonlinear relationship between scattering coefficients at 900 nm and firmness of Conference pears. Vangdal et al. (2008) found that the TRS absorption at 670 nm was closely related to firmness of Jubileum plums at the moment of picking. Less-ripe plums had higher absorption coefficients at 670 nm than riper plums. Zerbinni et al. (2009) concluded that the maturity of nectarines at harvest can be assessed by measuring the absorption coefficient at 670 nm, near the chlorophyll peak, in the fruit flesh. They used TRS to assess the maturity stage of nectarines and their ability to be shelf life stored and transported without losing their textural quality. They developed a kinetic model linking the absorption coefficient measured at the moment of harvest to firmness decrease during the ripening of the nectarines. Using this methodology, it would be possible to assign differently sorted nectarines on board a truck depending on the final market (different destinations or different market segments), so facilitating fruit management.

3.2.1.8 NMR Spectroscopy and Magnetic Resonance Imaging

Nuclear magnetic resonance (NMR) technique measures internal features based on the magnetic properties of the nucleus of atoms making up a material. NMR techniques have been used for examining foods since 1950. NMR is sensitive to the presence of mobile water, oil, and sugar, which are major components of agricultural materials. NMR devices are very expensive and their use is so far limited to research applications.

Thybo et al. (2004) used NMR imaging, the so-called magnetic resonance imaging (MRI), to predict sensory quality attributes of raw potatoes. They concluded that MRI, in addition to giving well-known information about water distribution, also provides information about anatomic structures within raw potatoes. This information is of importance for the perceived textural properties of the cooked potatoes. Zhou and Li (2007) applied texture analysis (TA) of magnetic resonance images to predict firmness of Huanghua pears (*Pyrus pyrifolia Nakai*, cv. Huanghua) during storage using an ANN. The optimal ANN model was able to predict the firmness of the pears with a mean absolute error (MAE) of 0.539 N and $R = 0.969$. Quevedo and Aguilera (2008) used computer vision and a stereoscopy technique to characterize and detect changes in the capacity of the salmon fillet surface to recover its original form after a constant weight was applied. A curvature index associated with fillet firmness, calculated by means of 3D information data obtained with the stereoscopy technique from the fat stripe on the fillets, was estimated over 6 months and was suggested as a characteristic of the recovery property of the fresh salmon fillet surface.

3.2.2 Novel Techniques for Crispness Measurement

Crispy and crunchy textures are desirable qualities for many foods, mainly for dry foods. Sterling and Simon (1954) reported that crispness as applied to the texture of a food product is not a well-defined term because crispness in one commodity does not mean the same textural quality as crispness in another food (e.g., crispness in fresh lettuce as compared with crispness in potato chips). They proposed a definition for crispness as the quality of fracturing under relatively slight distortion. Jowitt (1974) recognized two types of crispness: one type in fruits and vegetables such as apples, celery, and lettuce, and the other type in porous, dry foods, such as crackers and potato chips. He defined crispness perceived in fruits and vegetables as "the textural property manifested by a tendency to yield suddenly with a characteristic sound when subjected to an applied force" p. 52. He preferred the term *brittleness* to denote crispness in dry foods. Szczesniak (1987) reported that 94% of consumers classified crisp foods in three categories: raw fruits and vegetables, farinaceous products, and fried products. She concluded that responses from consumers suggest that a crisp food may be defined as one that is firm and snaps easily when deformed emitting a crunchy/crackly sound. Szczesniak and Kahn (1971) reached the conclusion that "Crispness appears to be the most versatile single texture parameter." It is particularly good as an appetizer and as a stimulant to active eating. It is notable as a popular accent-contributing or dramatizing characteristic. Crispness is very prominent in texture combinations that mark excellent cooking and is nearly synonymous with freshness and wholesomeness. Crispness is particularly appreciated in Oriental cuisine, and in contrast to the English language, which uses essentially only three words, crisp, crunchy, and brittle, Japanese and Chinese languages use up to forty and eleven words, respectively (Szczesniak, 1987). Fillion and Kilkast (2002) reported that they are complex concepts that combine a wide range of perceptions such as fracture characteristics, sound, density, and geometry, and also that there is better agreement on the meaning and use of crunchy terms by consumers and panelists than for crispy terms. They reported that *crispy* would refer to a light thin texture producing a sharp clean break with a high-pitched sound when a force is applied, mainly during the first bite with the front teeth. In contrast, *crunchy* would be associated with a hard and dense texture that fractures without prior deformation, producing a loud, low-pitched sound that is repeated over several chews.

Although sensory analysis gives a more complete description of the texture of tested products, there has been great interest in developing instrumental techniques for the assessment of the crispness of foods. These techniques are based mainly on mechanical, acoustic, and ultrasonic tests.

3.2.2.1 Mechanical Measurement of Crispness

The most commonly used tests can be categorized into three groups: flexure, shear, and compression tests. Sterling and Simone (1954) used a mechanical test to evaluate almond's crispness and proposed the amount of deflection of a bar from the first contact to the fracture of the almond as an index of the crispness. Katz and Labuza (1981) evaluated the crispness of several foods using a sensory test and an Instron machine. They used different types of cutting devices and concluded that shear force could be used as an indicator of crispness in expanded cereals. Stanley (1986) used a Universal Texture Analyzer and proposed the number of peaks before fracture for the determination of crispness in puffed cereal. Valles Pàmies et al. (2000) used a puncture test for crispness assessment. They calculated the average of the puncturing force (integral of force–time), the number of spatial ruptures (NSRs; ratio of the total number of peaks to the distance of puncturing), the average specific force of structural ruptures (ratio of the sum of force drops per peak to the number of peaks), and the crispness work (ratio of the average puncturing force to the NSR). NSR and puncturing force were found to correlate with sensory crispness and hardness, respectively. Onwulata et al. (2001) determined the breaking strength index (BSI) of some extruded foods using a texture analyzer (TA-XT2 from Stable Micro Systems, Surrey, England). This index is defined as the ratio between the peak breaking force (*N*) and the extrudate radium (mm). Veronica et al. (2006) demonstrated that the results from texture profile analysis (TPA) correlated well with those from sensory analysis. These mechanical methods to measure food crispness are destructive and cannot be installed on-line.

3.2.2.2 Acoustic Measurement of Crispness

As crispness has an auditory component, it is not surprising that some methods, developed to study crispness, have focused on the sounds generated at fracture, the sound being recorded during instrumental crushing or during mastication. Acoustic methods to characterize crispness of foods have recently been developed. Liu and Tan (1999) used sound signal features to accurately predict sensory crispness of snack food products, which were sensory evaluated by a trained sensory panel. They used an audio recording system that consisted of a super omnidirectional dynamic microphone, a Creative Sound Blaster 16 sound card, and a Waver Studio hosted by a Pentium personal computer. To mimic human biting with molars, a pair of pliers with two parallel plates was used to crush the samples. Principal component regression and neural network techniques were used to determine the usefulness of the sound signal features as predictors of sensory crispness. They developed a model able to predict sensory crispness with an R^2 of 0.89 and concluded that it is possible to use sound signals for crispness evaluation. Srisawas and Jindal (2003) used acoustic signals for predicting crispness of snack products. A 9-V DC-powered microphone connected to an analog-to-digital converter was used for sound signal acquisition. The voltage output of the microphone in the form of amplitude-time signal was converted into a power spectrum of frequencies using Pico FFT built-in software, and displayed on a PC. To simulate biting with incisors, a pair of pincers having a cutting area of 13 mm in length was used. They used neural networks to analyze the frequency domain spectra of acoustic patterns and developed a probabilistic neural network (PNN) model for classifying snack foods into four grades of sensory crispness with a prediction accuracy of approximately 96% to 98%. Sakurai et al. (2005) evaluated the texture of persimmon fruits by an acoustic measurement of crispness (AMC) and by a laser Doppler vibrometer (DLV). Laser Doppler measurement was carried out by placing the fruit on a vibrator stage (model 513-B, Emic, Tokio), and vibrating by sinusoidal signals from 0 to 2 kHz generated by a fast Fourier transformation (FFT) program. The signal was sent to a vibrator amplifier through a soundboard of a personal computer. A laser beam was focused on a reflecting tape on the top of the fruit. The signal response at the top of the fruit was measured by an LDV. Vibration at the vibrator stage was monitored by an accelerometer. The signal from the LDV was sent to the computer through channel A of a signal separator, and that from the accelerometer through channel B. Frequency response of the fruit to the vibration was obtained with the FFT program after subtracting the signal of the vibrator stage (B) from that at the top of the fruit (A). A second resonant peak was used for the calculation of elasticity. An AMC probe with conical tip to which a piezoelectric film was attached to its proximal end was inserted into the mesocarp tissue and the acoustic vibration up to 10 kHz was recorded. The signal was transformed to a frequency spectrum by a fast Fourier transformation. They calculated a texture parameter, called the *sharpness index*, by summation of signal intensities over the spectrum of AMC data; the high-frequency intensities were enhanced to compensate for the limited human hearing capacity. The sharpness index significantly correlated with sensory crispness. Using this methodology, Taniwaki et al. (2006) developed a device that enables direct measurement of food texture. The device is equipped with three parts because humans use three organs to evaluate food texture: teeth, nerves, and brain. A probe represents the teeth, a piezoelectric sensor the nerves, and a computer the brain. The probe was inserted into a food sample and the sensor detected the vibrations produced by the fracture of the sample. The device can measure acoustical vibrations that are created during the fracture of a food sample over a wide range of audio frequencies (0–25,600 Hz). The obtained signals are filtered using a half-octave multifilter to calculate the texture index (TI) for each frequency band. Taniwaki et al. (2009a) investigated the time-course changes in the elasticity index (EI), calculated from the results of a nondestructive vibrational method employing an LDV, and the TI calculated from the results of the acoustic test of two persimmon (*Diospyros kaki Thunb.*) cultivars (Fuyu and Taishuu) during the postharvest period. They defined the EI index as $EI = f^2 m^{2/3}$, where f_2 is the second resonance frequency of a sample, and m is the mass of the sample. The TI was calculated by summing up the amplitude of the texture signals and dividing by the data length $TI = 1/T)\Sigma|Vi|$, where T (s) is the sampling period and Vi (V), the amplitude of each data point. They concluded that both indexes can be used to determine the optimum eating ripeness of persimmons. The texture index reflects the level of sound generated per second when a sample is masticated and has been used to evaluate the texture of persimmons, apples, pears, potato chips, and blanched bunching onions. Taniwaki et al. (2009b) quantified the texture of pears during ripening with TI and found that TI declined gradually over a wide frequency range as the pear samples ripened. Taniwaki et al. (2010) calculated the TI for four kinds of potato chips using a probe with a 5-mm width, a 20-mm-long wedge, and with a 30° chisel tip angle. The probe speed was 22 mms^{-1}. This speed was estimated to be within the range of a typical human's mastication speed (Roudaut et al. 2002). The probe tip was perpendicular to the curvature of a potato chip so that the edge of the probe tip contacts the surface of the sample in unison. Vibration amplitudes (V), which reflect the variation in load applied to the probe, are detected by a piezoelectric sensor that is 1 mm thick and 10 mm in diameter (2Z10D-SYX, Fuji Ceramics Corp., Fuji, Japan). The signal detection covers a wide audio frequency range (0–25, 600 Hz). A sample was placed with its edge on the acrylic plate so that the friction between the sample and the plate helped to minimize the movement of the sample during the measurement. The data sampling rate was 80 kHz. The texture signals obtained were filtered into 19 bands using a half-octave multifilter for analyses in the frequency domain. Three selected parts of the acoustic vibration signals were used to evaluate potato chip texture: the anterior part between the contact of the probe with a sample, the major fracture part, and the full texture signals. The results showed that TI, which was calculated using the major fracture part, is the index that most clearly distinguished the texture characteristics of potato chips with differing degrees of crispness and thickness. The filtered data of each frequency band was subjected to quantification using a texture index (TI) defined by Equation (3.3) (Taniwaki and Sakurai, 2008):

$$TI = (f_l \times f_u) \times 1/n \sum_{i=1}^{n} V_i^2 \tag{3.3}$$

where f_l (Hz) represents the lower limit and f_u (Hz) the upper limit of each frequency band as determined by the half-octave multifilter, V_i (V) is the amplitude of the texture signal, and n is the number of data points. They concluded that a TI above 1600 Hz reflected the crispness of potato chips.

Chen et al. (2005) used an acoustic envelope detector (AED) attached to the Texture Analyzer to assess the crispness of six kinds of biscuits. The AED has been recently developed by the manufacturer of the Texture Analyzer, Stable Micro Systems (Surrey, U.K.) with a 5-kg load cell. The device can be easily attached to the Texture Analyzer as an additional fixture, so that both mechanical force and acoustic signals can be detected at the same time. The background noise was screened out by the filter function of the device, removing mechanical noise and acoustic noise below 1 kHz. This gives significantly enhanced signal-to-noise ratios because of the preponderance of environmental noise below the chosen cutoff. The Brüel and Kjær free-field microphone (Bruel and Kjaer, Naerum, Denmark) (8-mm diameter) was calibrated using the acoustic calibrator type 4231 (94 and 114 dB sound pressure level [SPL],1000 Hz). A three-point bending device was used for the breakup of biscuits. The span width was 60 mm for all the biscuits. The motion of the load cell was set at three different speeds: 0.01, 0.1, and 1.0 mm/s. Each test speed was repeated at least eight times and the average of the test results was used for data analysis. The integration time for acoustic signal analysis was set at two different values (1.25 and 0.25 ms) by changing a regulating capacitor connected to the AED. The force resolution is 0.1 g and the range resolution is 0.01 mm. The data acquisition rate was 500 points per second for both force and acoustic signals. The acoustic and the force–displacement signals were recorded simultaneously during the breakup of the biscuits. For each detected acoustic signal, there was a sudden drop in the compression force. The analysis of the force–displacement curve demonstrated the links between the second derivative of the force curve and the acoustic event, indicating the energy released through the air of these crack events. The acoustic behavior of the biscuits was assessed in terms of maximum sound pressure level and the number of acoustic events, which were further interpreted as the acoustic events per unit area of newly created surface area and the acoustic event per unit time. The acoustic ranking of biscuits from instrumental assessment was in very good agreement with that from sensory panel tests. The normal integration time (1.25 ms) for the AED was generally effective in detecting acoustic signals for crisp biscuits, but a shorter integration time (0.25 ms) was found advantageous in detecting acoustic signals that occur within a very short time period and gave better differentiation of crisp biscuits. Varela et al. (2006) used the AED attached to the Texture Analyzer to assess the crispness of roasted almonds, concluding that sensory crispness can be evaluated both acoustically and mechanically and that both measures are necessary to most effectively describe crispness.

Chaunier et al. (2005) studied the mechanical behavior of cereal flakes and concluded that a Kramer cell, on an Instron testing machine, made it possible to simultaneously record force in the range 0–5 kN and sound in the frequency domain 0–22 kHz during the crushing of 20-mm-thick beds of cereal flakes tested in bulk. It was shown to be the most reliable and efficient method for discrimination. Acoustic emission has been shown to correlate with crispness perception. Products perceived as not being crispy emitted signals with lower average amplitude and higher peaks, at low frequencies, less than 3 kHz, and opposed a high mechanical resistance to compression. Conversely, the crispiest flakes emitted sounds with larger average amplitude, fewer high peaks, and uniformly distributed in the frequency domain with a moderated mechanical resistance. Cheng et al. (2007) investigated the relationships between sensory perception of crispness of cornstarch-whey protein isolate extrudates and their mechanical and acoustic properties during crushing. They found high correlations between sensory attributes and instrumentally determined mechanical properties, including crushing force and crispness work. They developed a novel voice recognition technique utilizing frequency spectrograms, and used them successfully for understanding the differences in the sensory properties of various products. This mechanical-acoustic technique has the potential of reducing the time, costs, and subjectivity involved in evaluation of new foods by human panels, and can be a useful tool in the overall product development cycle. Arimi et al. (2010) developed an acoustic measurement system for analyzing crispness during mechanical and sensory testing. The acoustic recording system was used to capture sound during puncture and sensory tests. From the sound data, the maximum sound pressure, number of sound peaks, sound curve length, and area under the amplitude–time curve were obtained. The number of force and sound peaks, spatial ruptures, sound curve length, and area under the sound curve correlated well ($R^2 > 0.77$) with sensory crispness data. The system is designed in such a way that it can be used as an attachment to any food texture–analyzing system. In addition, the acoustic system can be used independently from a texture-analyzing instrument especially during sensory evaluation. Zdunek et al. (2010) developed a new contact acoustic emission detector (CAED) for instrumental texture evaluation of apples. They developed calibration models for sensory crispness and hardness that were able to measure these properties more accurately than puncture tests in terms of variance explained, root-mean-square, and bias.

3.2.2.3 Ultrasonic Measurement of Crispness

Povey and Harden (1981) studied the possibility of developing an objective method of measuring the crispness of biscuits using the ultrasonic pulse echo technique. They found an encouraging relationship between sensory crispness and the velocity of longitudinal sound. Antonova et al. (2003) determined ultrasonic parameters and studied the relationship between these parameters and sensory crispness in breaded fried chicken nuggets, reaching the conclusion that the ultrasonic velocity could predict sensory crispness.

3.2.3 Novel Techniques for Mealiness and Woolliness Identification

Mealiness is a negative textural attribute that appears in apples, tomatoes, and other fruits. It is characterized by a lack of juiciness and a sandy texture that is perceived when chewing (Arana et al., 2007). It is related to the relative strength of the cell wall and the middle lamella and usually appears after cold storage. *Wooliness* is a physiological alteration of peaches and nectarines characterized by dry and mealy texture, soft and dry fiber, lack of taste and aroma, diminished pulp brightness, and the impossibility of obtaining juice from the fruit (Boyes, 1952).

Given the sensory character of these attributes, it is impossible to detect them in the selection chain, so mealy and woolly fruits without sufficient quality reach the market. Therefore, it would be desirable to develop objective, nondestructive methods to evaluate fruit mealiness and woolliness on-line.

Mealiness and woolliness have been sensory evaluated by means of sensory panels. This method is subjective, destructive, expensive, time consuming, and needs expert tasters. Destructive mechanical tests have been used to characterize apple mealiness (Paoletti et al., 1993). Arana et al. (2007) used a texture analyzer (TA. XT2/25, Stable Micro Systems [SMS], London, U.K.) to perform mechanical tests on apples, peaches, and nectarines. A penetration test and a puncture test were performed on whole fruits. A shear stress rupture test was performed on fruit flesh test cylinders 13 mm in diameter and 50 mm long. The test is made by means of a special device made of methacrylate as shown in Figure 3.5. The texture analyzer pushes the

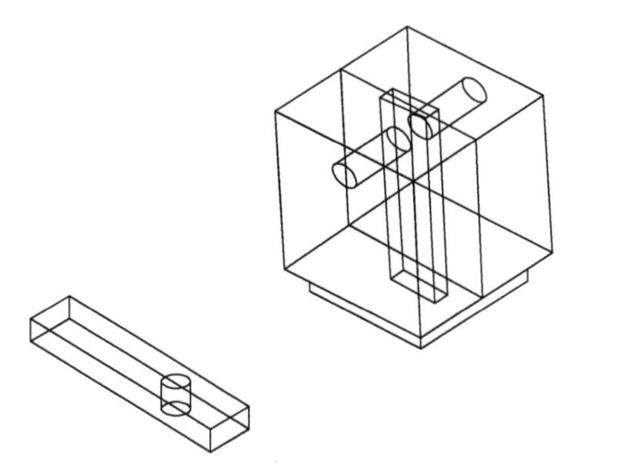
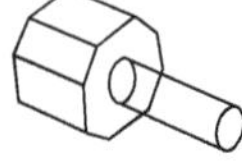

FIGURE 3.5 Special device made of methacrylate.

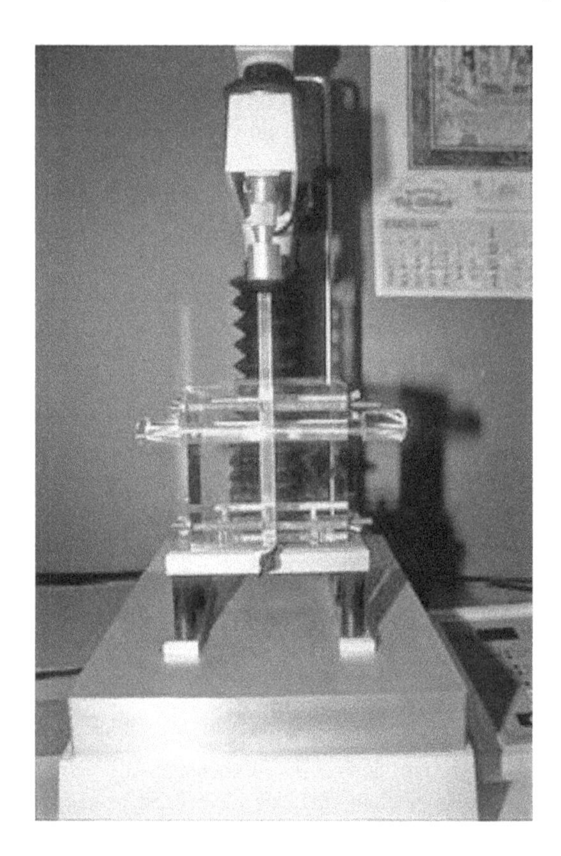

FIGURE 3.6 Shear stress rupture test.

guillotine until the rupture of the test tube, as shown in Figure 3.6. The test speed was 0.3 mm/min and the test distance was 15 mm.

They concluded that the best mechanical test to characterize apple mealiness individually was the shear stress rupture test. The correlation coefficient between mealiness and maximum resistance to shear stress rupture test (N) was higher than 0.8 with a significance level lower than 0.01. This test can be also used to evaluate the woolliness of peaches and nectarines.

In recent years, nondestructive tests were developed to evaluate mealiness and woolliness of fruits. These tests are based on impact response, quantitative measure of fruit juice, ultrasonic wave propagation, imaging, NMR spectroscopy and magnetic resonance imaging, time resolved reflectance spectroscopy, chlorophyll fluorescence, and modeling.

3.2.3.1 Impact Response–Based Techniques

Arana et al. (2004) used a nondestructive impact test to detect mealy apples. The fruits were sensory analyzed and classified on healthy or mealy categories according to their mealiness degree. They found that mealiness was intensely related to the maximum resistance to impact from 2 cm high (*F2*), which is a nondestructive test. By means of discriminate analysis, using variables from the impact response test, they classified apple samples as healthy and mealy. They always obtained healthy samples having a lower percentage of mealiness than original samples. These authors did not achieve samples without mealy fruits. They used signal detection tools, in particular receiver operate curves (ROCs), to achieve the whole spectrum of discriminate analyses produced by just varying the cutoff point (Gómez de la Cámara, 1998). They also used these tools to select the type of statistical analysis according to the goal, which could be to get samples without mealy apples, to maximize the textural quality of the sample, or to minimize the additional classification cost. They defined, in terms of the results of the discriminate analysis, a quality index (Q), and a cost index (C). The quality index can be defined as the percentage of mealy apples in the classified sample, while the cost index is the additional cost of the classification. They defined Equation (3.4) to calculate the quality of the apple sample classified as healthy depending on the statistical parameters of the sample.

$$\begin{aligned} Q &= 1 - VPP = FP/(VP + FP) \\ &= (1 - Pr) \times (1 - Sp)/[(\times (1 - Sp) \times (1 - Pr) + Se \times Pr] \end{aligned} \tag{3.4}$$

They defined Equation (3.5) to calculate the additional cost of the classification, depending on the cited parameters,

$$C = FN/(VP + FP) = Pr \times (1 - Se)/[(1 - Sp) \times (1 - Pr) + Se \times \mathrm{Pr}] \tag{3. 5}$$

where VPP is the positive predictor value, FP is the positive false, VP is the positive true, Pr is the prevalence, Sp is the specificity, and Se is the sensitivity.

As the textural quality of the sample increases, Q tends to zero and the cost improves if sensitivity improves and specificity decreases.

Using ROCs and nondestructive impact variables, it was possible to obtain samples of Golden Delicious and Top Red apple cultivars free of mealy fruits. The authors concluded that the lower limit for maximum resistance in the nondestructive test that ensures the lack of mealy fruits was 16 N for Golden Delicious apples and 19 N for Top Red apples.

Using the same methodology, Arana et al. (2005) obtained samples lacking woolly nectarines from Fairlane and Festina cultivars. The lower limit for maximum resistance in the nondestructive test, from 5 cm height ($F5$) that ensures the lack of woolly fruits was 25 N for Fairlane nectarines and 20 N for Festina nectarines.

Figures 3.7 and 3.8 show the quality of the classified sample and the additional cost of classification depending on the minimum value allowed for the maximum resistance opposed by the fruit during the nondestructive impact test. The quality and the cost increase when the minimum allowed value for $F5$ increases.

3.2.3.2 Quantitative Methods Based on Free Juice

Mealiness and woolliness are related to the lack of free juice during mastication. Therefore, there should be an inverse relationship between the amount of free juice and mealiness and woolliness. Crisosto and Labavitch (2002) extracted the juice of samples of tissue of peaches and nectarines by subjecting fruit tissue to a pressing

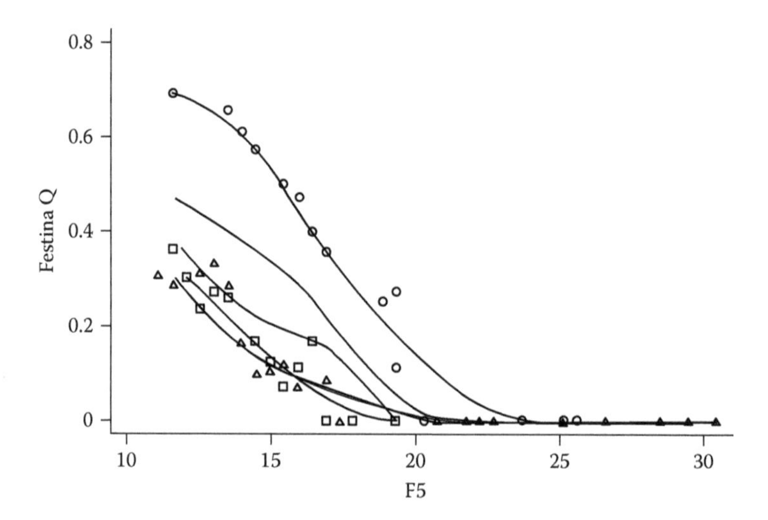

FIGURE 3.7 Textural quality of the apple sample achieved depending on the lower limit chosen for the maximum resistance exerted by the apple during the impact test.

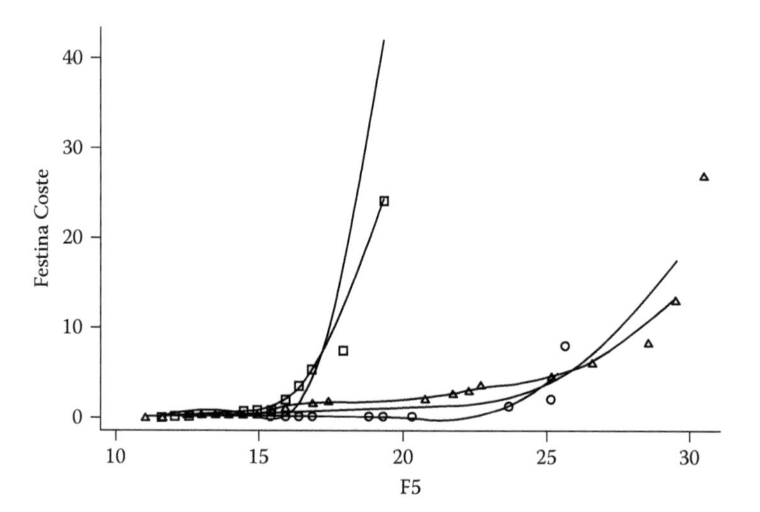

FIGURE 3.8 Additional cost of the classification method depending on the lower limit chosen for the maximum resistance exerted by the apple during the impact test.

force of 667 N for 1 minute after centrifugation. The supernatant was weighed and used as a measure of the amount of free juice. The percentage of fruit juice was more sensitive and it had a higher correlation to sensory woolliness. Arana et al. (2007) performed a confined compression test on apple, peach, and nectarine flesh cylinders 17 mm in diameter and 17 mm long. The fruit flesh cylinders are confined in a special device located on drying paper. The texture analyzer pushes the load cylinder with a speed of 0.3 mm min^{-1} until the maximum deformation of the fruit pulp cylinder of 2.5 mm is reached, as shown in Figure 3.9. As the cylinder of fruit pulp is

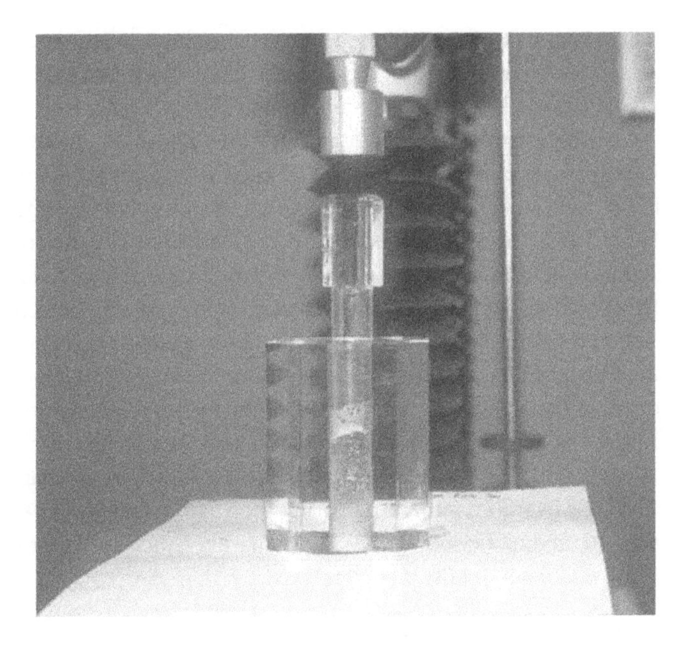

FIGURE 3.9 Confined compression test.

compressed, an amount of juice comes off and is gathered in the drying paper. The area (cm^2) stained by the juice is calculated by an image analysis as a measure of the fruit juiciness and is related to mealiness and woolliness.

3.2.3.3 Ultrasonic Wave Propagation Methods

Mizrach et al. (2003) used ultrasonic energy absorbance for nondestructive determination of three mealiness levels (fresh, ripe, and overripe) in Jonagold and Cox apples. Their results suggested that it is possible to distinguish among three mealiness levels in Jonagold apples. Bechar et al. (2005) developed an ultrasonic system comprising a high-power generator and a pair of 80-kHz ultrasonic transducers for nondestructive measurement of the three cited mealiness levels. One transducer, acting as a transmitter, sends a pulse through the apple tissue, which absorbs part of its energy depending on its internal textural attributes, and the transmitted pulse is received as an emerging signal by the other transducer. The detected ultrasound waves and the determination of the mealiness level, using a destructive confined compression test, were analyzed in parallel. The results obtained for Cox apples showed a high correlation between the ultrasound measurements and the confined-compression destructive tests for each mealiness level. Verlinden et al. (2004) evaluated the ability of ultrasonic propagation waves to measure chilling injury in tomatoes. They found that the attenuation of 50-kHz ultrasonic waves increased when chilling injury developed, but only when tomatoes were harvested at a mature stage.

3.2.3.4 Techniques Based on Imaging

De Smedt et al. (1998) visualized and quantified the differences between fresh and mealy apples using images made with a light microscope. They were able to identify fresh and mealy apples using four cell parameters such as area and perimeter and two roundness parameters measured from the images. They also found that the amount of broken cells at the surface of a fractured sample after a tensile loading test was significantly lower for mealy fruits than for fresh fruits. Zhu et al. (2010) proposed a supervised locally linear embedding (SLLE) coupled with support vector machines (SVM) to detect mealiness in apples using hyperspectral scattering images. They concluded that SLLE-SVM method would provide an effective classification method for apple mealiness detection using the hyperspectral scattering technique. Huang and Lu (2010) used the hyperspectral scattering technique for detecting mealy apples. The spectral scattering profiles at individual wavelengths were quantified by relative mean reflectance for a 10-mm scattering distance for the test apples. The reference mealiness of the apples was determined by the instrumental hardness and the juiciness measured by means of a confined compression test using the cited special device on drying paper. PLS discriminant models were built for two classes (mealy and non-mealy). The overall accuracies were between 74.6% and 86.7% when all the fruits were studied, and increased up to 93% when only fruits stored for 4–5 weeks at 20°C and 95% relative humidity were analyzed.

3.2.3.5 Mealiness and Woolliness Detection Using NMR Spectroscopy and Magnetic Resonance Imaging

Mealiness and wooliness are related to the lack of juiciness, and nuclear magnetic resonance is sensitive to the presence of mobile water. Therefore, NMR could be a useful tool to identify mealy and woolly fruits.

Sonego et al. (1995) detected woolly breakdown in nectarines using NMR spectroscopy. Woolly breakdown was detectable by NMR imaging as dark areas corresponding to low proton density. Barreiro et al. (1999) used MRI techniques to identify mealy apples. They identified mealy and nonmealy apples and recorded multi-slice, multi-echo magnetic resonance images (64–64 pixels) with an 8-ms echo time. Minimum T2 values obtained for mealy apples were significantly lower ($F = 13.21$) when compared with nonmealy apples. This seems to indicate that a more desegregated structure and a lower juiciness content leads to lower T2 signal. Also, there is a significant linear correlation ($r = -0.76$) between the number of pixels with a T2 value below 35 ms within a fruit image and the deformation parameter registered during the Magness-Taylor firmness test. In addition, all T2 maps of mealy apples show a regional variation of contrast that is not shown for nonmealy apples. Significant differences ($F = 19.43$) between mealy and nonmealy apples are found in the histograms of the T2 maps. Mealy apples show a skew histogram combined with a "tail" in their high T2 extreme, which is not shown in the histograms of nonmealy apples. Barreiro et al. (2000) validated these results using mathematical features from the histograms of the T2 maps and used the MRI technique to identify woolly peaches. The presence of a clear halo in the T2 maps of woolly peaches indicates variations in water mobility. Using the parameter "number of pixels below 60 ms," woolly peaches can be segregated. Marigheto et al. (2008) used two-dimensional NMR relaxation and diffusion techniques to characterize apple mealiness. They found that the T1 of the peak associated with the cell wall in mealy apples is much longer than that of fresh apples, which could be used to identify mealy apples.

3.2.3.6 Time-Resolved Reflectance Spectroscopy (TRS) Techniques

Valero et al. (2005) used time-resolved laser reflectance spectroscopy to determine apple mealiness. They instrumentally classified Golden Delicious and Cox apples on mealy and nonmealy fruits using a confined compression test. The optical coefficients were used as explanatory variables to build discriminant functions for mealiness. They used fifteen TRS variables to develop a calibration model that classified correctly 98% of the apples into two categories: mealy and nonmealy. A validation test was performed using independent samples showing that 80% of the samples were correctly classified. When trying to estimate three texture stages (fresh, nonmealy, and mealy) with TRS, the performance of the new models decreased from 98% to 71% and the percentage of validation decreased to 51%.

3.2.3.7 Methods Based on Chlorophyll Fluorescence

Fluorescence is the emission of light by a substance that has absorbed light or other electromagnetic radiation of a different wavelength. The main application of fluorescence is chlorophyll fluorescence. Fluorescence measurements of chlorophyll-containing biological materials give information about photosynthesis activity. Chlorophyll loss and decreasing photosynthesis rate are usually coupled to progressive ripening of vegetables. Tissue chlorophyll content is impaired by injuries and stress.

Moshou et al. (2003) stored Jonagold and Cox apples under different storage conditions to develop three different mealiness levels: not mealy, moderate, and strongly mealy. The mealiness of apples was evaluated using the results of destructive tests. The firmness of the apples was estimated by the maximum force during penetration of a plunger into the fruit. The hardness was determined by the slope of the force–deformation curve measured during the compression test. Juice content was determined by measuring the moistened area of the filter paper, which was placed under the sample during the confined compression test. Soluble solid content (SSC) was measured by means of a digital refractometer. Chlorophyll fluorescence kinetics of the apples were measured; fluorescence values were found to decrease with the mealiness level. Classification according to the mealiness level based on fluorescence measurements was more accurate than the one based on destructive measurements. They suggested that fluorescence can be used in an automatic sorting line to assess mealiness. By continuing the cited research, Moshou et al. (2005) fitted the fluorescence curves by a tenth-order polynomial and the regressions coefficients were used as input parameters in quadratic discriminant analysis to assign the fruits into different mealiness levels. The achieved classification performance was around 85%. Having an extremely short measurement time, a classification model using only two parameters such as the slope of the fluorescence curve at the origin and the normalized fluorescence at 1 ms, achieved a classification performance of about 80%.

3.2.3.8 The Use of Modeling for Predicting Mealiness

Multiple linear regression models with instrumental parameters were calculated to predict sensorial attributes such as tomato mealiness (Verberke et al. 1998). De

Smedt (2000) developed a mechanistic model to explain the resulting perception of mealiness during storage of apples. De Smedt et al. (2002) built a mathematical model to link changes in texture attributes related to mealiness to the development of the turgor pressure of the tissue as well as the degree of hydrolysis of the middle lamella.

3.3 CONCLUSIONS

There is an increasing need to characterize food firmness in a nondestructive way that could be used on-line. Novel techniques used to characterize firmness have to be objective, rapid, and not expensive, and also able to characterize the most important physical properties that characterize the quality of each food.

There are many textural characteristics, but for most foods, firmness and crispness are the ones that best characterize them. In addition, fruits have to be marketed with sufficient quality, which makes it necessary to identify and remove fruits having negative textural attributes such as mealiness and woolliness.

During recent years, nondestructive methods have been developed to measure firmness and crispness and to classify fruits into categories according to their firmness or crispness. Methods to detect fruits so mealy or wooly that they are unmarketable have been developed too.

Novel techniques developed to characterize firmness, crispness, mealiness, and woolliness are similar for these four attributes and are based on nondestructive impact response, microdeformation, vibration measurements, NIR spectroscopy, ultrasonic wave propagation, imaging analysis and NMR spectroscopy, TRS spectroscopy, and magnetic resonance imaging. These novel techniques require more development and improvement, but are currently able to measure textural properties of many fruits with enough accuracy.

REFERENCES

Ababneh, H.A.A. 2002. Development of a mechanical probe for nondestructive apple firmness evaluation. Unpublished Ph.D. diss., Michigan State University, East Lansing, MI.

Abbott, J.A., A. Henry, and L.A. Liljedahl. 1992. Firmness measurement of stored Delicious apples by sensory methods, Magness Taylor, and sonic transmission. *Journal of American Society of Horticulture Science* 117(4):590–595.

Abbott, J.A., and D.R. Massie. 1998. Nondestructive sonic measurement of kiwifruit firmness. *Journal of American Society Horticulturae Science* 123(2):317–322.

Antonova, I., P. Mallikarjunan, and S. Duncan. 2003. Correlating objective measurements of crispness in breaded fried chicken nuggets with sensory crispness. *Journal of Food Science* 68:1308–1315.

Arana, I., C. Jarén, and S. Arazuri. 2004. Apple mealiness detection by non-destructive mechanical impact. *Journal of Food Engineering* 62(4):399–408.

Arana, I., C. Jarén, and S. Arazuri. 2005. Nectarine woolliness detection by nondestructive mechanical impact. *Biosystems Engineering* 90(1):37–45.

Arana, I., C. Jarén, and S. Arazuri. 2007. Sensory and mechanical characterization of mealy apples and woolly peaches and nectarines. *Journal of Food, Agriculture & Environment* 5(2):101–106.

Arazuri, S., C. Jarén, J.I. Arana, and J.J. Perez De Ciriza. 2007. Influence of mechanical harvest on the physical properties of processing tomato (*Lycopersicon esculentum Mill.*). *Journal of Food Engineering* 80:190–198.

Arimi, J., E. Duggan, M. O´Sullivan, J. Ling, and E. O´Riordan. 2010. Development of an acoustic measurement system for analyzing crispiness during mechanical and sensory testing. *Journal of Texture Studies* 41:320–340.

Armstrong, H. 2001. Non-destructive online firmness test. *Fruit & Vegetables Technology* 1(1), 2./ www.HortiWorld.nl.

Barreiro, P., C. Ortiz, M. Ruiz Altisent, J. Ruiz Cabello, et al. 2000. Mealiness assessment in apples and peaches using MRI techniques. *Magnetic Resonance Imaging* 18(9):11755–1181.

Barreiro, P., J. Ruiz Cabello, M.E. Fernández-Valle, C. Ortiz, and M. Ruiz Altisent. 1999. Mealiness assessment in apples using MRI techniques. *Magnetic Resonance Imaging* 17(2):275–281.

Bechar, A., A. Mitzrach, P. Barreiro, and S. Landdahl. 2005. Determination of mealiness in apples using ultrasonic measurements. *Biosystems Engineering* 91:329–334.

Boyes, W.W. 1952. The development of wooliness in South Africa peaches during cold storage. IX International Congress Refrigeration, Purdue USA, 4, 533–537.

Cantwell, M. 2004. Fresh market Tomato Statewide Uniform variety Trial Report Field and Postharvest Evaluations. South San Joaquin Valley. UCCE. University of California Cooperative Extension. Michelle Le Strange, UCCE Farm Advisor, Tulare & Kings Counties.

Chaunier, L., P. Courcoux, G. Della Valle, and D. Lourdin. 2005. Physical and sensory evaluation of cornflakes crispness. *Journal of Texture studies* 36(1):93–118.

Chen, J., C. Karlsson, and M. Povey. 2005. Acoustic envelope detector for crispness assessment of biscuits. *Journal of Texture Studies*, 36(2):139–156.

Chen, P., and M. Ruiz Altisent. 1996. A low-mass impact sensor for high-speed firmness sensing of fruits. Paper 96F-003. Presented at AgEng96, Madrid, Spain, September 22–26.

Chen, P., Y. Sarig, and J.F. Thompson. 2000. A hand-held impact sensor for firmness sensing of fruits. Proc. Postharvest Congress, Jerusalem. March 26–31.

Chen, P., S. Tang, and S. Chen. 1985. Instrument for testing the response of fruits to impact. Proceedings of the ASAE Winter Meeting. Chicago, IL. 85-3357.

Cheng, E., S. Alavi, T. Pearson, and R. Agbisit. 2007. Mechanical-acoustic and sensory evaluations of cornstarch-whey protein isolate extrudates. *Journal of Texture Studies* 38:473–498.

Cooke, J.R., and R.H. Rand. 1973. A mathematical study of resonance in intact fruits and vegetables using a 3-media elastic sphere model. *Journal of Agricultural Science* 18:141–157.

Crisosto, C.H., and J.M. Labavitch. 2002. Developing a quantitative method to evaluate peach (*Prunus persica*) flesh mealiness. *Postharvest Biology and Technology* 25:151–158.

Cubeddu R., A. Pifferi, P. Taroni, et al. 1999. Non-destructive measurements of the optical properties of fruits by means of time-resolved reflectance. In *Optical Tomography and Spectroscopy of Tissue, III*, ed. B. Chance, R.R. Alfano, and B.J. Tromberg, 445–449. SPIE Press, Bellingham, WA.

De Baerdemaeker, J., L. Lemaitre, and R. Meire. 1982. Quality detection by frequency spectrum analysis of the fruit impact force. *Transactions of the ASAE* 25(1):175–178.

De Belie, N., K. Pedersen, M. Martens, R. Bro, L. Munk, and J. De Baerdemaeker. 2003. The use of visible and near-infrared reflectance measurements to assess sensory changes in carrot texture and sweetness during heat treatment. *Biosystems Engineering* 85(2):213–225.

De Belie, N., S. Schotte, P. Coucke, and J. De Baerdemaeker. 2000a. Development of an automated monitoring device to quantify changes in firmness of apples during storage. *Postharvest Biology and Technology* 18(1):1–8.

De Bellie, N., S. Schotte, J. Lammertyn, and B. Nicolăi. 2000b. Firmness changes of pear before and after harvest with acoustic impulse response technique. *Journal of Agricultural Engineering Research* 77(2):183–191.

De Ketelaere, B., M.S. Howard, L. Crezee, et al. 2006. Postharvest firmness changes as measured by acoustic and low-mass impact devices: a comparison of techniques. *Postharvest Biology and Technology* 41:151–158.

Delwiche, M.J., H. Arévalo, and J. Mehlschau. 1996. Second generation impact force response fruit firmness sorter. *Transactions of the ASAE* 39(3):1025–1033.

Delwiche, M.J., and Y. Sarig. 1991. A probe impact sensor for fruit texture measurements. *Transactions of the ASAE* 32(1):321–326.

De Smedt, V. 2000. Measurement and modeling of mealiness in apples. PhD Thesis no. 417. Catholic University of Leuven, Leuven, Belgium.

De Smedt, V., P. Barreiro, B.E. Verlinden, E.A. Veraverbeke, J. De Baerdemaeker, and B. Nicolaï. 2002. A mathematical model for the development of mealiness in apples. *Postharvest Biology and Technology* 25(3):273–291.

De Smedt, V., E. Pauwels, J. De Baerdemaeker, and B. Nicolaï. 1998. Microscopic observation of mealiness in apples: A quantitative approach. *Postharvest Biology and Technology* 14(2):275–284.

Desmet, M., J. Lammertyn, N. Scheerlinken, and B.E. Verlinden. 2002. A pendulum for testing puncture injury susceptibility of tomatoes. Proceedings Intnl. Conf. Agricultural Engineering. Budapest, June 30–July 4. Paper 02-PH-038.

Diezma-Iglesias, B., C. Valero, F.J. García-Ramos, and M. Ruiz-Altisent. 2006. Monitoring of firmness evolution of peaches during storage by combining acoustic and impact methods. *Journal of Food Engineering* 77(4):926–935.

Duprat, H., H. Chen, M. Grotte, D. Loonis, and E. Pietri. 1995. Laser light based machine vision system for non-destructive ripeness sensing of golden apples. Paper presented at IFAC/CIGR Workshop on Control Applications in Postharvest and Processing Technology. Ostende, Belgium, June 1–2,117–123.

Durofel (1996). Technical information. In Operator Instructions.

Farabee, L.M., and M.L. Stone. 1991. Determination of watermelon maturity with sonic impulse testing. ASAE Paper No. 91-3013, ISSN 0149-9890.

Fekete, A., and J. Felfôldi. 2000. Systems for fruit firmness evaluation. Proc. Intnl. Conf. Agricultural Engineering. Warwick, UK, July 2–7. Paper 00-PH-034.

Fillion, L., and D. Kilkast. 2002. Consumer perception of crispness and crunchiness in fruits and vegetables. *Food Quality and Preference* 13(1):23–29.

Finney, E.E., and D.R. Massic. 1975. Instrumentation for investigating dynamic mechanical properties of fruits and vegetables. *Transactions of the ASAE* 11(1):94–97.

Fu, X., Y. Ying, Y. Zhou, L. Xie, and H. Xu. 2008. Application of NIR spectroscopy for firmness evaluation of peaches. *Journal of Zhejiang University* 9(7):552–557.

García, C., M. Ruiz-Altisent, and P. Chen. 1988. Impact parameters related to bruising in selected fruits. Proc Summer Meeting of the ASAE. Rapid City, South Dakota. Paper No. 88-6027.

García Ramos, F.J., P. Barreiro, M. Ruiz Altisent, J. Ortiz Cañavate, J. Gil Sierra, and I. Homer. 2002. A procedure for testing padding materials in fruit parking lines using multiple logistic regression. *Transactions of the ASAE* 45(3):751–757.

García Ramos, F.J., J. Ortiz Cañavate, M. Ruiz-Altisent, et al. 2003. Development and implementation of an online impact sensor for firmness sensing of fruits. *Journal of Food Engineering* 58(1):53–57.

García Ramos, F.J., C. Valero, I. Homer, J. Ortiz Cañavate, and M. Ruiz-Altisent. 2005. Nondestructive fruit firmness sensors: A review. *Spanish Journal of Agricultural Research* 3(1):61–73.

Gómez de la Cámara, A. 1998. Caracterización de las pruebas diagnósticas. (Characterization of diagnostic tests). *Medicine* 7(104):4872–4877.

Gutierrez, A., J.A. Burgos, and E. Moltó. 2003. Estimación de la firmeza de los melocotones en una línea de confección precomercial. (Assessement of the peach firmnes in a premarket parking line). Proc .III Congreso de Agroingeniería, Cordoba. Spain, September 24–27, 1158–1163.

Hernández Gómez, A., Y. He, and A. García Pereira. 2006. Non-destructive measurement of acidity, soluble solids and firmness of Satsuma mandarin using Vis/NIR-spectroscopy techniques. *Journal of Food Engineering* 77:313–319.

Homer, I., F.J. García-Ramos, J. Ortiz Cañavate, and M. Ruiz-Altisent. 2010. Evaluation of a non-destructive impact sensor to determine on-line fruit firmness. *Chilean Journal of Agricultural Research* 70(1):67–74.

Horwarth, M.S., and Y. Ioannides. 2002. Sinclair IQ-firmness tester. Proceedings of the International Conference on Agricultural Engineering. Budapest. June 30–July 4. Paper No. 02-IE-0006.

Huang, M., and R. Lu. 2010. Apple mealiness detection using hyperspectral scattering technique. *Postharvest Biology and Technology* 58(3):168–175.

Hung, Y.C., and S.E. Prussia. 1995. Firmness measurement using a non destructive laser-puff detector. In *Proceedings FPAC IV Conference*, Nov 3–5, Chicago, IL. Published by ASAE.

Jarén, C., and E. García-Pardo. 2002. Using non-destructive impact testing for sorting fruits. *Journal of Food Engineering* 53(1):89–95.

Jowitt, R. 1974. The terminology of food texture. *Journal of Texture Studies* 5:351–358.

Karaoui, R., L. Pillonel, E. Schaller, J.O. Bosset, and J. De Baerdemaeker. 2007. Prediction of sensory attributes of European Emmental cheese using near-infrared spectroscopy: A feasibility study. *Food Chemistry* 101(3):1121–1129.

Katz, E.E., and T.P. Labuza. 1981. Effect of water activity on the sensory crispness and mechanical deformation oh snack food products. *Journal of Food Science* 46:403–409.

Kavdira, I., R. Rub, D. Ariana, and M. Ngouajioc. 2007. Visible and near-infrared spectroscopy for nondestructive quality assessment of pickling cucumbers. *Postharvest Biology and Technology* 44(2):165–174.

Krautkramer, J., and H. Krautkramer. 1990. *Ultrasonic testing of materials*. Heidelberg, Germany: Springer-Verlag.

Kuriyati, N., T. Matsuoka, and S. Kawano. 2004. Precise near infrared spectral acquisition of intact tomatoes in interactance mode. *Journal of Near Infrared Spectroscopy* 12:391–395.

Lammertyn, J., A. Peirs, J. De Baerdemaeker, and B.M. Nïcolai. 2000. Light penetration properties of NIR radiation in fruit with respect to non-destructive quality assessment. *Postharvest Biology and Technology* 18:121–132.

Lapsley, K.G. 1989. Texture of fresh apples: Evaluation and relationship to structure. PhD thesis, Zurich, Swiss Federal Institute of Technology, Switzerland.

Lesage, P., and M.F. Destain. 1996. Measurement of tomato firmness by using a non-destructive mechanical sensor. *Postharvest Biology and Technology* 8(1):45–55.

Liu, X., and J. Tan. 1999. Acoustic wave analysis for food crispness evaluation. *Journal of Texture Studies* 30:397–408.

Lovász, T., P. Merész, and A. Salgó. 1994. Application of near infrared transmission spectroscopy for the determination of some quality parameters of apples. *Journal of Near Infrared Spectroscopy* 2:213–221.

Lu, R. 2004. Multispectral imaging for predicting firmness and soluble solids content of apple fruit. *Postharvest Biology and Technology* 31(2):147–157.

Lu, R., and Y. Peng. 2006. Hyperspectral scattering for assessing peach fruit firmness. *Biosystems Engineering* 93(2):161–171.

Lu, R., K. Srivastava, and R.M. Beaudry. 2005. A new bioyield tester for measuring apple fruit firmness. *Applied Engineering in Agriculture* 21(5):893–900.
Macnish, J.A., D.C. Joyce, and A.J. Shorter. 1997. A simple non-destructive method for laboratory evaluation of fruit firmness. *Australian Journal of Experimental Agriculture* 37:709–713.
Manley, M., and A.E.J. McGill. 1996. Whole wheat grain hardness measurement by near infrared spectroscopy. Near Infrared Spectroscopy: the future waves. In *Proceedings of the 7th International Conference on Near Infrared Spectroscopy*, ed. A.M.C. Davies and P. Willians, 466–470. Montreal. Canada.
Marigheto, N., L. Venturi, and B. Hills. 2008. Two-dimensional NMR relaxation studies of apple quality. *Postharvest Biology and Technology* 48(3):331–340.
McGlone, V.A., H. Abe, and S. Kawano. 1998. Kiwifruit firmness by near infrared light scattering. *Journal of Near Infrared Spectroscopy* 5(2):83–89.
Mitzrach, A. 2000. Portable ultrasonic non-destructive fruit quality analyzer. The XIV Memorial CIGR World Congress 2000. Paper No. R6230. Tokyo, Japan.
Mizrach, A. 2008. Ultrasonic technology for quality evaluation of fresh fruit and vegetables in pre- and postharvest processes. *Postharvest Biology and Technology* 48:315–330.
Mitzrach, A., A. Bechar, Y. Grinshpon, A. Hofman, H. Egozi, and L. Rosenfeld. 2003. Ultrasonic classification of mealiness in apples. *Transactions of the ASAE* 46:397–400.
Mitzrach, A., U. Flitsanov, Z. Schmilovitch, and Y. Fuchs. 1999. Determination of mango physiological indices by mechanical wave analysis. *Postharvest Biology and technology* 16:179–186.
Mitzrach, A., U. Flitsanov, M. Akerman, and G. Zauberman. 2000. Monitoring avocado softening in low-temperature storage using ultrasonic measurements. *Computers and Electronics in Agriculture* 26:199–207.
Mohsenin, N. 1970. *Physical properties of plant and animal materials*. Boca Raton, FL: Gordon and Breach Sciences Publishers.
Moons, E., A. Dubois, P. Dardenme, and M. Lindic. 1997. Non destructive visible and NIR spectroscopy for the determination of internal quality in apple. In *Sensors for Nondestructive Testing*. 122–131. Ithaca, NY: Northeast Regional Agricultural Engineering Service (NRAES).
Moshou, D., S. Wahlen, R. Strasser, A. Schenk, and H. Ramon. 2003. Apple mealiness detection using fluorescence and self-organism maps. *Computers and Electronics in Agriculture* 40(1–3):103–114.
Moshou, D., S. Wahlen, R. Strasser, A. Schenk, J. De Baerdemaeker, and H. Ramon. 2005. Chlorophyll fluorescence as a tool for online quality sorting of apples. *Biosystems Engineering* 91(2):163–172.
Nicolaï, B.M., K. Beullens, E. Bobellyn, et al. 2006. Systems to characterize internal quality of fruit and vegetables. *Acta Horticulturae* 712, ISHS 2006.
Nicolaï, B.M., K. Beullens, E. Bobelyn, et al. 2007. Nondestructive measurement of fruit and vegetable quality by means of NIR spectroscopy: A review. *Postharvest Biology and Technology* 46:99–118.
Nicolaï, B.M., B.E. Verlinden, M. Desmet, et al. 2008. Time-resolved and continuous wave NIR reflectance spectroscopy to predict soluble solids content and firmness of pear. *Postharvest Biology and Technology* 47:68–74.
Onwulata, C.I., R.P. Konstance, P.V. Smith, and V.H. Holsinger. 2001. Co-extrusion of dietary fiber and milk proteins in expanded corn products. *Lebensmittel Untersuchung Forschung* 34:424–429.
Paoletti, F., E. Moneta, A. Bertone, and F. Sinesio. 1993. Mechanical properties and sensory evaluation of selected apple cultivars. *Lebensmittel-Wissenschaft und-Technologie* 26:264–270.
Park, B., J.A. Abbott, K.J. Lee, C.H. Choi, and K.H. Choi. 2003. Near-infrared diffuse reflectance for quantitative and qualitative measurement of soluble solids and firmness of Delicious Gala apples. *Transaction of the ASAE* 46:1721–1731.

Peirs, A., N. Scheerlink, K. Touchant, and B.M. Nicolaï. 2002. Comparison of Fourier transform and dispersive near infrared reflectance spectroscopy for apple quality measurements. *Biosystems Engineering* 81(3):305–311.

Peleg, K. 1993. Comparison of non-destructive measurement of apple firmness. *Journal of Agricultural Engineering Research* 55(3):227–238.

Povey, M.J.W. 1998. Ultrasonics of food. *Contemporary Physics* 39(6):467–478.

Povey, M.J.W., and C.A. Harden. 1981. An application of the ultrasonic pulse echo technique to the measurement of crispness of biscuits. *International Journal of Food Science and Technology* 16(2):167–175.

Prussia, S.E., J.J. Astleford, Y.C. Humg, and R. Hewlett. 1994. Non-destructive firmness measuring device. US Patent No. 5,372,030.

Quevedo, R., and J.M. Aguilera. 2008. Computer vision and stereoscopy for estimating firmness in the salmon (Salmon salar) fillets. *Food and Bioprocess Technology* 3(4):561–567.

Ragni, L., A. Berardinelli, and A. Guarnieri. 2010, Impact device for measuring the flesh firmness of kiwifruits. *Journal of Food Engineering* 96(4):591–597.

Roudaut, G., C. Dacremont, B.V. Pamies, B. Colas, and M. Le Meste. 2002. Crispness: A critical review on sensory and material science approaches. *Trends in Food Science and Technology* 13(6–7):217–227.

Sakurai, N., S. Iwatani, and R. Yamamoto. 2005. Evaluation of Fuyu persimmons texture by a new parameter, "Sharpness index ." *Journal of Japanese Society Horticulture Science* 74:150–158.

Schaare, P.N., and V.A. MacGlone. 1997. Design and performance of a fruit firmness grader. *Acta Horticulturae (ISHIS)* 464:417–422.

Shao, Y., Y. He, A.H. Gómez, A.G., Pereir, Z. Qiu, and Y. Zhang. 2007. Visible/near infrared spectrometric technique for non destructive assessment of tomato "Heatwave" (*Lycopersicum esculentum*) quality characteristics. *Journal of Food Engineering* 81:672–678.

Shmulevich, I., N. Galili, and M.S. Howarth. 2003. Nondestructive dynamic testing of apples for firmness evaluation. *Postharvest Biology and Technology* 29:287–299.

Slaughter, C., M. Ruiz-Altisent, J.F. Thompson, P. Chen, Y. Sarig, and M. Anderson. 2009. A handheld, low-mass, impact instrument to measure nondestructive firmness of fruit. *Transactions of the ASABE* 52(1):193–199.

Sonego, L., R. Ben-Arie, J. Rainal, and J.C. Pech. 1995. Biochemical and physical evaluation of textural characteristics of nectarines exhibiting woolly breakdown: NMR imaging, x-ray computed tomography and pectin composition. *Postharvest Biology and Technology* 5:187–198.

Srisawas, W., and V. Jindal. 2003. Acoustic testing of snack food crispness using neural networks. *Journal of Texture Studies* 34(4):401–420.

Stanley, D.W. 1986. Chemical and structural determinants of texture of fabricated foods. *Food Technology* 12:65–76.

Steinmetz, V., M. Crochon, V. Bellon, J.L. García, P. Barreiro, and L. Verstreken. 1996. Sensors for fruit firmness assessment: Comparison and fusion. *Journal Agricultural Engineering Research* 64(1):15–27.

Sterling, C., and M.J. Simone. 1954. Crispness in almonds. *Journal of Food Science* 19(1–6):276–281.

Sugiyama, J., T. Katsuray, J. Hong, H. Koyama, and K. Mikuriya. 1997. Portable melon firmness tester using acoustic impulse transmission. In *Proc. Sensors for Nondestructive Testing International Conference*, 3–12. Ithaca, NY: Northeast Regional agricultural Engineering Service Cooperative Extension.

Szczesniak, A.S. 1963. Classification of textural characteristics. *Journal of Food Science* 28:385–389.

Szcesniak, A.S. 1987. The meaning of textural characteristics: Crispness. *Journal of Texture Studies* 19:51–59.

Szcesniak, A.S., and E.L. Kahn. 1971. Consumers awareness and attitudes to food texture. I Adults. *Journal of Texture Studies* 2(3):280–295.

Taherian, A.R., S. Hosahalli, and S. Ramaswamy. 2009. Kinetic considerations of texture softening in heat treated root vegetables. *International Journal of Food Properties* 12(1):114–128.

Taniwaki, M., T. Hanada., and N. Sakurai. 2006. Device for acoustic measurement of food texture using a piezoelectric sensor. *Food Research International* 39(10):1099–1105.

Taniwaki, T., T. Hanada, and N. Sakurai. 2009a. Postharvest quality evaluation of Fuyu and Taishuu persimmons using a nondestructive vibrational method and an acoustic vibration technique. *Postharvest Biology and Technology* 51(1):80–85.

Taniwaki, T., T. Hanada, M. Tohro, and N. Sakurai. 2009b. Non-destructive determination of the optimum eating ripeness of pears and their texture measurements using acoustical vibration techniques. *Postharvest Biology and Technology* 51(3):305–310.

Taniwaki, M., and N. Sakurai. 2008. Texture measurement of cabbages using an acoustical vibration method. *Postharvest Biology and Technology* 50(2–3):176–181.

Taniwaki, T., N. Sakurai, and H. Kato. 2010. Texture measurement of potato chips using a novel analysis technique for acoustic vibration measurements. *Food Research International* 43(3):814–818.

Terasaki, S., N. Wada, N., Sakurai, N., Muramatsu, R. Yamamoto, and D.J. Nevins. 2001. Nondestructive measurement of kiwifruit ripeness using a laser Doppler vibrometer. *Transactions of the ASABE* 44(1):81–87.

Thybo, A.K., P.M. Szczpy, A.H. Karlsson, S. Donstrup, H.S. Stokilde-Jorgensen, and H.J. Andersem. 2004. Prediction of sensory texture quality attributes of cooked potatoes by NMR-imaging (MRI) of raw potatoes in combination with different image analysis methods. *Journal of Food Engineering* 61(1):91–100.

Tu, K., R. Be Busscher, J. De baerdemaeker, and E. Schrevens. 1995. Using laser beam as light source to study tomato and apple quality non-destructively. Proceeding Food Processing Automation IV Conference, 3–5 November. Chicago IL.

Valero, C. 2001. Aplicación de la espectroscopía láser de reflectancia difusa (ERDT) a la medida de la calidad interna de frutas y hortalizas (Application of the Laser diffuse re flectance spectroscopy to the assessement of the internal quality of fruits and vegetables). PhD Thesis. Universidad Politécnica de Madrid.

Valero, C., P. Barreiro, M. Ruiz-Altisent, et al. 2005. Mealiness detection in apples using time resolved spectroscopy. *Journal of Texture Studies* 36(4):439–458.

Vàlles Pámies, B., G. Roudaut, C. Dacremont, M. Le Meste, and R. Mitchell. 2000. Understanding the texture of low moisture cereal products. Part I: Mechanical and sensory measurements of crispness. *Journal of the Science of Food and Agriculture* 80:1679–1685.

Vangdal, E., M. Vanoli, P.E. Zerbinni, S. Jacob, A. Torricelli, and L. Spinelli. 2008. TRS-measurements as a non-destructive method assessing stage of maturity and ripening in plum (*Prunus Domestica* L.). *Acta Horticulturae (ISHIS)* 858:443–448.

Varela, P., J. Chen, S. Fiszman, and M.J.W. Povey. 2006. Crispness assessment of roasted almonds by an integrated approach to texture description: Texture, acoustics, sensory and structure. *Journal of Chemometrics* 20(6–7):311–320.

Verberke, W., J. Janse, and M. Kersten. 1998. Instrumental measurement and modeling of tomato fruit taste. *Acta Horticulturae (ISHS)* 456:199–206.

Verlinden, B,E., V. De Smedt, and B.M. Nicolaï. 2004. Evaluation of ultrasonic wave propagation to measure chilling injury in tomatoes. *Postharvest Biology and Technology* 32(1):109–113.

Veronica, A.O., O.O. Olusola, and O.A. Adebowale. 2006. Quality of extruded puffed snacks from maize/soybean mixture. *Journal of Food Science* 29:149–161.

Williams, P.C. 2001. *Near-infrared technology in the agricultural and food industries.* Saint Paul, MN: American Association of Cereal Chemists.

Windham, W.R., B.G. Lyon, E.T. Champagne, et al. 1997. Prediction of cooked rice texture quality using near infrared reflectance analysis of whole-grain milled samples. *Cereal Chemistry* 74:626–632.

Xie, F., F.E. Dowell, and X.S. Sun. 2003. Non-destructive tests on the prediction of apple fruit flesh firmness and soluble solids content on tree and in shelf life. *Cereal chemistry* 80(1):25–29.

Zdunek, A., J. Cybulska, D. Konopacka, and K. Rutkowski. 2010. New contact acoustic emission detector for texture evaluation of apples. *Journal of Food Engineering* 99(1):83–910.

Zerbinni, P.E., M. Vanoli, A. Rizzolo, et al. 2009. Time-resolved reflectance spectroscopy as a management tool in the fruit supply chain: an export trial with nectarines *Biosystem Engineering* 102(3):360–363.

Zhou, R., and Y. Li. 2007. Texture analysis of MR image for predicting the firmness of Huanghua pears (*Pyrus pyrifolia Nakai*, cv. Huanghua) during storage using an artificial neural network. *Magnetic Resonance Imaging* 25(5):727–732.

Zhu, Q., M. Huang, and G. Zhao. 2010. Mealiness detection using supervised locally linear embedding and support vector machine. Pittsburgh, Pennsylvania, June 23. 7008915. ASABE, St. Joseph, Michigan.

Zude, M., B. Herold, J.M. Roger, V. Bellon-Maurel, and S. Landahl. 2006. Non-destructive tests on the prediction of apple fruit flesh firmness and soluble solids content on tree and in shelf life. *Journal of Food Engineering* 77(2):254–260.

4 Optical Properties of Foods

Begoña Hernández Salueña and
Carlos Sáenz Gamasa

CONTENTS

4.1 INTRODUCTION

By optical properties of a material we understand all those properties that describe how the geometrical, spectral, and chromatic characteristics of light are affected or modified after its interaction with that material. Within the context of optics, we are mainly interested in phenomena restricted to the visible range (i.e., between 380 and 780 nm), the wavelength interval where the human visual system is sensitive to electromagnetic radiation.

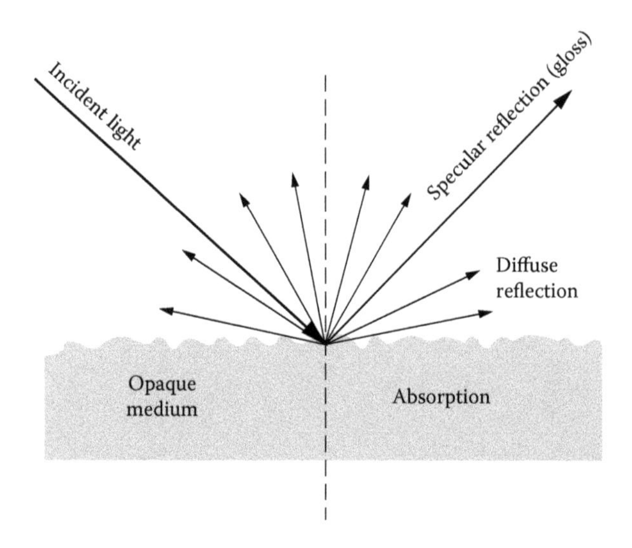

FIGURE 4.1 Light interaction in opaque media.

Visible light, hereafter just light, has a fundamental role in our perception of the appearance of objects. More than 80% of the information we gather from the environment through our senses is collected by the eyes. Not surprisingly, visual assessment has a central role in the food industry. If the visual appearance of a food product is not acceptable, then other attributes like odor or flavor will not even be considered. This leading role of visual appearance makes the study and analysis of the optical properties of foods the subject of extensive research. Optical properties are generally studied in connection with other parameters and objectives like shelf life, quality descriptors, characterization of varieties, visual appearance, consumer preferences, effect of storage and processing techniques, or the use of additives.

From a physical perspective, the interaction of light with an object is a complex process. Light falling on an object will be partially reflected, partially absorbed and scattered inside the object, and partially transmitted. Reflected light will have, in general, a specular or mirrorlike component and a diffuse component. Objects that reflect light predominantly in the specular direction will have a glossy appearance. If most of the light is diffusely reflected, the object will have a matte appearance. An object is said to be opaque if the fraction of transmitted light is negligible (Figure 4.1). Conversely, an object is said to be transparent if transmitted light cannot be neglected. Transmittance can be regular (i.e., in the same direction) that incident light, like in a transparent window glass (Figure 4.2), or diffuse as in turbid liquids (Figure 4.3).

Reflectance, absorption, and transmittance of light are wavelength-dependent processes. The spectral composition of light is modified in a geometrically dependent way and as a result, objects appear colored. The geometrical dependence of the spectral composition and intensity of reflected and transmitted light, and therefore of color, makes necessary the definition and even the standardization of the measurement geometries. Standard radiometric, photometric, and lighting vocabulary is

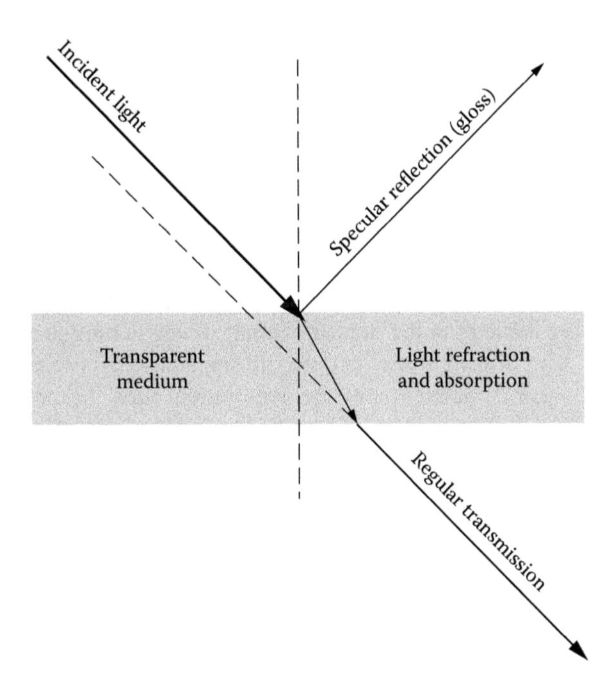

FIGURE 4.2 Light interaction in transparent media.

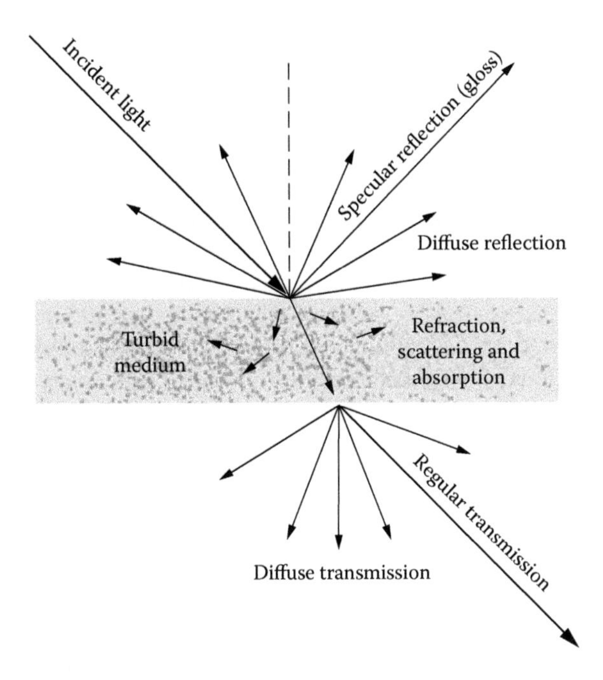

FIGURE 4.3 Light interaction in turbid media.

available to avoid confusion between similar terms with different meanings (CIE, 1987).

Foods can be found in a great variety of aggregation states: solids, liquids, dusts, granules, gels, and so on. In general, food products are inhomogeneously pigmented substances, with irregular shapes, sizes, and textures. The enormous variety of possible conditions in their light reflecting, absorbing, and transmitting properties frequently require specific sample preparation and measurement procedures for each particular food product. In general, these particularities make them difficult to measure with standardized procedures so common in other industries. Measurements become even more difficult in the industrial context, where on-line applications are necessary. Not only measurements, but also all the necessary processing to obtain the quality or control parameters must be done in real time. In these applications, measurement equipment must work in hostile conditions, very different from those found in laboratories.

In foods, physical and optical properties are always related to other chemical properties, structural characteristics, or quality factors. Optical properties are, in particular, fundamental to the description of the total appearance of foods. In this sense, the most important property is color, but gloss or translucency can also have a great influence on the visual aspect of a food product (Hunter and Harold, 1987; Hutchings, 1999; Hutchings, 2003; Calvo, 2004).

In this chapter we will first study color and gloss, magnitudes linked with our visual experience and completely meaningful only in the context of human observers. These magnitudes can be measured with adequate instrumentation, but they can also be evaluated by visual assessment and are directly related to the appearance of food products. We will then study other optical properties like light scattering, refractive index, or fluorescence that are not directly evaluable by visual assessment. Applications concerning computer vision and spectral imaging will be treated in separate sections at the end of the chapter.

According to the spirit of this volume, we have focused our attention on recent technical developments and novel applications where optical properties have a central importance. We have included a brief, and hopefully sufficient, explanation of each optical property in order to give a minimum background and maintain the coherence within each section. Nonetheless the detailed description and analysis of all optical properties exceeds the scope of this chapter. Due to limitations of available space, some well-known traditional measurement techniques like absorbance or transmittance in liquids will not be specifically treated, although references to them appear in several sections. For the same reason, only a selection of published works has been included, but we have included references to textbooks and recent reviews that the interested reader can consult to get a deeper insight in specific subjects.

4.2 COLOR

If we rank the several appearance attributes of foods we must recognize that color is always rated in the top positions. Color can inform us about the ripeness of a fruit, the state of preservation of a vegetable, or the time a cheese was cured. In general, the color of a food product allows us to make a first evaluation of its quality. Like other sensory attributes, the color of a particular food product is interpreted in terms of color–quality relationships obtained from previous experiences and serves as a first and fast clue of its suitability for consumption. This interpretation has a subjective component that may depend on consumer-related factors like sex, age, or education, but also on how a product is packed or displayed. On the other hand, color can be altered during processing, storage, or transport, and therefore it must be carefully controlled to ensure the quality and acceptability of food products.

Accordingly, colorimetry is used in the industry from two different perspectives:

- The standardization of the measurement procedures and color values in quality control processes
- The measurement of color as an indicator of other correlated attributes like the nutritional quality in the case of foods

In the first case we are mainly interested in the perception of color from the human (consumer) perspective. Materials must accomplish certain colorimetric specifications within some established tolerances. This is the objective of textile, plastic, or paint industries, and sometimes it is also the objective of the food industry, particularly in relation with the use of colorants (production of juices, jams, sauces, etc.). In general, the role of colorimetry in the food industry and food research corresponds to the second case. In color science, two different approaches are used: subjective color evaluation, based on color assessment by some observers, and the objective or instrumental color evaluation, based on the measurement of the necessary magnitudes according to established methods or standards. Both approaches are necessary and complement each other and will be analyzed here.

4.2.1 Color Measurement and Instrumentation

The instrument that is most frequently used for objective color measurement is the *spectrocolorimeter*. This apparatus measures the light reflected or transmitted by a sample and provides the spectral distribution of reflected or transmitted light. Using the instrument's built-in software or offline analysis programs we can then calculate color coordinates in different color spaces. A simpler color measurement device is the *tristimulus colorimeter*, which consists of an instrument that uses three filters to simulate the spectral sensitivity of the human visual system. Colorimeters directly provide color coordinates, although typically with poorer precision than spectrocolorimeters. Normally these instruments implement one of the standardized measurement geometries and have a built-in illumination source of known characteristics. In general, measurements require contact with the sample, but there are other instruments like spectroradiometers, which can be used to perform measurements at some distance from the sample. There are portable and benchtop instruments adapted for the measurement of solids, liquids, or both.

More recently, image analysis based in calibrated color imaging has stepped in to increase the available color measurement techniques, providing not only color, but also the spatial distribution of color in a product. Incorporating several visual

TABLE 4.1
Summary of the Necessary Information for the Correct Interpretation of Color Measurements

Parameter/Condition	Data to Specify
	Instrument
Type of instrument	Colorimeter, spectrocolorimeter, spectroradiometer, etc.
Spectral range	Spectral range and wavelength interval between data points.
Measurement mode	Reflectance, transmittance, etc.
Light source	Fluorescent, halogen, etc.
Geometry	Standard geometries (d/0°, 0°/45°, etc.)
	Nonstandard geometry: complete specification.
Specular component	Included or not included.
Calibration	Procedure and calibration elements (tiles, diffuse reflectance standard, reference transparent liquid, etc.)
	Reporting Data
Observer	2° or 10° standard observer
Illuminant	Standard illuminant (A, D65, D75, etc.)
	Non standard illuminant (spectral information, CCT, CRI, etc.)
Color space	CIELAB, Hunter Lab, etc.

attributes can bring us closer to the concept of food total appearance (MacDougall, 2002). We will review this particular subject in a specific section.

Color is basically specified by the geometry and spectral characteristics of three elements: the light source, the reflectance or transmittance of the sample, and the visual sensitivity of the observer. When color measurements are reported, there are several experimental parameters concerning these elements that must be specified to allow the correct interpretation of the data. A summary of these parameters can be found in Table 4.1.

Some of the parameters in Table 4.1 are fixed for a particular instrument (measurement geometry, light source, spectral range, etc.) and others that can be selected using the manufacturer's software (color space, illuminant, observer, etc.) depending on the application or product. In general, the use of standardized measurement geometries, illuminants, and observers, and the use of well-established colorimetric spaces is highly recommended.

4.2.2 Evaluation and Interpretation of Color Measurements

Sticking to recommended measurement procedures permits an easier interpretation and understanding of color data. The lack of some of the information stated in the previous section makes it difficult to evaluate the results and compare them with results from other works, seriously limiting their usefulness.

Let us suppose that we measure the color of a food sample and report the CIELAB color values. Color values are different for different illuminants (say illuminants A and D65) and therefore, if we do not specify the illuminant, the reported color values are meaningless. Special care must be taken if non-standard illuminants are used. For instance, we might be interested in the color of food products under display conditions. It is well known that special lamps are used to enhance the color characteristics of some products and make them more attractive. Meat products, for instance, are frequently illuminated with red loaded lamps to enhance their reddish hues, noticeably changing their color with respect to other types of illuminants (Saenz, Hernandez, et al. 2005). In these cases, the spectral characteristics of the illuminants should also be reported.

If spectral information is not available, the color-correlated temperature (CCT) can be a useful parameter. The CCT is the temperature of a black body radiator with similar chromaticity. For example, the D65 standard illuminant has an approximate CCT of 6500 K. However, not all light sources can be characterized by their CCT. Another useful parameter is the color rendering index (CRI). This parameter is intended to be a measure of the source's ability to reproduce colors in a *true* or *natural* way compared to a reference source. The CRI can be found in the technical information of most light sources.

The most common light sources used in the food industry are fluorescent lamps, which are cool sources that produce negligible heating of the food products. However, other technologies like those based in light-emitting diode (LED) are becoming available. Recently, the effect of CCT and CRI of halogen lamps (CCT = 3050 K), fluorescent lamps (CCT = 3950 K), and several combinations of LEDs in the color, attractiveness, and naturalness of fruits and vegetables has been studied. Results indicate that some combinations of white and red LEDs can produce a more attractive color than fluorescent and halogen lamps (Jost-Boissard et al., 2009).

We humans are trichromats; any color that we perceive can be described by only three attributes. Our visual system encodes these three dimensions using an opponent encoding along one lightness channel from black to white and two chromatic channels, the red-green and the yellow-blue channels.

In the CIELAB system, one of the most important color spaces, colors are described with three color coordinates L^*, a^* and b^*. Lightness L^* goes from 0 (black) to 100 (white), a^* is the red-green axis and b^* is the yellow-blue axis. They form a Cartesian coordinate system where any color can be represented. In the a^*b^* plane we can describe colors using the polar coordinates C^*_{ab} (chroma) and h_{ab} (hue angle), defined by Equations (4.1) and (4.2), respectively:

$$C^*_{ab} = \sqrt{a^{*2} + b^{*2}} \tag{4.1}$$

$$h_{ab} = \arctan\left(\frac{b^*}{a^*}\right) \tag{4.2}$$

The *chroma* indicates the vividness of one color; for instance, a pale pink has less chroma than a strong red. The *hue angle* is a continuous variable that indicates if a color is yellow, red, green, and so on. Sometimes these quantities are represented by symbols different than those shown here, which correspond to the CIE definitions

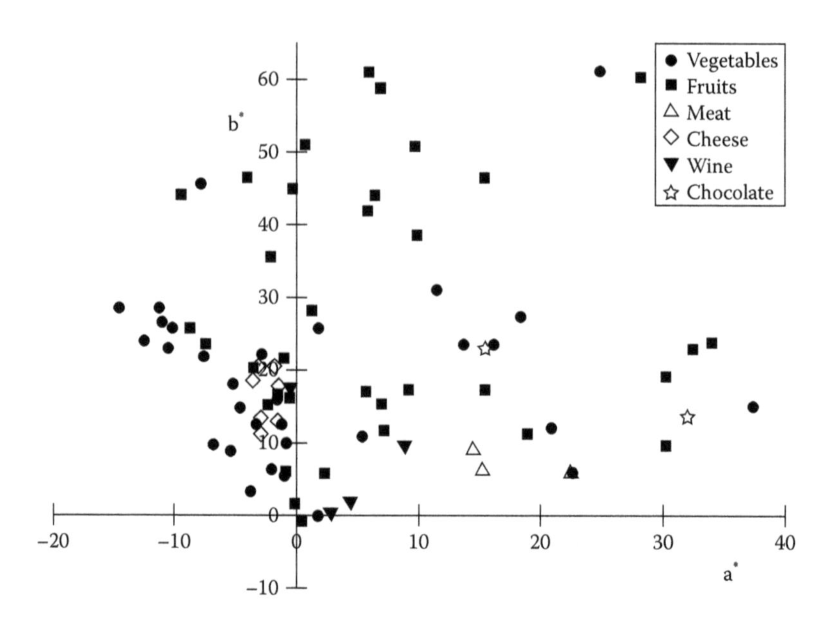

FIGURE 4.4 Typical CIE a*, b* color coordinate values of some representative food products.

(CIE 2004). An example of typical color coordinate values of a variety of food products can be seen in Figure 4.4.

One of the major advantages of colorimetric spaces is the possibility of making quantitative comparisons between two colors. In CIELAB, the color difference ΔE^*_{ab} between two colors with coordinates L^*_1, a^*_1, b^*_1 and L^*_2, a^*_2, b^*_2 is computed as the Euclidean distance defined by Equation (4.3):

$$\Delta E^*_{ab} = \sqrt{\left(L^*_2 - L^*_1\right)^2 + \left(L^*_2 - a^*_1\right)^2 + \left(b^*_2 - b^*_1\right)^2} \tag{4.3}$$

We should not oversimplify the meaning of the chromatic coordinates a^* and b^* with affirmations like "a^* is the measure of the amount of redness or greenness." For instance, samples having identical a^* values may exhibit purple, orange, or red appearance. Similarly, a pink and a dark red may have identical values of chroma C^*_{ab} (Wrolstad, Durst, et al. 2005).

For the interested reader, there are excellent publications that can be consulted to get a deeper insight into general colorimetry (Wyszecki and Stiles 2000) and also in the particular field of food color measurement (Hunter and Harold 1987; Hutchings 1999; MacDougall 2002; Calvo 2004).

4.2.3 Examples and Applications of Food Colorimetry

Controlling the effect that new treatments or processes have on food color is of great importance, and there are many examples in the literature where color is measured and related to other parameters. We will here mention only a few recent works that are innovative with respect to color measurement or application and that are also good examples of how to report color measurement procedures and colorimetric data.

4.2.3.1 Solid Products

Present lifestyles makes consumers demand more and more ready-to-eat products and minimally processed fruits and vegetables. In products such as fresh cut carrot slices (Villalobos-Carvajal, Hernandez-Munoz, et al. 2009), fresh cut cantaloupe (Ukuku, Fan, et al. 2006), or minimally processed shiitake mushrooms (Santana, Vanetti, et al. 2008), or asparagus (Sanz, Olarte, et al. 2009), color is a critical parameter directly related to freshness and quality.

Fish roes are an example of a food product where color measurement difficulties arise because of their irregular shape, small size, and translucency. Their color depends on the carotene content, but also on the fish species, diet, age, and maturity stage, among other factors. In fact, it has been shown that color, in particular a^*, can discriminate between roes from different species, and that color and the spectral reflectance of this product are differently affected by different types of processing (Bekhit, Morton, et al. 2009).

The attractive color of fresh strawberries is also affected by processing and storage. A recent study compared the color values obtained with three different instruments in fresh, chilled, and canned strawberries, and in strawberry jam and syrup. It was found that color values depended on the instrument's measurement geometry and that in some cases, for instance between fresh and chilled strawberries, color values did not reproduce the observed visual differences (Ngo, Wrolstad, et al. 2007).

4.2.3.2 Liquid Products

In general, the color of liquids is traditionally measured in transmittance mode pouring the liquid in glass or quartz cells with 1- to 10-mm light path lengths depending on the liquid opacity. Not only color coordinates, but also spectral transmittance or absorbance can be obtained with commercial cuvette spectrophotometers, specially designed for the measurement of liquid samples. Actually the determination of absorbance values has a long tradition due to the relationship between the absorbance and the concentration of light-absorbing substances within the sample. Many liquid food products are turbid to some extent, and they are centrifuged and filtered before measurement. In this way, however, we are changing an important characteristic of the visual appearance of the product (Melendez-Martinez, Vicario, et al. 2005). It must also be noted that these measurement conditions are very different from the conditions in which consumers assess the color of the product, and it is known that color coordinates vary with the measurement method (transmittance or reflectance) and the measurement geometry, as it has been recently reported in olive oil (Gomez-Robledo, Melgosa, et al. 2008).

Experimental conditions during transmittance measurements are very different from the viewing conditions in which liquid products are seen by consumers or evaluated during tasting. Some departure from traditional techniques seems to be necessary in order to reproduce visual color assessment scores. The use of calibrated color imaging systems has been tested, for instance in wines (Gonzalez-Miret, Ji, et al. 2007). Very good agreement between visual color assessment and instrumental color measurements has been obtained in red wines (Hernández, Sáenz, et al. 2008) and in white and rosé wines (Sáenz, Hernández, et al. 2009). These authors used a spectroradiometer and a color measurement procedure that carefully reproduces the tasting process with respect to illumination, background, standardized wine glass taster, and relative position of the sample and the measuring instrument.

4.2.4 Food Color and Pigments

Pigments are the most important cause of food color. Each pigment has an absorption spectrum with peaks at characteristic wavelengths. Absorbance or absorbance ratios at those wavelengths can be used to estimate pigment concentrations. Transmittance and reflectance values can be used in a similar way. Although characteristic wavelengths may be optimal to compute pigment concentrations, the presence of a pigment affects the entire reflectance or transmittance spectrum, and therefore the color of the sample. Since color depends on pigment concentrations, and concentrations can be obtained from characteristic wavelengths, we expect that color can be directly predicted from pigment concentrations and also from a reduced number of characteristic wavelengths.

Using absorption at 455 nm (carotene) and 670 nm (chlorophyll), the effect of the concentration of these pigments in the color of olive oil has been reported (Escolar, Haro, et al. 2007). Similarly the reflectance spectrum of beef meat can be reconstructed, and the color predicted, using reflectance values at 480, 580, and 620 nm, the characteristic wavelengths of the three chemical states of myoglobin (Mb, MbO_2, and MMb) (Aporta, Hernández, et al. 1996).

4.2.5 Sensory Color Evaluation and Color Scales

Sensory evaluation is a fundamental tool during product development or in quality control processes. During the evaluation, the organoleptic characteristics and the appearance attributes, color in particular, are judged and rated. Sensory evaluation is intrinsically subjective and care must be taken to control product preparation and display, the physical conditions in which evaluation is performed, and the characteristics of the panelists, including previous training if necessary and screening for color vision deficiencies. During the evaluation, it may be necessary or convenient to have adequate standards or references to aid panelists. In the case of color evaluation, this is particularly important since human observers show great ability in paired color comparison but poor color memory. An exception to this approach is when we are interested in the free, subjective evaluation of consumers. In this case, any training or visual aid that could bias the panelist scores must be avoided.

It must also be noted that there are situations in which the color of the product is concealed using adequate illumination or opaque recipients to prevent the influence of color in other sensory attributes. In olive oil tasting for instance, blue-tinted cups are used to reduce the effect of oil color, although recent research shows that these cups, which are not completely opaque, still influence observers (Melgosa, Gomez-Robledo, et al. 2009).

When possible, color scales can be conveniently used to provide some anchor points during color assessment. They provide a stable reference, helping to maintain consistent criteria among different tasting sessions and even over different years, as in the case of seasonal products like fruits or vegetables. It is known that without appropriate references, the color rating of individual samples will be influenced by the range of colors present in the sample set. A recent example has been reported in the evaluation of strawberry nectars (Gossinger, Mayer, et al. 2009).

Color scales constructed with plain color chips from color atlases and color collections have a reduced usefulness because they do not reproduce the color inhomogeneity and textural characteristics of most food products. Instead of plain color chips, it is possible to combine color and texture features simultaneously using color-processed images and calibrated color printing (Hernández, Alberdi, et al. 2005). In this method, color images are modified to obtain a collection of images resembling the visual appearance of the product at different degrees of quality. These scales are intended to make the work of panelists easier and accurate, and have been tested with piquillo peppers (Hernandez, Saenz, et al. 2004) and with beef meat (Goni, Indurain, et al. 2008) obtaining good agreement between visual assessment and instrumental color measurement.

4.2.6 Color Influence on Flavor and Odor

Color can have a halo effect that modifies subsequent flavor perceptions. A recent corroboration of this effect has been reported from a study about the influence of color clues on flavor discrimination and flavor intensity rating on fruit-flavored colored solutions (Zampini, Sanabria, et al. 2007). Without tasting the solutions, observers related specific colors to specific flavors like orange color with orange flavor, yellow with lemon flavor, and blue with spearmint flavor, whereas red, for instance, was not specially associated with any particular flavor. In a second experiment, observers were asked to taste, without swallowing, colored solutions. They were informed that the colors of the solutions could be misleading. In this case it was found that unexpected coloration of the solution affected the flavor discrimination. On the other hand, increasing the colorant concentration did not have noticeable effects in the perceived flavor intensity.

Color–flavor association can modulate flavor discrimination abilities even in the case of familiar food products such as sugar-coated chocolates (Levitan, Zampini, et al. 2008). In fruit-flavored yoghurts it has been found that, even with the same content of each fruit flavor (strawberry, orange, and fruit of the forest) and sugar, the greater the concentration of colorant the greater the perceived flavor intensity (Calvo, Salvador, et al. 2001).

Color also affects the identification of odors and aromas. It has been shown that observers chose odor–color combinations in a nonrandom manner (Dematte, Sanabria, et al. 2006). Caramel odor is associated with brown or yellow, cucumber to green, lemon to yellow, strawberry to red, and spearmint to blue. Interestingly, we can find, for instance, blue-colored sport drinks with fruit odors and flavors. Color can be used to evoke specific odors and flavors, but also to suggest particular sensations or ideas.

4.3 GLOSS

Gloss is a psychophysical phenomenon associated with the way in which light is reflected by an object in or near the specular direction. Glossy objects exhibit mirrorlike spots of reflected light and frequently even reflected images can be perceived in their surface. All these phenomena are barely perceptible in matte objects.

Gloss is an important perceptual attribute in many food products. It can be natural in origin or the result of fabrication processes. Gloss can be added by means of waxes and edible coatings, which are used to improve the appearance characteristics and preserve the quality of the product, protecting it during handling, since they reduce drying, UV damage, and fungi development (Hutchings 1999).

From a psychological perspective, several perceptual attributes related to gloss have been proposed: specular gloss, contrast gloss or luster, reflection haze, distinctness-of-reflected-image gloss, and sheen (CIE 1995). All of them are related to the directional properties of the reflected light. Physically, gloss is a surface phenomenon highly dependent on the change in the refractive index in the interface. Although refractive index is a wavelength-dependent quantity, for many common materials, the small variations of the refractive index values in the visible range have little influence in the spectral characteristics of the reflected light. This means that the spectral characteristics of incident and reflected light are quite similar. On the other hand, the fraction of light that is reflected in the specular direction depends on the angle of incidence. Therefore, illuminating and viewing geometry must be carefully specified for gloss measurement. The polarization of light also changes in the reflection, and although the human visual system is not sensitive to polarization, this must be taken into account in the design of gloss meters. The conditions for reliable gloss measurement have been standardized by the American Society for Testing and Materials (ASTM). The standard considers measurements at three angles of incidence (20°, 60°, and 85°) and defines the gloss unit (GU) so that a black glass with refractive index $n = 1.567$ has 100 GU for all angles of incidence (ASTM 1999).

Commercial gloss meters work according to this norm and are the basic instrument used for gloss determination in food products. However, the norm is for application to flat samples exclusively. In fact, deviations from perfect flatness have important consequences in gloss readings. Food products rarely possess a perfectly flat, homogeneous surface, and in most cases their surface is curved and inhomogeneous, making gloss measurements difficult and unreliable. In order to have flat samples, some foods, like fruits or vegetables, can be peeled and their skin flattened. Apart from being a destructive technique, gloss readings may not represent the appearance characteristics of the original, curved surface.

Not surprisingly, most recent developments in this field try to overcome this problem by using new instruments that measure gloss directly on the food product, without the requirement of having a flat sample. In existing alternatives, the (curved) sample is illuminated by a light (laser) beam and the spatial spread of the nearly specular reflected light is recorded using diode arrays or imaging devices.

The interpretation of the spatial distribution of reflected light in terms of psychophysical gloss is still an open question. Nussitovich et al. designed a gloss meter for the measurement of curved surfaces of fruits and vegetables. They used monochromatic light to illuminate the sample at the desired angle of incidence and reflected light was recorded with a video camera (Nussinovitch, Ward, et al. 1996). The spatial pattern of the reflected light had a distorted bell shape and authors used the full width at half maximum (FWHM) of the light profile, in pixels, as an indication of gloss. There are, however, other possible parameters like peak area, maximum intensity, or average of the reflected luminance flux, that could be better suited for this purpose (Mendoza, Dejmek, et al. 2010).

Imaging systems for gloss measurement have not been restricted to laboratory conditions. A machine vision system has been designed for real-time gloss grading of eggplants, in automatic grading conveyor lines (Chong, Nishi, et al. 2008). In this work, a gloss index was derived from the distribution of pixel intensities after image processing. Visual assessment classifies eggplants in high-, medium-, and low-gloss specimens. Gloss prediction was consistent and accurate in classifying high- and low-gloss samples. Some misclassification results were observed for the medium-gloss samples.

Attempts to include the spectral information of the reflected light have been also made. Jha et al. used a spectroradiometer to record the light reflected by the curved surface of eggplants illuminated by a halogen light source (Jha, Matsuoka, et al. 2002). The ratio between reflected and incident light at each wavelength was used to define a spectral gloss index which, when integrated in the visible range, gave the gloss index of the sample. It was found that the gloss index was correlated with the weight loss during storage of eggplants.

However, a gloss index defined in this way does not immediately translate into gloss units given by conventional gloss meters. Mizrach et al. have constructed and analyzed the capabilities of a spectrometer-based prototype gloss meter for curved surfaces that includes automatic sample positioning through image analysis (Mizrach, Lu, et al. 2009). Using a commercial catalog of flat samples of different sheen values and the measurements of a conventional gloss meter, the authors calibrate the system so that the gloss index can be converted into gloss units.

Gloss estimation using pixel intensity analysis can be performed using commercial imaging instruments designed for surface or appearance analysis. These measurements have been used to study the influence of different edible coatings in the gloss of chocolate-covered peanuts (Dangaran, Renner-Nantz, et al. 2006) and chocolate samples (Lee, Dangaran, et al. 2002). In chocolate, gloss and color vary due to fat blooming during storage at different temperatures (Pastor, Santamaria, et al. 2007). Chocolate has been also used to analyze the relationship between gloss and color and topographical and textural parameters obtained from scanning laser microscope images of samples with controlled induced surface roughness (Briones, Aguilera, et al. 2006).

4.4 LIGHT SCATTERING AND ABSORPTION

Many food products have a colloidal nature. In colloidal media such as gels, emulsions, foams, and turbid media, which may be practically opaque, light is basically scattered or absorbed. The appearance of such media depends on the way that it scatters and absorbs light. Scattering is related to attributes like turbidity, cloudiness, opacity, or lightness of an emulsion, whereas absorption determines its color

(McClements, Chantrapornchai, et al. 1998). Scattering depends on the concentration, size, and refractive index of the dispersed particles, whereas absorption depends on the concentration and type of existing chromophores (Hutchings 1999). When a beam of light passes through a colloidal dispersion, the particles or droplets scatter some of the light in all directions. When the particles are very small compared with the wavelength of the light, the intensity of the scattered light is uniform in all directions, but for larger particles, with diameters larger than approximately 250 nm, light intensity depends on the scattering angle.

The great majority of viscous liquid foods are emulsions. Beverage emulsions are primarily used to give opacity to clear beverages or to enhance their juicelike appearance (Taherian, Fustier, et al. 2006). Milk, yoghurts, and sauces are typical examples of low-volume fraction emulsions, and butter and margarine are high volume-fraction emulsions. Foams are the air-liquid analogue of emulsions and constitute another large category of foods. Finally, foods can be found in highly viscoelastic forms or even in glassy states (Mezzenga, Schurtenberger, et al. 2005). There are several optical characteristics and spectrophotometric techniques that can be used to characterize emulsions (Chanamai and McClements 2001; McClements 2002; Wackerbarth, Stoll, et al. 2009). These properties are important because of their relationship with the product appearance and also with the rheological properties (Gonzalez-Tomas and Costell 2006; Tarrega and Costell 2007; Sanz, Salvador, et al. 2008).

4.4.1 Dynamic Light Scattering and Diffuse Wave Spectroscopy

In the applications described previously, light illuminating the sample comes from noncoherent light sources. We will now briefly discuss two techniques that use monochromatic coherent light, from a laser for example, to produce interference effects in the scattered light.

Dynamic light scattering (DLS) is a technique based on the scattering of light by moving particles. Using coherent and monochromatic light it is possible to observe time-dependent fluctuations in the scattered intensity pattern, due to constructive and destructive interference of light scattered by neighboring particles. DLS requires a low concentration of scatter centers so that each incident photon of light is scattered only once within the sample (i.e., multiple scattering must be absent). This technique cannot be applied to concentrated suspensions like milk, but provides interesting information in dilute suspensions. Analysis of the time dependence of the intensity fluctuation yields the diffusion coefficient of the particles from which, knowing the viscosity of the medium, the hydrodynamic radius or diameter of the particles can be calculated (Dalgleish and Hallett 1995). This technique has been used to obtain the particle size distribution in a study of the stability and opacity of cloudy emulsions having Arabic gum and whey protein isolate (Klein, Aserin, et al. 2010). Particle size distributions were then related with absorbance measurements and with visual assessment.

Diffuse wave spectroscopy (DWS) is the extension of DLS to opaque samples, where light suffers multiple scattering. DWS uses the interference pattern generated by backscattered light from the sample. Due to particle motions, interference takes the form of a dynamic speckled pattern that can be recorded and analyzed. The rheological properties are deduced from dynamic changes in the speckled pattern using a well-established formalism (Alexander and Dalgleish 2006). DWS is especially well suited for turbid media and has been recently used to study the interactions of dairy proteins and the kinetics of aggregation in gelling systems under different destabilizing conditions (Corredig and Alexander 2008).

4.4.2 Translucency

Many food products are translucent, neither totally opaque nor completely transparent. In this kind of product light is reflected, absorbed, transmitted, and scattered, and the objective measurements of optical properties present serious difficulties (Calvo 2004).

The Kubelka-Munk (KM) theory is frequently used in these cases in relation to optical properties like spectral reflectance or color. The KM theory is a two-flux theory that describes the propagation of diffuse light through a material. Within this theory it is possible to relate the values of the absorption coefficient K and the scattering coefficient S with the reflectance R of the sample. In case of a sufficiently thick sample, so that it can be considered opaque, the reflectance is denoted as R_∞ and given by Equation (4.4):

$$\frac{K}{S} = \frac{(1 - R_\infty)^2}{2R_\infty} \tag{4.4}$$

Other, less simple expressions exist when the sample is not optically thick. The K/S ratio has the advantage of being approximately a linear function of the pigment concentration. This property makes it useful for many food products like juices (Melendez-Martinez, Vicario, et al. 2005), fruits (Contreras, Martin-Esparza, et al. 2008), meat (Aporta, Hernández, et al. 1996; Saenz, Hernandez, et al. 2008), and emulsions (McClements, Chantrapornchai, et al. 1998; Chantrapornchai, Clydesdale, et al. 1999). A detailed analysis of the KM theory and its applications to foods is beyond the scope of this chapter, but the interested reader can consult general works (Wyszecki and Stiles 2000) or works devoted to food science (MacDougall 2002; Calvo 2004).

Reflectance and color of translucent samples will be affected by the background where they are placed during measurements. A proposed measure of translucency based in the CIE tristimulus Y obtained from two measurements performed over a white ($CIE\ Y_{\text{white}}$) and a black ($CIE\ Y_{\text{black}}$) background and given by Equation (4.5)

$$translucency = 100 - (CIE\,Y_{Black} \,/\, CIE\,Y_{White}) \times 100 \tag{4.5}$$

has been used to evaluate translucency in white salted noodles (Solah, Crosbie, et al. 2007). In this work the relationship between several optical parameters and appearance attributes are studied: translucency, gloss measured with a gloss meter, color measured with a spectroradiometer, and the results of a sensory evaluation by panelists.

We cannot end this section without noting that translucency is considered an undesired property in some food products, negatively affecting their quality, mainly in the fresh-cut industry (Artes, Gomez, et al. 2007; Montero-Calderon, Rojas-Grau, et al. 2008).

4.5 FLUORESCENCE

In a fluorescent material, radiation absorbed at some wavelength is reemitted at a different wavelength after a very short period of time. In most of the applications in the food industry, absorbed ultraviolet radiation is emitted as visible light. Although it has innumerable applications, we will focus here on a reduced number of examples that combine fluorescence with other optical properties like color, absorbance, or scattering. Examples related to spectral imaging will be treated later. Notice that the general appearance of a product is affected by the presence of fluorescence and that natural light, as well as most artificial light sources, have a nonnegligible emission in the UV region.

Fluorescence is in general measured with an instrument called a spectrofluorometer. A complete characterization of the sample will require the recording of the emission spectrum corresponding to each excitation wavelength. However, many applications are based in measuring a single pair of excitation and emission wavelengths. For instance, excitation at 347 nm and emission at 415 has been used to follow the formation of Maillard reaction products (Morales and Jimenez-Perez 2001). Imaging spectrographs have been used to record the fluorescence emission spectrum and analyze it in relation to apple fruit quality (Noh and Lu 2007) and the influence of storage in the photo oxidation of cheese (Wold, Veberg, et al. 2006).

Finally, the effect of fluorescence in the appearance, in color in particular, has been studied for different products like concentrated milk systems (Rozycki, Pauletti, et al. 2007), meat emulsions (Allais, Viaud, et al. 2004), biscuits (Allais, Edoura-Gaena, et al. 2006), and infant formulas (Ferrer, Alegria, et al. 2005; Bosch, Alegria, et al. 2007).

4.6 REFRACTIVE INDEX

When light falls from one medium to another, it changes its speed and trajectory. These changes are determined by the value of the refractive index, which is a wavelength-dependent magnitude that can be derived from the dielectric permittivity, which in turn depends on other parameters like chemical composition, density, and so on. The refractive index can be determined rapidly and easily with a refractometer and a typical application in food science is the determination of alcohol or sucrose content in fruit juices, soft drinks, or wines.

A more modern technique to measure the refractive index is surface plasmon resonance (SPR). A surface plasmon is a light wave trapped in the surface of a metal. The propagation of such waves is very sensitive to tiny changes in the refractive index of the substance in contact with the metal surface. In this way, the presence of particles or chemical compounds can be determined by their influence in the refractive index of the substance. There are commercial biosensors that

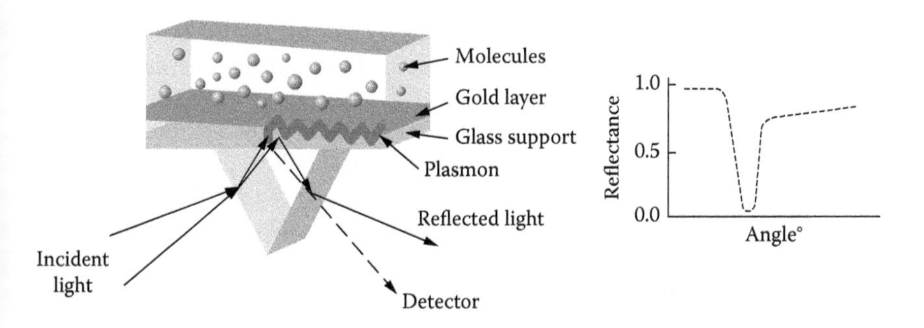

FIGURE 4.5 Typical configuration of a surface plasmon resonance (SPR) system measuring the refractive index.

implement this principle as it is schematically shown in Figure 4.5. A thin layer of metal, typically gold, having a thickness of only few nanometers, is deposited on a glass prism. The gold film is in contact with the substance to be analyzed. On the other side, a light beam enters the prism and suffers total reflection in the prism side in contact with gold and exits the prism, reaching the detector. Under appropriate circumstances, which depend on the refractive index of the substance, and for particular angles of incidence and wavelengths, a plasmon wave is generated in the metal layer. This destroys the total reflection process, drastically reducing the intensity of the reflected light that reaches the detector. In a calibrated device, the analysis of the reflected light gives us information about the refractive index of the substance in contact with the metal layer. It is even possible to make real-time measurements and follow time-dependent changes in the refractive index due to chemical reactions in the analyzed substance. Examples of the uses of this technique in food science are the detection of *E. Coli* O157:H7 in different food products (Waswa, Irudayaraj, et al. 2007) or the detection of insecticide residues (Yang and Cho 2008). A review of the application of SPR in food can be found in Pattnaik and Srivastav (2006).

4.7 COMPUTER VISION

Traditional instruments for measuring optical properties (spectrophotometer, spectroradiometer, colorimeter, gloss meter, etc.) measure over a relatively small area of the product. Since food products are rarely homogeneous, measurements must be repeated over different locations in the same product to obtain an average, representative value of the measured parameter. Similarly, liquids are frequently centrifuged and filtered to obtain clean, homogeneous samples. There are, however, products—a pizza, for instance—that are intrinsically inhomogeneous, presenting defects or discontinuities in their visual and optical properties (Yam and Papadakis 2004; Du and Sun 2005). In such cases, averaging measurements over different areas of the sample is obviously not a valid option.

In those products, shape, size, color, or texture taken independently are not adequate for an overall quality evaluation. The attractiveness of the product, the consumer preference, depends on its total appearance, a global characteristic difficult to obtain using traditional measurement techniques. This kind of evaluation is made by visual assessment by trained inspectors, which is a tedious, laborious, and subjective task.

In recent years, computer vision systems and image processing have been growing in importance as a tool to perform all these evaluations in an objective way, maintaining accuracy and consistency while eliminating the subjectivity of human inspectors (Du and Sun 2004). A typical application implements an image acquisition system, an image processing stage, and an image analysis stage. The color information in the image can be used to identify different constituents or elements and quantify their sizes, shapes, and distribution throughout the image.

In digital color cameras, color is given as red, green, and blue (RGB) values, usually in the range 0–255, that encode the light intensity recorded through three broadband filters. This procedure is based on the trichromacy of the human visual system and permits reasonable color reproduction in computer screens or printer hard copies. Imaging techniques need not be necessarily restricted to RGB filters. Images can be taken at multiple wavelengths combining the spatial resolution of conventional imaging with the spectral information conveyed in the reflected or transmitted light from an object. Spectral imaging will be analyzed later in a specific section.

The RGB encoding is a device-dependent process, but it is possible to obtain colorimetric information from an image after an adequate color characterization and calibration of the imaging system. In order to convert RGB values to a device-independent color space like CIELAB, several algorithms have been proposed for images obtained with digital cameras (Du and Sun 2005; Leon, Mery, et al. 2006; Mendoza, Dejmek, et al. 2006; Kang and Sabarez 2009; Valous, Mendoza, et al. 2009) and images obtained with color scanners (Hatcher, Symons, et al. 2004; Kihc, Onal-Ulusoy, et al. 2007).

During image acquisition, illumination deserves special attention. Uniformity, spectral composition, color rendering, and color temperature of the illuminating system must be considered. Fluorescent lamps are commonly used, but other solutions may be more adequate for a particular application, like the use of optical fiber to illuminate small regions of interest in the study of bubble size distribution in beer (Hepworth, Hammond, et al. 2004) or in the study of dried fish processes (Louka, Juhel, et al. 2004). Generally, samples are imaged over achromatic backgrounds: white, black, or neutral grays. Other alternatives, like cyan backgrounds, have been tested to increase the contrast in the image (Abdullah, Mohamad-Saleh, et al. 2006).

Calibrated color imaging systems can be arranged combining their different elements according to particular needs, but there are also complete systems that integrate all the elements necessary for calibrated color imaging (Luo, Cui, et al. 2001; MacDougall 2002).

A simple application of color images is the identification and segmentation of different objects or regions in the image that would otherwise be very similar in a gray image, as in the study of bicolor foods (Kang and Sabarez 2009) or in presliced hams (Valous, Mendoza, et al. 2009), but there are many others (Brosnan and Sun 2004). We will focus here on recent applications related to the measurement and evaluation of optical and visual properties like translucency, color, or appearance. Interesting examples are the study of maturity in star fruit (Abdullah, Mohamad-Saleh, et al.

2006), changes in color due to the migration of fat to the surface of chocolate during storage (Briones, Aguilera, et al. 2006), translucency of fresh tomato slices during refrigerated storage (Lana, Tijskens, et al. 2006), characterization of the appearance of sliced ham (Valous, Mendoza, et al. 2009), comparison of its color image characteristics with consumer evaluation (Iqbal, Valous, et al. 2010), enzymatic browning kinetics in pear slices (Quevedo, Diaz, et al. 2009), mushrooms (Vizhanyo and Felfoldi 2000), beer haze (Hepworth, Hammond, et al. 2004), color and speckness of oriental noodles (Hatcher, Symons, et al. 2004), study of relations between color and firmness in tomatoes (Schouten, Huijben, et al. 2007), and automatic grading systems for fruits and vegetables (Chong, Nishi, et al. 2008; Kondo 2010).

4.8 SPECTRAL IMAGING: MULTISPECTRAL AND HYPERSPECTRAL

By spectral imaging we understand all those imaging techniques that provide spectral information at each pixel in the image (i.e., at each point of the imaged object). Spectral information is very important since spectral characteristics are linked to the physical and chemical properties of foods. Spectral imaging provides information that is not available in conventional color imaging and can also be extended to other spectral ranges like ultraviolet (UV) or near-infrared (NIR).

The output of a spectral imaging system is a collection of images. Each image has been taken through a filter that can be broad band or narrow band, centered at some wavelength. The term *hyperspectral imaging* is normally used when data contains hundreds or even thousands of images, each at one wavelength. When fewer images are produced, we refer to the system as a *multispectral* imaging system. Notice that conventional RGB color imaging (treated in the previous section) is a particular case of multispectral imaging where images are taken through three broad-band filters.

Hyperspectral systems have the advantage of providing exhaustive spectral information on each image pixel. The analysis of this information requires complex and time-consuming algorithms, making this technique useful in off-line applications. If only a reduced number of wavelengths or wavebands (2–4) are necessary or sufficient, then simpler and faster multispectral systems can be devised that would be suitable for on-line applications. The identification of this reduced set of wavelengths can be done with traditional spectroscopic techniques, but as we will see, it is also one of the most frequent objectives of the analysis of hyperspectral imaging data.

A typical hardware configuration of a hyperspectral system consists of three basic elements: the illuminating system, the image acquisition system, and the sample holder. In general, the characteristics of the illuminating system are similar to those already described for conventional imaging. There are applications, like fluorescence imaging or transmittance imaging, which require special light sources or configurations. The imaging system is composed of a lens, a spectrograph, and a charge-coupled device (CCD) (or other type of imaging sensor). Light is collected through a narrow slit and focused by the lens into the spectrograph, which disperses the light into its spectral components along in specific direction by means of a combination of diffraction gratings and prisms. In this way, light coming from a narrow strip of the object is converted into a two-dimensional map, one direction corresponding to the spatial direction determined by the slit and the other

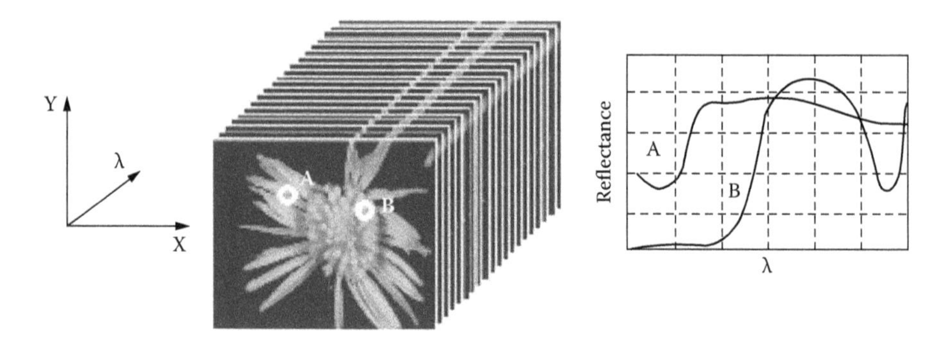

FIGURE 4.6 The hypercube. Spatial and spectral information from a hyperspectral system. For every pixel on the image (A and B in the picture) the complete spectral information can be recovered (right).

direction corresponding to the wavelength axis. In a single shot the CCD camera records all the spectral information of a narrow straight region of the object. In order to obtain a complete spatial scan, the process must be repeated, continuously scanning consecutive strips of the object. This requires a controlled movement of the object, usually accomplished by means of computer-controlled motorized sample holders. Taking successive images, the complete data set is formed. Data is usually referred to as the *hypercube*, since it has two spatial dimensions and a third, spectral dimension (Figure 4.6).

Spectral imaging has been applied to a great variety of foods. Table 4.2 summarizes some recent works organized by food product. Studies in the meat industry include poultry, beef, and pork. In the poultry industry, spectral imaging has been oriented to quality and safety inspection of poultry products in on-line applications. Fecal contamination and infectious conditions have deserved much attention due to zero-tolerance safety policies. Detection of fecal contaminants can be done using only the ratio of the reflectance at two wavelengths (Park, Lawrence, et al. 2005), but contaminants can be further classified depending on their origin (duodenum, caecum, colon, and ingesta) using hyperspectral data at 512 wavelengths (Park, Windham, et al. 2007).

With respect to infectious conditions, reasonable detection efficiencies for skin tumors have been obtained with spectral data at three wavelengths (Chao, Mehl, et al. 2002) and also with hyperspectral fluorescence imaging at ten wavelengths (Kong, Chen, et al. 2004). In this application, the importance of optimizing the wavelength selection algorithms has also been studied (Nakariyakul and Casasent 2009). Septicemia and toxemia, which affect the entire bird carcass, can be detected with only a few wavelengths, making possible an on-line multispectral system capable to identify unwholesome carcasses in the processing line working at 140 birds per minute with a detection efficiency close to 96% (Chao, Yang, et al. 2010).

Structural and textural information is of great importance in beef meat products. Major structural components like fat, connective tissue, and myofibers can be identified using the autofluorescence properties of naturally occurring fluorophores in meat. Using narrow interference filters for both the illuminating source and the CCD

TABLE 4.2
Examples of the Application of Spectral Imaging to Different Food Products

Commodity	Objective	Reference
	Meat Products	
Poultry	Fecal contamination	(Park, Lawrence et al. 2005; Park, Windham et al. 2007)
	Skin tumors	(Chao, Mehl et al. 2002; Kong, Chen et al. 2004; Nakariyakul and Casasent 2009)
	Wholesomeness high-speed inspection	(Chao, Yang et al. 2010)
Beef	Tenderness prediction	(Naganathan, Grimes et al. 2008; Wu, Peng et al. 2009)
	Structural components	(Skjervold, Taylor et al. 2003)
	Myoglobin concentration profiles	(Saenz, Hernandez et al. 2008)
Pork	Marbling classification	(Qiao, Ngadi et al. 2007)
	Drip-loss, pH, color	(Qiao, Wang et al. 2007)
	Fruits and Vegetables	
Apples	Fecal contamination	(Kim, Lefcourt et al. 2005; Lefcout, Kim et al. 2006; Liu, Chen et al. 2007)
	Bruise detection	(Xing, Bravo et al. 2005; Xing, Saeys et al. 2007; ElMasry, Wang et al. 2008)
	Chilling injury	(ElMasry, Wang et al. 2009)
	Optical properties	(Qin and Lu 2008; Qin and Lu 2009)
	Quality	(Noh and Lu 2007)
Citrus	Rottenness detection	(Gómez-Sanchis, Gómez-Chova et al. 2008)
	Skin defects	(Blasco, Aleixos et al. 2007)
	Canker detection	(Qin, Burks et al. 2009)
	Green citrus detection in trees	(Okamoto and Lee 2009)
	Corrections spherical objects	(Gomez-Sanchis, Molto et al. 2008)
Cantaloupe	Fecal contamination	(Vargas, Kim et al. 2005)
Peach	Firmness and maturity	(Lleo, Barreiro et al. 2009)
Strawberry	Quality attributes	(ElMasry, Wang et al. 2007)
Pickles	Internal defects	(Ariana and Lu 2010)
Mushrooms	Freeze damage	(Gowen, Taghizadeh et al. 2009)
Wheat	Grain cleanness	(Wallays, Missotten et al. 2009)

camera, a reduced number of illumination–detection wavelength pairs are necessary to identify them (Skjervold, Taylor, et al. 2003). Texture analysis using the images obtained from the most significant principal components of the hyperspectral data has been used to classify the tenderness of beef longissimus dorsi muscle into three categories based in shear force values with 96.4% accuracy (Naganathan, Grimes, et al. 2008). Direct multilinear regression with six wavelengths has also been used to predict the classification of tenderness into two categories, although with less accuracy (Wu, Peng, et al. 2009).

Hyperspectral imaging can be also used as an analytical tool. In unoxigenated meat myoglobin is found in the reduced form (Mb). In contact with air, oxygen diffuses through the surface of meat, converting Mb into oxymioglobin (MbO_2). With time MbO_2 will be further oxidized to form metmyoglobin (MMb). As oxygen penetrates inside meat it is consumed in these processes and, as a result, the concentration of each pigment becomes a function of time and the distance (depth) to the oxygenated surface. Relative concentrations of each myoglobin form can be obtained from reflectance data at some specific wavelengths (474, 525, 572, and 610 nm) as a direct application of the Kubelka-Munk theory (Sáenz and Hernández 1999). Using a multispectral system to capture a time sequence of spectral images transverse to the direction of penetration of oxygen, the concentration profiles of each pigment as a function of the oxygenation time have been obtained with a spatial resolution of 0.1 mm. Authors used Fick's second law to describe oxygen diffusion through the meat sample and a first-order kinetics for the Mb consumption. The model fitted the experimental evolution of the concentration profile of myoglobin inside meat with a root mean squared error (RMSE) of only 0.253% (Saenz, Hernandez, et al. 2008).

Spectral imaging in fruits has deserved considerable attention in recent years. In apples, efforts have been mainly directed toward the implementation of on-line applications for early detection and removal of damaged or contaminated fruits. Fecal contamination can be identified using the ratio of the reflectance at two different wavelengths using conventional halogen illumination during image acquisition (Liu, Chen, et al. 2007) or ultraviolet illumination and fluorescent emission (Kim, Lefcourt, et al. 2005). Bruise detection in mechanically damaged apples can also be performed with only 3 or 4 wavelengths, although different apple varieties may require a different set of wavelengths (Xing, Saeys, et al. 2007; ElMasry, Wang, et al. 2008).

The combination of hyperspectral imaging with laser-induced fluorescence has also been used to predict several apple quality parameters (Noh and Lu 2007). In this work, spatial images were not actually obtained, but spatial profiles of the intensity of fluorescence emission along a single scanning line situated a controlled distance from the laser spot.

Skin defects have also been studied in citrus fruits. The possibility of identifying skin damage in mandarins and oranges by their origin (green mould, chilling injury, phytotoxicity, etc.) has been studied with broad-band multispectral imaging combining a single UV image (200–400 nm), a single NIR image (700–1800 nm), and an induced fluorescence image FI at 560 nm. Reported detection accuracies depend on the nature of the skin defect and vary from 50% to 100% (Blasco, Aleixos, et al. 2007). Hyperspectral systems have also been applied to detect mandarin rottenness (Gómez-Sanchis, Gómez-Chova, et al. 2008) and citrus canker in grapefruits (Qin, Burks, et al. 2009). A multispectral system working with natural light has also been developed to detect green citrus in trees (Okamoto and Lee 2009).

The problems associated with the measurement of nonplanar objects, which we mentioned in the case of gloss, have also been studied in the context of spectral imaging. Using citrus fruits as a working example, a reflectance model for a spherical object has been developed with application to spectral and nonspectral imaging (Gomez-Sanchis, Molto, et al. 2008).

Among recent applications to vegetables, it is worth mentioning an on-line classification system that combines reflectance images in the visible and a transmittance image in the NIR spectrum that outperforms human inspectors in the detection of internal defects in whole pickles (Ariana and Lu 2010). Freeze damage in mushrooms (Gowen, Taghizadeh, et al. 2009), shelf life in mushrooms packed with different films (Taghizadeh, Gowen, et al. 2010), or the determination of grain cleanness in wheat (Wallays, Missotten, et al. 2009) are other recent examples of the possibilities of this technique.

4.9 CONCLUSIONS

Coexisting with traditional measurement techniques, present-day technologies offer new and promising procedures for determining the optical properties and appearance attributes of foods. Availability of advanced hardware and software solutions is undoubtedly spurring on new applications. Imaging technologies in particular hold a prominent position in number and diversity of implementations. Fast imaging systems based on conventional color imaging or multispectral imaging can be used for on-line applications. Hyperspectral systems, combining precise spatial and spectral information, are a powerful tool for many purposes. Presently limited to off-line applications due to the time-consuming analysis of the data, they will probably be used in real-time applications with the future development in computing systems and analysis algorithms.

The use of imaging systems is also very promising in the assessment of total appearance attributes of foods, combining color and color distribution, gloss, and texture in a unique global description. A better understanding of perceptual phenomena will be surely required, and in this quest, food science, optics, and psychophysics will have to keep walking side by side.

REFERENCES

Abdullah, M. Z., J. Mohamad-Saleh, et al. (2006). Discrimination and classification of fresh-cut starfruits (*Averrhoa carambola L.*) using automated machine vision system. *Journal of Food Engineering* 76(4): 506–523.

Alexander, M., and D. G. Dalgleish (2006). Dynamic light scattering techniques and their applications in food science. *Food Biophysics* 1(1): 2–13.

Allais, I., R. B. Edoura-Gaena, et al. (2006). Characterisation of lady finger batters and biscuits by fluorescence spectroscopy: Relation with density, color and texture. *Journal of Food Engineering* 77(4): 896–909.

Allais, I., C. Viaud, et al. (2004). A rapid method based on front-face fluorescence spectroscopy for the monitoring of the texture of meat emulsions and frankfurters. *Meat Science* 67(2): 219–229.

Aporta, J., B. Hernández, et al. (1996). Veal colour assessment with three wavelengths. *Meat Science* 44(1–2): 113–123.

Ariana, D. P., and R. F. Lu (2010). Evaluation of internal defect and surface color of whole pickles using hyperspectral imaging. *Journal of Food Engineering* 96(4): 583–590.

Artes, F., P. A. Gomez, et al. (2007). Physical, physiological and microbial deterioration of minimally fresh processed fruits and vegetables. *Food Science and Technology International* 13(3): 177–188.

ASTM (American Society for Testing and Materials). (1999). Standard test method for specular gloss, ASTM D523-89. In *ASTM Standards on color and appearance measurement*, 7th ed. West Conshohocken, PA: ASTM.

Bekhit, A., J. D. Morton, et al. (2009). Optical properties of raw and processed fish roes from six commercial New Zealand species. *Journal of Food Engineering* 91(2): 363–371.

Blasco, J., N. Aleixos, et al. (2007). Citrus sorting by identification of the most common defects using multispectral computer vision. *Journal of Food Engineering* 83(3): 384–393.

Bosch, L., A. Alegria, et al. (2007). Fluorescence and color as markers for the Maillard reaction in milk-cereal based infant foods during storage. *Food Chemistry* 105(3): 1135–1143.

Briones, V., J. M. Aguilera, et al. (2006). Effect of surface topography on color and gloss of chocolate samples. *Journal of Food Engineering* 77(4): 776–783.

Brosnan, T., and D. W. Sun (2004). Improving quality inspection of food products by computer vision - a review. *Journal of Food Engineering* 61(1): 3–16.

Calvo, C. (2004). Optical properties. In *Handbook of food analysis*, ed. L. M. L. Nollet. New York, Marcel Dekker.

Calvo, C., A. Salvador, et al. (2001). Influence of colour intensity on the perception of colour and sweetness in various fruit-flavoured yoghurts. *European Food Research and Technology* 213(2): 99–103.

Chanamai, R., and D. J. McClements (2001). Prediction of emulsion color from droplet characteristics: Dilute monodisperse oil-in-water emulsions. *Food Hydrocolloids* 15(1): 83–91.

Chantrapornchai, W., F. Clydesdale, et al. (1999). Theoretical and experimental study of spectral reflectance and color of concentrated oil-in-water emulsions. *Journal of Colloid and Interface Science* 218(1): 324–330.

Chao, K., P. M. Mehl, et al. (2002). Use of hyper- and multi-spectral imaging for detection of chicken skin tumors. *Applied Engineering in Agriculture* 18(1): 113–119.

Chao, K., C. C. Yang, et al. (2010). Spectral line-scan imaging system for high-speed non-destructive wholesomeness inspection of broilers. *Trends in Food Science and Technology* 21(3): 129–137.

Chong, V. K., T. Nishi, et al. (2008). Surface gloss measurement on eggplant fruit. *Applied Engineering in Agriculture* 24(6): 877–883.

CIE (Commission Internationale de L' Eclairage). (1987). *International lighting vocabulary*. Geneve: Bureau Central de la Commission Electrotechnique Internationale.

CIE (Commission Internationale de L' Eclairage). (1995). *CIE collection in colour and vision*. Vienna: Commission Internationale de L' Eclairage.

CIE (Commission Internationale de L' Eclairage). (2004). *Colorimetry*. Vienna: International Commission on Illumination.

Contreras, C., M. E. Martin-Esparza, et al. (2008). Influence of microwave application on convective drying: Effects on drying kinetics, and optical and mechanical properties of apple and strawberry. *Journal of Food Engineering* 88(1): 55–64.

Corredig, M., and M. Alexander (2008). Food emulsions studied by DWS: Recent advances. *Trends in Food Science & Technology* 19(2): 67–75.

Dalgleish, D. G., and F. R. Hallett (1995). Dynamic light-scattering: Applications to food systems. *Food Research International* 28(3): 181–193.

Dangaran, K. L., J. Renner-Nantz, et al. (2006). Whey protein-sucrose coating gloss and integrity stabilization by crystallization inhibitors. *Journal of Food Science* 71(3): E152–E157.

Dematte, M. L., D. Sanabria, et al. (2006). Cross-modal associations between odors and colors. *Chemical Senses* 31(6): 531–538.

Du, C. J., and D. W. Sun (2004). Recent developments in the applications of image processing techniques for food quality evaluation. *Trends in Food Science & Technology* 15(5): 230–249.

Du, C. J., and D. W. Sun (2005). Comparison of three methods for classification of pizza topping using different colour space transformations. *Journal of Food Engineering* 68(3): 277–287.

ElMasry, G., N. Wang, et al. (2007). Hyperspectral imaging for nondestructive determination of some quality attributes for strawberry. *Journal of Food Engineering* 81(1): 98–107.

ElMasry, G., N. Wang, et al. (2008). Early detection of apple bruises on different background colors using hyperspectral imaging. *LWT-Food Science and Technology* 41(2): 337–345.

ElMasry, G., N. Wang, et al. (2009). Detecting chilling injury in Red Delicious apple using hyperspectral imaging and neural networks. *Postharvest Biology and Technology* 52(1): 1–8.

Escolar, D., M. R. Haro, et al. (2007). The color space of foods: Virgin olive oil. *Journal of Agricultural and Food Chemistry* 55(6): 2085–2093.

Ferrer, E., A. Alegria, et al. (2005). Fluorescence, browning index, and color in infant formulas during storage. *Journal of Agricultural and Food Chemistry* 53(12): 4911–4917.

Gomez-Robledo, L., M. Melgosa, et al. (2008). Virgin olive oil color in relation to sample thickness and the measurement method. *Journal of the American Oil Chemists Society* 85(11): 1063–1071.

Gómez-Sanchis, J., L. Gómez-Chova, et al. (2008). Hyperspectral system for early detection of rottenness caused by *Penicillium digitatum* in mandarins. *Journal of Food Engineering* 89(1): 80–86.

Gomez-Sanchis, J., E. Molto, et al. (2008). Automatic correction of the effects of the light source on spherical objects. An application to the analysis of hyperspectral images of citrus fruits. *Journal of Food Engineering* 85(2): 191–200.

Goni, V., G. Indurain, et al. (2008). Measuring muscle color in beef using an instrumental method versus visual color scales. *Journal of Muscle Foods* 19(2): 209–221.

Gonzalez-Miret, M. L., W. Ji, et al. (2007). Measuring colour appearance of red wines. *Food Quality and Preference* 18(6): 862–871.

Gonzalez-Tomas, L. and E. Costell (2006). Relation between consumers' perceptions of color and texture of dairy desserts and instrumental measurements using a generalized Procrustes analysis. *Journal of Dairy Science* 89(12): 4511–4519.

Gossinger, M., F. Mayer, et al. (2009). Consumer's color acceptance of strawberry nectars from puree. *Journal of Sensory Studies* 24(1): 78–92.

Gowen, A. A., M. Taghizadeh, et al. (2009). Identification of mushrooms subjected to freeze damage using hyperspectral imaging. *Journal of Food Engineering* 93(1): 7–12.

Hatcher, D. W., S. J. Symons, et al. (2004). Developments in the use of image analysis for the assessment of oriental noodle appearance and colour. *Journal of Food Engineering* 61(1): 109–117.

Hepworth, N. J., J. R. M. Hammond, et al. (2004). Novel application of computer vision to determine bubble size distributions in beer. *Journal of Food Engineering* 61(1): 119–124.

Hernández, B., C. Alberdi, et al. (2005). Elaboración y uso de patrones para la evaluación visual de alimentos (Elaboration and use of standards for visual evaluation of foodstuffs) Patent. Universidad Pública de Navarra. Spain. P200201089.

Hernández, B., C. Saenz, et al. (2004). Design and performance of a color chart based in digitally processed images for sensory valuation of piquillo peppers (*Capsicum annuum*). *Color Research and Application* 29(4): 305–311.

Hernández, B., C. Sáenz, et al. (2009). Assessing the color of red wine like a taster's eye. *Color Research and Application* 34(2): 153–162.

Hunter, R. S., and R. W. Harold (1987). *The measurement of appearance*. New York: Wiley.

Hutchings, J. B. (1999). *Food color and appearance*. Gaithersburg, MD: Aspen Publishers.

Hutchings, J. B. (2003). *Expectations and the food industry: The impact of color and appearance*. New York: Kluwer Academic/Plenum Publishers.
Iqbal, A., N. A. Valous, et al. (2010). Classification of pre-sliced pork and turkey ham qualities based on image colour and textural features and their relationships with consumer responses. *Meat Science* 84(3): 455–465.
Jha, S. N., T. Matsuoka, et al. (2002). Surface gloss and weight of eggplant during storage. *Biosystems Engineering* 81(4): 407–412.
Jost-Boissard, S., M. Fontoynont, et al. (2009). Perceived lighting quality of LED sources for the presentation of fruit and vegetables. *Journal of Modern Optics* 56(13): 1420–1432.
Kang, S. P., and H. T. Sabarez (2009). Simple colour image segmentation of bicolour food products for quality measurement. *Journal of Food Engineering* 94(1): 21–25.
Kihc, K., B. Onal-Ulusoy, et al. (2007). Scanner-based color measurement in L*a*b* format with artificial neural networks (ANN). *European Food Research and Technology* 226(1–2): 121–126.
Kim, M. S., A. M. Lefcourt, et al. (2005). Automated detection of fecal contamination of apples based on multispectral fluorescence image fusion. *Journal of Food Engineering* 71(1): 85–91.
Klein, M., A. Aserin, et al. (2010). Enhanced stabilization of cloudy emulsions with gum Arabic and whey protein isolate. *Colloids and Surfaces B-Biointerfaces* 77(1): 75–81.
Kondo, N. (2010). Automation on fruit and vegetable grading system and food traceability. *Trends in Food Science & Technology* 21(3): 145–152.
Kong, S. G., Y. R. Chen, et al. (2004). Analysis of hyperspectral fluorescence images for poultry skin tumor inspection. *Applied Optics* 43(4): 824–833.
Lana, M. M., L. M. M. Tijskens, et al. (2006). Modelling RGB colour aspects and translucency of fresh-cut tomatoes. *Postharvest Biology and Technology* 40(1): 15–25.
Lee, S. Y., K. L. Dangaran, et al. (2002). Gloss stability of whey-protein/plasticizer coating formulations on chocolate surface. *Journal of Food Science* 67(3): 1121–1125.
Lefcout, A. M., M. S. Kim, et al. (2006). Systematic approach for using hyperspectral imaging data to develop multispectral imagining systems: Detection of feces on apples. *Computers and Electronics in Agriculture* 54(1): 22–35.
Leon, K., D. Mery, et al. (2006). Color measurement in L*a*b* units from RGB digital images. *Food Research International* 39(10): 1084–1091.
Levitan, C. A., M. Zampini, et al. (2008). Assessing the role of color cues and people's beliefs about color-flavor associations on the discrimination of the flavor of sugar-coated chocolates. *Chemical Senses* 33(5): 415–423.
Liu, Y. L., Y. R. Chen, et al. (2007). Development of simple algorithms for the detection of fecal contaminants on apples from visible/near infrared hyperspectral reflectance imaging. *Journal of Food Engineering* 81(2): 412–418.
Lleo, L., P. Barreiro, et al. (2009). Multispectral images of peach related to firmness and maturity at harvest. *Journal of Food Engineering* 93(2): 229–235.
Louka, N., F. Juhel, et al. (2004). A novel colorimetry analysis used to compare different drying fish processes. *Food Control* 15(5): 327–334.
Luo, M. R., G. H. Cui, et al. (2001). Apparatus and method for measuring colour (DigiEye System), DigiEye Plc, 4 British Patent (Application No. 0124683.4).
MacDougall, D. B. (2002). *Colour in food: Improving quality*. Boca Raton, FL: CRC Press; Cambridge: Woodhead Publishing.
McClements, D. J. (2002). Theoretical prediction of emulsion color. *Advances in Colloid and Interface Science* 97(1–3): 63–89.
McClements, D. J., W. Chantrapornchai, et al. (1998). Prediction of food emulsion color using light scattering theory. *Journal of Food Science* 63(6): 935–939.
Melendez-Martinez, A. J., I. M. Vicario, et al. (2005). Instrumental measurement of orange juice colour: A review. *Journal of the Science of Food and Agriculture* 85(6): 894–901.

Melgosa, M., L. Gomez-Robledo, et al. (2009). Color measurements in blue-tinted cups for virgin olive oil tasting. *Journal of the American Oil Chemists Society* 86(7): 627–636.

Mendoza, F., P. Dejmek, et al. (2006). Calibrated color measurements of agricultural foods using image analysis. *Postharvest Biology and Technology* 41(3): 285–295.

Mendoza, F., P. Dejmek, et al. (2010). Gloss measurements of raw agricultural products using image analysis. *Food Research International* 43(1): 18–25.

Mezzenga, R., P. Schurtenberger, et al. (2005). Understanding foods as soft materials. *Nature Materials* 4(10): 729–740.

Mizrach, A., R. F. Lu, et al. (2009). Gloss evaluation of curved-surface fruits and vegetables. *Food and Bioprocess Technology* 2(3): 300–307.

Montero-Calderon, M., M. A. Rojas-Grau, et al. (2008). Effect of packaging conditions on quality and shelf-life of fresh-cut pineapple (*Ananas comosus*). *Postharvest Biology and Technology* 50(2–3): 182–189.

Morales, F. J., and S. Jimenez-Perez (2001). Free radical scavenging capacity of Maillard reaction products as related to colour and fluorescence. *Food Chemistry* 72(1): 119–125.

Naganathan, G. K., L. M. Grimes, et al. (2008). Visible/near-infrared hyperspectral imaging for beef tenderness prediction. *Computers and Electronics in Agriculture* 64(2): 225–233.

Nakariyakul, S., and D. P. Casasent (2009). Fast feature selection algorithm for poultry skin tumor detection in hyperspectral data. *Journal of Food Engineering* 94(3–4): 358–365.

Ngo, T., R. E. Wrolstad, et al. (2007). Color quality of Oregon strawberries: Impact of genotype, composition, and processing. *Journal of Food Science* 72(1): C25–C32.

Noh, H. K. and R. Lu (2007). Hyperspectral laser-induced fluorescence imaging for assessing apple fruit quality. *Postharvest Biology and Technology* 43(2): 193–201.

Nussinovitch, A., G. Ward, et al. (1996). Nondestructive measurement of peel gloss and roughness to determine tomato fruit ripening and chilling injury. *Journal of Food Science* 61(2): 383–387.

Okamoto, H., and W. S. Lee (2009). Green citrus detection using hyperspectral imaging. *Computers and Electronics in Agriculture* 66(2): 201–208.

Park, B., K. C. Lawrence, et al. (2005). Detection of cecal contaminants in visceral cavity of broiler carcasses using hyperspectral imaging. *Applied Engineering in Agriculture* 21(4): 627–635.

Park, B., W. R. Windham, et al. (2007). Contaminant classification of poultry hyperspectral imagery using a spectral angle mapper algorithm. *Biosystems Engineering* 96(3): 323–333.

Pastor, C., J. Santamaria, et al. (2007). Gloss and colour of dark chocolate during storage. *Food Science and Technology International* 13(1): 27–34.

Pattnaik, P., and A. Srivastav (2006). Surface plasmon resonance: Applications in food science research: A review. *Journal of Food Science and Technology-Mysore* 43(4): 329–336.

Qiao, J., M. O. Ngadi, et al. (2007). Pork quality and marbling level assessment using a hyperspectral imaging system. *Journal of Food Engineering* 83(1): 10–16.

Qiao, J., N. Wang, et al. (2007). Prediction of drip-loss, pH, and color for pork using a hyperspectral imaging technique. *Meat Science* 76(1): 1–8.

Qin, J. and R. Lu (2009). Monte Carlo simulation for quantification of light transport features in apples. *Computers and Electronics in Agriculture* 68(1): 44–51.

Qin, J. W., T. F. Burks, et al. (2009). Detection of citrus canker using hyperspectral reflectance imaging with spectral information divergence. *Journal of Food Engineering* 93(2): 183–191.

Qin, J. W. and R. F. Lu (2008). Measurement of the optical properties of fruits and vegetables using spatially resolved hyperspectral diffuse reflectance imaging technique. *Postharvest Biology and Technology* 49(3): 355–365.

Quevedo, R., O. Diaz, et al. (2009). Quantification of enzymatic browning kinetics in pear slices using non-homogenous L* color information from digital images. *LWT-Food Science and Technology* 42(8): 1367–1373.

Rozycki, S. D., M. S. Pauletti, et al. (2007). The kinetics of colour and fluorescence development in concentrated milk systems. *International Dairy Journal* 17(8): 907–915.

Sáenz, C., and B. Hernández (1999). Analytical demonstration of the K/S ratio methods that give the relative proportions of myoglobin derivatives in meat. Paper presented at the 1st International Congress on Pigments in Food Industry, Sevilla, Spain.

Sáenz, C., B. Hernandez, et al. (2005). Meat color in retail displays with fluorescent illumination. *Color Research and Application* 30(4): 304–311.

Sáenz, C., B. Hernandez, et al. (2008). A multispectral imaging technique to determine concentration profiles of myoglobin derivatives during meat oxygenation. *European Food Research and Technology* 227(5): 1329–1338.

Sáenz, C., B. Hernández, et al. (2009). Measurement of the colour of white and rose wines in visual tasting conditions. *European Food Research and Technology* 229(2): 263–276.

Santana, C. C., M. C. D. Vanetti, et al. (2008). Microbial growth and colour of minimally processed shiitake mushroom stored at different temperatures. *International Journal of Food Science and Technology* 43(7): 1281–1285.

Sanz, S., C. Olarte, et al. (2009). Evolution of quality characteristics of minimally processed asparagus during storage in different lighting conditions. *Journal of Food Science* 74(6): S296–S302.

Sanz, T., A. Salvador, et al. (2008). Yogurt enrichment with functional asparagus fibre. Effect of fibre extraction method on rheological properties, colour, and sensory acceptance. *European Food Research and Technology* 227(5): 1515–1521.

Schouten, R. E., T. P. M. Huijben, et al. (2007). Modelling quality attributes of truss tomatoes: Linking colour and firmness maturity. *Postharvest Biology and Technology* 45(3): 298–306.

Skjervold, P. O., R. G. Taylor, et al. (2003). Development of intrinsic fluorescent multispectral imagery specific for fat, connective tissue, and myofibers in meat. *Journal of Food Science* 68(4): 1161–1168.

Solah, V. A., G. B. Crosbie, et al. (2007). Measurement of color, gloss, and translucency of white salted noodles: Effects of water addition and vacuum mixing. *Cereal Chemistry* 84(2): 145–151.

Taghizadeh, M., A. Gowen, et al. (2010). Use of hyperspectral imaging for evaluation of the shelf-life of fresh white button mushrooms (*Agaricus bisporus*) stored in different packaging films. *Innovative Food Science and Emerging Technologies* 11(3): 423–431.

Taherian, A. R., P. Fustier, et al. (2006). Effect of added oil and modified starch on rheological properties, droplet size distribution, opacity and stability of beverage cloud emulsions. *Journal of Food Engineering* 77(3): 687–696.

Tarrega, A., and E. Costell (2007). Colour and consistency of semi-solid dairy desserts: Instrumental and sensory measurements. *Journal of Food Engineering* 78(2): 655–661.

Ukuku, D. O., X. T. Fan, et al. (2006). Effect of vacuum-steam-vacuum treatment on microbial quality of whole and fresh-cut cantaloupe. *Journal of Food Protection* 69(7): 1623–1629.

Valous, N. A., F. Mendoza, et al. (2009). Colour calibration of a laboratory computer vision system for quality evaluation of pre-sliced hams. *Meat Science* 81(1): 132–141.

Vargas, A. M., M. S. Kim, et al. (2005). Detection of fecal contamination on cantaloupes using hyperspectral fluorescence imagery. *Journal of Food Science* 70(8).

Villalobos-Carvajal, R., P. Hernandez-Munoz, et al. (2009). Barrier and optical properties of edible hydroxypropyl methylcellulose coatings containing surfactants applied to fresh cut carrot slices. *Food Hydrocolloids* 23(2): 526–535.

Vizhanyo, T. and J. Felfoldi (2000). Enhancing colour differences in images of diseased mushrooms. *Computers and Electronics in Agriculture* 26(2): 187–198.

Wackerbarth, H., T. Stoll, et al. (2009). Carotenoid-protein interaction as an approach for the formulation of functional food emulsions. *Food Research International* 42(9): 1254–1258.

Wallays, C., B. Missotten, et al. (2009). Hyperspectral waveband selection for on-line measurement of grain cleanness. *Biosystems Engineering* 104(1): 1–7.

Waswa, J., J. Irudayaraj, et al. (2007). Direct detection of E-Coli O157: H7 in selected food systems by a surface plasmon resonance biosensor. *LWT-Food Science and Technology* 40(2): 187–192.

Wold, J. P., A. Veberg, et al. (2006). Influence of storage time and color of light on photo-oxidation in cheese: A study based on sensory analysis and fluorescence spectroscopy. *International Dairy Journal* 16(10): 1218–1226.

Wrolstad, R. E., R. W. Durst, et al. (2005). Tracking color and pigment changes in anthocyanin products. *Trends in Food Science & Technology* 16(9): 423–428.

Wu, J., Y. Peng, et al. (2009). Hyperspectral scattering profiles for prediction of beef tenderness. *Nongye Jixie Xuebao/Transactions of the Chinese Society of Agricultural Machinery* 40(12).

Wyszecki, G., and W. S. Stiles (2000). *Color science: Concepts and methods, quantitative data, and formulae*. New York: John Wiley & Sons.

Xing, J., C. Bravo, et al. (2005). Detecting bruises on Golden Delicious apples using hyperspectral imaging with multiple wavebands. *Biosystems Engineering* 90(1): 27–36.

Xing, J., W. Saeys, et al. (2007). Combination of chemometric tools and image processing for bruise detection on apples. *Computers and Electronics in Agriculture* 56(1): 1–13.

Yam, K. L., and S. E. Papadakis (2004). A simple digital imaging method for measuring and analyzing color of food surfaces. *Journal of Food Engineering* 61(1): 137–142.

Yang, G., and N. H. Cho (2008). Development, validation, and application of a portable SPR biosensor for the direct detection of insecticide residues. *Food Science and Biotechnology* 17(5): 1038–1046.

Zampini, M., D. Sanabria, et al. (2007). The multisensory perception of flavor: Assessing the influence of color cues on flavor discrimination responses. *Food Quality and Preference* 18(7): 975–984.

5 Electrical Properties of Foods

Satyanarayan R. S. Dev and G. S. Vijaya Raghavan

CONTENTS

5.1 INTRODUCTION

Electrical properties of food materials are of great importance in various scientific and industrial applications. Electrical properties are necessary when processing foods using electric fields, electric current conduction, or heating through electromagnetic waves. Electrical properties are also essential in processing foods with pulsed electric fields, ohmic heating, induction heating, radio frequency, and microwave heating.

Some popular applications of the knowledge of electric properties of food materials include rapid quality assessments, such as the measurement of moisture content in grain and other food products and to determine the absorption of energy in high-frequency dielectric heating and microwave heating applications that are useful in the processing of food materials to a large extent. These properties are also useful in establishing real-time monitoring of processing conditions (De Alwis and Fryer, 1992).

The processing of food always imparts changes to the food material treated; some are desirable and some are undesirable. For starchy foods, gelatinization makes the starch available for digestion (Angersbach et al., 1999). This is the key effect desired in heat processing. However, in the same temperature range, the integrity of the cell membrane is affected; this is accompanied by changes in textural properties through the loss of turgor pressure. Loss of compartmentalization leads to further changes (Dejmek and Miyawaki, 2002). The state of the cell membrane plays a major role in the electrical properties of food.

There are two main electrical properties in food engineering: electrical conductivity and electrical permittivity. From the electrical circuit point of view (Nelson, 1965), the electrical properties of different food materials that are of interest in research and commercial applications are conductivity (σ), the dielectric constant (ε'), the dielectric loss factor (ε''), and the loss tangent ($\tan \delta$). The last three of the previously mentioned properties together can be represented by the complex electrical permittivity of the food material. Predominantly, most of these properties are of considerable importance for implementation of electromagnetic field concepts (Nelson, 1973).

Three complex constitutive parameters, G, C, C_M, respectively, the conductance, capacitance, and magnetic capacitivity, universally describe the electromagnetic properties of all matter. In most biological materials, however, the magnetic capacitivity is the same as that of free space. This leaves us with only two variable parameters, namely, electrical conductance and capacitance, affecting the electrical properties of a food material. In order to be able to specify these properties irrespective of the quantity of the food material, both conductance and capacitance are expressed per unit length in the direction of flow of electric current, giving rise to conductivity (σ) and permittivity (ε) properties of matter. In Figure 5.1 the relationships among the major electrical properties of food materials are shown.

In this chapter, some of the electrical properties of foods, such as electrical conductance and capacitance, along with their properties and relationships will be highlighted. Their importance in certain food process situations will also be described.

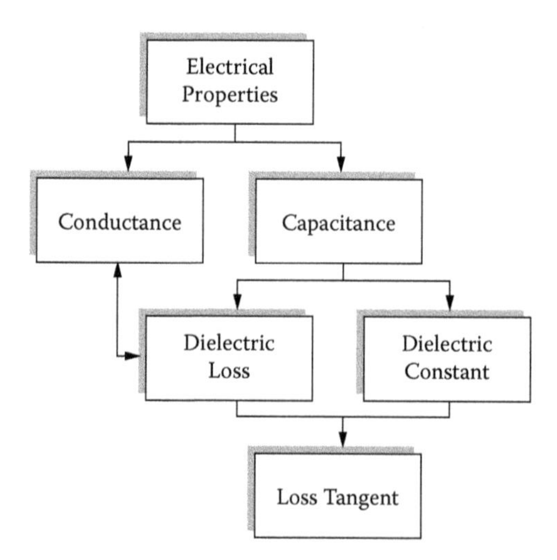

FIGURE 5.1 Major electrical properties of food materials and their relationships.

5.2 ELECTRICAL CONDUCTANCE

Electrical conductance of a food material is a measure of its ability to conduct an electric current between two points. It is dependent on the electrical conductivity and the geometric dimensions of the conducting object.

Electrical conductivity is a measure of how well a material accommodates the movement of an electric charge or, in other words, the property of the ease with which an electric current can be transmitted through a material. It is the ratio of the current density to the electric field strength. Electrical conductivity is a very useful property since its values are affected by such things as a substance's chemical composition and the stress state of crystalline structures. Therefore, electrical conductivity information can be used for measuring the purity of water, sorting materials, checking for proper heat treatment of metals, and inspecting for heat damage in some materials (Castro et al., 2003).

Food materials that contain positively charged or negatively charged electrolytes, or charged molecules or macromolecules, are capable of transmitting an electric current. In this context, positively charged ions are called *cations*, and negatively charged ions are called *anions*.

Carriers are those mobile components that are necessary to transmit electric current in the form of charged ions present in the food material. Unlike in metals, the charge carriers for the conduction of electricity in foods are ions, instead of electrons. In ionic materials, the band gap is too large for thermal electron promotion. Cation vacancies allow ionic motion in the direction of an applied electric field; this is referred to as *ionic conduction*.

High temperatures produce more vacancies and thus higher ionic conductivity. At low temperatures, electrical conduction in insulators is usually along the surface, due to the deposition of moisture that contains impurity ions. Under normal applications, ions carry the charges as the mass of ions moves along the electrical field. The concentration and mobility of ions determine the electrical conductivity. Temperature and other ingredients in foods affect the ion mobility. Under an extreme electric field, electron hopping takes place between the ions or molecules (Barbosa-Cánovas et al., 2006). This is the precursor of dielectric breakdown of foods, in which case an arc is the observed result.

Factors influencing the electrical conductivity of foods are the concentration of charge carriers' salinity, formulation charge, and number of charge carriers (mainly singly charged or doubly charged ions), mobility of charge carriers, their aggregate state, molecular mass, and type of bonding.

In general, if there are charged ions with mobile carriers present within a sample of food material, then if we apply a voltage potential across the food sample, an electric current will flow through the sample, as part of an electric circuit. The strength of this electric current depends on the electrical resistance R of the food sample. The electrical resistance impedes the flow of electric current through the sample. The reciprocal of resistance is conductance, G. Therefore, the resistance in an electric circuit is inversely proportional to its conductance. According to Ohm's law (Equation 5.1), there is a linear relationship between voltage, current, and resistance within an electric circuit:

$$V = IR \tag{5.1}$$

where

V is the potential difference in volts
I is the electric current in amperes
R is the electrical resistance of the material in ohms

Electrical conductivity of a food material is a measure of the extent of the flow of electric current through the material of cross-sectional area A, length L, and resistance R; in other words it is the inverse value of electrical resistivity of the food material of unit cross-sectional area and unit length. It can also be defined as the reciprocal of the specific volume resistivity, which is the resistance between opposite faces of a cube of unit dimensions of the material. The conductivity of a material σ is expressed in SI units S/m in Equation (5.2).

$$\sigma = L/(AR) \tag{5.2}$$

The electrical conductivity of the food material when subjected to direct current (DC) results from the motion of free charges and ions, which is not very different from the total conductivity of the material at low frequencies of alternating current (AC) as well. Figure 5.2 shows the ionic conductivity of food materials under DC and low-frequency AC environments.

Conductivity is the basic physics behind ohmic heating, in which electricity is transformed to thermal energy when an alternating current (AC) flows through a food material. It is important to know the effective conductivity or the overall resistance of liquid–particle mixtures for its potential use in fluid pasteurization. Furthermore, liquid–particle mixtures can be pasteurized using pulsed electric field technology, where products with low electrical conductivity are better and more energy efficient to process. Electrical conductivity varies with acidity of the food material, and therefore a good tool to monitor the change in acidity in real-time in processes like fermentation. Crystallization processes (for example, in sugar solutions) can also be monitored with conductivity measurements. Conductivity has also been found to be inversely proportional to viscosity, and viscosity in turn

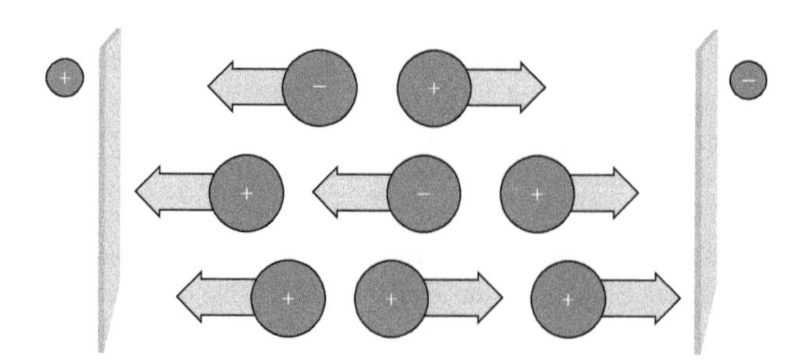

FIGURE 5.2 Ionic conductivity of food materials in DC and low-frequency AC environments.

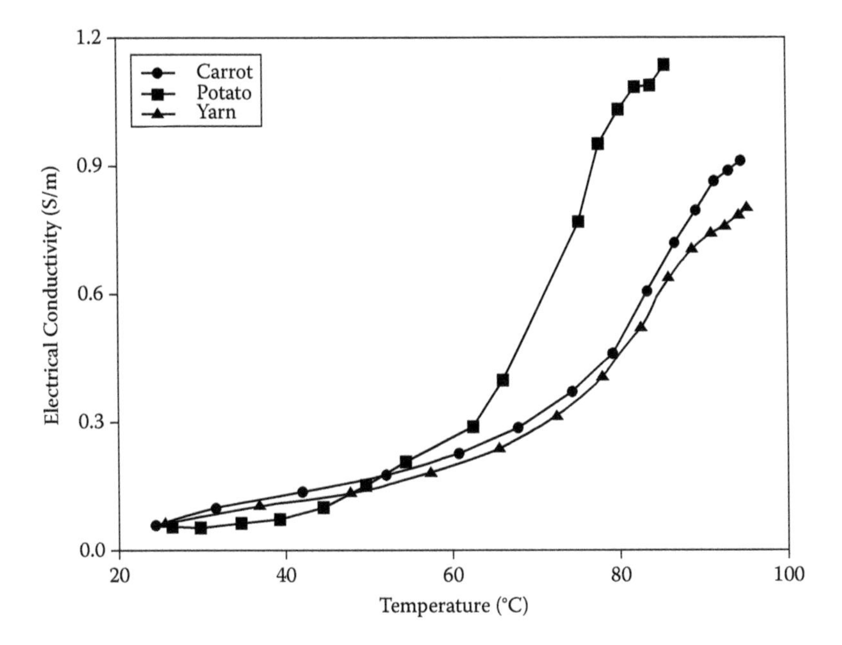

FIGURE 5.3 Electrical conductivity of vegetable tissue during conventional heating. (From Palaniappan, S., and S.K. Sastry. 1991. Electrical conductivity of selected solid foods during ohmic heating. *Journal of Food Process Engineering* 14:221–236. Reprinted with permission from John Wiley and Sons. Wiley Press, License number 2620901126146, dated March 02, 2011.)

is directly proportional to supersaturation. Conductivity measurements have also been used to measure moisture content in materials, particularly grain products and particulates.

The electrical conductivity of foods has been found to increase with temperature (linearly), and with water and ionic content. Mathematical relationships have been developed to predict the electrical conductivity of food materials: for example, for modeling heating rates through electrical conductivity measurements, or for probability distribution of conductivity through liquid–particle mixtures. Electrical conductivity shows different behaviors during ohmic and conventional heating. At freezing temperatures, electrical conductivity increases with temperature, due to the poor conductivity of ice compared to water. Starch transitions and cell structural changes affect electrical conductivity, and fat content decreases conductivity. Analogous to the thermal properties, the porosity of the food plays an important role in the conduction of electrons through the food. Electrical conductivity of different fruits and vegetables also follow a similar trend with temperature. Figure 5.3 gives the change in electrical conductivity of three different vegetable materials during conventional heating and Table 5.1 gives the electrical conductivity of various common food materials.

TABLE 5.1
Electrical Conductivity of Common Food Materials

Sample	Electrical Conductivity (mS/cm)
Pear	0.41
Apple	0.23
Potato	0.38
Turnip	0.26
Onion	0.22
Tomato	0.45
Pepper	0.48
Mushrooms	0.22
Carrot	0.25–0.42
Chicken	0.8
Egg White	0.71
Egg Yolk	0.37

Source: Amiali, M., Ngadi, M.O., Raghavan, G.S.V., and D.H. Nguyen. 2006. Electrical conductivities of liquid egg products and fruit juices exposed to high pulsed electric fields. *International Journal of Food Properties* 9(3):533–540; Sanjay S., Sastry, S.K., and L. Knipe. 2008. Electrical conductivity of fruits and meats during ohmic heating. *Journal of Food Engineering* 87(3):351–356.

5.3 ELECTRICAL CAPACITANCE

The electrical capacitance of any food material is its capacity to store and dissipate electrical energy. It is a factor of its electrical permittivity, which is a dielectric property used to explain interactions of foods with electric fields.

The cell membrane is nonconducting, but in intact cells, the presence of an internal electrolyte gives rise to an induced capacitance in an alternating electric field. This effect has been used to measure the cell population density (biomass) in a cell culture. The effects of freezing on cell integrity are also reflected in the dielectric properties of foods. Dielectric properties can also be utilized to monitor the change in cell size and to measure the cell membrane permeability.

5.4 DIELECTRIC PROPERTIES

Dielectric properties determine the interaction of electric fields and in turn the electromagnetic waves with matter, and defines the charge density under an electric field. The electrical field inside the food is determined by the dielectric properties and the geometry of the load, and by the oven configuration. These properties are also useful in detecting processing conditions, or the quality of foods. The major uses for dielectric properties are measurement of the current status of the food material and in electromagnetic heating applications.

Historically, the interest in dielectric properties of materials has been associated with the design of electrical equipment, where various dielectrics are used for insulating conductors and other components of electric equipment. Measurement of the bulk dielectric properties (dielectric constant, dielectric loss factor) is not an end unto itself. Rather, these properties are an intermediary vehicle for understanding, explaining, and empirically relating certain physicochemical properties of the food material. Measurements of the dielectric properties of materials are finding increasing application as new electrotechnology is adapted for use in food and bioprocessing industries.

As dielectric properties describe how the material interacts with electromagnetic radiation, they are also important in the selection of proper packaging materials and cooking utensils, and in the design of microwave and radio frequency heating equipment.

Knowledge of dielectric properties in partially frozen material is critical in determining the rates and uniformity of heating in microwave thawing. As the ice in the material melts, absorption of energy increases tremendously. Thus, the portions of material that thaw first absorb significantly more energy and heat at increasing rates, which can lead to localized boiling temperatures while other areas are still frozen. Salt affects the situation through freezing point depression, leaving more water unfrozen at a given temperature.

In solids, liquid, and gases, the dielectric properties depend on two values, namely the dielectric constant (ε') and the dielectric loss factor (ε'').

5.5 DIELECTRIC CONSTANT

Dielectric constant ε' is related to the capacitance of a substance and its ability to store electrical energy.

5.6 DIELECTRIC LOSS

The dielectric loss factor is related to energy losses when the food is subjected to an alternating electrical field (i.e., dielectric relaxation and ionic conduction). The loss component of the complex conductivity is included in the dielectric loss component of the complex permittivity (Equation [5.3]).

$$\varepsilon = \varepsilon' - j\varepsilon'' \tag{5.3}$$

5.7 RELATIONSHIP BETWEEN ELECTRICAL CONDUCTANCE AND CAPACITANCE

In practical applications, the parameter customarily used to define the electric property of any material is the permittivity of materials relative to the free-space permittivity, ε_o.

The complex relative permittivity is, therefore, given by Equation (5.4):

$$\varepsilon_r = \varepsilon_r' - j\varepsilon_r'' \text{ (in other form) } \varepsilon/\varepsilon_o = \varepsilon'/\varepsilon_o - j\varepsilon''/\varepsilon_o \tag{5.4}$$

where the real component is the relative dielectric constant, and the imaginary component is the relative dielectric loss factor. The extent of the dielectric loss that can happen in a material in relation to its dielectric constant at any given frequency is a good quantifier of the material's ability to convert electrical energy to heat energy. The *loss tangent* (tangent of the loss angle of the dielectric), also called the *dissipation factor*, is related to the dielectric constant and loss factor as given by Equation (5.5):

$$\tan\delta = \varepsilon_r''/\varepsilon_r' \tag{5.5}$$

The AC conductivity of a material is given by Equation (5.6):

$$\sigma = \omega\varepsilon_o\,\varepsilon_r'' \tag{5.6}$$

where $\omega = 2\pi f$ is the angular frequency of the alternating fields. The dielectric constant, loss factor, and loss tangent are dimensionless quantities, but the conductivity can be expressed as given by Equation (5.7):

$$\sigma = 0.556\,f\varepsilon_r'' \tag{5.7}$$

where the unit of conductivity would be nmhos per cm when f is in kHz, in μmhos per cm when f is in MHz, and in mmhos per cm when f is in GHz, which is equal to a conductivity of 10^{-1} siemens per meter (S/m).

5.8 REACTION TO ELECTROMAGNETIC RADIATION

Permittivity and moisture are closely correlated when the water content is high. Properly designed electrical instruments can be used to determine moisture content or water activity. At high frequencies that are too high for the ionic conductivity, alternating electric fields produce an additional component of conductivity leading to the capacitance of the food material, and food material will act like a dielectric material for a capacitor in an AC circuit. Figure 5.4 represents the dipole rotation of a water molecule, which results in molecular friction and heating at high-frequency dielectric heating.

Basically, the dielectric constant of a material is related to its capability for storing energy in an electric field in the material, whereas the loss factor is related to the material's capability for absorbing energy from the field. The ratio of the capacitance of a capacitor, with the food material as its dielectric, to the capacitance of the same capacitor with air (or more properly, with a vacuum as its dielectric), is known as the *dielectric constant* of a material. When the material is exposed to alternating electric fields, the index of the material's energy dissipation characteristics is known as its *dielectric loss tangent*.

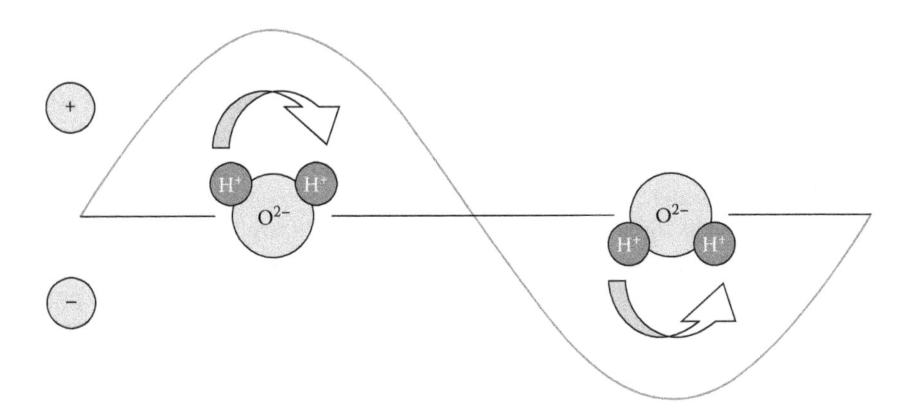

FIGURE 5.4 Dipole rotation of a water molecule subjected to electromagnetic radiation.

In foods, permittivity can be related to chemical composition, physical structure, frequency, and temperature, with moisture content being the dominant factor. Dielectric properties (ε', ε'') are primarily determined by their chemical composition (presence of mobile ions and permanent dipole moments associated with water and other molecules) and, to a much lesser extent, by their physical structure. The influence of water and salt (or ash) content largely depends on the manner in which they are bound or restricted in movement by other food components. Free water and dissociated salts have a high dielectric activity, while bound water-associated salts and colloidal solids have low activity.

Power dissipation is directly related to the dielectric loss factor ε'' and depends on the specific heat of the food, density of the material, and changes in moisture content (for example, vaporization). Permittivity also depends on the frequency of the applied alternating electric field. Frequency contributes to the polarization of molecules such as water. In general, the dielectric constant increases with temperature, whereas the loss factor may either increase or decrease depending on the operating frequency. Both the dielectric constant ε' and loss factor ε'' decrease significantly as more water freezes.

Studies of heating uniformity and temperature elevation rate involve dielectric properties. Large internal hot and cold areas, internal focusing effects, and the edge-heating phenomenon are some of the typical features of power density patterns of a load. For example, when a raw egg is heated in a microwave environment it may explode because the power density near its center is much higher than in other parts, causing violent shattering as the interior becomes superheated. The dielectric properties of materials are very important in evaluating the penetration depth of energy that can be achieved in a certain food.

By understanding and fine tuning all these parameters, raw eggs can be successfully pasteurized in a microwave environment.

5.9 NOVEL MEASURING TECHNIQUES AND APPLICATIONS

The measurement of the conductivity of a material is generally done by passing a known amount of current at constant voltage through a known volume of the material and by determining resistance. The total conductivity is then calculated by taking the inverse of the total resistivity. These basic measurements consist of bridge networks (such as the Wheatstone bridge circuit) or a galvanometer. There are other devices that measure electrical conductivity of foods under ohmic or conventional heating conditions, using thermocouples and voltage and current transducers to measure voltage across and current through the samples.

Known methods for measuring dielectric properties are cavity perturbation, open-ended coaxial probe, and transmission line methods. Since modern microwave network analyzers have become available, the methods of obtaining dielectric properties over frequency ranges have become more efficient. Computer control of impedance analyzers and network analyzers has facilitated the automatic measurement of dielectric properties over wide frequency ranges, and special calibration methods have also been developed to eliminate errors caused by unknown reflections in the coaxial-line systems. Distribution functions can be used in expressing the temperature dependence of dielectric properties (Venkatesh and Raghavan, 2005).

The measurement of electrical conductivity in foods is used in food quality testing laboratories for the testing of fats and oils to determine the oil stability index (OSI), which is an indicator of their shelf stability or stability against oxidation (oxidative rancidity). The fat or oil is made to oxidise and the short-chain fatty acids, which are products of oxidation, are dissolved in water; the change in electrical properties of the water is directly proportional to the extent of oxidation, and therefore provides an index of stability for the oil or fat (Pike, 2001).

Electrical conductivity can be measured using conductivity or an inductance, capacitance, resistance (LCR) meter. Ruhlman et al. (2001) used an electrical conductivity meter to measure electrical conductivities of different products, including apple and orange juices, at temperatures between 4 and 60°C. Other authors have also used conductivity meters to determine electrical conductivities of different food products, including orange, apple, and pineapple juices, at a constant temperature of 20°C (Raso et al., 1998). However, most commonly available conductivity meters may not be suited for measurement of high-conductivity products. Marcotte et al. (2000) used a static ohmic heating cell to measure electrical conductivity of some hydrocolloid solutions.

For pulsed electric field (PEF) applications, it is more suitable to determine electrical conductivity on-line while the product is treated in a treatment chamber. This will allow measurement of conductivity and monitoring of changes in the product during a PEF treatment. This approach has not been reported in the literature. Data on the relationship between temperature and electrical conductivity of liquid foods is scarce and has not been investigated using PEF technology as a measurement device (Zhang et al., 1994; Dunn et al., 1987).

Measurements of electrical properties through nonmagnetically permeable metals using directed magnetic beams and magnetic lenses (Amini, 2003) is one of the buzzing, up-to-date technologies. Real-time monitoring of fresh fruits and vegetables for freezing/chilling injury by measuring the change in electrical properties continuously helps prevent chilling injuries (Zhang et al., 1992; Ishikawa et al., 1997). Electric property measurements help on-line assessment and determination of accretion in fermentation reactors (Davey, 1991).

5.10 CONCLUSIONS

Electrical properties of naturally occurring food materials and food products vary widely depending on several factors. Frequency and temperature are the basic factors that influence values of electrical properties. Knowledge of the frequency dependence can be helpful in predicting the dielectric behavior of food materials. The dielectric constant of a material either decreases or remains constant as frequency increases, and the increase or decrease in the loss tangent and the dielectric loss factor with frequency depends upon the frequency range and the nature of the absorption process. Conductivity generally increases with increasing frequency, but is also dependent upon the loss factor. Temperature dependence is closely related to frequency dependence in theory, and the nature of the dispersion can be explored by varying either of these factors. For every food material, the relaxation frequency increases as temperature increases, as a consequence of the decrease in relaxation time within a certain range of frequencies, the dielectric constant increases with increasing temperature whereas outside the range it decreases with increasing temperature. This range is termed as the dielectric dispersion region. Measurement of electrical properties of food materials and correlating them to different quality and processing parameters would help innovation and push real-time quality monitoring and process control to its limits.

REFERENCES

Amiali, M., M.O. Ngadi, G.S.V. Raghavan, and D.H. Nguyen. 2006. Electrical conductivities of liquid egg products and fruit juices exposed to high pulsed electric fields. *International Journal of Food Properties* 9(3):533–540.

Amini, B.K. 2003. Measurements of electrical properties through non magneticially permeable metals using directed magnetic beams and magnetic lenses. US Patent No. 6,630,831 B2. Issued Oct 7, 2003. Assignee: Em-Tech Sensors LLC, Houston, TX, USA.

Angersbach, A., V. Heinz, and D. Knorr. 1999. Electrophysiological model of intact and processed plant tissues: cell disintegration criteria. *Biotechnology Progress* 15:753–762.

Barbosa-Cánovas, G.V., P. Juliano, and M. Peleg. 2006. Engineering properties of foods. In *Food engineering*, ed. G.V. Barbosa-Cánovas, *Encyclopedia of Life Support Systems* (EOLSS), Developed under the Auspices of UNESCO. Oxford, UK: Eolss Publishers.

Castro, I., J.A. Teixeira, S. Salengke, S.K. Sastry, and A.A. Vicente. 2003. The influence of field strength, sugar, and solid content on electrical conductivity of strawberry products. *Journal of Food Process Engineering* 26(1):17–29.

Davey, C.L., W. Penaloza, D.B. Kell, and D.N., Hedger. 1991. Real-time monitoring of the accretion of Rhizop oligosporus biomass during the solid substrate temperature fermentation. *World Journal of Microbiology and Biotechnology* 7:248–259.

De Alwis, A.A.P., and P.J. Fryer. 1992. Operability of the ohmic heating process: Electrical conductivity effects. *Journal of Food Engineering* 15:21–48.

Dejmek, P., and O. Miyawaki. 2002. Relationship between the electrical and rheological properties of potato tuber tissue after various forms of processing. *Bioscience, Biotechnology and Biochemistry* 66(6):1218–1223.

Dunn, J.E., and J.S. Pearlman. 1987. Methods and apparatus of extending the shelf-life of fluid food products. US Patent 4,695,472, Filed May 31, 1985.

Ishikawa, E., S.K. Bae, O. Miyawaki, K. Nakamura, Y. Shiinoki, and K. Ito. 1997. Freezing injury of cultured rice cells analyzed by dielectric measurement. *Journal of Fermentation and Bioengineering* 83:222–226.

Marcotte, M., M. Trigui, and H.S. Ramaswamy. 2000. Effect of salt and citric acid on electrical conductivities and ohmic heating of viscous liquids. *Journal of Food Processing and Preservation* 24:389–406.

Nelson, S.O. 1965. Dielectric properties of grain and seed in the 1 to 50-mc range. *Transactions of the ASAE* 8:38.

Nelson, S.O. 1973. Electrical properties of agricultural products: A critical review. *Transactions of the ASAE* 16:384. Special Publication SP-05-73, American Society of Agricultural and Biological Engineers, St. Joseph, MI.

Palaniappan, S., and S.K. Sastry. 1991. Electrical conductivity of selected solid foods during ohmic heating. *Journal of Food Process Engineering* 14:221–236.

Pike, O.A. 2001. *Current protocols in food analytical chemistry*, D2.3.1–D2.3.5. New York: John Wiley & Sons, Inc.

Raso, J., M.L. Calderón, M.M. Góngora-Nieto, G.V. Barbosa-Cánovas, and B.G. Swason. 1998. Inactivation of *Zygosaccharomyces bailii* in fruit juices by heat, hydrostatic pressure and pulsed electric fields. *Journal of Food Sciences* 63(6):1042–1044.

Ruhlman, K.T., Z.T. Jin, and Q.H. Zhang. 2001. Physical properties of liquid foods for pulsed electric fields treatment. In *Pulsed electric field in food processing: Fundamental aspect and application*, ed. G.V. Barbosa-Cánovas and Q.W. Zhang, 45–56. Lancaster, PA: Technomic Publishing.

Sanjay S., S.K. Sastry, and L. Knipe. 2008. Electrical conductivity of fruits and meats during ohmic heating. *Journal of Food Engineering* 87(3):351–356.

Venkatesh, M.S., and G.S.V. Raghavan. 2005. An overview of dielectric properties measuring techniques. *Canadian Biosystems Engineering / Le génie des biosystèmes au Canada* 47:7.15–7.30.

Zhang, M.I.N., and J.H.M. Willison. 1992. Electric impedance analysis in plant tissues: In vivo detection of freezing injury. *Canadian Journal of Botany* 70:2254–2258.

Zhang, Q.H., F.J. Chang, G.V. Barbosa-Cánovas, and B.G. Swason. 1994. Inactivation of microorganisms in a semisolid model food using high-voltage pulsed electric fields. *Lebensmittel-Wissenschaft und-Technologie* 27:538–543.

6 Thermodynamic Properties of Agricultural Products Processes

Paulo Cesar Corrêa, Gabriel Henrique Horta de Oliveira, and Emílio de Souza Santos

CONTENTS

6.1 POSTHARVEST PROCESSES

With the increasing production of crops, the study of new technologies and solutions to problems related to postharvesting has become indispensable, particularly with regard to storage options. In order to guarantee the quality and conservation of grain, important postharvest procedures should be followed, including safe transport and the storage of products in dry, well-ventilated facilities, with low levels of moisture content (X).

Drying of foodstuff is one of these postharvest processes, and is indispensable to control and maintain the quality of these products. The main objective of drying is to reduce the X and water activity (a_w) to certain levels, since these properties are key quality factors. Too high X and a_w may lead to microorganism growth and fermentation of the product, which deteriorates the quality, while too low X and a_w may lead to excessive energy consumption and product quality damages.

The development of drying equipment, the calculation of dehydration process energy requirements, the determination of water properties on the food surface, and the determination of sorption kinetic parameters are some of the items for which thermodynamic properties provide useful information (Corrêa et al., 2007). The thermodynamic concept of a_w (Lewicki, 2004) is used in drying applications through the sorption isotherms.

6.2 SORPTION ISOTHERMS

In order to correctly conduct drying and storage operations, it is necessary to know the relationship between air temperature (T) and relative humidity (RH), and desirable conditions for preserving the product. To obtain this information, sorption isotherms are indispensable.

Simal et al. (2007) reports that engineering design is an important application of sorption isotherms in separation processes when interface water transport is involved. Thermodynamics in sorption processes is used to understand water properties and for calculation of the required energy associated with heat and mass transfer in biological systems, properties of water, food microstructure, physical phenomena on food surfaces, and sorption kinetic parameters (Corrêa et al., 2010a). The thermodynamic properties of sorption provide a better understanding of the equilibrium state of water under certain T and RH conditions (Fasina, 2006). They offer information regarding sorbent affinity to water and the spontaneity of the sorption process; they also define the concept of an order/disorder state in water-sorbent systems. Thermodynamic functions can be obtained through sorption isotherms, facilitating the interpretation of thermodynamic parameters (Rizvi and Benado, 1984).

Moisture sorption isotherms for materials describe the equilibrium relationship between X and RH of the surrounding environment, which is equal to a_w at the equilibrium state. The equilibrium moisture content (X_{eq}) is reached when the partial pressure of water vapor in the product is equal to the partial pressure of water vapor of the surrounding air. The X_{eq} of a hygroscopic product under given T and RH conditions depends on the path used to achieve that equilibrium. At a given RH, two different isotherms, called *adsorption* and *desorption* isotherms, can thus be obtained as a function of the initial experimental conditions. The difference between desorption and adsorption is known as *hysteresis* (Wolf et al., 1972).

Water sorption isotherms are unique for individual foodstuffs and are a useful tool for understanding the moisture relationship of a material and consequently its stability problems, leading to a better understanding of moisture variations during storage. This information is then used directly to solve food processing design problems and to predict energy requirements.

Sorption isotherms can be expressed by mathematical models. There are over 200 models proposed in the literature to represent hygroscopic equilibrium of agricultural models. These models differ in their theoretical or empirical basis and in the number of parameters involved (Mulet et al., 2002). The description of sorption isotherms of agricultural products can be made by more than one model of X_{eq}. The main criteria to select models are the adjustment degree to experimental data and the simplicity of the model (Furmaniak et al., 2007).

Several models currently used to describe moisture sorption isotherms can be separated into different groups: kinetic models based on a monolayer (BET [Brunauer, Emmett, and Teller] model), kinetic models based on a multilayer and condensed film (GAB [Guggenheim, Anderson, and De Boer] model), semi-empirical (Ferro-Fontan, Henderson, and Halsey models) and empirical models (Smith and Oswin models). The GAB equation has been recommended by the European Project Group COST 90 on Physical Properties of Foods (Wolf et al., 1985) as the fundamental equation for the characterization of water sorption of food materials. The major advantages of the GAB model (Equation [6.1]) are its theoretical background, its physically meaningful parameters, and its adequacy in describing experimental data of water activity up to 0.90.

$$X_{eq} = \frac{X_m C K a_w}{(1 - K a_w)(1 - K a_w + C K a_w)} \tag{6.1}$$

where X_{eq} is the equilibrium moisture content, % d.b.; a_w is the water activity, decimal; X_m is the monolayer moisture content, % d.b.; and C and K are GAB constants and are related to monolayer and multilayer properties (van den Berg, 1984).

6.3 DRYING

Drying is defined as a simultaneous process of heat and mass transfer (moisture) among the product and drying air. Drying diminishes the respiration rate of the product and increases the storage time with the minimum possible loss (Corrêa et al., 2010c).

According to Incropera and Dewitt (2002), drying conditions during drying of agricultural products are considered to be isothermal and moisture transfer is restricted to the product surface. However, the knowledge of water molecule movement within the product is of vital importance to the correct study of interactions between water and agricultural components. Moisture can move inside agricultural products during drying in different ways, among them liquid diffusion, capillary diffusion, surface diffusion, hydrodynamic flow, and vapor and thermal diffusion (Martinazzo et al., 2007).

In addition to sorption isotherms, thermodynamic properties during the drying procedure can be studied. Corrêa et al. (2010b) accomplished an analysis of the drying rate coefficient, obtained from drying curves. The methodology used in this work is explained later in this chapter. These authors used coffee grain dried at three different temperatures (35°, 45°, and, 55°C), concluding that enthalpy values decreased along with temperature increase.

6.4 THERMODYNAMIC PARAMETERS

An understanding of the water properties in relation to a biological system can be classified into three categories: structural, dynamic, and thermodynamic (Rizvi and Benado, 1984). As stated before, the thermodynamic approach provides an understanding of water equilibrium with its surroundings at a certain *RH* and *T* condition.

The following thermodynamic parameters have been studied over the past years by different authors, including isosteric heat of sorption or differential enthalpy (ΔH), differential entropy (ΔS), Gibbs free energy (ΔG), activation energy, enthalpy–entropy theory or isokinetic theory.

The information that water sorption isotherms and ΔH (which is the energy required of the process being studied) provides is essential in modeling several food industry processes (Goneli et al., 2010a). Besides required energy, these properties can provide important information in respect to sorption mechanisms and interactions between product components and water molecules (Tolaba et al., 2004). Also, it affects the equipment design procedures for dehydration processes and the qualitative understanding of the state of water on the food surface.

With drying on the scope, ΔH is used as an indicator of the binding strength of the water to the solid (Moreira et al., 2008), meaning the higher the magnitude of this parameter, the more tightly water is bound to the product. According to Wang and Brennan (1991), for water removal from hygroscopic materials, the required energy is greater than that required to vaporize the same amount of free water, at the same temperature and pressure conditions. Because of the strong bonds between water and the surface of the adsorbent substance, this additional energy is denoted as the *isosteric heat of sorption* in the drying processes and is a good parameter for estimating the minimal amount of heat required to remove a quantity of water from the product, as well as to provide data on the state of moisture in the product (Oliveira et al., 2010). Alterations of this parameter along with moisture content indicates the availability of polar sites to water vapor as sorption (desorption or adsorption) proceeds.

The sorption processes in systems exhibiting hysteresis involve thermodynamically irreversible phenomena. Different authors have explained this phenomenon and Al Hodali (1997) correlated the hysteresis effect with the rigidity of the structure of a pore connected to a small capillary. During the adsorption process, the capillary begins to swell as a consequence of the rise in RH when the pore was still empty. The *X* moves to the interior of the pore when the partial vapor pressure of the air becomes higher than the capillary pressure. In desorption, the pore is saturated at the beginning of the process. The water diffusion occurs surrounding the grain surface when the partial vapor pressure of the surrounding air is lower than the vapor pressure inside the capillary. The differences between the adsorption and desorption can be observed because of the exposure to the elevated capillary diameters present in the pores (Lahsasni et al., 2003).

The differential entropy production arises from this irreversibility due to hysteresis. Entropy is associated with the forces of attraction or repulsion of water molecules to product components and is linked with the spatial arrangement of the water–sorbent relationship. Thus, it defines the number of available desorption sites corresponding to a specific energy level as proportional to ΔS, which describes the degree of disorder and motion randomness of water molecules. It is a measurement of the mobility of the adsorbed water molecules, indicating the level at which the water and substrate interaction is greater than the interaction between the water molecules, as stated by Moreira et al. (2008). Also, the entropy variation is useful for the interpretation of procedures at food processing such as dissolution, crystallization, and swelling.

The feasibility and extent of a chemical reaction is best determined by measuring the changes in Gibbs free energy (ΔG), that is, the vaporization of water during the dehydration process. Gibbs function of a fixed mass is also correlated to be a measure of its potential to perform optimum work in a steady flow reactor (Annamalai and Puri, 2002). According to Telis et al. (2000), ΔG is indicative of the affinity of sorbents for water and provides a criterion of whether water sorption occurs as a spontaneous process. Based on its sign, this parameter is indicative of whether water sorption is a spontaneous ($\Delta G < 0$) or nonspontaneous ($\Delta G > 0$) process.

Activation energy (E_a) is defined as the energy that must be overcome or the minimum energy necessary to initiate a chemical reaction (Oliveira et al., 2010). It may also be defined as the minimum energy required for starting a chemical reaction. Activation energy can be thought of as the height of the potential that separates two minima of potential energy. For a chemical reaction to proceed at a reasonable rate there should exist an appreciable number of molecules with energy equal to or greater than the activation energy.

Enthalpy–entropy compensation, primarily applied by Bell (1937), is a theory that has been extensively considered in research regarding the physical and chemical phenomena involved in the water sorption process. This theory states that when a chemical reaction exhibits a linear relationship between enthalpy and entropy, it provides a strong form of compensation; the weak form of compensation is described when ΔH and ΔS have the same sign because a change occurs in some thermodynamic quantity excluding temperature. According to Laidler (1959), the compensation arises from changes in the nature of the interaction between the solute and solvent causing the reaction, and the relationship between enthalpy and entropy for a specific reaction is linear.

For a linear or strong enthalpy–entropy compensation, the isokinetic temperature (T_b) can be determined from the slope of the line, and if the theory is valid, should be constant at any point (Heyrovsky, 1970). This author states that at this temperature all the reactions in the series proceed at the same rate. According to Lai et al. (2000), the existence of the compensation theory implies that only one reaction mechanism is followed by all members of the reaction series, and therefore, a reliable evaluation of the isokinetic relationship aids elucidation of the reaction mechanisms.

6.5 EXPERIMENTAL PROCEDURES

6.5.1 Sorption Isotherms

There are two methods to determine the equilibrium moisture content of agricultural products: the static or dynamic (gravimetric) method. In the dynamic or gravimetric method, the air is mechanically moved and therefore requires less time to reach the equilibrium between the product and the air. The static method may require several weeks depending of the initial conditions of moisture content of the product. Thus, the dynamic method is preferred.

To obtain the isotherms, specific saturated saline solutions and acid solutions for each desired relative humidity at a certain temperature can be used inside desiccators or dryers (static method). Several authors used saturated saline solutions in order to acquire sorption isotherms (Goneli et al., 2010a, 2010b; Corrêa et al., 2010a; Mujumdar, 2006). Saturated saline solutions may also be used along with air flow in order to minimize the required time to reach equilibrium (dynamic method).

The dynamic or gravimetric method is simple and yields reliable results, and it can be a fully automated measuring system or it can be measured using weight measurements. In order to attain equilibrium, three consecutive weight measurements are commonly used, and consecutive readings yield variations that must be lower than 0.01 g.

As stated before, several mathematical models can be used to represent the experimental data of sorption of agricultural products. However, there are few models that possess a theoretical background, and over the past years, the Guggenheim-Anderson-De Boer (GAB) model has been successfully applied to different agricultural products. Its parameters are meaningful and allow the study of heat and mass transfer in a more detailed analysis, as explained below.

The monolayer moisture content (X_m) indicates the quantity of water molecules that are strongly adsorbed to specific sites at the food surface. Alterations of this parameter with temperature are a result of modifications in the physical and chemical characteristics of the product due to temperature (Perdomo et al., 2009). Components of agricultural products also affect the behavior of X_m; Oliveira et al. (2010), studying different corn cultivars, reported that the cultivar with the highest sugar content presented lower binding strength of water to solid. Also, the monolayer moisture content is recognized as the safest X, providing the longest time period with minimum quality loss at a certain temperature during preservation, and values below X_m lead to a decrease of deteriorative reactions, pest attacks, and respiration rate of the product.

The C constant in the GAB model is associated with the difference in chemical potential between the monolayer and the upper layers. This parameter allows us to classify sorption isotherms according to Brunauer's classification (Brunauer et al., 1940). The constant K corresponds to the difference in chemical potential between the multilayer state and the water's pure liquid state.

Using the results from the GAB parameters, enthalpy values for each coefficient can be calculated using ln X_m, ln C, and ln K versus the inverse of absolute temperature, as described by Simal et al. (2000) and shown in Equations (6.2), (6.3), and (6.4):

$$C = C_0 \exp\left[\frac{\Delta H_1}{R(T + 273,16)}\right] \tag{6.2}$$

$$K = K_0 \exp\left[\frac{\Delta H_2}{R(T + 273,16)}\right] \tag{6.3}$$

$$U = U_0 \exp\left[\frac{\Delta H_3}{R(T + 273,16)}\right] \tag{6.4}$$

In which C_0 and K_0 are equation parameters, dimensionless; U_0 is the equation parameter, % d.b.; T is temperature, °C; ΔH_1, ΔH_2, and ΔH_3 are enthalpy values from GAB model coefficients, kJ kg^{-1}; and, R is the water vapor constant, 0.462 kJ kg^{-1} K^{-1}.

6.5.2 Isosteric Heat of Sorption, Differential Entropy, and Gibbs Free Energy

The isosteric heat of sorption is a differential molar quantity derived from the temperature dependence of the isotherm at a constant amount of sorbed water moles (n_w) (Equation [6.5]).

$$\left(\frac{\partial \ln(p)}{\partial T}\right)_{n_w} = \frac{\Delta H}{RT_K^2} \tag{6.5}$$

in which p is the partial pressure of water vapor; T is temperature, °C; ΔH is the isosteric heat of sorption, kJ kg^{-1}; R is the water vapor constant, 0.462 kJ kg^{-1} K^{-1}; and T_K is temperature, K.

The net isosteric heat of sorption is derived from moisture sorption data using the Clausius-Clapeyron equation (Equation [6.6]):

$$\frac{\partial \ln(p_0)}{\partial T} = \frac{\Delta H_{vap}}{RT_K^2} \tag{6.6}$$

in which p_0 is saturation pressure; ΔH_{vap} is the latent heat of vaporization of pure water, kJ kg^{-1}. The ΔH_{vap} is obtained for the average temperature (T_c) of the sorption isotherms using Equation (6.7):

$$\Delta H_{vap} = 2502.2 - 2.39\ T_c \tag{6.7}$$

Subtracting Equation (6.6) from Equation (6.5) results in Equation (6.8):

$$\left(\frac{\partial \ln(p/p)}{\partial T}\right)_{n_w}\left(\frac{\partial \ln a_w}{\partial T}\right)_{n_w} = \frac{\Delta H_{st}}{RT_K^2} \tag{6.8}$$

in which ΔH_{st} is the net isosteric heat of sorption, kJ kg^{-1}.

The net isosteric heat of desorption is obtained by subtraction ΔH_{vap} from ΔH. Integrating Equation (6.5) and assuming that ΔH_{st} is independent of T, the isosteric heat of sorption is calculated for each X_{eq}, according to Equation (6.9) (Wang and Brennan, 1991).

$$\frac{\partial \ln(a_w)}{\partial(1/T)} = -\frac{\Delta H - \Delta H_{vap}}{R} = -\frac{\Delta H_{st}}{R} \tag{6.9}$$

Alterations in the differential entropy of sorption (ΔS) are calculated by means of the Gibbs-Helmholtz equation (Equation 6.10):

$$\Delta S = \frac{\Delta H_{st} - \Delta G}{T_K} \tag{6.10}$$

in which ΔS is the differential entropy of sorption, kJ kg^{-1} K^{-1}; and ΔG is the Gibbs free energy, kJ kg^{-1} mol^{-1}.

Gibbs free energy is calculated through the Van't Hoff equation, which relates the change in temperature to the change in the equilibrium constant given the standard enthalpy change for the process. ΔG is calculated by means of Equation (6.11) (Moran and Shapiro, 2006), and it is a function of temperature and relative humidity. For desorption, the changes are positive, while they are negative in the case of adsorption (Atkins, 2000).

$$\Delta G = \pm\, RT_K \ln a_w \tag{6.11}$$

The influence of water sorption variations over free energy is normally accompanied by enthalpy and entropy changes. Thus, substituting Equation (6.11) into Equation (6.10) and rearranging, Equation (6.12) can be obtained:

$$\ln a_w = \pm\left(\frac{\Delta H_{st}}{RT_K} - \frac{\Delta S}{R}\right) \tag{6.12}$$

ΔH_{st} and ΔS are calculated using Equation (6.12) and plotting ln (a_w) values versus to the respective values of X_{eq}. The magnitudes of a_w, T_K and X_{eq} are acquired through the equation that best fits the sorption experimental data.

6.5.3 Enthalpy–Entropy Compensation Theory

According to the enthalpy–entropy compensation theory explained previously, the linear relationship between enthalpy and entropy for a specific reaction is given by Equation (6.13):

$$\Delta H_{st} = T_B\,(\Delta S) + \Delta G_B \tag{6.13}$$

in which, T_B is the isokinetic temperature, K; and ΔG_B is the Gibbs free energy at the isokinetic temperature, kJ kg^{-1} mol^{-1}.

The T_B represents the T in which all reactions occur at the same rate. Once the enthalpy and entropy are highly correlated, it is assumed that the compensation theory can be valid for the sorption process. To confirm the existence of the compensation, the T_B is compared with the harmonic mean temperature (T_{hm}), which is defined by Equation (6.14):

$$T_{hm} = \frac{n_t}{\sum_{i=1}^{n_t}\left(\frac{1}{T_i}\right)} \tag{6.14}$$

in which T_{hm} is the harmonic mean temperature, K; and n_t is the number of temperatures utilized.

An approximate confidence interval, (1 – α)100%, to the T_B can be calculated using Equation (6.15):

$$T_B = \hat{T}_B \pm t_{m-2,\alpha/2}\sqrt{Var(T_B)} \tag{6.15}$$

in which and $Var\ (T_B)$ are calculated using Equations (6.16) and (6.17), respectively.

$$\hat{T}_B = \frac{\sum(\Delta\hat{H}_{st})(\Delta\hat{S} - \Delta\bar{S})}{(\Delta\hat{S} - \Delta\bar{S})^2} \tag{6.16}$$

$$Var\left(T_B\right) = \frac{\sum(\Delta\hat{H}_{st} - \Delta\bar{G}_B - \hat{T}_B\Delta\hat{S})^2}{(m-2)(\Delta\hat{S} - \Delta\bar{S})^2} \tag{6.17}$$

in which m is the number of data pairs of enthalpy and entropy; $\Delta\bar{H}_{st}$ is the mean enthalpy, kJ kg^{-1}; and $\Delta\bar{S}$ is the mean entropy, kJ kg^{-1} K^{-1}.

A linear compensation pattern only exists if $T_B \neq T_{hm}$. Moreover, if $T_B > T_{hm}$, the process is enthalpy driven, while the opposite condition is considered to be controlled by entropy.

The well-known Arrhenius equation (Equation 6.18) and the X_m values obtained through statistical procedures are used to obtain the values of E_a. This equation gives the relationship between activation energy and the velocity at which the reaction occurs:

$$X_m = A_0 \exp\left(-\frac{E_a}{RT_K}\right) \tag{6.18}$$

in which A_0 is the pre-exponential factor, % d.b.; and E_a is the activation energy, kJ kg^{-1}.

6.5.4 Drying

Corrêa et al. (2010b) obtained the thermodynamic properties regarding the drying rate for coffee fruits. These authors obtained the k values (drying rate) through fitting of drying curves (moisture ratio) using known models such as Page, Midili, and Verna, and applied the method described by Jideani and Mpotokwana (2009) (Equations [6.19], [6.20], and [6.21]) to acquire the thermodynamic parameters of drying process:

$$\Delta H^* = E_a - RT \tag{6.19}$$

$$\Delta S^* = R\left(\ln A_0 - \ln\frac{k_B}{h_p} - \ln T_K\right) \tag{6.20}$$

$$\Delta G^* = \Delta H^* - T\Delta S^* \tag{6.21}$$

in which ΔH^* is enthalpy, J mol^{-1}; ΔS^* is entropy, J mol^{-1}; ΔG^* is Gibbs free energy, J mol^{-1}; k_B is the Boltzmann constant, 1.38×10^{-23} J K^{-1}; and h_P is the Planck constant, 6.626×10^{-34} J s^{-1}.

REFERENCES

Al Hodali, R. 1997. Numerical simulation of an agricultural foodstuffs drying unit using solar energy and adsorption process. PhD Thesis, Universite´ Libre de Bruxelles, Belgium.

Annamalai, K., and I.K., Puri. 2002. *Advanced thermodynamics engineering*. Boca Raton, FL: CRC Press.

Atkins, W.P. 2000. *Chimie physique*, 6th ed., De Boeck Université, Paris, Bruxelles (Chapter 28).

Bell, R.P. 1937. Relations between the energy and entropy of solution and their significance. *Transactions of the Faraday Society* 33: 496–501.

Brunauer, S., L.S. Deming, W.E. Deming, and E. Teller. 1940. On a theory of the van der Waals adsorption of gases. *Journal of American Chemistry Society* 62: 1723–1732.

Corrêa, P.C., A.L.D. Goneli, C. Jarén, D.M. Ribeiro, and O. Resende. 2007. Sorption isotherms and isosteric heat of peanut pods, kernels and hulls. *Food Science and Technology International* 13: 231–238.

Corrêa, P.C., A.L.D. Goneli, P.C. Afonso Jr., G.H.H. Oliveira, and D.S.M. Valente. 2010a. Moisture sorption isotherms and isosteric heat of sorption of coffee in different processing levels. *International Journal of Food Science and Technology* 45: 2016–2022.

Corrêa, P.C., Oliveira, G.H.H., Botelho, F.M., Goneli, A.L.D., and F.M. Carvalho. 2010b. Modelagem matemática e determinação das propriedades termodinâmicas do café (Coffea arabica L.) durante o proceso de secagem. Revista Ceres 57(5): 595-601.

Corrêa, P.C., G.H.H. Oliveira, A.P.L. Rodrigues, S.C. Campos, and F.M. Botelho. 2010c. Hygroscopic equilibrium and physical properties evaluation affected by parchment presence of coffee grain. *Spanish Journal of Agricultural Research* 8(3): 694–702.

Fasina, O., S. Sokhansanj, and R. Tyler. 1997. Thermodynamics of moisture sorption in alfalfa pellets. *Drying Technology* 15: 1553–1570.

Furmaniak, S., A.P. Terzik, and P.A. Gauden. 2007. The general mechanism of water sorption on foodstuffs: Importance of the multitemperature fitting of data and the hierarchy of models. *Journal of Food Engineering* 82: 528–535.

Goneli, A.L.D., P.C. Corrêa, G.H.H. Oliveira, C.F. Gomes, and F.M. Botelho. 2010a. Water sorption isotherms and thermodynamic properties of pearl millet grain. International Journal of *Food Science and Technology* 45: 828–838.

Goneli, A.L.D., P.C. Corrêa, G.H.H. Oliveira, and F.M. Botelho. 2010b. Water desorption and thermodynamic properties of okra seeds. *Transactions of the ASABE* 53(1): 191–197.

Heyrovsky, J. . 1970. Determination of isokinetic temperature. *Nature* 227: 66–67.

Incropera, F.P., and D.P. Dewitt. 2002. *Fundamentals of heat and mass transfer*. 5th ed. New York: John Wiley and Sons.

Jideani, V.A., and S.M. Mpotokwana. 2009. Modeling of water absorption of botswana bambara varieties using Peleg's equation. *Journal of Food Engineering* 92(2): 182–188.

Lahsasni, S., M. Kouhila, M. Mahrouz, and M. Fliyou. 2003. Moisture adsorption-desorption isotherms of prickly pear cladode (*Opuntia ficus indica*) at different temperatures. *Energy Conversion and Management* 44: 923–936.

Lai, V.M., C. Lii, W. Hung, and T. Lu. 2000. Kinetic compensation in depolymerisation of food polysaccharides. *Food Chemistry* 68: 319–325.

Laidler, K.J. 1959. Thermodynamics of the ionization process. Part. 1: General theory of substituent effects. *Transactions of the Faraday Society* 55: 1725–1730.

Lewicki, P.P. 2004. Water as the determinant of food engineering properties. A review. *Journal of Food Engineering* 61: 483–495.

Martinazzo, A.P., P.C. Corrêa, O. Resende, and E.C. Melo. 2007. Análise e descrição matemática da cinética de secagem de folhas de capim limão. *Revista Brasileira de Engenharia Agrícola e Ambiental* 11(3): 301–306.

Moran, M.J., and H.N. Shapiro. 2006. *Fundamentals of engineering thermodynamics*. New York: John Wiley and Sons.

Moreira, R., F. Chenlo, M.D. Torres, and N. Vallejo. 2008. Thermodynamic analysis of experimental sorption isotherms of loquat and quince fruits. *Journal of Food Engineering* 88(4): 514–521.

Mujumdar, A.S. 2006. Handbook of industrial drying, 3rd ed. Boca Raton, FL: CRC Press.

Mulet, A., P. García-Pascual, N. Sanjuán, and J. García-Reverter. 2002. Equilibrium isotherms and isosteric heats of morel (*Morchella esculenta*). *Journal of Food Engineering* 53: 75–81.

Oliveira, G.H.H., P.C. Corrêa, E.F. Araujo, D.S.M. Valente, and F.M. Botelho. 2010. Desorption isotherms and thermodynamic properties of sweet corn cultivars (*Zea mays* L.). *International Journal of Food Science and Technology* 45: 546–554.

Perdomo, J., A. Cova, A.J. Sandoval, L. García, E. Laredo, and A.J. Müller. 2009. Glass transition temperatures and water sorption isotherms of cassava starch. *Carbohydrate Polymers* 76: 305–313.

Rizvi, S.S.H., and A.L. Benado. 1984. Thermodynamic properties of dehydrated foods. *Food Technology* 38: 83–92.

Simal, S., Femenia, A., Llul, P., and C. Roselló. 2000. Dehydration of aloe vera: Simulation of drying curves and evaluation of functional properties. *Journal of Food Engineering* 43:109–104.

Simal, S., A. Femenia, A. Castell-Palou, and C. Roselló. 2007. Water desorption thermodynamic parameters of pineapple. *Journal of Food Engineering* 80: 1293–1301.

Telis, V.R.N., A.L. Gabas, F.C. Menegalli, and J. Telis-Romero. 2000. Water sorption thermodynamic properties applied to persimmon skin and pulp. *Thermochimica Acta* 343: 49–50.

Tolaba, M.P., M. Peltzer, N. Enriquez, and M.L. Pollio. 2004. Grain sorption equilibria of quinoa grains. *Journal of Food Engineering* 61: 365–371.

Van den Berg, C. 1984. Description of water activity of foods for engineering purposes by means of the GAB model of sorption. In Engineering and food, ed. B.M. McKenna, 311–321. New York: Elsevier Applied Science.

Wang, N., and J.G. Brennan. 1991. Moisture sorption isotherm characteristics of potato at four temperatures. *Journal of Food Engineering* 14: 269–287.

Wolf, M., J.E. Walker, and J.G. Kapsalis. 1972. Water sorption hysteresis in dehydrated food. *Journal of Agricultural and Food Chemistry* 20: 1073–1077.

Wolf, W., W.E.L. Spiess, and G. Jung. 1985. Standardization of isotherm measurements (COST-project 90 and 90 bis). In *Properties of water in foods in relation to quality and stability*, ed. D. Simatos and J.L. Multon, 661–679. Dordrecht: Martinus Nijhoff.

7 Flow Properties of Foods

Paulo Cesar Corrêa, Emílio de Souza Santos, and Pedro Casanova Treto

CONTENTS

7.1 INTRODUCTION

Most operations of separation, conservation, and processing of agricultural products use air or water as a transportation medium of mass (impurity and water) and/or energy. Therefore, we need to know how the flow passes through the product, in addition to the transfer phenomena and thermodynamics of the fluid and the product, in order to correct handling and simulation of these operations.

7.2 DRAG OF SOLID FOODS

The external flow passing through the product is important in separation operations, which comprise studies of the boundary layer, drag forces, and terminal velocity. Knowledge of terminal velocity and the impurities associated with a product is of fundamental importance in the development of separation machinery that uses a fluid to remove impurities.

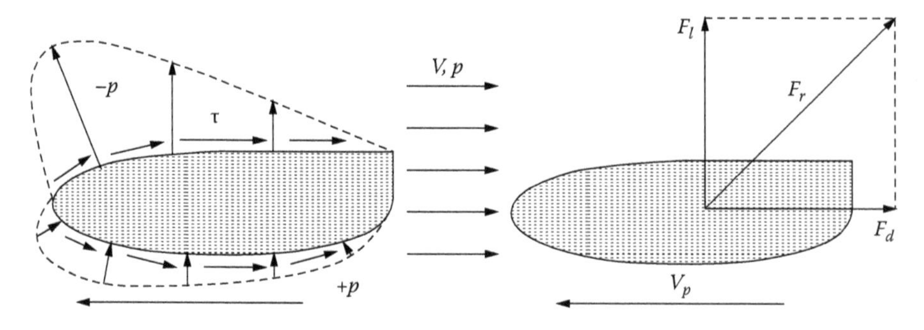

FIGURE 7.1 Flow through an immersed body.

7.2.1 Mechanics

Whenever there is relative motion between a solid body and a viscous fluid surrounding it, the body will experience tangential stresses at its surface due to viscous action and normal stresses as a result of local pressure. The resultant force, F_r, at this surface is resolved into the drag force, F_d, acting parallel to the direction of relative motion, and lift force, F_1, acting perpendicular to the motion. A scheme is shown in Figure 7.1.

The equation to calculate the drag force has been derived by dimensional analysis, considering the force function of the surface area of the product, A; the relative velocity, V; the fluid density, ρ_f; and the fluid viscosity, η. Thus, the drag force is given by Equation (7.1).

$$F_d = \frac{CA_p\rho_f V^2}{2} \tag{7.1}$$

in which

F_d is the drag force, N
C is the drag coefficient, dimensionless
A_p is the projected area of the object, m^2
ρ_f is the fluid density, kg m^{-3}
V is the relative velocity between the object and the fluid, m s^{-1}

In common separation systems, the particles are separate when one particle moves in a different direction than other particles, due to the difference between its drag forces. As discussed, the drag force is function of the relative velocity between the body and the fluid, which turn the measure of this force hard to be done. For this reason, terminal velocity (V_t) is used as a parameter in designing separation systems.

When a body is moving though a fluid, due to the action of constant force (e.g., centripetal or gravitational force), a drag force arises as a reaction. When the vectors of these forces are in equilibrium, the body acceleration is annulated, and begins movement at a constant velocity called *terminal velocity* (Mohsenin, 1986). A scheme is shown in Figure 7.2.

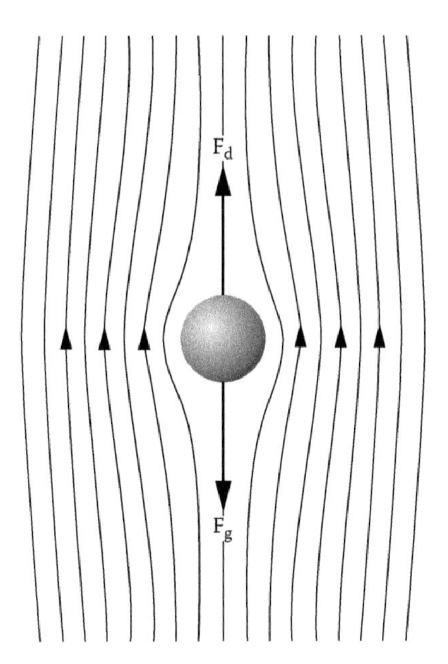

FIGURE 7.2 Force diagram of a body in a flow when it reaches terminal velocity.

In that way, when the drag force and the force due to gravity are in equilibrium, the relative velocity between the fluid and the object is equal to the terminal velocity of the object. Thus, from Equation (7.1), Equations (7.2) and (7.3) can be obtained. These equations relate the drag coefficient and the terminal velocity with the remaining parameters:

$$F_g = m_p g \left[\frac{(\rho_p - \rho_f)}{\rho_p} \right]$$

$$F_d = \frac{C A_p \rho_f V_t^2}{2}$$

in which

ρ_p is the particle density (or real), kg m^{-3}
g is the gravity acceleration, m s^{-2}
m_p is the particle mass, kg
V_t is the terminal velocity, m s^{-1}

Thus:

$$m_p g \left[\frac{(\rho_p - \rho_f)}{\rho_p} \right] = \frac{C A_p \rho_f V_t^2}{2}$$

Rearranging this equation:

$$V_t = \left[\frac{2W(\rho_p - \rho_f)}{\rho_p \rho_f A_p C} \right]^{1/2} \tag{7.2}$$

or

$$C_t = \frac{2W(\rho_p - \rho_f)}{V_t^2 A_p \rho_p \rho_f} \tag{7.3}$$

in which W is the particle weight, N.

Notice that the drag coefficient and terminal velocity are intimately connected and cannot be determined individually through these relationships. The drag coefficient is a dimensionless fluid dynamic that depends upon the shape and size of a particle immersed in a flow. Two objects with the same area moving at the same velocity are submitted to a drag force proportional to their drag coefficients; solid objects can obtain higher values than hollow objects. In order to estimate the terminal velocity of a grain, Equation (7.2) can be used along with average values of the drag coefficient obtained experimentally.

The drag coefficient is a function of shape, roughness, Reynolds number, orientation, and other parameters. In general, determination of the drag coefficient must be experimental. Many models have been presented in the literature, most of them validated with experimental values for each product and conditions.

7.2.2 Experimental Techniques

For practical purposes, in airflow separation systems, the terminal velocity can be experimentally obtained when the grain is floating in air at a constant height inside an acrylic tube connected to a fan, by measuring the air velocity. An example of such equipment is shown in Figure 7.3. Most agricultural products present a decrease of terminal velocity with a decrease in moisture content. Table 7.1 shows some examples of drag coefficient values.

According to Couto et al. (2003), determination of experimental values of terminal velocity of a body is frequently accomplished in two different ways: (a) determining the required velocity so that a body is able to float in an ascendant air flow; and (b) through the displacement measurement, as a function of time of the free fall of a particle. This movement can be defined by Equation (7.4):

$$m\frac{d^2x}{dt^2} = mg - F_d \tag{7.4}$$

in which

- m is the particle mass, kg
- x is the particle displacement, m
- g is the gravity acceleration, m s^{-2}
- F_r is the resistant force, N

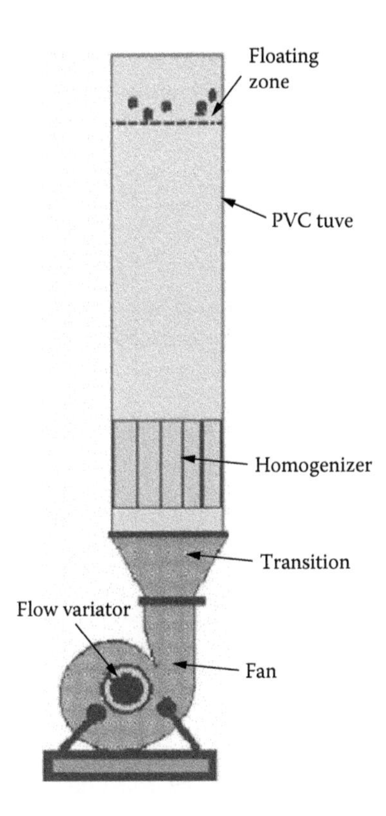

FIGURE 7.3 Scheme of the system used to determine the terminal velocity of solid foods.

TABLE 7.1
Some Examples of the Drag Coefficient

Airplanes, Buildings, and Agricultural Products	Drag Coefficient
Learjet 24	0.022
Boeing 747	0.031
Empire State Building	1.3–1.5
Eiffel Tower	1.8–2.0
Corn	0.56
Soybeans	0.45
Alfalfa seeds	0.50
Cotton seeds	0.52
Barley seeds	0.50

Source: Bilanski et al., 1962

When the particle reaches terminal velocity, the analytic solution for Equation (7.4) is Equation (7.5). This equation can be used to calculate the displacement when the product is submitted to an ascendant flow.

$$x = \frac{V_t}{g} \ln\left[\cosh\left(\frac{g}{V_t} t\right)\right] \tag{7.5}$$

Another method to estimate the terminal velocity and drag coefficient of grains is by using an abacus developed by Mohsenin (1986), shown in the Appendix of this chapter. They relate the products CN_R^2 along with N_R to different shapes of the object when the following are known:

1. Physical properties of the product:
   - A_p is the projected area of the object, m^2
   - ρ_p is the particle density, kg m^{-3}
   - W is the particle weight, N
   - D is the equivalent diameter, m
2. Fluid properties:
   - η is the fluid viscosity, kg m^{-1} s^{-1}
   - ρ_f is the fluid density, kg m^{-3}

It is known that the relationship between the Reynolds number and terminal velocity of a sphere of diameter D is given by Equation (7.6). The projected area of a sphere is given by Equation (7.7).

$$N_R = \frac{V_t D \rho_f}{\eta} \tag{7.6}$$

$$A_p = \frac{\pi D^2}{4} \tag{7.7}$$

Thus, it is possible to obtain Equation (7.8), which relates the products CN_R^2 to a sphere by combining Equations (7.3), (7.6), and (7.7), as illustrated here:

$$CN_R^2 = \frac{2W(\rho_p - \rho_f)}{V_t^2 A_p \rho_p \rho_f}\left(\frac{V_t D \rho_f}{\eta}\right)^2$$

$$CN_R^2 = \frac{2W(\rho_p - \rho_f)}{V_t^2 \frac{\pi D^2}{4} \rho_p \rho_f}\left(\frac{V_t D \rho_f}{\eta}\right)^2 \tag{7.8}$$

$$CN_R^2 = \frac{2W \rho_f (\rho_p - \rho_f)}{\pi \eta^2 \rho_p}$$

The values of CN_R^2 of objects that present a high sphericity index (Equation [7.9]) can be obtained by Equation (7.12), which considers the object's weight (Equation [7.10]) and its equivalent diameter (Equation [7.11]).

$$\Phi = \frac{(abc)^{1/3}}{a} \tag{7.9}$$

$$W = \frac{\pi \rho_p g D^2}{6} \tag{7.10}$$

$$D = (abc)^{1/3} \tag{7.11}$$

$$CN_R^2 = \frac{4gD^3\rho_f\left(\rho_p - \rho_f\right)}{3\eta^2} \tag{7.12}$$

in which

Φ is the sphericity, dimensionless

a, b, and c are the characteristic dimensions of the particle

After determining the CN_R^2 value, the Reynolds number is estimated through the use of the abacus mentioned previously. Utilizing this Reynolds number, the terminal velocity of the object is obtained by means of Equation (7.6). The drag coefficient is estimated by dividing the CN_R^2 by the Reynolds number squared.

Several researchers obtained different values of terminal velocity as a function of moisture content (M), as shown in the Appendix at the end of this chapter. The following is an example of the method used in this work to determine values of the terminal velocity of agricultural products.

7.2.2.1 Example

The projected area, terminal velocity, and drag coefficient of a certain agricultural product, having moisture content ranging from 10.7 to 26.6 % w.b, has been calculated from the data shown in Table 7.2, which have been experimentally obtained. This has been done using the abacus and reflects air viscosity equal to 1.83×10^{-5}, air density equal to 1.185 kg m^{-3}, and gravity acceleration equal to 9.81 m s^{-2}.

7.2.2.2 Solution

The projected area of the product can be approximated to the area of a sphere with diameter

$$D\left(A_p = \frac{\pi D^2}{4}\right)$$

Thus, it is possible to obtain the data shown in Table 7.3. Using Equation (7.14), the products CN_R^2 are obtained, as shown in Table 7.4. Through the abacus shown

TABLE 7.2
Experimental Data for a Certain Agricultural Product

Moisture Content (% w.b.)	Sphericity (%)	D (mm)	ρ_p (kg m^{-3})	A_p (m^2)
26.6	85	10.79	932.53	
22.7	87	10.69	888.90	
15.1	83	10.58	874.44	
13.6	82	10.55	738.62	
10.7	82	10.53	690.02	

TABLE 7.3
Projected Area Calculated from Data Shown in Table 7.2

Moisture Content (% w.b.)	Sphericity (%)	D (mm)	A_p (x 10^{-5} m^2)
26.6	85	10.79	9.14
22.7	87	10.69	8.98
15.1	83	10.58	8.79
13.6	82	10.55	8.74
10.7	82	10.53	8.71

TABLE 7.4
CN_R^2 Calculated from Data Shown in Table 7.2

Moisture Content (% w.b.)	D (m)	CN_R^2 (× 10^7)
26.6	0.01079	5.41
22.7	0.01069	5.01
15.1	0.01058	4.78
13.6	0.01055	4.00
10.7	0.01053	3.72

in Figure 7.4, N_R values are determined and are shown in Table 7.5. Using N_R in Equation (7.8), V_t can be estimated (see Table 7.6).

The division between CN_R^2 and N_R^2 results in the drag coefficient (C). Table 7.7 shows the obtained values.

The behavior of terminal velocity of the product as a function of moisture content is shown in Figure 7.5.

Drag coefficient can be calculated based on the Wadell equation, cited by West (1972). This author established drag coefficient curves versus the Reynolds number

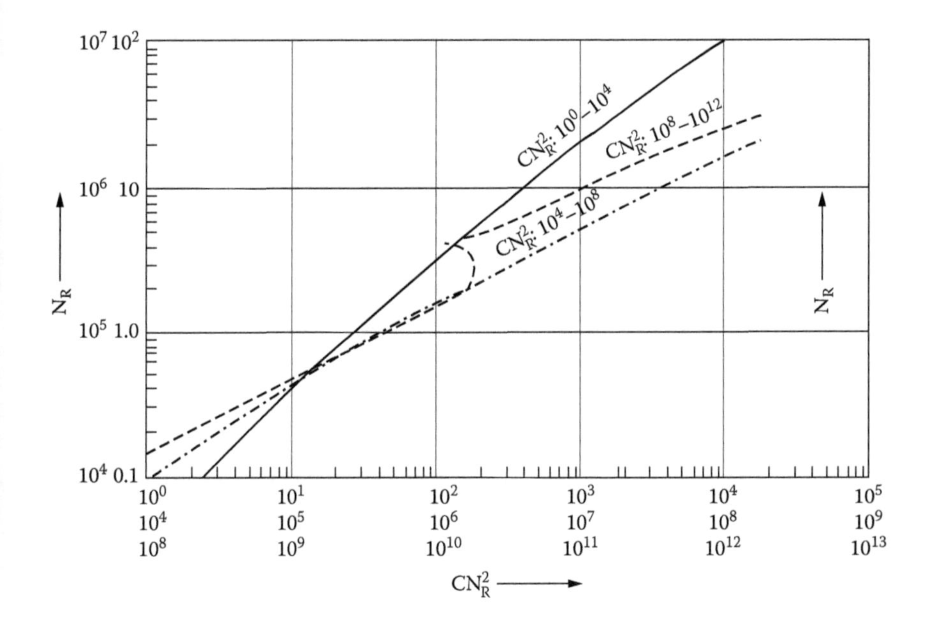

FIGURE 7.4 Reynolds number, NR, versus CNR2 for spheres (From Mohsenin, N.N. 1986. *Physical properties of plant and animal materials*. New York: Gordon and Breach Publishers.)

TABLE 7.5
N_R Obtained through the Abacus

Moisture Content (% w.b.)	CN_R^2 (× 10^7)	N_R
26.6	5.41	12860
22.7	5.01	11430
15.1	4.78	11070
13.6	4.00	10280
10.7	3.72	10000

TABLE 7.6
V_t Values Calculated Using Equation 7.8

Moisture Content (% w.b.)	CN_R^2 (× 10^7)	N_R	V_t (m s^{-1})
26.6	5.41	12860	18.4
22.7	5.01	11430	16.5
15.1	4.78	11070	16.2
13.6	4.00	10280	15.1
10.7	3.72	10000	14.7

TABLE 7.7
Drag Coefficient of the Product

Moisture Content (% w.b.)	CN_R^2 (× 10^7)	N_R	C
26.6	5.41	12860	0.33
22.7	5.01	11430	0.38
15.1	4.78	11070	0.39
13.6	4.00	10280	0.38
10.7	3.72	10000	0.37

(N_R), in different sphericities. The result was an empirical equation with N_R values between 2,000 and 20,000, as a function of sphericity (Equation [7.13]).

$$C = 5.31 - 4.88\ \Phi \tag{7.13}$$

Becker (1959) proposed an equation for the drag coefficient that depends on shape, orientation, and the Reynolds number. For a Reynolds number above of 2,000, the relationship is expressed by Equation (7.14).

$$C = 2.53 - 2.83\ e^{2\Phi} \tag{7.14}$$

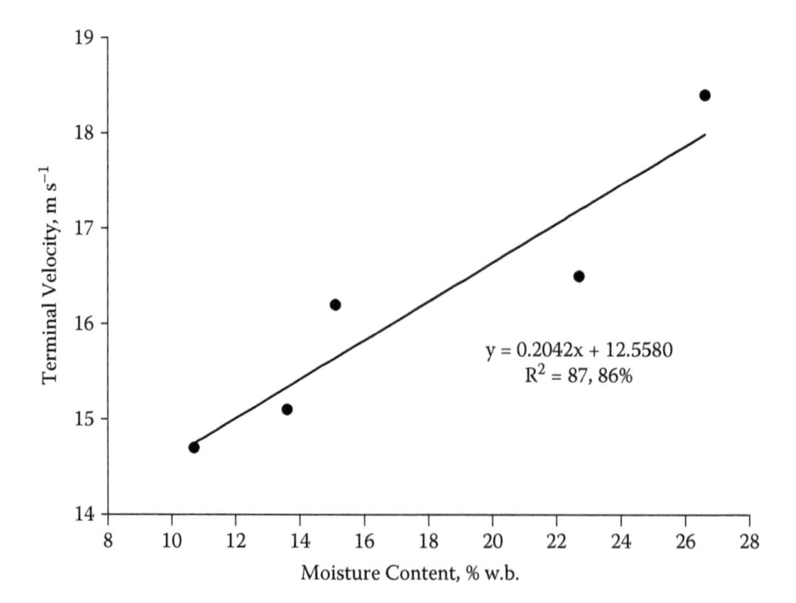

FIGURE 7.5 Terminal velocity of the product as a function of moisture content.

7.3 FRICTION LOSSES ON FOOD PROCESSES

The internal flow passing through pipes and porous media is important in fluid transport, especially in the design of pumps, fans, and blowers. The mechanics between the solid surface and the fluid decrease the energy associated with the flow, which is called *friction loss*. As will be shown, the friction loss is related to solid surface and fluid properties. So, depending on the operation and the food state, the product can be considered as a fluid (oils, juices) or a solid media (grains, powders).

7.3.1 Fluid Dynamics

Any fluid moves from levels of higher energy to levels of lower energy. This concept is based on energy conservation. In a fluid system, the energy in a point has to be equal in another point, adding the transferred energy to another system. This is the Bernoulli concept and is defined in Equation (7.15) for an incompressible flow:

$$h_1 + \frac{P_1}{\gamma} + \frac{V_1^2}{2g} + W - F = h_2 + \frac{P_2}{\gamma} + \frac{V_2^2}{2g} \quad (7.15)$$

in which h is the fluid height, m; p is the pressure, Pa; γ is the fluid specific weight, N m^{-3}; V the velocity, m s^{-1}; g is the gravity, 9,81 m s^{-2}; W is the energy added to the system, J N^{-1}; and F is the energy loss due to friction, J N^{-1}.

The friction term F in the Bernoulli equation represents the fluid dynamic energy loss due to effects such as internal fluid resistance, turbulence, and resistance of the flow retainer. The Darcy equation, shown as Equation (7.16), is the most widely used equation for determining friction loss in a circular pipe:

$$F = f \frac{L}{D_p} \frac{V^2}{2g} \quad (7.16)$$

in which f is the friction factor; L is the length of pipe, m; and D_p is the internal diameter of pipe, m.

The friction factor is a function of the Reynolds number (N_R), representing the fluid resistance and the specific roughness (e/D_p), representing the external resistance offered by the retainer. If $N_R < 2130$, then the flow is laminar, and the friction factor must be calculated using Equation (7.3). If $N_R > 2130$, the flow is turbulent, and the friction factor must be calculated by more complex equations, like the Colebrook-White equation (Equation [7.17]). The roughness factor (e) for various materials are given in Table 7.8:

$$f = \frac{64}{N_R}(3) \frac{1}{\sqrt{f}} = -2\log_{10}\left(\frac{e}{3,7D_p} + \frac{2,51}{N_R\sqrt{f}}\right) \quad (7.17)$$

in which e is the roughness factor, mm.

TABLE 7.8
Roughness Factor for Various Types of Pipes

Material	e (mm)
Riveted steel	0.9–9.0
Concrete	0.3–3.0
Wood stave	0.2–0.9
Cast iron	0.26
Galvanized iron	0.15
Asphalted cast iron	0.12
Commercial steel	0.046
Drawn tubing	0.0015

Source: Adapted from Henderson, S.M., R.L. Perry, and J.H. Young. 1997. *Principles of process engineering*, 4th ed., 353. St. Joseph, MI: ASAE.

When the pipe is not circular, the hydraulic diameter (Equation [7.18]) is used instead of the simple diameter, as an approximation. The friction loss due to valves, fittings, and contractions is generally given by the manufacturer as an equivalent length (L_e/D). This parameter must be added to the Darcy equation to calculate the total friction losses of the system:

$$D_h = \frac{4A}{P} \tag{7.18}$$

in which A is the transversal area of the pipe, m^2; and P is the perimeter of the pipe, m.

As previously discussed, the friction factor is dependent on the Reynolds Number, which is dependent on density and viscosity. The greatest challenge for food technologists is determining these properties for all types of products and processes, since these properties vary greatly depending on the product constitution and the process changes. In Chapters 1 and 2, many techniques were presented to measure density and viscosity of foods, in order to fill this gap.

7.3.2 Fluid Dynamics through Porous Media

When the product is a solid material and it is processed as a mass inside a volume, this product acts as a porous media that offers resistance to any fluid that crosses it. Therefore, the friction losses for this fluid increase as compared to the same empty volume.

7.3.2.1 Fundamentals

In 1856, Darcy published the first experimental work on flow through porous media. Based on a large amont of experimental data, he proposed a modification on the momentum equation of flow through a pipe, which is defined by Equation (7.19) (De Wiest, 1969):

$$\frac{dp}{dx} = \frac{\eta}{k}V \tag{7.19}$$

in which *dp/dx* is the gradient pressure; Pa m^{-1}; and *k* is the medium permeability, m^2.

This relationship is valid only for a laminar flow through a porous media. It was verified that for a turbulent flow, this relationship is not valid. Scheideger (1973) suggested the use of an expression given by the quadratic form of the Forchheimer equation (Equation [7.20]), which can be applied to both laminar and turbulent flow:

$$\frac{dp}{dx} = \frac{\eta}{k}\left[1 + \frac{f_g \rho_f \sqrt{k}}{\eta}|V|\right]V \tag{7.20}$$

in which f_g is the particle geometric factor.

The medium permeability is the most important property; it shows the fluid flow facility through a porous media. Kozeny (quoted by Ramesh and Moshenin, 1980) developed an equation to estimate the permeability (Equation [7.21]). He based it on an analogy that the flow through a porous media is similar to a flow through various pipes, with a depth equal to the media depth and volume equal to the media void volume. Subsequently, Càrman (quoted by Ramesh and Moshenin, 1980) fitted this equation to experimental data and observed an average value of β of 5:

$$k = \frac{\varepsilon_p^3}{\beta(1-\varepsilon_p)^2 s_p^2} \tag{7.21}$$

in which β is the Kozeny constant; s_p is the particle specific surface; $m^2\ m^{-3}$; and ε_p is the external porosity of product mass.

Ergun (1952) related the coefficient f_g with porosity (Equation [7.22]), based on experiment data. In addition, he suggested a general expression (Equation [7.23]) for friction loss on porous media, modifying the Darcy equation for friction loss in pipes:

$$f_g = \frac{0{,}143}{\varepsilon_p^{3/2}} \tag{7.22}$$

$$f = \frac{1-\varepsilon_p}{\varepsilon_p^3}\left(\frac{300(1-\varepsilon_p)}{N_R} + 3{,}5\right) \tag{7.23}$$

The particle equivalent diameter is used to calculate both friction loss on the Darcy Equation (Equation [7.16]) and the Reynolds number (Equation [7.6]). When the particle has a regular shape, Equation (7.13) is recommended. If the particles have irregular shapes, the equivalent diameter must be calculated using Equation (7.24):

$$D = \frac{6v_p}{S_p} \tag{7.24}$$

in which v_p is the particle volume, m³.

Bakker-Arkema et al. (1969) compared the experimental data of friction loss of a cherry mass to the values given by the Ergun equation, observing a constant relation. They suggested multiplying the Ergun equation by an empiric coefficient, function of the product type. Such difference is probably due to the product roughness, which is neglected in the Ergun equation, and the different spatial arrangement, which varies by product. Both assumptions could modify the coefficients c and β. More studies about the influence of shape, roughness, and porosity of irregular particles should be done in order to develop a general theoretical equation that represents the flow through an irregular porous media.

7.3.2.2 Empiric Equations

Many researchers suggested using empiric equations due to the facility for working with these equations and the nonexistence of a general equation to describe the flow through porous media. Hukill and Ives (1955), examining the experimental data of Shedd (1951, 1953), estimated the nonlinearity of it and suggested Equation (7.25) to describe the relation between air velocity and the pressure gradient:

$$\frac{dp}{dx} = \frac{aV^2}{\ln(1+bV)} \tag{7.25}$$

in which a and b are the empiric coefficients.

Resistance data observed by various researchers, and the coefficient of the Hukill and Ives equation adjusted to these data, was included as ASAE Data D272 in the *ASAE Standards* (2001). These coefficients are shown in Table 7.9.

The resistance data presented in the *ASAE Standards* are for a cleaning and nondense layer of product. However, in reality the friction loss is higher due to the layer densification caused by the handling operation and the product's impurity. Therefore, it is common practice to apply a correction factor in order to more accurately predict the pressure drop in a real situation. These factors normally have a value between 1.1 and 1.5. Thus, the total friction loss of a grain layer can be calculated by Equation (7.26):

$$F = \frac{aC_f LV^2}{\gamma \ln(1+bV)} \tag{7.26}$$

in which C_f is the correction factor.

The Hukill and Ives empiric equation is limited for specific conditions of the experiment, and it is not possible to vary the other properties like porosity, shape, size, or fluids. Since some of the variation in these properties is important, the use of a theoretical equation, like the Ergun equation, is more indicated. An example is

TABLE 7.9
Values for Constants in Airflow Resistance Equation

Material	*a* (10^3 Pa s^2 m^{-3})	*b* (m^2 s m^{-3})	Range of *V* (m^3 m^{-2} s^{-1})
Alfalfa	64	3.99	0.0056–0.152
Alfalfa cubes	1.27	22.99	0.13–3.15
Alfalfa pellets	18	68.72	0.0053–0.63
Barley	21.4	13.2	0.0056–0.203
Bromes Grass	13.5	8.88	0.0056–0.152
Canola, Tobin	52.2	7.27	0.0243–0.2633
Canola, Westar	45.5	9.72	0.0243–0.2633
Clover, alsike	61.1	2.24	0.0056–0.101
Clover, crimson	53.2	5.12	0.0056–0.203
Clover, red	62.4	3.55	0.0056–0.152
Corn, ear	10.4	325	0.051–0.353
Corn, shelled	20.7	30.4	0.0056–0.304
Corn, shelled low airflow	9.77	8.55	0.00025–0.0203
Fescue	31.5	6.7	0.0056–0.203
Flax	86.3	8.29	0.0056–0.152
Lentils	54.3	36.79	0.0028–0.5926
Lespedeza, Kobe	19.5	6.3	0.0056–0.203
Lespedeza, Sericea	64	3.99	0.0056–0.152
Lupine, blue	10.7	21.1	0.0056–0.152
Milkweed pods	2.11	4.65	0.06–0.4
Oats	24.1	13.9	0.0056–0.203
Peanuts	3.8	111	0.03–0.304
Peppers, Bell	0.544	868	0.03–1
Popcorn, White	21.9	11.8	0.0056–0.203
Popcorn, yellow	17.8	17.6	0.0056–0.203
Potatoes	2.18	824	0.03–0.3
Rescue	8.11	11.7	0.0056–0.203
Rice, rough	25.7	13.2	0.0056–0.152
Rice, long brown	20.5	7.74	0.0055–0.164
Rice, long milled	21.8	8.34	0.0055–0.164
Rice, medium brown	34.9	10.9	0.0055–0.164
Rice, medium milled	29	10.6	0.0055–0.164
Sorghum	21.2	8.06	0.0056–0.203
Soybeans	10.2	16	0,0056–0.304
Sunflower, confectionery	11	18.1	0.055–0.178
Sonflower, oil	24.9	23.7	0.025–0.570
Swett Potatoes	3.4	6.1 10^8	0.05–0.499
Wheat	27	8.77	0.0056–0.203
Wheat, low airflow	8.41	2.72	0.00025–0.0203

Source: Adapted from ASAE (American Society of Agricultural Engineers). *ASAE Standards 2001: Standards engineering practices data*. St. Joseph, MI: ASAE.

the case of drying operations, during which the porosity can vary a lot; or a cleaning operation, during which water is used as the fluid.

The friction loss due to a perforated floor or wall is very common in agricultural operations, such as cleaning operations and aeration. Henderson (1943) developed the experimental Equation (7.27) for friction loss due to perforated floors:

$$F = \frac{1{,}071\, Pa\, s^2 m^{-2}}{\gamma} \left(\frac{V}{\varepsilon_c \varepsilon_p} \right)^2 \tag{7.27}$$

in which ε_c is the ratio of the perforate area to the total area.

REFERENCES

Afonso, P.C., Jr., P.C. Corrêa, F.A.C. Pinto, and D.M Queiroz. 2007. Aerodynamic properties of coffee cherries and beans. *Biosystems Engineering* 98(1): 39–46.

ASAE (American Society of Agricultural Engineers). *ASAE Standards 2001: Standards engineering practices data*. St. Joseph, MI: ASAE.

Aydin, C. 2003. Physical properties of almond nut and kernel. *Journal of Food Engineering* 60(3): 315–320.

Bakker-Arkema, F.W., R.J. Patterson, and W.G. Bicker. 1969. Static pressure-airflow relationships in packed beds of granular biological materials such as cherry pits. *Transaction of ASAE* 12(1): 134–136.

Baryeh, E.A. 2002. Physical properties of millet. *Journal of Food Engineering* 51(1): 39–46.

Baryeh, E.A., and B.K. Mangope. 2003. Some physical properties of QP-38 variety pigeon pea. *Journal of Food Engineering* 56(1): 59–65.

Becker, H.A. 1959. The effects of shape and Reynolds number on drag in the motion of a freely oriented body in an infinite fluid. *The Canadian Journal of Chemical Engineering* 37: 85–100.

Bilanski, W.K., Collins, S.K., and Chu, P. 1962. Aerodynamic properties of seed grains, *Transactions of the American Society of Agricultural Engineers* 8(1), pp. 49–52.

Çalişir, S., and C. Aydın. 2004. Some physico-mechanic properties of cherry laurel (*Prunus lauracerasus* L.) fruits. *Journal of Food Engineering* 65(1): 145–150.

Cetin, M. 2007. Physical properties of barbunia bean (Phaseolus vulgaris L. cv. 'Barbunia') seed. *Journal of Food Engineering* 80(1), pp. 353–358.

Coşkun, M.B., I. Yalçın, I. and C. Özarslan. 2006. Physical properties of sweet corn seed (*Zea mays saccharata Sturt.*). *Journal of Food Engineering* 74(4): 523–528.

Couto, S.M., A.C. Magalhães, D.M. Queiroz, and I.T. Bastos. 2003. Parâmetros relevantes na determinação da velocidade terminal de frutos de café. *Revista Brasileira de Engenharia Agrícola e Ambiental* 7(1): 141–148.

De Wiest, R.J.M. 1969. *Flow through porous media*, 530. New York: Academic Press.

Dursun, E., and I. Dursun. 2005. Some physical properties of caper seed. *Biosystems Engineering* 92(2): 237–245.

Dursun, İ., K.M. Tuğrul, and E. Dursun. 2007. Some physical properties of sugarbeet seed. *Journal of Stored Products Research* 43(2): 149–155.

Ergun, S. 1952. Fluid flow through packed columns. *Chemical Engineering Progress* 48(9): 89–94.

Gupta, R.K., and S.K. Das. 1997. Physical properties of sunflower seeds. *Journal of Agricultural Engineering Research* 66(1): 1–8.

Henderson, S.M. 1943. Resistance of shelled corn and bin walls to airflow. *Journal of Agricultural Engineering* 24(11): 367–369.

Henderson, S.M., R.L. Perry, and J.H. Young. 1997. *Principles of process engineering*, 4th ed., 353. St. Joseph, MI: ASAE.

Hukill, W.V., and N.C. Ives. 1955. Radial airflow resistance of grain. *Journal of Agricultural Engineering*. 36(5): 332–335.

Işik, E. and H. Ünal. 2007. Moisture-dependent physical properties of white speckled red kidney bean grains. *Journal of Food Engineering* 82(2): 209–216.

Joshi, D.C., S.K. Das, and R.K. Mukherjee. 1993. Physical properties of pumpkin seeds. *Journal of Agricultural Engineering Research* 54(3): 219–229.

Kashaninejad, M., A. Mortazavi, A. Safekordi, and L.G. Tabil. 2006. Some physical properties of Pistachio (*Pistacia vera* L.) nut and its kernel. *Journal of Food Engineering* 72(1): 30–38.

Masoumi, A.A., A. Rajabipour, L. Tabil, and A.A. Akram. 2003. Terminal velocity and frictional properties of garlic (*Allium sativum* L.). CSAE/SCGR 2003 Meeting. Paper No. 03-330.

Mohsenin, N.N. 1986. *Physical properties of plant and animal materials*. New York: Gordon and Breach Publishers.

Rajabipour, A., A. Tabatabaeefar, and M. Farahani. 2006. Effect of moisture on terminal velocity of wheat varieties. *International Journal of Agriculture & Biology* 8(1): 10–13.

Ramesh, P., and N. N. Moshenin. 1980. Permeability of porous media as a function of porosity and particle size distribution. *Transaction of ASAE* 23(3): 742–745.

Sacilik, K., R. Öztürk, and R. Keskin. 2003. Some physical properties of hemp seed. *Biosystems Engineering* 86(2): 191–198.

Scheideger, A.E 1973. *The physics of flow through porous media*, 3rd ed., 353. Toronto: University of Toronto Press.

Shedd, C.K. 1951 Some new data on resistance of grains to air flow. *Journal of Agricultural Engineering* 32(9): 493–495.

Shedd, C.K. 1953. Resistance of grains and seeds to air flow. *Journal of Agricultural Engineering* 34(9), 616–619.

Silva, D.J.P., S.M. Couto, A.B. Peixoto, A.E.O. Santos, and S.M.J. Vieira. 2006. Determinação de características fluidodinâmicas do café. *Revista Brasileira de Armazenamento* Café (9): 56–66.

Silva, F.S., P.C. Corrêa, P.C. Afonso Jr., and A.L. Goneli. 2003. Influência do teor de umidade na velocidade terminal de grãos de sorgo e milheto. *Revista Brasileira de Milho e Sorgo* 2(3): 143–147.

Singh, K.P., H.N. Mishra, and S. Saha. 2010. Moisture-dependent properties of barnyard millet grain and kernel. *Journal of Food Engineering* 96(4): 598–606.

Suthar, S.H., and S.K. Das. 1996. Some physical properties of karingda (*Citrullus lanatus* [Thumb] *Mansf*) seeds. *Journal of Agricultural Engineering Research* 65(1): 15–22.

West, N.L. 1972. Aerodynamic force predictions. *Transactions of the ASAE* 15(3): 584–587.

Yalçın, I. 2007. Physical properties of cowpea (*Vigna sinensis* L.) seed. *Journal of Food Engineering* 79(1): 57–62.

Yalçın, İ., and C. Özarsla. 2004. Physical properties of vetch seed. *Biosystems Engineering* 88(4): 507–512.

Yalçın, İ., C. Özarslan, and T. Akbaş. 2007. Physical properties of pea (*Pisum sativum*) seed. *Journal of Food Engineering* 79(2): 731–735.

Zewdu, A.D. 2007. Aerodynamic properties of tef grain and straw material. *Biosystems Engineering* 98(3): 304–309.

APPENDIX

Product		Variety	Terminal Velocity	R^2	Citation
Almond	Nut		$V_t = 0.0209M + 7.5042$	0.930	Aydin (2003)
	Kernel		$V_t = 0.0726M + 5.596$	0.900	
Barbunia	Seed	Barbunia	$V_t = 11.258 + 0.1641M$	0.990	Centin (2007)
Caper	Seed		$V_t = 5.234 + 0.158M$	0.975	Dursun & Dursun (2005)
Coffee fruits	Green	Catauí	$V_t = 12.26 - 0.04539M - 0.001386M^2$	0.954	Couto et al. (2003)
	Cherry	Catauí	$V_t = 0.09432M + 10.10$	0.977	
	Green	Timor	$V_t = 9.391 + 0.08439M$	0.956	
	Cherry	Timor	$V_t = 12.01 - 0.02304M + 0.001366M^2$	0.930	
Coffee fruits	Cherry	Catuaí	$V_t = -3.293 - 0.361M + 0.0415\rho_p - 0.265 \times 10^{-4} \rho_p^2 + 0.411 \times 10^{-3}M\rho_p$	0.984	Afonso Júnior (2007)
	Bean	Catauí	$V_t = 35.904 - 0.508M + 0.0103M^2 + 0.102\rho_p - 0.519 \times 10^{-4}\rho_p^2$	0.986	
	Cherry	Conilon	$V_t = 7.837 + 0.0538M + 0.373 \times 10^{-2} \rho_p$	0.986	
	Bean	Conilon	$V_t = -17.948 - 0.539M + 0.0664 \rho_p - 0.408 \times 10^{-4} \rho_p^2 + 0.615 \times 10^{-3}M\rho_p$	0.990	
Coffee fruits	Pulpless	Catauí	$V_t = 0.04215M + 6.713$	0.764	Silva et al. (2006)
	Over ripe	Catauí	$V_t = 0.07647M + 10.406$	0.940	
Cowpea	Seed		$V_t = 9.2074 + 0.0108M$	0.970	Yalçın (2007)
Garlic	Cloves	White	$V_t = 0.09M + 8.14$	0.983	Masoumi et al. (2003)
	Cloves	Pink	$V_t = 0.14M + 6.75$	0.987	
Hemp	Seed		$V_t = \mathbf{4.9 + 0.074M}$	0.990	Sacilik et al. (2003)
Karingda	Seed		$V_t = 4.15 + 0.06M$	0.990	Suthar & Das (1996)
	Kernel		$V_t = 3.28 + 0.04M$	1.000	
Laurel Fruits	Cherry		$V_t = 0.0148M + 7.3076$	0.924	Çalısır & Aydın (2004)
Millet	Grain		$V_t = 2.251 + 0.101M$	0.978	Baryeh (2002)
Millet	Grain	Barnyard	$V_t = 3.99 + 2.36M + 27.60M^2$	0.990	Singh et al. (2010)
	Kernel	Barnyard	$V_t = 3.62 + 2.27M + 22.31M^2$	0.980	

Pea	Seed		$V_t = 8.8771 + 0.0149M$	0.982	Yalçın et al. (2007)
Pearl millet	Grain	BRS 1501	$V_t = 2.0551 + 0.0498M$	0.882	Silva et al. (2003)
Pigeon pea	Grain	QP-38	$V_t = 5.321 + 0.251M$	0.996	Baryeh & Mangope (2003)
Pistachio	Nut		$V_t = 0.0186M + 7.29$	0.791	Kashaninejad et al. (2006)
	Kernel		$V_t = 0.0210M + 6.31$	0.692	
Pumpkin	Seed		$V_t = 4.59 + 0.05M$	0.992	Joshi et al. (1993)
	Kernel		$V_t = 4.25 + 0.02M$	0.946	
Sorghum	Grain	BR 304	$V_t = 4.6063 + 0.0668M$	0.854	Silva et al. (2003)
Sugarbeet	Seed		$V_t = 4.32 + 0.163M$	0.970	Dursun et al. (2007)
Sunflower	Seed		$V_t = 5.37 + 0.112M$	0.981	Gupta & Das (1997)
	Kernel		$V_t = 2.86 + 0.15M$	0.993	
Sweet corn	Seed		$V_t = 5.239 + 0.0263M$	0.920	Coşkun et al. (2006)
Speckled Red Kidney Bean	Grain		$V_t = 7.0427 + 0.0974M$	0.900	Işik & Ünal (2007)
Teg	Grain		$V_t = 0.0363M + 2.8858$	0.980	Zewdu (2007)
Vetch	Seed		$V_t = 9.57 + 0.038M$	0.960	Yalçın et al. (2004)
Wheat	Seed	Mahdavi	$V_t = 6.65 + 0.03M$	0.950	Rajabipour et al. (2006)
	Seed	Marvdashat	$V_t = 5.67 + 0.04M$	0.910	
	Seed	Pishtaz	$V_t = 5.53M^{0.09}$	0.920	

8 The Acoustic Properties Applied to the Determination of Internal Quality Parameters in Fruits and Vegetables

Belén Diezma and Margarita Ruiz-Altisent

CONTENTS

8.1 INTRODUCTION

Acoustics can be defined as the generation, transmission, and reception of energy in the form of vibrational waves in matter. As the atoms or molecules of a fluid or sound are moved from their normal configurations, restorative elastic force arises. The elastic restoring force, coupled with the inertia of the system, allows the material to participate in oscillatory vibrations and consequently generate and transmit acoustic waves. Acoustics cover a wide range of scientific and engineering disciplines (Recuero López, 2000).

The best-known acoustic phenomenon is related to the sensation of sound. For young people, a disturbance is interpreted as a vibrational sound if its frequency is in the range of about 20 to 20,000 Hz, however, in its broadest sense, sound also includes ultrasonic frequencies above 20,000 Hz and infrasonic frequencies, below 20 Hz. The nature of the vibrations associated with acoustics are very different—from those produced by stringed instruments to those generated by an explosion.

The study of materials has developed dynamic methods for determining the elastic properties of samples homogeneously formed as an alternative to compression static methods or similar. The techniques for determining the resonant frequencies are tests used to characterize the elastic properties of metals, ceramics, alloys, and compounds (Lemmens, 1990). For homogeneous objects with simple geometries (cylinders, discs, spheres, rings, etc.), it is possible to establish expressions relating the resonant frequencies of the material properties (elastic modulus, Poisson ratio, density, etc.) and geometric properties (size and shape) (Blevins, 1993). The energy applied to the structure is amplified to a certain frequency, it is the resonant frequencies, the value of each resonant frequency is dependent on the geometry, density, and elastic properties of the sample. The translation of the possibilities of these techniques to horticultural products must overcome some difficulties inherent in the products, such as heterogeneity in their structures or irregular shapes and variables. In the following paragraphs, we review the models described for these products, the devices used, and the applications developed.

8.2 APPLICATION OF VIBRATIONS IN THE AUDIBLE RANGE

Dull (1986) reviewed nondestructive methods for assessing the quality of fruits and vegetables, and distinguishes between *acoustic impulses*, in which a body hits the sample, and the techniques of *resonance frequency*, in which the sample is subjected to vibration frequencies. Chen (1996) also distinguishes between *sonic vibration*, and the acoustic response to the impact (*acoustic response*). This classification serves as the structure for studying works aimed at implementing sound properties to food products.

8.2.1 Determination of Textural Properties of Fruits and Vegetables: Vibrational Excitation

Texture is a complex and variable qualitative parameter of great importance in the evaluation of fruits and vegetables for fresh consumption. Different factors, such as water status, the physical properties of cell walls, and tissue structure, act together to determine the strength, firmness, and elasticity, which are characteristics that make up the texture. The importance of this attribute has led numerous research groups to develop destructive and nondestructive methods to objectively determine parameters for characterizing texture. In this context, the development of techniques and devices that analyze the vibrational response of horticultural products has been largely dedicated to the study of texture.

8.2.1.1 Definition and Application of the Coefficient of Rigidity and the Acoustic Parameter for Determining Textural Characteristics of Fruits and Vegetables

Traditional and popular practices have been used to hit some horticultural products, such as watermelons, and attend to the resonance that occurs to get an idea of its maturity. Beginning in 1968 (Abbott et al., 1968; Bachman et al. 1968), researchers stated that this resonance property is characteristic of all fruits and vegetables, and is a good indicator of the internal conditions of texture of these products. These authors used two techniques to measure the textural characteristics of fruit using acoustic energy, which considered several basic principles:

- The resonant frequency depends on the size, shape, and texture of the product: large products and soft textures have lower resonant frequencies.
- The response can be analyzed in terms of configuration and elastic properties of the resonating body: A steel sphere 6 cm in diameter rings to about

45,000 Hz, while a spherical fruit of the same size makes it around 420 Hz; it is often easily detected by the human ear.

- The consideration of damping is also important: Fast damping is characteristic of high internal friction materials that absorb vibrational energy. In fruits and vegetables, this damping of vibration is very fast. The internal friction of a material is measured by the damping ratio: η.

The elastic modulus of a simple configuration object with uniform texture can be determined by the resonance frequency and the damping ratio. Young's modulus, E, can be expressed in terms of the flexural resonant frequency (see the following explanation of flexural vibration) of a bar of known length, diameter, and density. The *complex modulus* E^*, is expressed in relation to E. The complex modulus is an important factor in the evaluation of texture. The Equation (8.1) is the expression of the complex modulus as a function of Young´s modulus and the damping ratio.

$$E^* = E(1 - i\eta) \tag{8.1}$$

The firmness of the fruit is indicative of the consistency of the material. There are different methods to measure fruit firmness, although the most universal is the Magness-Taylor test. Abbott (1968) stated that the total resistance against the penetration material is composed of two factors: resistance to compression and resistance to shear, which leads to great variability. We propose an alternative method based on the vibrational properties of products to characterize texture. Two experimental procedures are discussed:

1. *Study of the behavior of a whole apple subjected to vibration*: An acoustic spectrometer causes a metal rod to vibrate at defined frequencies ranging across the spectrum. One end of the rod rests on the apple, and the apple is suspended from a wire to allow it to vibrate freely. The vibration of the apple is picked up with another very light rod, one end resting on the upper region of the fruit on the opposite side of the excitation. Another contact is made with a piezoelectric material that transduces the mechanical signal to an electrical signal, which is represented as amplitude. A representation of the frequency versus amplitude is thus obtained. When the apple is in resonance with the frequency of waves generated, the spectrometer produces a maximum amplitude reading. The position, shape, and relative amplitude of these peaks are characteristic of each apple studied.
2. *Study of the textural characteristics of a cylinder of apple pulp through its acoustic behavior*: The device described above was applied to a cylinder of apple of known dimensions. The Young's modulus and the damping ratio were calculated based on the resonant frequencies obtained in the cylinders studied. Estimates of E and the damping coefficient were performed: f is the resonant frequency, L the length of the cylinder, d diameter, ρ density, and the frequency difference corresponds to the width of the resonance peak

taken at half of its maximum amplitude. Equation (8.2) and (8.3) are the expressions for calculating Young's modulus and the damping coefficient, respectively, according to Abbott et al. (1968).

$$E = 50.9\, f^2 L^4 d^2\, \rho \tag{8.2}$$

$$\eta = (f_2 - f_1) / f\sqrt{3} \tag{8.3}$$

The spectrum obtained between 20 and 2400 Hz was studied in the whole apple. Four resonance peaks were observed in the spectrum. The first one corresponds to a type of longitudinal vibration. The second resonance peak, attributed to a flexural vibration, was related to fruit firmness, although it was also affected by the mass of the apple. The extreme values were found at 1,488 Hz in a small apple (40.6 g) and firm, and at 370 Hz for a large apple (162.2 g) and unstable. The frequency of this peak was less affected by the change in the position of the rods' excitatory and signal pickup. It was noted that certain defects in the flesh of the apple (bruises, brown heart, etc.) affected the resonance frequency.

They proposed the determination of the coefficient of rigidity of the entire apple using Equation (8.4), where $f_{n=2}$ is the resonant frequency of the second peak and mass m. Decreases in this ratio in samples subjected to longer periods of storage were confirmed.

$$f_{n=2}^2 \cdot m \tag{8.4}$$

Finney (1970) also studied the mechanical resonance of Red Delicious apples in relation to the texture of the fruit. The device used consisted of an electromagnetic vibrator that excited the sample by generating a vibration between 5 and 10,000 Hz. The vibration amplitude of the fruit was measured with a miniature accelerometer attached to the fruit surface, and was stored as a function of frequency. The frequency spectrum showed several resonant peaks in the audible range. The second resonant frequency was used to calculate the stiffness coefficient as defined by Abbott (1968) (Equation [8.4]). Studying changes in the coefficient of rigidity and modulus of elasticity of the pulp over a storage period of 6 months showed a significant correlation between them (r greater than 0.8). Shackelford and Clark (1970) were the first to apply the technique of resonance to peaches. Finney (1971) found, in their experiments with peaches, that the resonant frequency of fruit decreased as the fruit matured. The coefficient of rigidity, called the "maturity index" by this author, as Equation (8.4), with the second resonant frequency f, obtained values of 343 × 106 Hz^2g in more immature peaches, up to 39 × 106 Hz^2g for the more mature fruit.

Cooke (Cooke and Rand, 1973; Cooke, 1972) addressed the definition of the coefficient of rigidity from theoretical considerations of the problem, modeling fruit and vegetables as elastic spheres consisting of three concentric layers with different mechanical characteristics and relating the frequencies' resonant vibration modes corresponding to the internal mechanical properties of intact samples.

The work of Cooke adopts the nomenclature used in geophysical surveys to define and name the modes of vibration. Thus, the vibration modes that can occur in an area are classified into two types: spheroidal modes of vibration and torsional vibration modes. In spheroidal modes, the volume of the body subjected to vibration is not constant and the displacement vector must be defined by three independent components; this type is identified with the letter *S*. The torsional vibration does not produce change in the volume of the body and there is no radial component of the displacement vector; this is identified by the letter *T*. Specific modes are identified with superscripts and subscripts: ${}_{i}T_{n}^{m}$, *n* and *m* characterize the deformation of surfaces, and *i* refers to the order of frequency for the modes. In spheroidal vibrations, $n = 0$ corresponds to purely radial vibration mode, the so-called breathing, $n = 2$ occurs at frequencies lower than $n = 0$ and is characterized by an oscillation of the field between the forms of an oblate spheroid and a spheroid elongated at the poles (oblate-prolate).

Abbott's frequencies f_1 and f_2 were identified as modes of spheroidal and torsional vibration, respectively: ${}_{0}S_{2}^{2}$ and ${}_{0}T_{2}^{2}$. So, what Abbott called longitudinal vibration modes are interpreted by Cooke and Rand as spheroidal vibrations, and the flexural modes are identified as torsional.

The spheroidal vibration mode of the lowest frequency is characterized by the movement of expansion and contraction of a diameter, or simultaneous movement of contraction and expansion of the perpendicular diameter.

Torsional vibration of the lowest frequency is characterized by no change in volume occurring during the oscillation; the radial component of displacement is 0. If 16 meridians are established in the sphere spaced at 22.5 degrees, for a given meridian, the points at the same distance from the sphere's equator to the north and south have the same magnitude as to the east and west, but in the opposite direction, while the north–south component remains the same in magnitude and direction. At the sphere's equator, there are 4 nodal points (that is, points with displacement of 0 during the oscillation).

According to Abbott, the use of resonance to infer the elastic properties of horticultural products requires special attention to the following factors: the mode of the excitation of the resonance, the identification of mode of resonance frequency, and the influence of the sample holder. Depending on the mode of vibration of interest it is more convenient a configuration of these factors or another.

Thus, based on the description of the vibration modes, it appears that the device used by Abbott in their experiments, favors the excitation of the torsional mode, where the receiver and its location (at the top of the apple) is sensitive to the north–south movement, although the larger north–south movement at the equator occurs at the point opposite the driver. The Finney device, with its radial movement, favors spheroidal excitation of the sample. Similarly, the Abbott device (apple suspended from the stalk) is the optimum positioning for torsional vibration, which is defined at the pole at a nodal point.

According to the elastic sphere model composed of three concentric layers, the shear modulus of layer 2 (G_2) (renumbered starting from the inside), comparable to the pulp, can be expressed as Equation (8.5), where ρ_2 is the density of layer 2 and *CR* is a dimensionless conversion factor that depends explicitly on the relationship between the radii and the densities of different layers, and implicitly includes the relationship between the shear modules.

$$G_2 = f^2 m^{2/3} \rho_2^{1/3} .CR \tag{8.5}$$

Thus, one can nondestructively calculate the shear modulus of the pulp of apples from the vibration frequency torsional mode, but assuming that the relationships among the radius, density, and shear modulus does not vary much from one apple to another.

If we consider Abbott's (1968) definition of the coefficient of rigidity, we can establish Equation (8.6), which shows the relationship between the stiffness coefficient defined by Abbott (1968) and the expression of the shear modulus of Cooke (1972).

$$f^2 m = (G_2 m^{1/3}) / CR.\rho_2^{1/3} \tag{8.6}$$

It can then be observed that Abbott's coefficient of rigidity is dependent on mass. If we apply the assumptions already identified, and assume a variation of the mass of apples that is small enough, we can relate the stiffness coefficient and the elastic properties of the fruit. Assuming that the overall density of the fruit is about the same as that of the pulp, the mass can be approximated to the mean radius, so you could say that although the stiffness coefficient depends on the size of the fruit, if this variation is small, you can get a high correlation between shear modulus and the coefficient of rigidity. Similarly, taking into account the relationship between G (shear modulus) and E (Young's modulus) it can be said that if variations in Poisson's ratio are small enough, there will be a high correlation between E and the coefficient of rigidity.

Applying the theoretical model for calculating the resonant frequencies of the modes of spheroidal and torsional vibration, a sensitivity study was conducted to define the variation produced in the resonant frequencies of each mode depending on the change in the relationships of radius, densities, and shear modulus. Spheroidal frequencies show higher sensitivity to these variations.

In apples, it was found experimentally that the expression $f^2.m^{2/3}$ reduced the error in the prediction of E, compared to the expression $f^2.m$ prediction: in 24 of the 38 groups chosen for the study, the mean square error was lower in the first regression.

Yong and Bilanski (1979) subjected spherically shaped fruit to vibrations; the output signal was acquired by an accelerometer attached to different points of the sample and the system itself. The sample was then placed on a vibrating plate. The accelerometer mounting points were the vibrating plate, the upper end of the sample, and a point on the equator. A comparison of the output signals corresponding to different positions of the accelerometer was performed. In the signal of the vibrating table, a resonant peak was observed between 100 and 200 Hz; for higher frequencies a decrease of amplitude below the signal amplitude levels of other signals was observed. In the curve from the upper part of the sample, there was another peak in the vicinity of those frequencies and a second, pronounced peak near 900 Hz. The signal from the equator, the peak of the first resonant frequency, was significantly lower than the same in the other two signals, producing the second resonant peak of the same intensity as on the polar curve of the fruit.

In addition to the direct reading of the signal, a ratio signal is proposed that takes into account the input signal. Thus, the ratio of the output when the accelerometer is at the top of the sample and the entrance, shows two resonant frequencies, the first at 110 Hz and the second at 900 Hz, but the same ratio when the accelerometer is in the equatorial zone shows only a clear resonant peak at 900 Hz. Waves corresponding to the resonant peaks of the signals from the equator of the sample and the top of the samples were 180° out of phase. The fact that the accelerometer only measured in the perpendicular direction allowed the vibration of the resonant peak of 900 Hz to be defined as a spheroidal mode oblate-prolate. As explained, the first resonant frequency is that for a vibrating system with one degree of freedom, with the skin contact plate and the pulp adjacent to the spring and damping system, and the rest of the fruit mass of the system. Thus, the first resonant frequency not is a spheroidal or torsional mode, but could be represented by the Kelvin model of one degree of freedom, as shown in Equation (8.7). The authors question the assertions of other studies (Cooke, 1972), which identified the first two resonant frequencies, the first of which was associated with a spheroidal vibration in apples and a torsional vibration in peaches.

$$mx'' + bx' + Kx = by' + ky \tag{8.7}$$

Clark and Shackelford (1973) studied (in peaches) the relationship between the expression $f^2.m^{2/3}$ (stiffness index, according to the authors), where f is the second resonant frequency of the spectrum, with the value of force in the flesh penetration Magness-Taylor test. Their paper does not refer to the mode of vibration that corresponds to the resonance frequency used. The different regressions established for 7 different varieties gave a correlation coefficient (R) of between 0.52 and 0.78.

In an attempt to better understand the relationship between the vibrational behavior and the elastic properties of the products, the dynamic modulus is considered. The determination of dynamic modulus requires a destructive test in which a cylinder of the sample is placed on a force transducer that is mounted on a vibrating system. The whole system can be likened to a mass-spring-damper and Young's modulus can be calculated from the transfer function. Van Woensel and De Baerdemaeker (1985) determined Young's modulus, a stiffness index defined as $f^2m^{2/3}$, and Young's modulus in a quasi-static uniaxial compression test.

The stiffness index was obtained using a vibrating device for excitation of the whole apple and an accelerometer placed on the surface of the fruit; f reflects the second resonant frequency, which is identified as a spheroidal-type vibration, which contradicts Abbott et al. (1968) and Cooke (1972). The first resonant frequency is ruled out because it is the frequency corresponding to rigid body motion. Uniaxial compression tests were performed on cylinders of apple pulp, and the center of the curve force versus deformation is considered to calculate Young's modulus. The three tests were applied to apples with different levels of consumer acceptability (sensory evaluation that considered various aspects such as manual pressure, mouthfeel and taste), at different times in its maturity, and during storage (150 days of storage). Both the stiffness index and the break point in the compression test declined throughout the storage period. A further decline was observed in the first parameter until apples reached the climacteric stage, then the trend became smoother. In regard to the relationship of the parameters measured by consumer acceptability, a better relationship is observed between the dynamic parameters (coefficient of strength and modulus of elasticity) and acceptability, than between acceptability and the parameters of the static test.

Chen and De Baerdemaeker (1990) tried to relate the changes of firmness in tomatoes produced throughout the ripening process with the second resonant frequency spectrum. The excitement of the samples was done using a vibrator that transmitted a pseudo-random noise signal with a bandwidth of 800 Hz. The vibration of the fruit was determined with an accelerometer. The stiffness index, defined as $f^2m^{2/3}$, was best correlated with the color (measured with color charts) and the Young's modulus, determined by analyzing quasi-static compression. The coefficient $f^2m^{2/3}$ decreases as the red color of tomatoes is accentuated and decreases E. Indexes $fm^{2/3}$ and fm^2, showed worse correlations with the ripeness of tomatoes, especially the latter. Similarly, the damping coefficient of the signal was not well correlated with color or the Young's modulus of the samples. The internal structure of the tomato can cause different vibration signals when measured at one point or another. In conclusion, the expression $f^2m^{2/3}$, used so far in apples and peaches, is valid also for tomatoes.

Abbott and Liljedahl (1994) dealt with an investigation of three varieties of apples with the aim of relating the sonic signal of apples with individual compression tests of probes and the classic Magness-Taylor test. The testing device consisted of a clay holder attached to an electromagnetic vibrator. An accelerometer is held in a light arm to get vertical and opposite to the vibrator. The sample was placed horizontally on the support and the accelerometer lay at the equator of the sample. A pulse (sin t)/t (where t is time) containing all frequencies up to 2000 Hz with the same energy density at each frequency generated the movement of the vibrator. In this case, the authors extracted the frequencies of the second and third resonant peak (f^2 and f^3) from the spectrum without specifying the corresponding vibration modes, their respective amplitudes (a_2 and a_3), and the index defined as and f^2m^2 y $f^2m^{2/3}$ with $f = f^2$ and f^3. In addition, in conjunction with researchers who had identified the first peak resonant with rigid body motion, the search for f^2 and f^3 are scheduled to obviate the spectrum located below 500 Hz to avoid choosing the first resonant frequency f^1. The compression tests allowed determination of the force/deflection curve (*F*/*D*) and identification of the following parameters: maximum force, area under the curve, and slope of the curve. The Magness-Taylor test was done with a rod 11 mm in diameter in two points of the apple's equator, which was considered the average of the maximum force obtained in each penetration.

Both the resonant frequencies and indexes, as the parameters of the *F*/*D* curve and Magness-Taylor force decreased over the storage months for the three apple varieties studied. However, it is clear that as the varieties have shapes that differ from the spherical shape, the correlations between nondestructive sonic parameters and parameters of the *F*/*D* curve decrease dramatically.

In all three varieties, the resonant frequencies are related more to the slope of the *F*/*D* curve than to the forces, which is consistent with the theory that the resonant frequency is proportional to the modulus of elasticity, not strength. Also, according to theory, the resonant frequency of a body is proportional to its size, density, and Poisson's ratio. In this study, these parameters were not measured, but an approach was performed using the mass and calculating the two accepted indexes, which include the resonant frequencies and the masses. However, although indexes were better predictors of the slope of the *F/D* curve, the correlations were only slightly higher than the correlations between the slope and resonant frequencies.

A replica of the work of Abbott and Lilje-dahl was carried out by Abbott et al. (1995), this time employing a type of Golden Delicious apple, Golden Delicious Stark, and two types of Delicious, Red, and Triple Ace. Again stored for 4 months the acoustic parameters were good predictors of destructive firmness variables in Golden Delicious apples ($r > 0.8$), but not in Red Delicious.

8.2.1.2 The Importance of Media and Means of Contact between Vibrating Structures and Samples

Affeldt and Abbott (1989) studied the importance of supports and materials of contact between vibrating structures and samples. Structures and transmission spectra with a low flat attenuation in the region of interest are needed. The effect of different supports and materials was studied. These elements were implemented in the structure of vibration of an electromagnetic vibrator collecting the response of apples of different maturity stages. Although the best response was obtained with a base of clay, different granular media (sand and powdered sugar) gave good results, reducing attenuation and variations in the resonant frequencies because the medium was thinner. The response of samples was collected with a miniature accelerometer placed at the equator at the point diametrically opposite to the excitement.

Each apple was tested acoustically 8 times by turning 45 degrees each time. They observed the appearance of distortions in the transmission of the signal when the signal is applied at locations with surface defects. The same effects could be due to minor bruises, but when bruising affected a significant volume of fruit, observed modal distortions were observed in almost all measurement points around the equator.

In addition to considering the correlations between the first two resonant frequencies, which were then considered as longitudinal and torsional modes, respectively, and the stiffness coefficients calculated with each of them (f^2m) and the Magness-Taylor penetration test, prediction models were established that included one or two fixed frequencies, amplitudes at fixed frequencies, or the ratio of fixed frequencies. Using these parameters, the sweep of the response signal looking resonant peaks was avoided. The maximum value of r in these regressions exceeded 0.6 in the model of two fixed frequencies (860 and 165 Hz), improving the correlation between the first torsional frequency and the Magness-Taylor test parameter.

8.2.1.3 Speed of Propagation of Acoustic Perturbation as a Parameter for Determining the Textural Characteristics of Fruits and Vegetables

Garrett and Furry (1972) considered the fact that the propagation of sound waves in a material depends on its modulus of elasticity, density, and coefficient of Poisson. The aim of this research was to use the propagation of mechanical disturbances in entire apples as a nondestructive method of determining the mechanical properties of the fruit flesh. The speed of wave propagation in whole apples, apple quarters, and core samples were measured.

By exciting whole apples, apple quarters (without the central region), and cylinders of pulp, the apparent speed of propagation (v_{WF}), the speed of propagation of waves of expansion (V_d), and the speed of propagation in a bar (V_b) were determined. For determining V_{WF} and V_d, a piezoelectric rod contacting the surface of the sample recorded the signal on the face opposite to the excitation zone and transmitted it to an oscilloscope; for determination of V_b, two detectors were placed on the sample cylinder and spaced a certain distance, and the time a signal peak took to pass from one detector to another was measured. Theoretically V_b, V_d, and their relationship are defined in Equation (8.8), where υ is the Poisson ratio.

$$V_b^2 = \frac{E}{\rho}; V_d^2 = \frac{E(1-\upsilon)}{\rho(1+\upsilon)(1-2\upsilon)}; V_b^2 = V_d^2\frac{(1+\upsilon)(1-2\upsilon)}{(1-\upsilon)} \tag{8.8}$$

The elastic modulus, E, was calculated using Equation (8.8) from V_b, and the results were consistent with those of other studies. However, the estimation of E from v_{WF} was 13.6% of the average value of E. According to the authors, in general, the determination of velocity in the whole apple cannot define the specific values of the elastic properties of the flesh of the apple. However, under conditions in which Poisson's ratio remains constant, changes in E can be detected, and if E remains constant, changes in υ can be detected.

8.2.1.4 Other Vibration Sensor Systems on the Surface of a Body Subjected to a Vibrational Excitation: Phase Shift of the Signal as a Parameter Related to the Textural Characteristics of Fruits and Vegetables

Muramatsu et al. (1999) used a laser Doppler vibrometer to capture the vibrations that occur on the surface of a fruit subjected to vibrational excitation; the frequency of the laser wave reflected from the surface varies depending on the speed of vibration. The method allows measurement without the need for contact between the sample and the sensing elements. Kiwis, pears, and peaches were measured to determine their state of maturity, and citrus to detect units affected by cold damage (collapse of the vesicles). The samples were placed in a holder through which were transmitted an excitation consisting of sine wave frequencies from 5 to 2000 Hz. The vibrometer measured the vibration transmitted to the upper surface of the fruit. The parameter used to determine the textural characteristics, or quality of the fruit, was the change in phase between the vibration signals of input and output. This is given by the frequency response function, defined as the ratio of Fourier transforms of input and output signals multiplied by the complex conjugation of the input signal.

The changes in phase at the frequencies 800, 1200, and 1600 Hz showed good correlation with the destructive measures of firmness ($r < 0.8$), and appeared as good predictors of disorders caused by cold in citrus.

The method is presented as an alternative to the use of accelerometers (which require contact between the sample and the sensing element) with respect to which is more sensitive to vibration. A disadvantage of laser technology is the duration of the measure. In Muramatsu et al., the duration is 3 minutes to make a sweep from 5 to 2000 Hz; it can be assumed that limiting the number of excitation frequencies implies a decrease of measurement time to 45 ms if reduced to a single frequency.

A noncontact laser vibrometer Doppler effect was also applied for the prediction of storage time and consequently of postharvest maturity and size of plums (Bengtsson et al., 2003). The excitation was performed with pulses of frequencies from 0 to 1600 Hz, which required a measurement time of approximately 10 seconds. Frequencies with the highest regression coefficients were in the 41–151 Hz range in relation to postharvest parameters (storage time), and 7–70 Hz for the parameters of size (weight and diameter).

More recently, Taniwaky optimized the nondestructive vibrational method by employing a laser Doppler vibrometer to define and calculate an index involving resonant frequencies and mass in order to determine the optimum eating ripeness of pears and melons and the postharvest quality evaluation of persimmons (Taniwaki et al., 2009a, 2009b, 2009c) and other products.

8.2.2 Determination of Textural Properties of Fruits and Vegetables: The Impact Acoustic Response

8.2.2.1 Development of Theoretical Models and Expressions of the Elastic Properties of Horticultural Products: Applications

Armstrong et al. (1990) propose an elastic sphere model to predict the firmness of apples. According to the spherical resonator theory, based on the expression of the displacement for small Navier movements (Equation [8.9]) expressed the modulus of rigidity (G) and the modulus of elasticity (E) of an elastic sphere in terms of density, Poisson's ratio, frequency, and mass (Equations [8.10] and [8.11]).

$$\mu\nabla^2 u + (\lambda+\mu)\nabla\nabla\cdot u = \rho\ddot{u}$$

$$\text{u} = \text{ displacement}, \rho = \text{ density}, \lambda \text{ y } \mu = \text{ Lamé constants} \tag{8.9}$$

$$G = \left[\frac{\rho^{1/3}(6\pi^2)^{2/3}}{\Omega^2}\right] f^2 m^{2/3} \tag{8.10}$$

$$E = \left[\frac{\rho^{1/3}(6\pi^2)^{2/3}2(1+\upsilon)}{\Omega^2}\right] f^2 m^{2/3} \tag{8.11}$$

Ω, corresponds to a frequency standard that is defined by Equation (8.12), where C_t is the transverse phase velocity, ω is the frequency in rad/s, and a is the radius of the sphere (Armstrong et al., 1990).

$$\Omega = \frac{\omega a}{C_t} \tag{8.12}$$

As cited in the work from Armstrong et al. (1990), Auld (1973) gives the vibration modes and their normalized frequencies that solve this scalar equation and the corresponding vectorial equation. The vibrational modes of lower orders and the classification of them are listed here:

- *Pure compressional modes:* S_{0n} ($n \leq 1$); S_{01} is the so-called breathing in which all particles move in the same radial.
- *Pure shear mode:* T_{nm}, T_{20} can be described as two connected hemispheres rotating in opposite directions, and the movement of particles located at the same distance from equator but in different hemispheres are equal; T_{11} represents surface of two concentric spheres rotating in opposite directions.
- The mode S_{20} is a mixed mode of vibration in which a diameter of the sphere expands while the perpendicular diameter is compressed.

In the experimental works (Armstrong et al., 1990) apples with different storage periods and treatments were used. The characterization of the textural properties of fruits is done by the Magness-Taylor penetration test, including determination of modulus of elasticity (cylindrical and cubic samples of fruit) and the determination of density. The excitement of the apple was achieved by hitting it with a ball of wax of 20 mm in diameter attached to a steel rod. The acoustic emission was collected with a microphone (2048 points acquired to a sampling frequency of 25 kHz), sending the signal to a digital oscilloscope. By applying the fast Fourier transform (FFT) function, the frequency spectrum was obtained to extract the natural frequencies (resolution of FFT for the range from 0 to 12.5 kHz. 12.2 Hz). Spectra with a single clear resonant frequency between 891 and 1013 Hz were obtained.

The regressions between the predicted elastic modulus according to Equation (8.11), and the elastic modulus measured at a cubical specimen that was included in the core of the apple, were relevant: r^2 between 0.7 and 0.8. Value of Ω^2 used in the expression of E was 4.9. Using an average density (a constant) rather than the density measured in each apple gave similar results, and did not worsen the results. Although this research did not determine the type of vibration corresponding to the resonant frequency introduced in the calculations of E, it was assumed that the resonant frequency of the model was the mode of vibration of pure compression S_{01}.

In subsequent works, Armstrong and Brown (1990) applied Equation (8.11) in an attempt to predict the firmness of apples after a storage period based on the same measurement made before such storage. The spectrum corresponding to excitation of the samples using a slight impact, showed a predominant resonant frequency, typically between 800 and 1200 Hz. It was this frequency that was introduced in the equation for calculating E; the density of the apple and Poisson's ratio is considered constant with values of 800 kg/m^3 and 0.3, respectively. The prediction of E was adequate for firmness in groups, but did not yield good results for individual apples.

Huarng et al. (1993) started afresh with the expression of the Navier equation (Equation [8.13]) governing small motions of an isotropic, elastic sphere, this time to determine the theoretical modes of vibration and check whether these are consistent with those measured with two microphones placed in the apple equator excited by an impact.

$$\mu\nabla^2\bar{u}+(\lambda+\mu)\nabla\left[\nabla.\bar{u}\right]=\rho\bar{u}'' \tag{8.13}$$

In Equation (8.13), λ and μ are Lamé constants, $\bar{u}$ is the displacement vector, and ρ is the density. The solution of this equation is obtained by expressing the displacement vector and a scalar potential and vector potential, as given by Equation (8.14):

$$\bar{u}=\nabla\phi+\nabla\times\bar{\psi} \tag{8.14}$$

Equation (8.14) satisfies the Navier equation if the conditions in Equations (8.15) and (8.16) are satisfied. These equations are expressions of the motion vector decomposition of an escalar potential and a vector potential, respectively.

$$\nabla^2\phi=\frac{1}{V_l^2}\ddot{\phi} \quad \text{con} \quad V_l=\left(\frac{\lambda+2\mu}{\rho}\right)^{1/2} \tag{8.15}$$

$$\nabla^2\bar{\psi}=\frac{1}{V_s^2}\overline{\psi''} \quad \text{con} \quad V_s=\left(\frac{\mu}{\rho}\right)^{1/2} \tag{8.16}$$

Either of the these equations can be solved by the method of separation of variables. It then briefly describes the process for resolving ϕ. Equation (8.17) is the expression of the scalar potential as a product of a function of displacement and a function of time.

$$\phi=\Phi(r,\theta,\delta)T(t) \tag{8.17}$$

where r, θ, and δ are the spherical coordinates. The following expressions are obtained.

Equation (8.18) expresses the scalar potential in terms of the Legendre polynomial and Bessel function. It is the expression of the solution of Equation (8.15), in the case of excitation by a pulse to the normal surface. According to this equation, the deformation of the surface of the sphere (with r equal to the radius of the sphere) is characterized by the Legendre polynomial, $P_n(\cos\theta)$.

$$\phi_n=z_n\left(\frac{\omega r}{V_l}\right)P_n\left(\cos\theta\right)e^{-i\omega t} \quad \text{con} \quad n=1,2,\ldots \tag{8.18}$$

If it is $R + Pn$ $(\cos\theta)$ against θ, for $n = 2$, 3, and 4, where R is the radius of the sphere ,and considering the value of the Legendre polynomial to these values (Equation [8.19]), we obtain the representative theoretical spheroidal modes of vibration S_{20}, S_{30}, and S_{40}, respectively.

$$P_0(\cos\theta)=1$$
$$P_1(\cos\theta)=\cos\theta$$

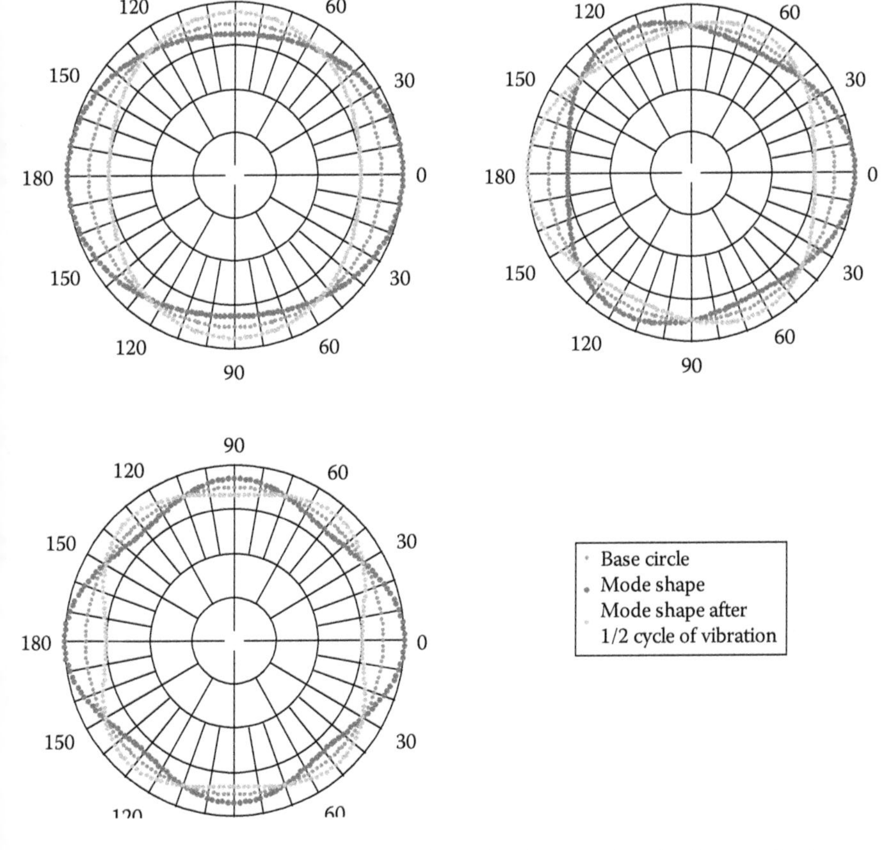

FIGURE 8.1 Representation of the theoretical vibrational modes S_{20}, S_{30}, and S_{40}. (Images courtesy of Prof. P. Chen.)

$$
\begin{aligned}
P_2(\cos\theta) &= (1/4)(3\cos 2\theta + 1) \\
P_3(\cos\theta) &= (1/8)(5\cos 3\theta + 3\cos\theta) \\
P_4(\cos\theta) &= (1/64)(35\cos\theta\, 4\theta + 20\cos 2\theta + 9)
\end{aligned}
\tag{8.19}
$$

In S_{20} mode, there are two nodal lines located at $\theta = 54.7°$ and $\theta = 125.3°$, which divide the sphere into 3 zones. In S_{30} mode there are three nodal lines $\theta = 39.2°$, 90°, and 140.8°. In S_{40} the nodal lines are given as $\theta = 30.6°$, 70.1°, 109.9°, and 149.9° (Figure 8.1).

In the experimental device the apples tested were excited on a point at the equator, the point 0°. A microphone, called a reference, was placed in all the impacts to 180° of the excitation point. A second microphone was moving for another 18 points on the equator, spaced at 30 degrees. For each measurement, the fruit is hit on the point 0° and the acoustic signal is collected simultaneously by the two microphones. The signal in the time spectrum was obtained by fast Fourier transform, as the sampling frequency and time of collection for each measurement signal; 256 pairs of Fourier coefficients are obtained, a_j and b_j, $j = 1, 2, \ldots 256$. The amplitude and phase are calculated according to the following terms using Equation (8.20):

$$A_j = \left(a_j^2 + b_j^2\right)^{1/2}$$
$$\phi = \tan^{-1}\left(b_j / a_j\right) \tag{8.20}$$

A representation similar to that made with the theoretical data was carried out using the experimental measurements. To this end, each measuring point's relative amplitude was calculated by dividing the amplitude at that point by the amplitude at the reference point; the phase was determined by subtracting a phase reference point for the measurement point.

The spectra showed three clear resonant peaks. Using each of them to perform the calculations described in the previous paragraph, it was observed that they corresponded to the vibration modes S_{20}, S_{30}, and S_{40}. Figure 8.2 shows the points obtained experimentally on the representation of the movement of the vibration modes. The level of agreement allows the authors to identify as spheroidal modes the real vibrations corresponding to the three resonant peaks appearing in the spectra of apples when they are excited by an impact at any point on the equator.

Rosenfeld et al. (1992) conducted a theoretical analysis based on numerical simulation of elastic and viscoelastic bodies subjected to dynamic boundary conditions, in order to model the response of the fruit and the possibilities of using it as a method to classify fruits according to firmness. To do this, the fruit is represented as a continuous three-dimensional object arbitrarily satisfying the wave equation elasto-dynamics and boundary conditions. The simulation characterizes the behavior of the surface points in terms of displacement and traction both in the time and frequency domains.

The mathematical model of the fruit is based on the equation of motion of an elastic body, homogeneous, linear, and isotropic defined in a field Ω and limited by the surface Γ is shown in Equation (8.21):

$$\left(c_1^2 - c_2^2\right) u_{i,ij}\left(x,t\right) + c_2^2 u_{i,ij}\left(x,t\right) + b_j = \frac{\partial^2 u_j\left(x,t\right)}{\partial t^2} \tag{8.21}$$

where $u\ (x,\ t)$ is the displacement vector associated with any point x of the body; c_1^2 and c_2^2 are the velocities of longitudinal and transverse waves, respectively, according to Lamé constants and density of the material; and b_j is the body forces unit mass.

The boundary conditions were established as a combination of displacement u_i and traction t_i in complementary regions of the boundary, as shown in Equation (8.22):

$$u_i = \overline{u_i}(x,t) \quad \text{en } \Gamma_1$$
$$t_i = \sigma_{ij} n_j = \overline{t_i}\left(x,t\right) \quad \text{en } \Gamma_2 \tag{8.22}$$

The initial conditions for all parts of the body were given by Equation (8.23):

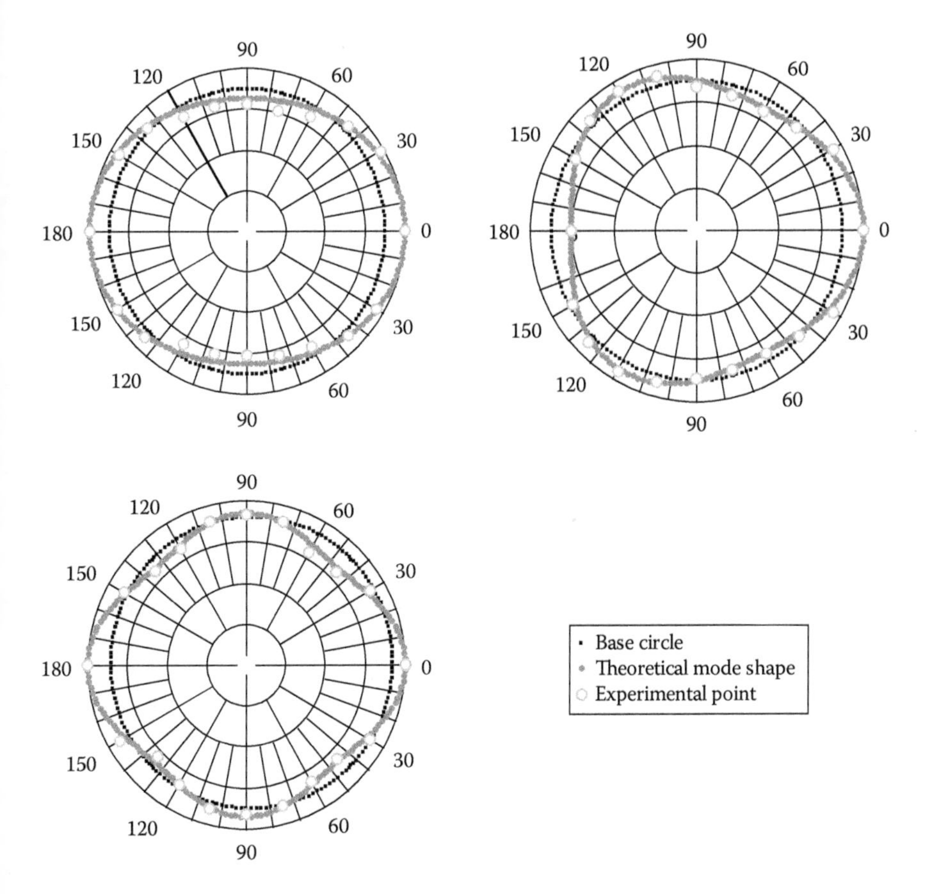

FIGURE 8.2 Representation of the vibration modes of theoretical and experimental S_{20}, S_{30}, and S_{40}. (Images courtesy of Prof. P. Chen.)

$$u_i(x,0) = u_{i0}(x); \quad \frac{\partial}{\partial t}\left[u_i(x,0)\right] = \dot{u}_{i0}(x) \tag{8.23}$$

The solution to an equation derived from Equations (8.21), (8.22), and (8.23) is obtained by the method of boundary element approximation.

In this framework, an ellipsoidal fruit was simulated. In different simulations the following values of elastic parameters were used: density of 780 to 880 kg/m^3, Poisson's ratio of 0.2 to 0.35, and Young's modulus of 3 to 9 MPa. In the simulation, the fruit was subjected to a sinusoidal perturbation amplitude of 1 mm at its bottom and the pull in the rest of the points of the fruit was 0.

In the simulation, the frequency response of several points on the surface and for different directions of motion were obtained, so the results of the work could be used to study the vertical displacement of the point located on the top of the body, the vertical displacement of a node on the equator, the horizontal displacement of this node, and the vertical traction of the excitation point (the lowest point of the body).

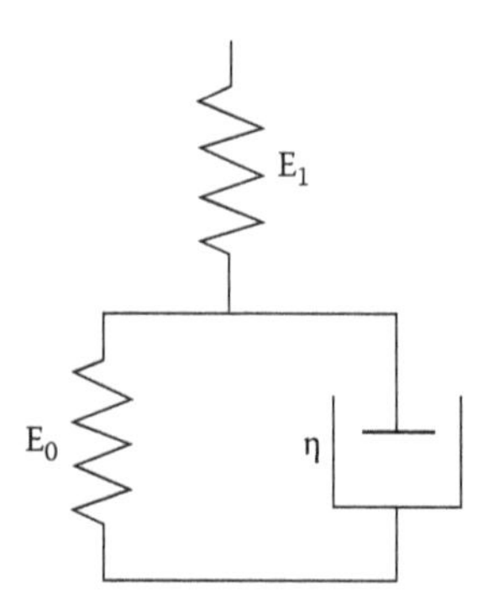

FIGURE 8.3 Three-dimensional viscoelastic solid model used to replace Young's modulus.

The representation for each point of its velocity divided by the traction at the point of excitation ($mm.s^{-1}/Pa$) versus frequency (Hz) showed the resonant frequencies of the fruit for certain values of the elastic parameters. Analyzing this representation for the vertical displacement of the top point and the horizontal and vertical displacements of the equatorial point, the authors describe the modes of vibration of the body: The first resonant frequency observed in the vertical displacements (below 300 Hz) was not in the horizontal displacement, which coincides with the results of Yong and Bilanski (1979) on the assimilation of the first resonance the motion of the system of one degree of freedom. The second resonant peak (above 500 Hz) occurred in both the vertical displacement of the top spot as the horizontal displacement of the equatorial point, and did not appear in the vertical movement of this node, so that this resonance is identified as the spheroidal vibration mode oblate-prolate.

The mechanical properties of fruits are frequently represented as viscoelastic models. Many studies have provided data for these models of static tests; however, in the absence of data in dynamic models, the authors defined a three-dimensional viscoelastic solid model to replace Young's modulus, so the Young complex modulus can be expressed as Equation (8.24), which represents the model in Figure 8.3.

$$E^*(-i\omega) = \frac{E_0 E_1 + E_1 \eta(-1\omega)}{E_0 + E_1 + \eta(-i\omega)} \tag{8.24}$$

The influence of mechanical properties in the frequency response was examined through the effect produced by the variation of the Poison ratio, density, fruit size, and parameters affecting the complex Young's modulus of the viscoelastic model (Figure 8.3). Horizontal speed was calculated at the highest point of the fruit; the result of dividing this rate by the amplitude of excitation is plotted versus frequency for different values of these parameters.

The variation of E_1 between 2 and 3 MPa caused an increase in the resonant frequencies. The increase of the damping ratio η caused a decrease in the amplitude of the resonant peaks, while the resonant frequencies appeared virtually unchanged; the increase in E_0 had the same effect. The decrease in fruit size simulated a decrease in mass, which produced an increase in the resonant frequencies; the decrease in density and consequently in the mass produced the same trend in the resonant frequencies.

Finally, the decrease in the Poisson ratio caused a strong damping in the second resonant frequency, while changes in the first were less pronounced. The authors explained the vibration modes corresponding to each of the two peaks as follows: the first mode corresponds to a "longitudinal" intuitively independent of the Poisson ratio, and the second relates to a spheroidal mode in which there is lateral movement that is directly related to this ratio.

In this study there were higher lags in the measurement of the signal between the upper and equatorial point in the resonant peaks, which could provide another tool for finding the resonant frequencies of the samples.

8.2.2.2 Collection of the Signal by Recording System: Use of Pattern Spectrum for the Extraction of Frequencies with Higher Discriminating Power between Stages of Maturity

One of the first applications of the acoustic response technique to impact was developed by Saltveit et al. (1975). This paper discusses the sound produced by a tomato when it is tapped with the finger as compared to the time the tomato needs to reach the state of green-red color. The sound produced by the samples was recorded. Using the algorithm of the fast Fourier transform, spectra signals were extracted. The region of interest in the spectrum was established at 2,000 Hz. The resulting average spectra of 9 tomatoes that were close to achieving a green-red color was used as the pattern spectrum of this state of maturity. The same procedures were followed to obtain the pattern spectrum for green tomatoes. By the difference (subtraction of the amplitudes at each frequency) of the two pattern spectra, the frequencies with greater distances between the two groups were obtained, assuming that the amplitudes at these frequencies could be correlated with the maturity of tomatoes. Seven frequencies were used to define the parameters for classification: mean amplitude of each of the 40-Hz-wide windows centered at the frequencies of higher differences. The window between 610 and 650 Hz showed the highest correlation with the stage of maturity, but the variability in the results was such that the classification of an individual showed a high probability of error.

The maximum amplitude frequency was also used as a parameter of classification, with similar results. However, although a tendency for the frequency of maximum amplitude to be greater for less-mature tomatoes was observed, the dispersion of the data was such that it was impossible the classification of tomatoes in maturity groups based on these data.

8.2.2.3 Signal Collection with a Microphone

8.2.2.3.1 Resonant Frequencies Used as Acoustic Parameters in Analysis

Chen et al. (1992) compared measurements of acoustic response to impact of 14 apples of different varieties, ranked in terms of sound perceived by three experts. The influence of the impact location on the fruit, the sample holder, and the method of generation of the impact were studied.

TABLE 8.1
Methods of Impact and Support Elements of the Sample in the Measurement of Acoustic Impact Response in Apples (Chen et al., 1992)

Methods of Impact	By Hand: Impact with the Finger	Flexible Rod with a Metallic Disc in the Extreme
Support Elements	Suspended	On a Plexiglas Cylinder

Expert classification was carried out by hitting the sample with the finger and hearing the sound with the ear close to the apple and evaluating the tone of the sound produced. Ten groups were established based on tone. For all tests, the microphone was placed on the opposite side of the point of impact (all around the equator) and separated from the surface of the fruit by about 3 mm. Accidental contact between the sample and the microphone sensing element caused significant distortions in the signal.

The impact methods analyzed are shown in Table 8.1. By analyzing the spectra it was observed that although most of them had several resonant frequencies, in some of them appeared only the lowest resonant frequency (as Yamamoto et al. [1980] and Armstrong et al. (1990) pointed out in their studies). According to preliminary studies, the contact time of impact affects the resonant frequencies that are excited. By changing the striking mass and using a more flexible rod, it was possible to increase the contact time; manual excitation also produces a lower impact speed. The results showed that the impact time affects the second and subsequent resonant frequencies: a contact time of T seconds excites the vibration corresponding approximately to the resonant frequencies $1/T$ Hz.

Variations in the location of the point of impact around the equator showed no significant differences in frequencies, although their amplitudes changed. The same effect was found when comparing different methods of support (Table 8.1): When the fruit is suspended, the restrictions on its vibration are minimal and the resonant frequency of greatest amplitude is first, followed by the second, third, and fourth. When the fruit is placed in a cylinder, the second resonant frequency shows the greater amplitude.

The classification into 10 groups based on the tone heard by experts correlated well with the first two resonant frequencies ($R^2 = 0.9$ for the first resonant frequency and $R^2 = 0.96$ for the second).

8.2.2.3.2 Stiffness Coefficients Used as Acoustic Parameters in Analysis

Duprat et al. (1997) applied the formula proposed by Cooke and Rand (1973) to estimate the modulus of elasticity ($E = f^2 m^{2/3} \rho^{1/3}$) of apples while on the tree and in storage, as well as tomatoes at different stages of maturity. The f in the above expression corresponds to the frequency with the higher amplitude. The acoustic test equipment consisted of an impact system and a microphone that picked up the signal at the point diametrically opposite to the impact.

The average firmness of the flesh in a penetration test at a depth of between 5 and 9 mm is considered the reference parameter. The decrease in modulus of elasticity calculated from the resonant frequency and the firmness while maturing on the tree for over one month followed similar patterns. This study states that an increase in the modulus was recorded on a mid-term date; however, on the same date, the firmness parameter, which was obtained destructively, continued the decreasing trend. The explanation may be found in the rainfall two days before the time of this test, which caused an increase in fruit turgor. Similar decreases in firmness and elasticity modulus were also observed throughout the storage period, and were especially pronounced until day 110, after which values stabilized.

To establish the degree of relationship between firmness and elasticity coefficient set with the acoustic instrument, a data set of the three seasons studied and the progression both in tree and in storage was employed; the correlation coefficient was 0.84. In the graph of firmness versus modulus of elasticity, this can be observed as a change of slope around 1.9 MPa. Two populations can be distinguished: one corresponding to the immature hard apple and the other consisting of the mature fruit. The coefficient of variation of the modulus of elasticity for the population of softer individuals is 8.5%, while for mature individuals it is 17.5%; in the case of firmness, coefficients of variation were 10% and 4.5%, respectively. It seems therefore that the firmness parameter would discriminate better between immature individuals, while the modulus of elasticity would be the best indicator for mature individuals. Seventy five apples of similar firmness (12,02 + 0,93 N) were selected to study the effect of different masses (measured range was from 89 to 216g) in the resonant frequencies. As expected, there was a clear relationship between the resonant frequency and mass (correlation coefficient 0.91), whereas there was poor correlation between the modulus of elasticity and mass (0.05).

The work that Schotte et al. (1999) did with tomatoes aimed to evaluate the correlation between the stiffness index $f^2m^{2/3}$ obtained in the measurements of the acoustic response to mechanical impact and the sensory evaluation panel of experts to determine the slightest difference in the stiffness index that evaluators are able to detect, and to use the technique of acoustic response to mechanical impact to study the changes in firmness produced during storage of tomatoes and the influence of factors such as maturity at time of harvest and in storage conditions.

In the acoustic test the tomatoes were placed horizontally on a support covered with a foam cushion. In the same support a microphone was placed perpendicular to the surface of the fruit and close to the cushion (a few mm). Tomatoes are excited by manual impact at three points on the equator with a head attached to a plastic rod; the average acoustic parameter of the three impacts characterized each fruit. The determination of the stiffness index was performed using the first resonant frequency, for which an arbitrary criterion was established that the frequency selected as the first resonance would have to arise at least 50% of the maximum amplitude of the spectrum.

Some discrepancies were found in the sensory tests conducted with different panels of experts, which means that results must be rescaled in order to compare the results of different panels. A logarithmic model explained the relationship between the stiffness index value and the scores of the experts ($R^2 = 0.79$). The experts showed greater power of segregation in low index value samples: For index values below 5 × 106 $Hz^2.g^{2/3}$, all the experts were able to accurately distinguish changes of 2.5×10^6 $Hz^2.g^{2/3}$, and only a few were able to find variations of 1.5×10^6 $Hz^2.g^{2/3}$. In regard to the index value limits, tomatoes with less than 2×10^6 Hz^2.g$^{2/3}$ had a 50% chance of being rejected by the experts.

Assuming that the changes of firmness, for a given temperature are a function of time, the firmness index can be expressed as a function of degradation in a first-order model (Equation [8.25]), where S is the rate of firmness, the index is S_0, the initial time is t, and the degradation constant is α (1/day).

$$\frac{dS}{dt} = -\alpha S \;\; ó \;\; S = S_0 e^{-(\alpha, t)} \tag{8.25}$$

Firmness index measurements were made in different batches for different combinations of storage temperature, harvest dates, early stages of firmness, and so on. The measured values of S were used to calculate the degradation coefficients for each of the combinations of factors, noting that for a storage temperature of 20°C, the degradation coefficient is not affected by harvest time or the state of maturity at harvest, but for temperatures of 12°C, constant degradation was higher in tomatoes collected at the point of color change.

Introducing the state of initial firmness of the samples and their degradation coefficient under certain conditions in Equation (8.25), the authors established the maximum storage time beyond which tomatoes reached a firmness index of 2.5×10^6 Hz^2.g$^{2/3}$.

De Belieet al. (2000) adapted the acoustic response device used by Schotteet al. (1999) for mechanical impact as an automated control of the evolution of the firmness within cold storage chambers.

The system of automatic monitoring of firmness, used in this study to monitor apples stored at 1°C and 95% relative humidity for 140 days, consisted of a line rotating with 13 holes in which the samples were placed. The system also included an optical sensor to detect the position of the samples, an electromagnetic excitation system whose activation depended on the optical sensor signal, a microphone placed in a fixed base under the revolving line and vertical to the excitation, and a data acquisition system with a PC located outside of the refrigerator. Making periodic measurements allowed the researchers to establish the degradation rate based on the firmness index according to Equation (8.25) and the maximum storage time allowable, after which the firmness was not suitable for consumption. The firmness index was calculated using the first resonant frequency of the response signal of the apples.

There was some discrepancy between the data obtained with the automated system and the fixed acoustic analysis of the response to mechanical impact. The automated system gave consistently higher resonant frequencies than the fixed system, which could be due to different coating materials and support.

In the last ten years, some researchers have applied the method of impact acoustic response as a reference method. In these cases, the stiffness coefficient has gone from being a parameter associated with a method under development to a reference measurement used to accept or reject hypotheses regarding other techniques or variables.

Roth et al. (2003) used this technique to measure changes during the storage of apples grown using ecologic practices and to compare different storage technologies. The only parameter used in this study to characterize the firmness of the samples was defined as the stiffness coefficient $f^2.m$ (N / mm). The use of this coefficient instead of the now most widely used coefficient $f^2.m^{2/3}$ ($s^{-2}kg^{2/3}$) is justified by the authors as being more accepted by the industry because it is expressed in N/mm units. There were differences in this parameter due to the grown practices, station, and storage practices. The variation along the storage time of f^2m fits well in an exponential model when using the average data for each day of measurement.

The hypothetical action of the enzyme polygalacturonase in the process of fruit softening was studied by correlating the cumulative activity of this enzyme and the stiffness index (f^2m). A significant correlation between the two parameters was observed.

Herppichet et al. (2003) designed an experiment to study the effect of water status in tissue elasticity. To this aim, the dehydration process by free transpiration of 18 radishes was controlled. During the dehydration process, several parameters were periodically measured: fresh mass, the stiffness index $f^2m^{2/3}$ where f is the first resonant frequency of the spectrum, the apparent modulus E (compression test quasi-static), and the potential of water using a pressure chamber. Finally, after the period of transpiration, dry weight (drying in an oven at 85°C) was determined for each unit tested.

The apparent elastic modulus E decreases linearly with an increase in water potential ($r^2 = 0.69$). The representation of E versus relative water deficit follows a curve ($r^2 = 0.8$) in which E increases by increasing the water deficit. The slope of these representations fall sharply when water losses are so large that the power is less than −1 MPa and the relative water deficit is greater than 0.12. In the inflection points of these representations (to −1.17 MPa water potential and water relative deficiency of 0,117) pressure potential (related to turgor) is almost 0, which may indicate that the influence of water on the mechanical properties of the tissue becomes very minor when there is no hydrostatic pressure component in the water potential.

However, the coefficient of rigidity does not show a good relationship with measures of water status in the product. Hardly a trend in the data is observed: stiffness coefficient decreased with decreasing water potential ($r^2 = 0.06$) and increasing relative water deficit ($r^2 = 0.12$). It could be concluded that the stiffness index is not a good predictor of the elastic characteristics of the radish, but this does not match what is concluded in the work done on other products. This could be explained if we consider that in many of these studies have involved long-term storage and/or growth, in which changes in the elastic properties of the product is due more to changes in the cell walls than variations in water status. Thus, the authors conclude that the stiffness index is more related to the texture component, which refers to the properties of cell walls. The structure and composition of the product studied in this work differs greatly from those of the fruits analyzed in other research. On the other hand, there are few tested units, so that the statements and conclusions in this paper should be viewed with some caution before being extended to all cases.

8.2.2.3.3 *Other Treatments of the Signals: Other Acoustic Parameters for Determining the Internal Quality of Horticultural Products*

Horticultural products are composed of tissues with high damping, in which the vibration after impact only lasts a few ms, typically 10 ms, so the number of points in the signal in time to be used in fast Fourier transform (FFT) is very small. This, coupled with (1) the asymmetry of these items and (2) the heterogeneity of their structures (which causes different resonant frequencies to be determined depending on the point of impact and point of measurement) cause the appearance of several resonant frequencies very close to each other and with similar amplitudes, resulting in low reproducibility in determining a resonant frequency in a specific sample (De Ketelaere and De Baerdemaeker, 2001). The application of nonparametric smoothing to the spectrum according to the results of these authors increases the robustness of the determination of resonant frequencies. The application of methods of smoothing the signal, based on Fourier series, can find optimal smoothing parameters that do not cause data loss, and eliminate as much noise as possible.

The estimated smoothing function is given in Equation (8.26):

$$\hat{r}(x;m)=\hat{\phi}_0+2\sum_{j=1}^{m}\hat{\phi}_j\cos(\pi jx) \qquad 0\le x\ge 1 \tag{8.26}$$

where the estimate of φ_j is defined in Equation (8.27), *m* is the number of points in the calculation of point *j*, the value of *m* is one that maximizes the so-called risk function, which involves the variance of spectrum, the estimated Fourier coefficients, and *m*.

$$\hat{\phi}_j=\frac{1}{n}\sum_{i=1}^{n}Y_i\cos(\pi jx_i) \qquad \mathrm{j=0,\dots\ n\text{-}1} \tag{8.27}$$

Nonparametric smoothing was applied to the signals of tomatoes subjected to impact. The benefit of softening of the spectra was evident in the softest fruits, in which a significant decrease in the coefficient of variation of the resonant frequency was achieved when the spectrum was smoothed (De Ketelaere and De Baerdemaeker, 2001).

The technique of acoustic impact response was used for detection of internal voids in Conference pears (Schrevens et al., 2001); new systems of discrimination and new acoustic parameters were proposed. The combination of certain conditions of collection and storage can cause physiological disorders in some species and is relatively common to find apples with internal browning and even holes. The spectra of pears with these internal problems had certain abnormalities with respect to healthy pears, as seen in the appearance of peaks near the peak of maximum amplitude. Two groups of parameters to characterize the spectra were established (regions of interest ranged between 300 and 1,300 Hz). On the one hand were considered the following parameters: frequency of maximum amplitude, the number of peaks with amplitudes greater than 33% of the higher amplitude, damping measured at the peak of greater amplitude, area under the spectrum, and momentum of order 0 of the spectral curve with the rotation axis passing through the peak of greater amplitude. On the other hand, the normalized spectra of pears were approached by Fourier series consisting of 24 parameters (12 sines and 12 cosines). A Fourier series expresses a function as the sum of cosines and sines with a period of $2\pi\lambda$. The multiple correlation coefficient (R^2) between the observed spectrum and that predicted by the Fourier series was 99%. Each of the groups of parameters was introduced in a statistical analysis based on classification and regression trees: The algorithm of classification and regression trees select variables with a significant level of prediction, and build the decision rules in a binary decision tree.

The decision tree based on the first parameter group presented two nodes: Pears whose spectrum has a peak with an amplitude higher than 33% of the maximum amplitude are considered pears with internal disorders; those that do not meet this condition are considered pears without internal quality problems if their momentum exceeds a threshold, and are classified as suffering some problem if their momentum is less than or equal to this threshold. In an analysis of 70 pears, 10% of pears can be well classified using this decision tree.

Another decision tree was established with 5 nodes, and using the coefficients of the Fourier series. With this decision tree the percentage of well-classified units rose to about 95%.

8.2.2.3.4 *Speed of Propagation of a Disturbance Used as an Acoustic Parameter in Analysis*

Sugiyamaet et al. (1994) questioned the use of resonant frequencies as a parameter for estimating the maturity of fruits and vegetables, although it has been shown that the resonant frequencies decrease as the fruits mature, the resonant frequencies are affected by the shape and size of the bodies. Some authors have proposed rates including resonant frequency, mass, and density to offset the effect of size, but then assume the characteristics of an elastic sphere.

In their experimental design, the equator of melons tested is divided in 24 equal parts. A pendulum hits one of the points of the equator and the signal is picked up with a microphone placed each time at one of the 21 remaining points, excluding the point impact and the two closest to it. The 21 signals per sample were included in a tridimensional diagram in which the axes were the time, location, and intensity of the microphone. These signals exhibited a pattern in X, so the authors concluded that the impact response signal is transmitted through the surface with a uniform speed in both directions.

The transmission of the impact response at the equator was considered as part of a spherical wave spreading across the surface. A simplification was taken into account considering only one dimension. The wave equation is expressed in Equation (8.28), where v is the transmission speed and u (x, t) is the displacement at a point x, in an instant t.

$$\frac{\partial^2 u(x,t)}{\partial t^2} = v^2 \frac{\partial^2 u(x,t)}{\partial x^2} \tag{8.28}$$

Considering a small volume of the sample, with a linear density ρ, subjected to a pure displacement of traction, the expression in Equation (8.29) is proposed, where E is the modulus of elasticity.

$$v^2 = E/\rho \quad (8.29)$$

The speed of transmission of the response to the impact was measured on 36 melons. The maximum penetration force and the score given by a panel of 48 tasters were the reference variables used to compare with the speed measurements.

The testing device consisted of a pendulum impact generator and two microphones located parallel and spaced 5 mm and 3 mm from the surface of the melon. The microphones were placed to record the signal at a point on the melon that formed an angle of less than 90° to the point of impact, because otherwise, the transmitted wave in the other direction would interfere with that which you wish to measure. With this microphone arrangement, the microphone nearest the impact point recorded the response first; the other microphone recorded this response 100 μs s later. The lag time between the two microphones corresponds to a distance of 5 mm (the distance between two microphones).

The delay time was measured by calculating the cross-correlation function between the signals in time. For this purpose, 256 points were selected in each of the microphone signals behind the point of maximum intensity of the first peak and 256 points later. The delay time is determined considering the peak of the cross-correlation function.

A correlation coefficient of 0.832 between the parameter transmission speed and maximum penetration force was obtained, observing a lower repeatability in the force of penetration than in the transmission speed.

Similar experimental design was applied in the work of Hayashi and Sugiyama (2000), in which melons, apples, and pears were measured. The velocity of the waves transmitted on the surface is the parameter that relates to firmness in these fruits.

Sugiyamaet et al. (1998) moved to a portable device for field use which determined the speed of transmission of the wave response to mechanical impulse. A more compact design was achieved as a result of including in a single carcass, the striker element, and microphones. A trigger attached to a spring system caused the impact rod to move. The microphones were spaced 16 mm apart.

A total of 72 melons of different varieties were measured with the portable device. The correlation found between the apparent elasticity measured in a compression test of probes, and the transmission speed, was higher than 0.9.

This device also monitors a melon in plant, collecting data starting 20 days after pollination and finishing 13 days after harvest, held 61 days after pollination. During the first 5 days an increase in the transmission speed was observed; from the beginning of the formation of the grid surface of the melon and the increase in its circumference, the transmission speed decreased until harvest. After harvest, the decrease of the speed was more pronounced.

8.2.2.4 Signal Collecting with an Accelerometer

De Belie et al. (2000) checked the viability of the acoustic impact response system on nonspherical fruit, particularly pears, measuring changes in the firmness index ($f^2m^{2/3}$) produced during the ripening of the fruit on the tree.

In this case, an accelerometer sensor element was placed at the end of the flower, and the excitation was produced by an impact near the stalk. Due to the difficulties of measuring the weight of each fruit on the tree, an estimate was made that took into account the diameter and length of each fruit. A decrease in the firmness index along the track of samples on the tree was observed (from 4 weeks before the date of commercial harvest, up to 2 weeks later), noting that between three weeks and one week before the commercial harvest date, there is a sudden drop in the firmness index, from which this parameter is set to Equation (8.25), where t is the time elapsed since the sharp decline occurs in the firmness index. However, the calculation of the degradation rate can be better estimated in the evolution of fruit after harvest rather than on the tree.

The method of measuring the response to mechanical impact was considered to be a good system to monitor the evolution of firmness of pears on the tree, and to determine the optimal harvest date, especially if we consider that the correlation coefficients between the firmness index and destructive variables indicative of maturity (maximum penetration force, soluble solids content, and modulus of elasticity) ranged between 0.59 and 0.82. Moreover, as a nondestructive method, it was possible to monitor the same fruits along the ripening process, giving a more accurate idea of the evolution of firmness than with destructive methods, which require different samples each time the measurements are made, giving a more irregular pattern of changes.

8.2.2.5 Signal Collecting with a Piezoelectric Sensor

8.2.2.5.1 Stiffness Coefficients as Acoustic Parameters Used in Analysis. Other Parameters: Damping, Frequency Spectrum Centroid

Shmulevich et al. (1994) developed a new system for assessing firmness based on the acoustic response of fruit produced by impacting the sample with a pendulum; however, the sensor element of the system consisted of a flexible piezoelectric film. The piezoelectric sensor was formed by a sheet of polyvinylidene fluoride (PVDF) coated with two layers of conductor material, all surrounded by a polyethylene foam to allow vibration of the fruit. The system included a load cell to record the weight of each fruit. For each measure, the firmness index ($f^2m^{2/3}$) was determined.

The research studied the evolution of the firmness of plums, nectarines, and mangoes during storage. The plums were placed in the holder of the device. The piezoelectric sensor measured at two positions, one below the sample and one on the side, coinciding with equator and 90 degrees from the point of impact. The sensor configuration was the same for nectarines, except that the fruit was placed in the holder with the stem horizontal, so that the sensor located under the sample was located at the equator and was at 90 degrees from the point of impact, while the lateral sensor measured in one shoulder of the sample. The mango, due to its shape and size, could only be measured with the piezoelectric sensor located in the bed of support.

A second-order polynomial function allowed the firmness index value to be modeled versus days of storage of the fruit. The best results, defined as less dispersion of data, were obtained with sensors that measured in the equator of plums and nectarines at 90 degrees from the point of impact. The results showed greater variability in mango, which seems to require a different support design adapted to the characteristics of shape and size of this fruit.

The use of the sensor element introduced in this work could broaden the perspectives of use in the agricultural industry of classification devices based on the vibration characteristics of products.

An exhaustive investigation in avocados, in which the technique and the device previously described was used (Galili et al., 1998), allowed proposal of new acoustic parameters in an attempt to relate the firmness and the maturity of the fruit. Regarding the resonant frequencies and the resulting firmness index ($f^2m^{2/3}$), two searching algorithms were used to determine the frequency of maximum amplitude (f_{max}) and the first resonant frequency (f_1). For this determination, the following conditions were set: it must be a local maximum, the amplitude had to be at least 50% of the maximum amplitude, and at least 10% greater than the value of the curve in points on the right side, its value should be greater than a threshold established experimentally (in this case 200 Hz). From the two determined resonant frequencies, the firmness indexes were calculated: FI_1 and FI_{max}.

An alternative method to determine damping was proposed (Equation [8.30]). The value of damping ζ for frequencies f_{max} and f_1 was calculated by averaging the values obtained by applying Equation (8.31) at 10 points to the left of the resonant frequency. Equation (8.30) is the expression of the damping in resonant peak frequencies based on the amplitude equal to half the maximum amplitude peak.

$$\zeta = \frac{\omega_2 - \omega_1}{2\omega_n} \tag{8.30}$$

Equation (8.31) is the expression of the damping in a resonant peak in terms of the relationship between frequency ω of the peak and the resonant frequency, and their respective amplitudes.

$$\zeta = \frac{\left|1 - \omega'^2\right|}{2\left[\left(\frac{H_m}{H}\right)^2 - \omega'^2\right]^{1/2}} \tag{8.31}$$

In these expressions, ω_n is the value of the resonant frequency, ω' is the relationship ω/ω_n, H is the amplitude at ω, and H_m is the amplitude of the resonant peak.

This study also introduced a parameter called the *centroid* of the frequency spectrum, which is calculated according to Equation (8.32).

$$f_c = \frac{\sum_{i=1}^{n}(f_i H_i)}{\sum_{i=1}^{n} H_i} \tag{8.32}$$

This parameter is easy and fast to compute, which gives potential in applications requiring high speed. In addition, it does not depend on specific resonance frequencies, but encompasses the spectrum of information as a whole, which might be of interest in samples containing numerous asymmetric resonant peaks close together.

The experimental design was made up of several lots of avocados from two different varieties: The fruits were tested on the day of harvest, and after a period of 18 days under refrigeration (4°C and 80% RH); then the fruits were measured every day for 6 days to check the shelf life.

In the process of ripening of avocados, variations in the resonant frequencies with maximum amplitude were observed, so not always the same frequency f_n presented the maximum amplitude of the spectrum. This, of course, affects the calculation of the firmness index based on the f_{max}: erratic tendencies were seen in the parameter value. During maturation it is important to keep track of the same resonant frequency, which was demonstrated by using the f_1 in the calculation of the firmness index, which showed a steady decline in its value over the days of shelf evolution. The damping increased during the shelf life of fruits.

The correlations between the maximum penetration force and the acoustic parameters based on f_1, were higher than those for f_{max} parameters: between 0.695 and 0.778 and between 0.58 and 0.72, respectively. The best correlation of the penetration force was obtained with the centroid of frequencies (0.806).

Despite having defined all the parameters described in the preceding paragraphs, the author uses only the stiffness coefficient ($f^2m^{2/3}$) in one of his last works (Shmulevich, 2003) to analyze the capabilities of the device developed in mango, avocados, melons, and nectarines, and to compare it with two prototypes designed by the Sinclair company. Both systems recorded the response to a low-intensity impact with a sensor element. The first of Sinclair devices has a small pendulum to produce excitation; in the other (Sinclair IQTM Firmness Tester) the sensor element is placed inside a bellows that expands to impact the test point in the sample. Through a system of acquisition and signal analysis, Sinclair indexes are generated: TIQ and SIQ respectively.

Other measurements were performed as references to characterize the state of the samples: compression between parallel plates for determining the apparent modulus E – (ASAE Standard S368.3) and penetration of the pulp according to species. Good correlations were obtained between the Sinclair SIQ index and the modulus (R greater than 0.7 in all species). The coefficient of firmness ($f^2m^{2/3}$) was correlated with the Sinclair indexes (R between 0.81 and 0.99). A multisensor classification system that incorporates a firmness index and the Sinclair SIQ index was proposed, which improved the results of each device.

8.2.2.6 Band Magnitude as an Acoustic Parameter Used in Analysis

Another system in which piezoelectric sensors were implemented as signal receiving elements was employed by Zhang et al. (1994). They compared the results with the acoustic device described by Farabee and Stone (1991) (explained in more detail in Section 8.2.3.3.) and the results of a drop test with the traditional parameters characterizing the firmness of peaches, such as background color through the Hunter Lab color space, in which the "a" coordinate ranges from values for green and positive values for red, and firmness Effegi. The impact test parameters in free fall were analyzed: time of contact, relationship between maximum force and time of maximum force, and absorbed energy.

The relationship between the amplitude of the second and the first peak was obtained from the acoustic signal on the domain of time, thus making an approach to the damping coefficient. From spectrum (obtained by applying the FFT) the parameter BM80-130, *band magnitude*, was extracted, calculated by summing the spectrum magnitudes contained between 80 and 130 Hz and dividing by the sum of the magnitudes of the spectrum contained between 0 and 500 Hz. BM values are thus proportional to the energy content between these frequencies.

All measured parameters showed a decrease over time in the laboratory (two days), but the percentage of variation was different for each variable and variety, with the Effegi firmness being the parameter with the highest percentage of variation between day 0 and day 2. Both impact parameters in freefall, as those for the acoustic test, showed a wide range of values of correlation with Effegi firmness (r ranged from 0.68 to 0.89). Differences occur depending on the variety studied. Effegi firmness turned out to be the most sensitive parameter to changes in firmness, and also proved to be the parameter with greater variations within a single fruit.

Stone et al. (1998) reapplied the device of Farabee and Stone (1991) on peaches in order to perceive changes in the evolution of firmness in different areas of the fruit. To accomplish this, each peach was beaten on two locations on the equator, on the suture, on the flower, and on the stem, with 8 impacts per fruit. In conducting these experiments, the peaches were placed on a support with a curved surface and adaptable to the shape of peach, allowing the point of impact to be matched with the center of the transducer. The reference parameter used was destructive Effegi firmness at each of the 8 test points of each peach. Based on previous studies, the only sound spectrum parameter studied was the BM150-200, calculated as the sum of the intensity between 150 and 200 Hz in the normalized spectrum.

In the Cresthaven variety, significant differences between the locations of test points were observed. The correlations between the acoustic parameters and Effegi firmness of local measures were less than 0.37 for all locations, rising to 0.44 if the mean values obtained in the two parts of the equator, the suture, the flower, or the stem are considered. However, in the Loring cultivar, variations due to location were lower for both Effegi firmness and the acoustic parameter; the correlations between acoustic parameters and reference of local measures reached a maximum of 0.82 for the equator location.

8.2.2.7 Comparison of Two Systems that Differ in the Element Sensor: Piezoelectric Sensor versus Microphone

Piezoelectric sensors and microphones have been presented as possible alternative sensor elements of the vibration signal in the design of different acoustic devices. Armstrong et al. (1997b) compared the two systems using two devices in studies with peaches. One of the systems used was described by Farabee and Stone (1991), which includes a piezoelectric sensor; the other system basically consists of an arm tapping one side of the sample and a microphone located opposite the point of impact.

The acoustic parameters extracted from the spectra were the same for two devices: CFN50 and BMX. Both types of variables are explained in detail in Section 8.2.3.3. to address these authors' previous work, the bandwidths associated with BM parameters, size-band, are the same as those used by Farabee and Stone (1991).

Two varieties of peaches, Red Haven and Cresthaven, were tested using the two acoustic devices and determining the strength Effegi. All measurements were made twice in each peach in points opposite each other and at 90° of suture. The average of the two measurements was the value adopted for the characterization of each unit.

Multiple linear regressions between the variables for each of the acoustic devices and Effegi firmness were computed for each variety. All BM parameters were included in the regressions. The inclusion of weight in the regression functions slightly improved their adjusted r^2. For both varieties, the microphone system provided better results: adjusted r^2 of 0.64 and 0.68 in functions where the weight was not included, and 0.64 and 0.74 in the regressions including the variable weight.

It could be said that the ability of any two devices to reliably predict the firmness Effegi is limited. However, the microphone seems to be a potential instrument to perform a preclassification to separate very hard fruit (> 50 N Effegi) and very soft fruit (<20 N).

8.2.3 Internal Quality of Watermelons

Knowledge of the textural characteristics of the pulp of the watermelon before cutting the fruit has been a concern of many researchers for decades. Special attention has been paid to this fruit with respect to the acoustic techniques.

Clark (1975) related the acoustic properties of whole fruit with the Magness-Taylor firmness of the pulp. The watermelon was hit by an arm at 4 points around the equator; for each hit, two signals were recorded by two sound-level meters, one on the side opposite the impact and another on the same side as the impact. The parameters extracted from this test were the damping time of the signal transmitted through the fruit (T_T) and the damping time of the reflected signal (T_R). The correlation coefficients between the damping times and Magness-Taylor firmness was less than 0.2.

8.2.3.1 Signal Collecting with an Accelerometer

Sasao (1985) developed an experiment on growing watermelons (on plant). During the fruit development (from day 20 after pollination until 57 days after pollination), acoustic vibration measurements were taken. The watermelon was placed on a rubber ring; it was excited by an impact and vibration response was collected with an accelerometer at the point diametrically opposite to the excitation point. The evolution of weight was regularly monitored. In the sample tested, consisting of watermelons of the same variety, 3 groups were distinguished according to the weight gain recorded between 20 and 40 days after pollination. The first group included watermelons whose weight increased 2.5 kg in that period (50%), the second group varied from 3.5 to 4 kg (30%), and the third group (20%) the ranged from 5 to 6 kg. This classification is of interest in observing the evolution of frequency spectra of each group. The group with less weight gain showed a spectrum in which the first resonant frequency decreased from 207 to 162 Hz; in the last days of monitoring (days 41–57) a second resonant frequency around 225 Hz was observed; that is the frequency of maximum amplitude on the last day. In the other two groups, although you can see a decreasing trend in the first resonant frequency (from 176 Hz and 188 Hz, respectively to 102 Hz), from days 41 and 34, the spectra become more irregular, and other peaks appear with higher amplitude than the first resonant frequency. The proliferation of resonant peaks (irregular spectrum) is related to the appearance of holes and internal cracks.

8.2.3.2 Signal Collecting with a Microphone

Yamamoto et al. (1980) presented a nondestructive technique for measuring the textural quality of apples and watermelons, based on the acoustic properties of these products. The advantage lay in the fact of using a microphone located a few millimeters from the surface of the sample, instead of an accelerometer attached to it.

The mechanism impactor consisted of a pendulum with a wooden ball on the end. A condenser microphone was positioned at 180 degrees from the point of impact, and transmitted the signal to an amplifier. The power spectrum is calculated by FFT. During the test, the watermelons were placed on a hard flat surface due to the difficulty of suspending them by their heavy weight. The samples were subjected to different periods of storage, after which they were measured. Sensory evaluation and determination of Magness-Taylor firmness were performed as reference tests.

In watermelons, the spectra showed a high number of frequency peaks. The frequencies of these peaks (considered the first three) had good reproducibility. It was a negative correlation between weight and peak frequencies, so the parameters proposed by other authors to correct this influence were applied. Table 8.2 shows these different *firmness indexes.*

All acoustic parameters were very poorly correlated with the Magness-Taylor firmness test and on the order of 0.7 to the sensory evaluation of firmness.

8.2.3.2.1 Using the Speed of Propagation of the Disturbance as an Acoustic Parameter Classification

In the work of Hayashi and Sugiyama (2000), watermelon and pineapple are included. According to the three-dimensional representations, the authors identified transmitted waves through the surface and other waves transmitted internally; it is the speed of the internally transmitted waves that is related to the firmness

TABLE 8.2
Definition of the Different Firmness Indexes Based on the Resonant Frequencies and Their Authors

Authors	Index
Abbott et al. (1968); Finney (1971, 1972)	mf^2
Cooke (1972)	$m^{2/3}\rho^{1/3}f^2$
Yamamoto (1980)	$m^{2/3}f^2$

of watermelon. A disruption of these internal waves was observed in watermelon with cracked pulp. The first part of the time signal is the zone that is identified with the transmission of disturbance through the internal part of the sample (time <5 ms).

8.2.3.3 Signal Collecting with a Piezoelectric Sensor

Farabee and Stone (1991) designed and developed a response to the impact sensor for watermelons with the aim of providing a nondestructive method for determining maturity.

It is important to have reference methods in these works that allow us to see the virtues of new developments. However, there are few references on destructive determination of physical properties that are good predictors of maturity; some researchers have used the color of tissue inside the watermelon, sugar content, firmness, and thickness of the cortex (measured as the thickness of the outer white ring in a cross-sectional or longitudinal cut of the fruit). These references were considered by the authors as indications of maturity.

The element that generates vibration and the piezoelectric sensor for the registration of the vibration were mounted in a plastic housing. A solenoid acted as the pulse generator; the piezoelectric sensor sent the signal to an oscilloscope, which transferred it to a PC for analysis.

Seven signals were collected for each watermelon: 3 at the equator of the watermelon, two below the equator (halfway between the equator and the flower), and two above the equator (halfway between the equator and the stem). The first 1,000 points of the signal are used to calculate the frequency spectrum. The 7 spectra of each watermelon were normalized to the maximum amplitude of energy, and the average spectrum characteristic of each watermelon was calculated. With normalization, the energy differences between spectra are eliminated, but the energy differences between the different frequencies of the same spectrum are preserved.

The acoustic parameters extracted from the normalized average spectra were:

- Frequency band center narrower containing 50% of the total energy spectrum: CFN50.
- Energy contained in the band bounded by 85 and 160 Hz frequency, band magnitude: BM85-160.
- Relationship between the second and third highest peaks of the signal at the time (an approach used by Clark (1975), who observed a relationship between signal attenuation and the ripeness of watermelon, estimated by the color of the flesh).
- Resonant frequencies; parameters that had been successful in determining the state of firmness of apples and peaches.

The best correlation coefficients were obtained between the refractive index (soluble solids) and spectral parameters CFN50 and BM85-160: 0.67 and 0.56, respectively.

Multiple regressions were used to calculate prediction models of parameters CFN50 and BM85-160 in terms of destructive measurements. The regression equation with the highest correlation (0.734) was the CFN50 calculated according to the refractive index and the firmness, measured as the slope of the linear region of the curve obtained in a Magness-Taylor penetration of 8 mm. The introduction of the weight of the samples in regressions did not improve the models, which suggests that the spectral parameters used are independent of sample weight.

The studies explained in the preceding paragraphs were also performed in 13 seedless watermelons corresponding to a variety that differs substantially in shape and size from the range studied in the heart of the investigation; it is larger and elongate, while the seedless watermelon is of a more spherical shape. However, the most significant regressions obtained reinstated the refractive index and the Magness-Taylor slope as independent variables to predict CFN50 and BM85-160, with correlation coefficients above 0.8, so that shape and weight did not appear to reduce the effectiveness of these two parameters.

Both CFN50 as BM85-160 showed good prospects as parameters to predict the state of ripeness of watermelons, but the simplicity of calculating the BM85-160 makes it suitable for possible application in control systems.

The same equipment was used to determine the ripeness of watermelons in the field (Stone et al., 1996). The influences of different support bases during the execution of the acoustic measurement were considered: the bed of earth where the watermelon grew (watermelon still attached to the plant), a wooden frame of 2 cm × 25 cm × 50 cm (watermelon separated from the plant).

Stone et al. extracted a greater number of parameters to characterize the normalized frequency spectrum: CFN50 and 7 different BMX. Each BM (band magnitude) is calculated by adding the magnitudes between x and y frequencies in the normalized spectrum and dividing this sum by the sum of the magnitudes of the spectrum between 0 and 500 Hz. The parameters studied were: BM85-160, BM40-90, BM60-110, BM70-120, BM80-130, BM100, and BM120-180-200.

The acoustic tests were conducted in three different locations of the fruit within the central area. The analysis of variance performed showed no significant differences at the 5% level different locations. However, the differences were significant between the acoustic measurements made with different supports.

The results were different for different varieties studied. In general, the correlations between destructive measurements of maturity (flesh color, soluble solids content, and Effegi firmness) were poor, which could be explained considering that these relationships are dependent on variety and a single parameter cannot be used as a good reference of maturity in all varieties.

Even the best correlations between acoustic parameters and a single destructive reference parameter not were high enough to think of them as reliable predictors of the state of firmness. The acoustic parameter BM40-90 showed the highest correlation with sugar content and color (0.494 and 0.636), and this result was reasonably consistent with the two supports tested. By contrast, in some varieties, the results between the different supports were not coincident, indicating the importance of support in the resonant response of a body.

These acoustic parameters were used to establish the relationship between them and the level of hollow heart in watermelons. In this study, the wooden support was used for implementation of the acoustic test. Watermelons were divided into 4 groups based on the level of hollowness by visual inspection: watermelons without hollowness were level 0, high volume of hole were level 4. For the Queen of Hearts variety ($n = 60$), the parameter BM60-110 showed the best correlation with the level of hollowness (0.584), but for the Black Diamond variety, BM80-130 was the most well-correlated parameter with a hollowness level (0.784).

The same authors (Armstrong et al., 1997a), years later returned to the Farabee device to evaluate its ability to detect internal bruises in watermelons. Different levels of bruising in the fruit were achieved with free-falling of samples from a height of 25 cm onto a hard surface repeatedly (up to 5 times). The acoustic and compression testing up to 2.5 cm between parallel plates (nondestructive measurements) were repeated before inflicting any damage to the fruit and after each fall. In the compression test, the watermelon was placed with the stem-flower line horizontal.

From the compression test, two variables were studied: maximum force to 2.5 cm and the slope of the stress–strain curve. Both variables decreased with increases in the level of a particular fruit bruise. However, means of all watermelons tested for each bruised level were not significantly different from each other, reducing the value of these parameters as criteria for the classification of the watermelons in terms of their internal state. The acoustic parameters extracted from the *t* were the same as those defined by Stone (1996). The best model to estimate the slope of compression from acoustic parameters ($r^2 = 0.7$) was a multiple linear regression that included BM60-110, BM70-120, BM80, and BM120-130-200 as dependent variables. Both the acoustic and compression settings were only effective in resolving major differences in the internal state of watermelons.

Waveband magnitude parameters, obtained by summing the magnitude of the spectrum between two frequencies in a specified band width have been also applied in devices including a microphone as a recording signal system (Diezma-Iglesias et al., 2004; Diezma-Iglesias et al., 2006).

8.3 OTHER APPLICATIONS OF SOUND FOR THE DETERMINATION OF QUALITY IN FOOD PRODUCTS

The analysis of the vibrational response to a nondestructive impact has been applied to other food products.

8.3.1 Surface Crack Detection in Eggs Using the Impact Acoustic Response Technique

This technique has been tried to solve the serious problem of detecting cracks in eggs. Coucke et al. (1999) calculated the dynamic stiffness of each egg and set the relationships between this parameter and other geometric and physical parameters that characterized the product. A modal analysis previously showed a spheroidal mode of vibration at lower resonant frequency; the points of excitation and collection of the signal were fixed at the equator of the eggs spaced 180 degrees, so that the system was optimized for the collection of the spheroidal mode. The excitation mechanism was a manual impact of low intensity; a microphone, located at few millimeters from the surface of the sample, picked up the signal. The first resonant frequency occurred at around 3,600 Hz.

To set the correction factor to compensate for the effect of mass differences in resonant frequencies, it is assumed that the dynamic behavior of the egg can be expressed as a linear mass-spring system. The motion of the system can be written in the second-order differential equation shown as Equation (8.33):

$$m.\frac{d^2x(t)}{d^2}+k\,x(t)=f(t) \tag{8.33}$$

where t is time, x (t) is the displacement of mass along the x axis, m is the mass of the spring, spring stiffness K (N/kg) and f (t) is the external force applied. If f (t) is 0, the solution of this equation can be written as given by Equation (8.34):

$$x(t) = A\cos(\omega t + \varphi) \tag{8.34}$$

where A is the amplitude of the periodic signal, ω is the natural frequency, and ϕ is the phase at the start of the vibration. Substituting Equation (8.34) into Equation (8.33) we obtain Equation (8.35):

$$\omega = 2\pi f = \sqrt{\frac{K_{din}}{m}} \tag{8.35}$$

So, knowing the f (resonant frequency) and the mass, K_{din} (dynamic) for each egg was calculated. K_{est} (static) was determined using a compression test between two parallel plates. Both constants were compared and correlated. There was good correlation between K_{din} and K_{est} (0.71); K_{din} was related to the thickness of the shell at the equator, and the egg shape was expressed as a ratio between the minor axis and major axis of the egg.

In a later paper, De Ketelaereet et al. (2000) studied the procedures for detecting cracks in the eggshell and the parameters and thresholds for classification. The testing device in this case used a microphone near the point of impact. On the motion that describes the first spheroidal vibration mode, any of the following locations of the microphone with respect to the impact point: 0°, 90°, 180°, and 270° are optimum. If using the device for on-line classification, the best option would be the location at 0°.

Each egg was hit and measured at 4 points on the equator, 90° apart. Eggs without discontinuities in the shell had a higher level of similarity between the spectra corresponding to the 4 impacts. The first resonant frequencies corresponding to the spheroidal mode in these eggs ranged from 3,000 Hz to 6,000 Hz, and the amplitude of the resonant peak values showed a range of about a factor of 20. Cracked eggs showed spectra with higher differences between them. To identify cracked eggs, spectra were compared 2 to 2 between the 4 spectra obtained from an egg. The relevant parameter to characterize and classify an egg was the lowest Pearson correlation coefficient obtained in comparisons of the spectra. A threshold of 0.9 in the correlation coefficient optimized the detection of cracks (90% of cracked eggs are rejected) but gives the worst levels of false rejection (sound eggs classified as cracked, 1.16%). The threshold of minimum correlation coefficient can be modified to meet the requirements of the potential user of the device.

In order to determine on-line the static stiffness as a characteristic quality of eggs, some works (De Ketelaere and De Baerdemaeker, 2000; De Ketelaere et al., 2000; De Ketelaere et al., 2003) propose a multivariate model that includes dynamic stiffness (determined by the method of acoustic response to the impact), damping signal, and an index defined as the ratio of the major axis to the minor axis. An alternative procedure for estimating the resonant frequency and damping of the signal was proposed, which is not based on the FFT algorithm. Research has shown that obtaining the frequency spectrum using the FFT does not result in very accurate estimators when short time signals are involved. In general, constructing the frequency spectrum using an FFT can be regarded as a way to reconstruct the power distribution of a time signal as a function of the frequency. In general, the parameters of interest are the resonant frequency and the corresponding damping of the vibration, rather than the whole energy distribution over all frequencies up to the Nyquist frequency (spectral line estimation). Starting from the frequency spectrum, the resonant frequency is calculated as the frequency with the largest power. The frequency resolution of the frequency spectrum is set by the ratio between the sample frequency and the number of points in the time signal. It is clear from this relation that highly damped specimens give rise to a low-frequency resolution. Altering the sample frequency has no effect on the resolution, since it also alters the number of points in the time domain. Since an impact is used as the excitation method, there is no way to increase the number of relevant points in the time signal: the length is a characteristic of the specimen under study. So, the only way to enlarge the number of points without changing the energy content of the time signal is to adopt zero padding. Indeed, in this way the denominator of the equation becomes larger, giving a better resolution (De Ketelaereet al., 2003). Due to those drawbacks of the FFT, another method was proposed to find robust and high-resolution estimators of the resonant frequency and damping. High-resolution estimators distinguish themselves from low-resolution estimators (such as the FFT) by the fact that their aim is not to construct the power distribution of a given time signal over all frequencies. Instead, their aim is to find precise knowledge about one line in a given spectrum (spectral line estimation). The term *high resolution* is justified since they are able to resolve spectral lines separated in frequency f by less than $1/N$ cycles per sampling interval, which is the resolution limit for classical FFT. High-resolution estimators exploit an exact parametric description of the signal (in this case a damped cosine). In the impulse–response measurements taken for the nondestructive quality assessment, this parametric form is easily derived by solving the equation of motion for a mass–spring–damper system, which leads to the parametric form of a damped cosine. In this study, the simplification toward a single degree-of-freedom system is made, since the setup is optimized in such a way that only the first modal shape is encountered in the response. The response follows the parametric expression of a damped cosine shown in Equation (8.36):

$$s_n = \alpha \cos(\omega n + \varphi) e^{-\delta n} \quad \text{para n} = 1.....\text{N} \tag{8.36}$$

where N is the number of points earned. In the simplification of the system to a single degree of freedom, it is assumed that the device optimized to collect the signal is only one mode of vibration (as modal analysis is required in advance). Denoting the noise term by z, the acquired signal y is the sum of the parametric forms and the noise z (Equation [8.37]):

$$y_n = s_n + z_n \quad \text{para n = 1.....N} \tag{8.37}$$

The determination of the unknown parameters that minimize Equation (8.38) is the way of estimating the spectral line, ω, A, ϕ and δ are the frequency, the amplitude, the phase, and the damping, respectively.

$$F = \sum_{i=1}^{n}\left(y_i - \hat{y}_i\right)^2 = \sum_{i=1}^{n}\left[y_i - \left(Ae^{-\delta t}\cos(\omega t + \varphi)\right)\right]^2 \tag{8.38}$$

A principal component analysis showed that the information provided by the resonant frequency calculated by either method was the same, which does not support the criticisms made by the authors with respect to determining the resonant frequency from the FFT. However, the damping itself differed significantly. It is shown that the damping determined by high-resolution estimation is the major factor in explaining eggshell strength.

Surface crack detection in eggs was also the main objective of the work of Cho and Choi (2000), which developed an experimental device in which each egg was impacted 4 times, twice in the area at the ends of the egg. The criteria for segregation between eggs with cracks on the surface and intact eggs were based on a series of measures from the frequency spectrum such as area under the curve of the spectrum from 0 to 10 kHz, coordinates of the centroid of the spectrum area, and resonant frequency. It was noted that the signals on the time and frequency domain corresponding to a cracked egg differed from each other significantly, while the curves in intact eggs showed a higher degree of similarity. The decision rule for identifying eggs with a cracked surface was generated by discriminant analysis, including the average area of the spectra of the 4 impacts, average of the coordinates of the centroids, difference between the maximum and minimum x-coordinate of the centroid, the difference between the extreme values of the y-coordinate of the centroid, and the average of the resonant frequencies. It achieved a level of success of 90%.

The monitoring of cheese maturity has also been the object of study by acoustic impulse response. Acoustic parameters extracted from spectra computed by FFT have been compared with textural parameters measured by traditional instrumental methods. The impact setup consisted of a free-falling impact probe; a microphone was located at the same face as the impact point. For the whole spectrum, the momentum M_0, the momentum M_1, the central frequency DF, and the variance Var, were calculated for the interval ranging from 10 to 70 Hz and from 70 to 400 Hz, and are shown in Equations (8.39), (8.40), and (8.41):

$$M_o = \sum_{f_1}^{f_2} X(f)\Delta f \tag{8.39}$$

$$CF = \frac{M_1}{M_0} \tag{8.40}$$

$$Var = \frac{\sum_{f_1}^{f_2} (f\_CF)X(f)}{\sum_{f_1}^{f_2} X(f)} \tag{8.41}$$

where f is the frequency and $X(f)$ is the amplitude. Changes in the frequency spectra took place as cheese matured, increasing higher frequencies and energy content. Multiple linear regression (MLR) and partial least squares regression (PLSR), considering the acoustical variables extracted from the spectrum, allowed for a good estimation of cheese texture. The textural characteristics of the cheese surface, and in particular the maximum force in compression experiments ($R^2 > 0.937$ for MLR and $R^2 > 0.852$ for PLSR) were accurately predicted by the acoustic method; however, the texture of the central layers of the cheese were poorly assessed ($R^2 < 0.720$) (Benedito et al., 2006; Conde et al., 2007).

8.3.2 Determination of Textural Properties of Different Foods by the Sound Produced in Mastication

The textural properties of food are used as key factors in the acceptability of the product. These properties are perceived by the consumer through a combination of visual, tactile, and auditory sensations (Duizer, 2001). The sound produced by chewing some foods has been studied for information about their texture. Various procedures of acquisition of sound have been presented by different researchers and for different applications: a microphone in front of the mouth of the individual who is chewing the sample (Vickers and Bourne, 1976), a microphone placed over the ear canal (Vickers, 1981; Vickers, 1983; De Belie et al., 2003), a microphone on the cheek near the jaw angle (Lee et al., 1988), or a microphone placed in the ear canal on the side on which the sample is chewed (De Belie et al., 2000b).

Any of the techniques used may provide an empirical measure of the difference between textures of food samples. In spite of this, there are some difficult problems to solve in technology. The soft structures of the mouth tend to dampen high-frequency sounds (Vickers, 1991), possible interference between chewing movements in general and the specific sound of the crunching of the food can occur, and the frequencies of the sounds of the opening movement of the mouth have been located at around 160 Hz (Drake, 1963; Kapur, 1971).

Different parameters have been used to judge the sound of chewing, such as amplitude of the signal in time (Drake, 1965) and intensity in defined frequency ranges (Drake, 1965; Seymour and Hamann, 1988).

(De Belie et al., 2000b) observed that the Royal Gala apple amplitudes between 700 and 900 Hz, and *t* between 1200 and 1400 Hz were due to the general movement of mastication and that the amplitudes near 4000 Hz were related to the first bite to the sample. According to these authors, combining the information contained in different spectrum bands could produce a good correlation with sensory evaluations. However, the significant frequencies in the analysis were found to be different for each consumer, making it difficult to use a combination of frequency bands as a parameter for measuring crispiness.

In a further attempt to use this technique to distinguish apples of different textures, mealiness was induced in Cox's Orange Pippin apples by storage times and conditions causing the disorder. Mealiness implies a loss of crispiness without causing an appreciable loss of moisture. The recording of the sounds produced by chewing a piece of apple with a defined shape and size was performed by placing the microphone in the ear canal on the side on which chewing occurred. The sound collected corresponds to the cut of the sample with the incisors and the successive chews made with the molars. FFT was applied to this signal to obtain the corresponding spectrum. The application of principal component analysis reduced the number of variables to consider for the generation of a classification model. In this study, 3 principal components explained 73% of the variance of the population. Frequencies between 100 and 500 Hz and between 800 and 1100 Hz contributed particularly to the principal components. Plotting the values of the two first principal components for mealy, crisp, and intermediate state apples, there was a good separation of groups of extreme characteristics. It was noted that the technique is more related to the sensory evaluation of the mealiness that to the mechanical tests. The development of a technique for determining crispiness requires a previous calibration for each person because each person emits different sounds in the process of mastication.

The same technique was applied to the differentiation of types of fries and snacks (De Belieet al., 2003). The sensitivity of human hearing is proportional to the logarithm of the intensity, so that high frequencies (7000 to 8000 Hz) may be important for assessing crispiness. Therefore, principal component analysis was applied to the logarithms of the spectra obtained in the process of chewing products. A satisfactory separation of chips of regular shapes and uniform appearance was obtained, while the less homogeneous samples were more often misclassified. Again, it was observed that the system should be calibrated for each person.

REFERENCES

Abbott, J.A., G.S. Bachman, R.F. Childers, J.V. Fitzgerald, and F.J. Matusik. 1968. Sonic techniques for measuring texture of fruits and vegetables. *Food Technology* 22:635–646.

Abbott, J.A., and L.A. Liljedahl. 1994. Relationship of sonic resonant frequency to compression test and Magness-Taylor firmness of apples during refrigerated storage. *Transactions of the ASABE* 37(4):1211–1215.

Abbott, J.A., Massie, D.R., Upchurch, B.L., and W.R. Hruschka. 1995. Nondestructive sonic firmness measurement of apples. *Transactions of the ASAE* 38(5):1461–1466.

Affeldt, H.A. and Abbott J.A. 1989. Apple firmness and sensory quality using contact acoustic transmission. In: V.A. Dodd and P.H. Grace, Editors, Proceedings of the Eleventh International Congress on Agricultural Engineering. Dublin, Ireland: 2037–2045.

Armstrong, P.R., M.L. Stone, and G.H. Brusewitz. 1997a. Nondestructive acoustic and compression measurements of watermelon for internal damage detection. *Applied Engineering in Agriculture* 13(5):641–645.

Armstrong, P.R., M.L. Stone, and G.H. Brusewitz. 1997b. Peach firmness determination using two different nondestructive vibrational sensing instruments. *Transactions of the ASAE* 40(3):699–703.

Armstrong, P.R., H.R. Zapp, and G. K. Brown. 1990. Impulsive excitation of acoustic vibrations in apples for firmness determination. *Transactions of the ASAE* 33(4):1353–1359.

ASAE Standards. 1995. S368.3. Compression test for food materials of convex shape. St. Joseph, Mich.: ASAE.

Auld, B.A. 1973. *Acoustic fields and waves in solids*. New York: John Wiley and Sons.

Benedito, J., T. Conde, G. Clemente, and A. Mulet. 2006. Use of the acoustic impulse-response technique for the nondestructive assessment of Manchego cheese texture. *Journal of Dairy Science* 89:4490–4502.

Bengtsson, G.B., F. Lundby, J.E. Haugen, B. Egelandsdal, and J.A. Marheim. 2003. Prediction of postharvest maturity and size of Victoria plums by vibration response. *Acta Horticulturae*.

Blevins, R.D. 1993. *Formulas for natural frequency and mode shapes*. Malabar, FL: Krieger Publishing Company.

Chen, P. 1996. Quality evaluation technology for agricultural products. In *Proceedings of the International Conference on Agricultural Machinery Engineering*, 12–15.

Chen, H., and J. De Baerdemaeker. 1990. Resonance frequency and firmness of tomatoes during ripening. FIMA, 90.

Chen, P., Z. Sun, and L. Huarng. 1992. Factors affecting acoustic responses of apples. *Transactions of the ASAE* 35(6):1915–1919.

Clark, J.R., and P.S. Shackelford. 1973. Resonance and optical properties of peaches as related to flesh firmness. *Transactions of the ASAE* 16(6):1140–1142.

Clark, R. L. 1975. An investigation of the acoustical properties of watermelon as related to maturity. ASAE Paper no. 75-6004.

Conde, T., J.A. Cárcel, J.V. García-Pérez, and J. Benedito. 2007. Non-destructive analysis of Manchego cheese texture using impact force-deformation and acoustic impulse-response techniques. *Journal of Food Engineering* 82:238–245.

Cooke, J. R. 1972. An interpretation of the resonance behavior of intact fruit and vegetables. *Transactions of the ASAE* 15:1075–1079.

Cooke, J.R., and R.H. Rand. 1973. A mathematical study of resonance in intact fruits and vegetables using a 3-media elastic sphere model. *Journal of Agricultural Engineering Research* 18:141–157.

Coucke, P., E. Dewil, E., Deguypere, and J. De Baerdemaeker. 1999. Measuring the mechanical stiffness of an eggshell using resonant frequency analysis. *British Poultry Science* 40:227–232.

De Belie, N., S. Schotte, P. Coucke, and J. De Baerdemaeker. 2000a. Development of an automated monitoring device to quantify changes of firmness of apples during storage. *Postharvest Biology and Technology* 18(1):1–8.

De Belie, N., V. De Smedt, and J. De Baerdemaeker. 2000b. Principal component analysis of chewing sounds to detect differences in apple crispness. *Postharvest Biology and Technology* 18:109–119.

De Belie, N., S. Schotte, J. Lammertyn, B. Nicolai, and J. De Baerdemaeker. 2000c. Firmness changes of pear fruit before and after harvest with the acoustic impulse response technique. *Journal of Agricultural Engineering Research* 77(2):183–191.

De Belie, N., M. Sivertsvik, and J. De Baerdemaeker. 2003. Differences in chewing sounds of dry-crisp snacks by multivariate data analysis. *Journal of Sound and Vibration* 266:625–643.

De Ketelaere, B., P. Coucke, and J. De Baerdemaeker. 2000. Eggshell crack detection based on acoustic resonance frequency analysis. *Journal of Agricultural Engineering Research* 76:157–163.

De Ketelaere, B., and J. De Baerdemaeker. 2000. Part I: Parameter estimation for the non-destructive quality assessment of agro-products using vibration measurements. ISMA 25: International Conference on Noise and Vibration Engineering, Leuven.

De Ketelaere, B., and J. De Baerdemaeker. 2001. Advances in spectral analysis of vibration for non-destructive determination of tomato firmness. *Journal of Agricultural Engineering Research* 78(2):177–185.

De Ketelaere, B., H. Vanhoutte, and J. De Baerdemaeker. 2003. Parameter estimation and multivariable model building for the non-destructive, on-line determination of eggshell strength. *Journal of Sound and Vibration* 266:699–709.

Diezma-Iglesias, B., M. Ruiz-Altisent, and P. Barreiro. 2004. Detection of internal quality in seedless watermelon by acoustic impulse response. *Biosystems Engineering* 88:221–230.

Diezma-Iglesias, B., C. Valero, F.J. García-Ramos, and M. Ruiz-Altisent. 2006. Monitoring of firmness evolution of peaches during storage by combining acoustic and impact methods. *Journal of Food Engineering* 77:926–935.

Drake, B. K. 1963. Food crushing sounds. An introductory study. *Journal of Food Science* 28:233–241.

Drake, B. K. 1965. Food crushing sounds: Comparisons of objective and subjective data. *Journal of Food Science* 30(3):556–559.

Duizer, L. 2001. A review of acoustic research for studying the sensory perception of crisp, crunchy and crackly textures. *Trends in Food Science and Technology* 12:17–24.

Dull, G. 1986. Nondestructive evaluation of quality of stored fruits and vegetables. *Outstanding Symposia in Food Science and Technology* 40(5):106–110.

Duprat, F., M. Grotte, E. Pietri, and D. Loonis. 1997. The acoustic impulse response method for measuring the overall firmness to fruit. *Journal of Agricultural Engineering Research* 66:251–259.

Farabee, M.L., and M.L. Stone. 1991. Determination of watermelon maturity with sonic impulse testing. ASAE Paper No. 91-3013.

Finney, E.E. 1970. Mechanical resonance within Red delicious apples and its relation to fruit texture. *Transactions of the ASABE* 13:177–80.

Galili, N., I. Shmulevich, and N. Benichou. 1998. Acoustic testing of avocado for fruit ripeness evaluation. *Transactions of the ASABE* 41(2):399–407.

Garrett, R.E., and R.B. Furry. 1972. Velocity of sonic pulses in apples. *Transactions of the ASABE* 15(4):770–774.

Hayashi, S., and J. Sugiyama. 2000. Nondestructive quality evaluation of fruits by acoustic impulse transmission. ISMA 25: International Conference on Noise and Vibration Engineering, Leuven.

Herppich, W.B., B. Herold, S. Landahl, and J. De Baerdemaeker. 2003. Interactive effect of water status and produce texture: an evaluation of non-destructive methods. *Acta Horticulturae* 599:281–288.

Huarng, L., P. Chen, and S. Upadhyaya. 1993. Determination of acoustic vibration modes in apples. *Transactions of the ASAE* 36(5):1423–1429.

Kapur, K. 1971. Frequency spectrographic analysis of bone conducted chewing sounds in persons with natural and artificial dentitions. *Journal of Texture Studies* 2(1):50–61.

Lee, W.B., A.E. Deibel, C.T. Glembin, and E.G. Munday. 1988. Analysis of food crushing sounds during mastication: Frequency-time studies. *Journal of Texture Studies* 21:165–178.

Lemmens, J.W. 1990. *Impulse excitation: A technique for dynamic modulus measurement.* Philadelphia: American Society for Testing and Materials.

Muramatsu, N., N. Sakurai, N. Wada, R.Yamamoto, T. Takahara, T. Ogata, K. Tanaka, T. Asakura, Y. Ishikawa-Takano, and D.J. Nevins. 1999. Evaluation of fruit tissue texture and internal disorders by laser Doppler detection. *Postharvest Biology and Technology* 15:83–88.

Recuero López, M. 2000. *Ingeniería acústica*. Madrid: Paraninfo.

Rosenfeld, D., I. Shmulevich, and G. Rosenhouse. 1992. Three-dimensional simulation of the acoustic response of fruit for firmness sorting. *Transactions of the ASAE* 35(4):1267–1274.

Róth, E., E. Kovács, and J. Felföldi. 2003. Investigating the firmness of stored apples by non-destructive method. *Acta Horticulturae* 599:257–263.

Saltveit, M.E., S.K. Upadhyaya, J.F. Happ, R. Cavaletto, and M. O'Brien. 1975. Maturity determination of tomatoes using acoustic methods. ASAE Paper No. 85-3536.

Sasao, A. 1985. Impact response properties of watermelons in growth process. *Journal of the Japanese Society of Agricultural Machinery* 47(3):335–358.

Schotte, S., N. De Belie, and J. De Baerdemaeker. 1999. Acoustic impulse-response technique for evaluation and modelling of firmness of tomato fruit. *Postharvest Biology and Technology* 17(2):105–115.

Schrevens, E., R. De Busscher, L. Verstreken, and J. De Baerdemaeker. 2001. Detection of hollow pears by tree based modelling on non-destructive acoustic response spectra. *Acta Horticulturae* 464:441–446.

Seymour, S.K., and D.D. Hamann. 1988. Crispness and crunchiness of selected low moisture foods. *Journal of Texture Studies* 19:79–95.

Shackelford, P.S., and R.L. Clark. 1970. Evaluation of peach maturity by mechanical resonance. *ASAE* Paper No. 70–552.

Shmulevich, I., N. Galili, and D. Rosenfeld. 1994. Firmness testing device based on fruit acoustic response. AgEng. Milan. Paper No. 94-6-080.

Stone, M.L., P.R. Armstrong, X. Zhang, G.H. Brusewitz, and D.D. Chen. 1996. Watermelon maturity determination in the field using acoustic impulse impedance techniques. *Transactions of the ASABE* 39(6):2325–2330.

Stone, M.L., P.R. Armstrong, P. Chen, G.H. Brusewitz, and N.O. Maness. 1998. Peach firmness prediction by multiple location impulse testing. *Transactions of the ASABE* 41(1):115–119.

Sugiyama, J., T. Katsurai, J. Hong, H. Koyama, and K. Mikuriya. 1998. Melon ripeness monitoring by a portable firmness tester. *Transactions of the ASABE* 41(1):121–127.

Sugiyama, J., K. Otobe, S. Hayashi, and S. Usui. 1994. Firmness measurement of muskmelons by acoustic impulse transmission. *Transactions of the ASABE* 37(4):1235–1241.

Taniwaki, M., T. Hanada, and N. Sakurai. 2009a. Postharvest quality evaluation of "Fuyu" and "Taishuu" persimmons using a nondestructive vibrational method and an acoustic vibration technique. *Postharvest Biology and Technology* 51:80–85.

Taniwaki, M., M. Takahashi, and N. Sakurai. 2009b. Determination of optimum ripeness for edibility of postharvest melons using nondestructive vibration. *Food Research International* 42:137–141.

Taniwaki, M., T. Hanada, M. Tohro, and N. Sakurai. 2009c. Non-destructive determination of the optimum eating ripeness of pears and their texture measurements using acoustical vibration techniques. *Postharvest Biology and Technology* 51:305–310.

Van Woensel, G., and J. De Baerdemaeker. 1985. Dynamic and static mechanical properties of apples and fruit quality. Paper presented at the 3rd International Conference Physical Properties of Agricultural Materials and Their Influence on Design and Performance of Agricultural Machineries and Technologies.

Vickers, Z. 1981. Relationships of chewing sounds to judgments of crispness, crunchiness and hardness. *Journal of Food Science* 47:121–124.

Vickers, Z. 1983. Crackliness: Relationships of auditory judgment to tactile judgments and instrumental acoustical measurements. *Journal of Texture Studies* 15:49–58.

Vickers, Z. 1991. Sound perception and food quality. *Journal of Food Quality* 14:87–96.

Vickers, Z., and M.C. Bourne. 1976. A psychoacoustical theory of crispness. *Journal of Food Science* 41:1158–1164.

Yamamoto, H., M. Iwamoto, and S. Haginuma. 1980. Acoustic impulse response method for measuring natural frequency of intact fruit and preliminary applications to internal quality evaluation of apples and watermelons. *Journal of Texture Studies* 11:117–136.

Yong, Y.C., and W.K. Bilanski. 1979. Modes of vibration of spheroids at the first and second resonant frequencies. *Transactions of the ASABE* 22(6):1463–1466.

Zhang, X., M.L. Stone, D. Chen, N.O. Maness, and G.H. Brusewitz. 1994. Peach firmness determination by puncture resistance, drop impact, and sonic impulse. *Transactions of the ASABE* 37(2):495–500.

9 Medical Diagnostic Theory Applied to Food Technology

Alejandro Arana and J. Ignacio Arana

CONTENTS

9.1 INTRODUCTION

Diagnostic systems are all around us. They are used in military science to detect threats of enemies, in meteorology to predict weather conditions, in civil engineering to reveal malfunctions in buildings and factories, and in medicine to detect diseases in people.

Other diagnostic systems are used to make rational selections from many objects; psychologists use selection algorithms to classify job or school applicants according to the likelihood of success, governments to evaluate fraudulent tax declarations, and policemen to detect criminal suspects.

In medicine, diagnosis is a main challenge. Diagnostic tests are used to discriminate between healthy and unhealthy subjects, and are of daily use because patients are central in clinical practice (*Harrison's Principles*, 2011). The methodology to evaluate the accuracy and characteristics of diagnostic tests is well established in the medical field (Sackett et al., 1985).

Quality is of increasing importance in hortofruticultural activities. With the introduction of the global quality concept, it is crucial for optimal marketing to evaluate each characteristic of fruit samples. Discriminant analysis has been used for this purpose in the past. The formal purpose of classification or discriminant analysis is to assign objects to one of several groups or classes based on a set of measurements obtained from each object or observation. Classification techniques are also used informally to study the separability of labeled groups of observations in the measurement space (Friedman, 1989). The objective of discriminant analysis is to maximize group classification accuracy, but this is not enough when the objective is to maximize the quality of the sample (i.e., no unhealthy fruits in the sample). This requires individual evaluation of the fruits by nondestructive methods.

In this chapter we present the rationale behind diagnostic tests, their characteristics, the standard methods used for the evaluation of those characteristics, and the application of these methods to the evaluation of a test meant to discriminate pathological fruits from a sample.

9.2 DIAGNOSTIC TESTS IN MEDICINE

A diagnostic test is any kind of medical test performed to aid in the diagnosis or detection of disease. Medical tests can be as traditional as the classical physical examination, or may be based in the most modern technology like the positron emission tomography (PET) scan, shown in Figure 9.1.

Diagnostic tests help physicians revise disease probability for their patients. The physician aims to get closer to certainty when diagnosing or discarding a disease. A diagnostic test can be performed to establish a diagnosis in symptomatic patients; for

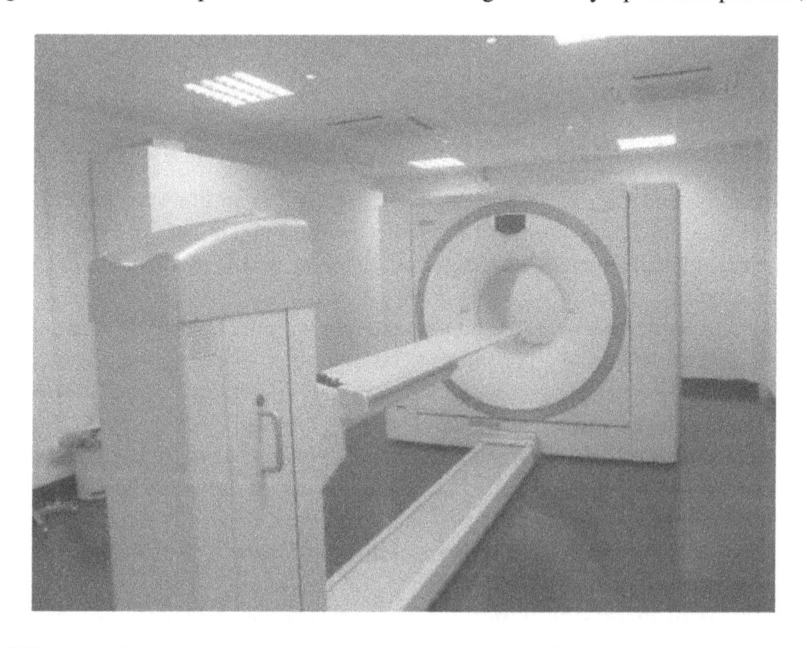

FIGURE 9.1 PET scan, courtesy of Mr. F.J. Sanz and Drs. M. Garcia-Miralles and R. Gaston.

example, an electrocardiogram (ECG) to diagnose myocardial infarction in patients with chest pain. It can also be used to screen for disease in asymptomatic patients; for example, a prostate-specific antigen (PSA) test to screen out prostate cancer in men older than 50 years. It also provides prognostic information in patients with established disease; for example, a CD4 lymphocyte count in patients with human immunodeficiency virus (HIV). It is used in therapeutics, to monitor either benefits or side effects of medicines; for example, measuring the international normalized ratio (INR) in patients taking Warfarin. Lastly, a test may be performed to confirm that a person is free from a condition; for example, a pregnancy test to exclude the diagnosis of pregnancy (*Harrison's Principles*, 2011).

In medicine, the reference standard test definitively decides either presence or absence of a disease and it is called the *gold standard test* (Sackett et al., 1985). Examples of gold standard tests include pathological specimens for malignancies and pulmonary angiography for pulmonary embolism. However, gold standard tests routinely come with drawbacks; they are usually expensive, less widely available, more invasive, and riskier. These issues usually compel most physicians to choose other diagnostic tests as surrogates for their criterion standard test.

For example, venography, the criterion standard for vein thrombosis, is an invasive procedure with significant complications including renal failure, allergic reaction, and clot formation. These risks make venography less desirable than the alternative diagnostic test—venous duplex ultrasonography (Carpenter et al., 1993). There are instances where the gold standard test is the result of an autopsy, but it is evident that these test cannot be performed upon a living patient. The price most diagnostic tests pay for their ease of use compared to their criterion standard is a decrease in accuracy. How to account for this trade-off between diagnostic accuracy and patient acceptability will be described in this chapter.

9.3 EVALUATION OF DIAGNOSTIC TESTS

9.3.1 The Concept

To understand the concept of evaluation or validation of diagnostic tests, we present the following example. A farmer is interested in the occurrence of frost over a period of 10 consecutive days. During these 10 days, three days actually had frost with no frost on the other seven. Over the same period, on each preceding day, a TV channel weather forecast announced frost for six of the days. Two of the days forecasted for frost actually had frost. Frost was wrongly forecasted on four days, and on one of the days frost occurred without having been forecasted.

This information can be expressed in what is known as a two-by-two table (Table 9.1). Note that the "truth" (whether or not there was frost) is expressed along the horizontal title row, whereas the TV forecast (which may or may not reflect the truth) is expressed on the vertical column.

These figures, if they are typical, reflect several features of this particular forecast:

- The TV channel correctly identifies two out of every three true frost days.
- It correctly forecasts three out of every seven frost-free days.

TABLE 9.1
Two-by-Two Table Showing the Outcome of 10 Consecutive Days

	True Weather	
TV Forecast	**Frost**	**No Frost**
Frost	Rightly forecasted (2 days)	Wrongly forecasted (4 days)
No Frost	Wrongly forecasted (1 day)	Rightly forecasted (3 days)

- If this TV channel forecasts a day with frost, there is still only a one in three (33%) chance that frost will actually occur.
- If this TV channel forecasts a day as frost free, there is a three in four chance (75%) of no frost.
- In five cases out of every 10 (50%), the TV forecast is right.

These five features constitute, respectively, the sensitivity, specificity, positive predictive value, negative predictive value, and accuracy of this TV channel forecast's performance.

9.3.2 Characteristics of a Diagnostic Test

We mentioned previously that the evaluation of a diagnostic test needs to be done against another test that is considered to provide the truth. The diagnostic test providing the truth is denominated, in medicine, the gold standard test (Sackett et al., 1985). In the previous example, we tested the forecasted weather conditions against the gold standard test, which is the registered weather.

Subjects, units of analysis, are tested against both the diagnostic test and the gold standard test. The results of the test are displayed in a two-by-two table (Table 9.2).

Subjects are then classified into four categories:

- *True positive (a):* Subject with the characteristic (disease, pathology, etc.) detected with the gold standard test whose test result is positive for the presence of the characteristic

TABLE 9.2
Two-by-Two Table Notation for Expressing the Results of Validation Study for a Diagnostic or Screening Test

	Result of Gold Standard Test	
Result of Diagnostic Test	**Disease Positive (a + c)**	**Disease Negative (b + d)**
Test positive (a + b)	True positive (a)	False positive (b)
Test negative (c + d)	False negative (c)	True negative (d)

- *False positive (b):* Subject without the characteristic according to the gold standard test whose test result is positive for the presence of the characteristic
- *False negative (c):* Subject with the characteristic present according to the gold standard test whose test result is negative for the presence of the characteristic
- *True negative (d):* Subject without the characteristic according to the gold standard test whose test result is negative for it

The following five features constitute the properties of a diagnostic test:

- *Sensitivity (SE):* How many true positives does the procedure detect as positive? How good is the test at detecting disease?
- *Specificity (SP):* How many true negatives does the procedure detect as negative? How good is the test to detect absence of disease?
- *Positive predictive value of the positive result:* How many of the tested positives are true positives? How reliable is the positive result of the test?
- *Negative predictive value of the negative result:* How many of the tested negatives are true negatives? How reliable is the negative result of the test?
- *Accuracy:* How many cases are correctly classified by the procedure?

Table 9.3 summarizes these properties.

TABLE 9.3
Features of Diagnostic Test That Can Be Calculated by Comparison with Gold Standard in Validation Study

Feature of the Test	Alternative Name	Question Addressed	Formula (see Table 9.2)
Sensitivity	True positive rate (positive in disease)	How good is this test at picking up individuals with the condition?	a/(a + c)
Specificity	True negative rate (negative in health)	How good is this test at identifying individuals without the condition?	d/(b + d)
Positive predictive value	Posttest probability of a positive test	If a test is positive, what is the probability that the condition is truly present?	a/(a + b)
Negative predictive value	Posttest probability of a negative test	If a test is negative, what is the probability that the condition is truly absent?	d/(c + d)
Accuracy	—	What proportion of all tests have given the correct result? (true positives and true negatives as a proportion of all results)	(a + d)/(a + b + c + d)

Once the tests are characterized by evaluating their features, they can be chosen according to the needs of the user.

9.3.2.1 Evaluation of a Diagnostic Test with a Continuous Result

Many tests do not result in dichotomous (yes/no) results, but in a value suitable to be measured in an ordinal or continuous scale. Take, for example, the use of the levels of tau protein in the cerebrospinal fluid (CSF-tau) test to screen for Alzheimer's disease (AD) (Andreasen et al., 1999). Most elderly will have some detectable protein in their cerebrospinal fluid (say, 200 pg/ml), and most of those with probable AD will have high concentrations (above about 600 pg/ml). But a concentration of, say, 300 pg/ml may be found either in a perfectly normal person or in someone with early AD. There simply is not a clean cutoff between normal and abnormal.

Results of a validation study of this test can be used against a gold standard for AD (AD diagnosed according to National Institute of Neurological and Communicative Disorders and Stroke-Alzheimer's Disease and Related Disorders Association criteria [McKhann et al., 1984]) to draw up a whole series of two-by-two tables. Each table would use a different definition of an abnormal test result to classify patients as *normal* or *abnormal*. For each of the tables, we could generate the sensitivity and specificity associated with a protein concentration above each different cutoff point. When faced with a test result in the *grey zone*, we would at least be able to say, "This test has not proved that the patient has Alzheimer's disease, but the probability of having it is x."

Another way to express the relationship between sensitivity and specificity for a given test is to construct a curve called a receiver operating characteristic (ROC) curve, shown in Figure 9.2 (Peterson et al., 1954; Schisterman et al., 2001). It is

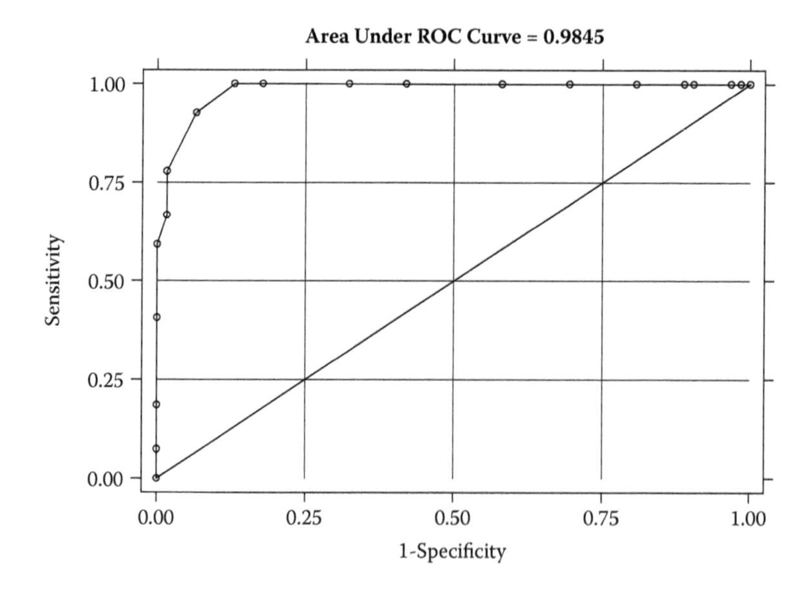

FIGURE 9.2 Receiver operating characteristics curve for the validation study of the nondestructive test to detect mealy fruits.

constructed by plotting the true positive rate (sensitivity) against the false positive rate (1, specificity). The values on the axes run from a probability of 0 to 1.0 (or, alternatively, from 0 to 100%).

Tests that discriminate well, group toward the upper left corner of the ROC curve. Tests that perform less well have curves that fall closer to the diagonal line running from lower left to upper right. This line describes a test that contributes no information.

The ROC curve is used to describe the accuracy of a test over a range of cutoff points. It can serve as a normogram for reading off the specificity that corresponds to a given sensitivity. It shows how severe the trade-off between sensitivity and specificity is for a test and can be used to help decide where the best cutoff point would be. The overall accuracy of a test can be described as the area under the ROC curve (AUC); the larger the area, or the closest to 1, the better the test. The AUC has an important statistical property: the AUC of a classifier is equivalent to the probability that the classifier will rank a randomly chosen positive instance higher than a randomly chosen negative instance (DeLong et al., 1988).

Although we are forced to make a trade-off between sensitivity and specificity for any given test, it is possible that a new test can be both more sensitive and more specific than its predecessors.

9.3.2.2 Combining Multiple Tests

Few diagnostic tests are both highly sensitive and highly specific. For this reason, patients are sometimes diagnosed using two or more tests. These tests may be performed either in parallel (i.e., at the same time and interpreted together) or in series (i.e., the results of the first test determine whether the second test is performed at all) (Zhou et al., 2002). The latter has the advantage of avoiding unnecessary tests, but the disadvantage of potentially delaying treatment for diseased patients by lengthening the diagnostic testing period.

Tests performed in parallel can be considered positive if any of the tests results is positive. This is called the *OR rule*. Alternatively, the combination can be considered positive if both the tests results are positive. This is called the *AND rule*. Under the OR rule, the sensitivity of the combined result is higher than that of either test alone, but the combined specificity is lower than that of either test. With the AND rule, this is reversed: The specificity of the combined result is higher than either test alone, but the combined sensitivity is lower than that of either test.

Serial testing is an alternative to parallel testing that is particularly cost-efficient when screening for rare conditions, and is often used when the second test is expensive and/or risky. Under the OR rule, if the first test is positive, the diagnosis is positive; otherwise, the second test is performed. If the second test is positive after a negative first test, then the diagnosis is also positive; otherwise, the diagnosis is negative. The OR rule, then, leads to a higher overall sensitivity than either test by itself. With the AND rule, if the first test is positive, the second test is performed. If the second test is positive, the diagnosis is positive; otherwise, the diagnosis is negative. The AND rule, then, leads to a higher overall specificity than either test by itself.

Further discussion on the methodology and interpretation of diagnostic tests can be found in the books by Galen and Bambino (1975), Sackett et al. (1985), and Zhou et al. (2002), the paper by Sheps and Schechter (1984), and the series of articles from the Department of Clinical Epidemiology and Biostatistics at MacMaster University (1983).

9.3.2.3 Examples of the Evaluation of a Diagnostic Test in Food Technology

Textural characteristics of fruits are of extreme importance for quality assurance and marketing. For apples, the most appreciated one is the absence of *mealy* or *farinaceous* texture (Arana et al., 2004).

To discriminate mealy apples is not yet possible without using destructive tests, mainly tasting the apple by a professional tester. The destructive test is equivalent to what in medicine is called the gold standard test.

A nondestructive test has recently been developed. It is called the impact test (IT), for the discrimination of mealy apples (Arana et al., 2004). The test measures the maximum resistance to the impact of a spherical object in free fall from a height of 2 cm onto an apple.

Eighty-nine apples were tested with the IT and subsequently tasted by the professional tester. Table 9.4 shows the results of the tests, IT and taster, for each of the apples studied.

The next step was to draw up a whole series of two-by-two tables. Each table would use a different definition of an abnormal test result, mainly each value of the IT test, and classifies apples as *normal* or *abnormal* (Tables 9.5a and 9.5b). Each cutoff point is characterized by its sensitivity and specificity. The predictive values for the cutoff point can be calculated.

The sensitivity for the cutoff point of 18.38 Newtons is 0.98, meaning that 98% of the mealy fruits will have values lower than 18.38 N in the maximum resistance test. The specificity is 0.66, meaning that 66% of nonmealy apples will have values greater or equal than 18.38 N in the test. If the result of the test was lower than 18.38 N, the probability of that apple being mealy was 87% (positive predictive value), and if the result was greater or equal to 18.38 N, there was a 94% probability of actually being nonmealy.

The sensitivity for the cutoff point 17.41 N is 0.94 and the specificity is 0.93.

The corresponding table and sensitivity and specificity calculations need to be performed for every IT test result value as a cut-off point (Table 9.6). These values are transformed into a ROC curve (Figure 9.2).

This analysis allows the decision maker to select the discriminator value of the IT test according to his business needs. If all mealy fruits need to be diagnosed and excluded from the sample, the cutoff point maximizing sensitivity will be selected. If, on the contrary, no healthy nonmealy fruits have to be excluded from the sample, the lowest cutoff point maximizing specificity will be selected. Arana et al. (2005) used the impact test for the discrimination of woolly nectarines.

Cataltepe et al. (2004) used transmittance images to classify insect-damaged and undamaged wheat kernels. They achieved a specificity of 80% and a sensitivity of 80%. The AUC was 0.86. Combining this and another test resulted in a specificity of 90%, a sensitivity of 80%, and an AUC of 0.92.

Ariana et al. (2006) used the AUC to compare the overall performance of three classification methods, based on near-infrared hyperspectral reflectance imaging, developed by them to identify bruised and normal pickling cucumbers. They used ratio of reflectance (R) images for two wavelengths (R = R988nm/R1085nm) and

TABLE 9.4
Results of the Two Tests. Mealy Apple as Tasted by a Professional Tester and Nondestructive Test (IT) for Each Fruit in the Sample

n	Mealy Diagnosis by Professional Taster	IT Result in Newtons	*n*	Mealy Diagnosis by Professional Taster	IT Result in Newtons	*n*	Mealy Diagnosis by Professional Taster	IT Result in Newtons
1	no	17.41	31	yes	16.44	61	yes	18.38
2	no	19.34	32	yes	15.48	62	yes	14.51
3	no	20.79	33	yes	14.51	63	yes	14.99
4	no	19.34	34	yes	14.02	64	yes	17.41
5	no	18.86	35	yes	15.96	65	yes	13.54
6	no	19.34	36	yes	16.44	66	yes	14.99
7	no	18.86	37	yes	14.99	67	yes	15.48
8	no	19.34	38	yes	12.57	68	yes	15.96
9	no	19.83	39	yes	15.96	69	yes	15.48
10	no	17.89	40	yes	14.99	70	yes	14.51
11	no	17.41	41	yes	15.96	71	yes	14.02
12	no	19.34	42	yes	15.96	72	yes	15.96
13	no	17.41	43	yes	17.41	73	yes	14.51
14	no	18.38	44	yes	12.57	74	yes	12.09
15	no	19.83	45	yes	14.99	75	yes	13.54
16	no	19.83	46	yes	14.02	76	yes	15.96
17	no	18.86	47	yes	15.48	77	yes	14.51
18	no	18.86	48	yes	16.93	78	yes	16.93
19	no	17.41	49	yes	14.51	79	yes	13.06
20	yes	16.93	50	yes	14.02	80	yes	14.99
21	no	16.93	51	yes	14.51	81	yes	15.48
22	no	18.38	52	yes	14.02	82	yes	13.54
23	no	19.34	53	yes	14.99	83	yes	13.54
24	no	17.89	54	yes	13.54	84	yes	11.61
25	no	18.86	55	yes	14.99	85	yes	17.41
26	no	16.93	56	yes	14.99	86	yes	15.96
27	no	20.79	57	yes	14.02	87	yes	15.96
28	no	17.89	58	yes	16.44	88	yes	12.57
29	yes	16.93	59	yes	12.57	89	yes	14.99
30	yes	15.48	60	yes	14.02	—	—	—

difference of reflectance images for two wavelengths (D = R1346 nm −R1425 nm), over a period of 0 to 6 days.

The accuracy based on the ratio of two wavelengths was slightly better than that based on the difference of two wavelengths for 0 and 1 days after bruising, whereas the difference of two wavelengths was superior for 3 and 6 days. Based on the AUC comparisons, the classification performance for the band difference method was slightly better than that of the band ratio at 0, 3, and 6 days.

TABLE 9.5A
Two-by-Two Table for the Results of Defining Positive "Mealy" through a Cutoff Point (Maximum Resistance Less than 18.38 Newtons)

	Result of Gold Standard Test	
Result of Diagnostic Test	**Disease Positive "Mealy" as Defined by Professional Tester**	**Disease Negative "Not Mealy" as Defined by Professional Tester**
Test positive (< 18.38 N)	61	9
Test negative (≥ 18.38 N)	1	18

TABLE 9.5B
Two-by-Two Table for the Results of Defining Positive "Mealy" through a Cutoff Point (Maximum Resistance Less than 17.41 Newtons)

	Result of Gold Standard Test	
Result of Diagnostic Test	**Disease Positive "Mealy" as Defined by Professional Tester**	**Disease Negative "Not Mealy" as Defined by Professional Tester**
Test positive (< 17.41 N)	58	2
Test negative (≥ 17.41 N)	4	25

Some authors used diagnostic theory for the fast and accurate detection of *Salmonella spp.* in chicken rinses, minced meat, fish, and raw milk (Malorny et al. 2004) and in spinach, tomatoes, and in both jalapeno and serrano peppers (González-Escalona et al., 2009).

Jordan et al. (2007) used ROC analysis to estimate and compare the errors that would be experienced if densities of coliform, enterobacteriaceae, or APC were used to predict the presence or absence of *E. coli* biotype I in samples from beef carcasses, sheep carcasses, frozen beef, and frozen sheep meat.

García-Rey et al. (2004) used ROC curves to classify hams according to normal pH and low pH in order to study the relationship between pH before salting and dry-cured ham quality.

Hung et al. (2011) efficiently differentiated ostrich meat from pork, beef, and chicken meat, and evaluated grades and freshness of ostrich meat using electrochemical (EC) profiling through copper nanoparticle-plated screen-printed electrode. Statistical analysis (ROC curve) demonstrated that peak ratios could be used to evaluate ostrich meat grades with high sensitivity (up to 95%) and specificity (up to 100%).

There are many situations in agriculture in which the evaluation of individual pieces by nondestructive tests is necessary. On those occasions and for the correct evaluation of the tests, the application of the methods described here is essential.

TABLE 9.6
Force in Newtons, and Associated Sensitivity and Specificity Values

Force in Newtons	Sensitivity	Specificity
11.61	0.016129	1.000000
12.09	0.032258	1.000000
12.57	0.096774	1.000000
13.06	0.112903	1.000000
13.54	0.193548	1.000000
14.02	0.306452	1.000000
14.51	0.419355	1.000000
14.99	0.580645	1.000000
15.48	0.677419	1.000000
15.96	0.822581	1.000000
16.44	0.870968	1.000000
16.93	0.935484	0.925926
17.41	0.983871	0.777778
17.89	0.983871	0.666667
18.38	1.000000	0.592593
18.86	1.000000	0.407407
19.34	1.000000	0.185185
19.83	1.000000	0.074074
20.79	1.000000	0.000000

In food technology, where some individuals from the sample can be sacrificed, the establishment of the gold standard is easier than in medicine.

In medicine, where diagnosis is one of the main challenges, there is a lot of experience with the application and evaluation of diagnostic tests. The theory has been developed and it is well established. It is also used in military science, in meteorology, in civil engineering, in behavioral research, and criminology (Swets, 1988). We are glad that it is applied to food technology as well.

A wider and deeper understanding of the needs and possibilities of measuring accuracy in diagnostic tests would be beneficial for the food technology field, and also for society. Scientists are increasingly aware that a successful science of accuracy testing exists, and that the fundamental factors in evaluation of diagnostic tests are the same across the different disciplines where it is applied.

REFERENCES

Andreasen, N., L. Minthon, A. Clarberg, P. Davidsson, J. Gottfries, E. Vanmechelen, H. Vanderstichele, B. Winblad, and K. Blennow. 1999. Sensitivity, specificity, and stability of CSF-tau in AD in a community-based patient sample. *Neurology* 53:1488–1494.

Ariana, D.P., R. Lu, and D.E. Guyer. 2006. Near-infrared hyperspectral reflectance imaging for detection of bruises on pickling cucumbers. *Computers and Electronics in Agriculture* 53:60–70.

Arana, I., C. Jarén, and S. Arazuri. 2004. Apple mealiness detection by non-destructive mechanical impact. *Journal of Food Engineering* 62(4):399–408.

Arana, I., C. Jarén, and S. Arazuri. 2005. Nectarine woolliness detection by non-destructive mechanical impact. *Biosystems Engineering* 90(1):37–45.

Carpenter, J.P., G.A. Holland, R.A. Baum, R.S. Owen, J.T. Carpenter, and C. Cope. 1993. Magnetic resonance venography for the detection of deep venous thrombosis: comparison with contrast venography and duplex Doppler ultrasonography. *Journal of Vascular Surgery* 18:734–741.

Cataltepe, Z., E. Cetin, and T. Pearson. 2004. Identification of insect damaged wheat kernels using transmittance images. ICIP 04 *International Conference on Image Processing* 5:2917–2920.

DeLong, E.R., D.M. DeLong, and D.L. Clarke-Pearson. 1988. Comparing the areas under two or more correlated receiver operating characteristic curves: a nonparametric approach. *Biometrics* 44(3):837–845.

Department of Clinical Epidemiology and Biostatistics, McMaster University. 1983. Interpretation of diagnostic data. *Canadian Medical Association Journal* 129: 429–432, 559–564, 586, 705–710, 832–835, 947–954, 1093–1099.

Friedman, J.H. 1989. Regularized discriminant analysis. *Journal of the American Statistical Association* 84:165–175.

Galen, R.S., and S.R. Bambino. 1975. *Beyond normality: The predictive value and efficiency of medical diagnosis*. New York: Wiley.

García-Rey, R.M.,J.A. García-Garrido, R. Quiles-Zafra, J. Tapiador, and M.D. Luque de Castro. 2004. Relationship between pH before salting and dry-cured ham quality. *Meat Science* 67(4):625–632.

González-Escalona, N., T.S. Hammack, M. Russell, A.P. Jacobson, A.J. De Jesús, E.W. Brown, and K. A. Lampel. 2009. Detection of live *Salmonella* sp. cells in produce by a TaqMan-based quantitative reverse transcriptase real-time PCR targeting invA mRNA. *Applied and Environmental Microbiology* 75(11):3714–3720.

Harrison's Principles of Internal Medicine 17th ed. Harrison's Online. http://accessmedicine.com/resourceTOC.aspx?resourceID=4 (accessed February 1, 2011).

Hung, C.J., H.P. Ho, C.C. Chang, M.R. Lee, C.A. Franje, S.I. Kuo, R.J. Lee, and C.C. Chou. 2011. Electrochemical profiling using copper nanoparticle-plated electrode for identification of ostrich meat and evaluation of meat grades. *Food Chemistry* 26(3):1417–1423.

Jordan, D., D. Philipps, J. Sumner, S. Morris, and I. Jenson. 2007. Relationships between the density of different indicator organisms on sheep and beef carcasses and in frozen beef and sheep meat. *Journal of applied Microbiology* 102(1):57–64.

Malorny, B., E. Paccassoni, P. Fach, C. Bunge, E. Martin, and R. Helmuth. 2004. Diagnostic real-time PCR for detection of *Salmonella* in food. *Applied and Enviromental Microbiology* 70(12):7046–7052.

McKhann, G., D. Drachman, M. Folstein, R. Katzman, D. Price, and E.M. Stadlan. 1984. Clinical diagnosis of Alzheimer's disease: Report of the NINCDS-ADRDA Work Group under the auspices of Department of Health and Human Services Task Force on Alzheimer's Disease. *Neurology* 34:939–944.

Peterson, W.W., T.G. Birdsall, and W.C. Fox. 1954. The theory of signal detection. Trans. IRE Professional Group of Information Theory, PGIT-4: 171–212. www.ee.kth.se/sip/courses/FEN3100/docs/Peterson1954.pdf

Sackett, D.L., R.B. Haynes, and P. Tugwell. 1985. *Clinical epidemiology: A basic science for clinical medicine*. Boston: Little, Brown and Company.

Schisterman, E.F., D. Faraggi, B. Reiser, and M. Trevisan. 2001. Statistical inference for the area under the receiver operating characteristic curve in the presence of random measurement error. *American Journal of Epidemiology* 154:174–179.

Sheps, S.B., and M.T. Schechter.1984. The assessment of diagnostic tests. A survey of current medical research. *Journal of American Medical Association* 252:2418–22.
Swets, J.A. 1988. Measuring the accuracy of diagnostic systems. *Science* 40:1285–1293.
Zhou, X.H., N.A. Obuchowski, and D.K. McClish. 2002. *Statistical methods in diagnostic medicine*. New York: Wiley & Sons.

10 Mechanical Damage of Foods

Silvia Arazuri

CONTENTS

10.1 INTRODUCTION

External appearance is the main quality aspect that each consumer is confronted with when buying food products (Nicolaï et al., 2009). Internal quality is important at the moment of product consumption. Apart from other characteristics, the absence of damage is one of the most important signs of quality of an agricultural product.

Mechanical damages appear due to impacts and compressions produced during harvesting, transport, and manipulation processes. Damages can appear at the moment at which the impact or compression takes place, or later, during storage.

These damages have a direct effect on loss of quality and reduce sale prices. External quality is considered of paramount importance in the marketing and sale of fruits. The appearance—size, shape, color, and presence of blemishes—influences consumer perceptions and therefore determines the level of acceptability prior to purchase. The consumer associates desirable internal quality characteristics with certain external appearance (Brosnan et al., 2004). The improvement of a food product's quality will make it more profitable and will open new sale markets.

This chapter seeks to give an overview of damages produced during manipulation processes as well as modern measuring techniques used for their detection.

10.2 MECHANICAL DAMAGES: DEFINITION

Mohsenin (1986) defined damage as the failure of the product under excessive deformation when it is forced through fixed clearance or excessive force when it is subjected to impact. Mechanical damages in agricultural products are due to external forces under static or dynamic conditions or to internal forces. Damages due to internal forces can be the result of physical changes, such as variation in temperature and moisture content, or chemical and biological changes. Mechanical damages due to external forces are mechanical injuries in fruits and vegetables, grains, and so on.

In terms of intact agricultural products, failure is usually manifested through a rupture in the internal or external cellular structure of the material.

Injuries could be classified as postharvest mechanical injuries (bruises, cuts, and punctures), preharvest mechanical injuries (healed lesions caused by rub injury or pest attacks), physiological disorders (badly misshapen fruit, growth cracks, or cracking), preharvest diseases (rust and shot-hole), and postharvest diseases (rots, including brown rot) (Amorim et al., 2008).

One of the main objectives of researchers is to know the behavior of agricultural products. Prediction of susceptibility to damage will improve preharvest and postharvest management. However, fruits and vegetables do not behave ideally because they are not ideal materials.

Gunasekaran and Mehmet (2000) described the particularity of agricultural products by comparison with ideal solids and fluids. These authors explained that an ideal solid material will respond to an applied load by deforming finitely and recovering that deformation upon removal of the load (elastic behavior). In contrast, an ideal fluid will deform and continue to deform as long as the load is applied and will not recover from its deformation when the load is removed. This response is called *viscous*.

From energy considerations, elastic behavior represents complete recovery of energy expended during deformation, whereas viscous flow represents complete loss of energy as all the energy supplied during deformation is dissipated as heat. Ideal elastic and ideal viscous behaviors present two extreme responses of materials to external stresses. However, real materials exhibit a wide array of responses between viscous and elastic. Most materials exhibit some viscous and some elastic behavior simultaneously and are called *viscoelastic*. Almost all foods, both liquid and solid, belong to this group.

In relation to this characteristic, in 1954, Bowden and Tabor divided the impact of colliding bodies into four phases:

1. Initial elastic deformation, during which the region of contact is deformed elastically and fully recovers without residual deformation. The time of impact, mean pressures, and deformation for this phase can be found using equations based on Hertz theory.
2. Onset of plastic deformation during which the mean pressure exceeds the dynamic yield pressure of the material and the resulting deformation will not be fully recovered. It has been shown that this condition occurs at extremely small impact energies. The fact that the indentation produced

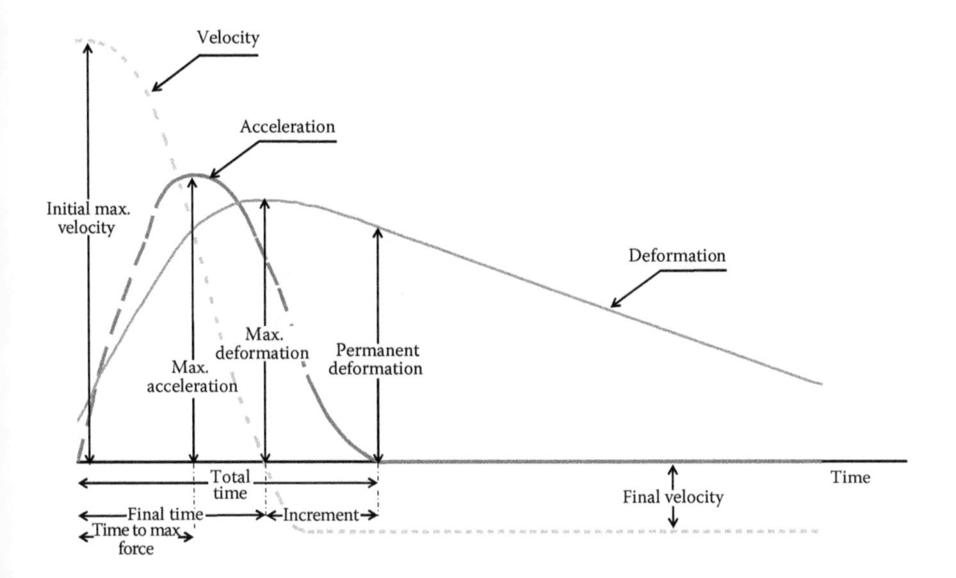

FIGURE 10.1 Impact characteristic curves from an impact test. (*Source:* Arana et al. 2005.)

by a spherical indenter is partially elastic at this stage has been proved by applying the original load several times and observing the recovered indentation, which remained essentially unchanged in diameter and in radius of curvature.

3. Full plastic deformation, during which the deformation continues from elastic-plastic to fully plastic until the pressure falls below the dynamic yield pressure.
4. Elastic rebound during which a release of elastic stresses stored in both bodies takes place.

Figure 10.1 shows an impact characteristic curve in which some of the phases previously described can be seen.

10.3 DETECTING DAMAGES DURING HARVEST AND MANIPULATION

There are several events during harvest and postharvest that can damage internal and external appearance of edible horticultural products, reducing quality by producing injuries, defects, or rot. These damages occur chiefly during harvest, transport, grading, and processing. Thus, reduction of fruit damages during harvest and postharvest manipulation is a vital step in order to increase quality. Qualitative and quantitative losses in fruits and tubers caused by mechanical damages as well as the increasing demand for quality products are of vital importance. This is the reason why devices known as sensors have been developed to quickly locate the points at which damages originate (Jaren et al., 2008). IS-100 (Figure 10.2), PMS-60, potato-shaped instrumented device (Figure 10.3) or Smart Spud (Figure 10.4) are some of the best-known

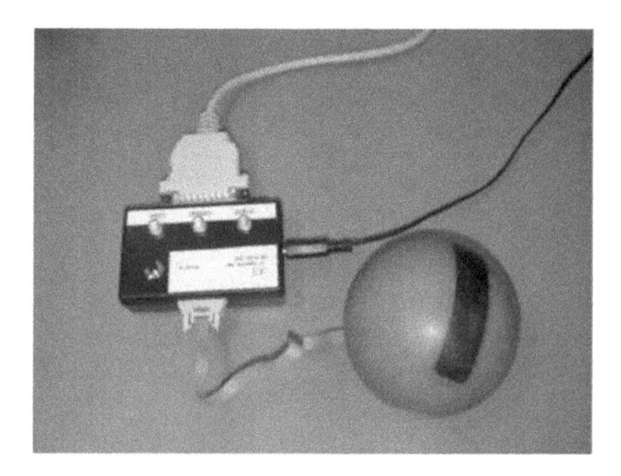

FIGURE 10.2 IS-100 instrumented sphere connected to communication interface.

devices for detection of those critical points. Further information about these sensors was published by Jaren et al. (2008).

In order to obtain good information about our machines, impact data must be related to bruise susceptibility of each fruit type by establishing impact damage thresholds of the products. Impact characteristics depend on different parameters: velocity, transfer height, padding materials, and transfer point design (Garcia-Ramos et al., 2003). Analyzing four different sizers using an IS-100 instrumented sphere, Garcia-Ramos et al. (2004) observed that most of the impacts recorded in the sizers showed acceleration values over 50 g and were produced against hard surfaces. Damage probability was high in the case of the stone fruits (peaches and apricots) tested. Moreover, most of the analyzed points were could be improved.

Arazuri et al. (2001) and Salvadores et al. (2001) evaluated different tomato harvesters in order to identify critical points where mechanical damages occur during harvesting. An IS-100 instrumented sphere was selected because of the similarity

FIGURE 10.3 Potato instrumented device or PTR200 and data logger.

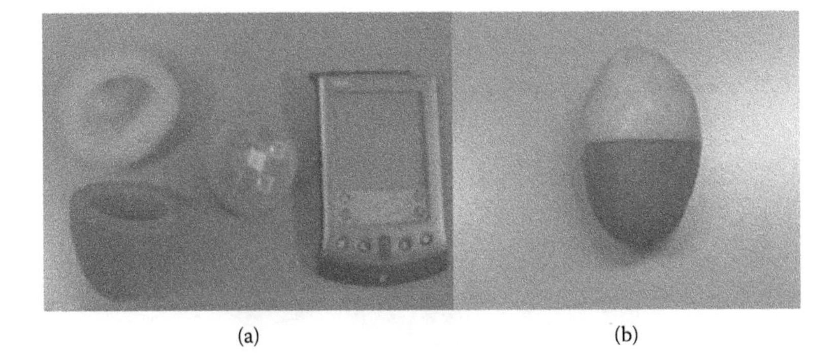

(a) (b)

FIGURE 10.4 Smart Spud instrumented device: (a) covers, sensor, and handheld, (b) Smart Spud ready to work.

FIGURE 10.5 IS-100 and tested tomato. (*Source:* Arazuri et al. 2010.)

between the shape of tomatoes and the device (Figure 10.5). The IS-100 was placed in tomato plants and picked up with them to follow the whole route with the tomatoes. The IS-100 recorded information about maximum impact acceleration, impact duration, and velocity change for every impact. These data provided information about impact intensity, location, and type of material against which the impact occurred—cushioning material or no cushioning material. In order to obtain as much information as possible about damage threshold, different samples were taken: manually harvested samples as control samples, and samples after shaking and after discharge elements. It was observed that shaking was the element in which damage risk was higher. Moreover, the influence of the machines on the firmness properties of the tomatoes were evaluated using impact tests and a texture analyzer TA-XT2 for compression, puncture resistance, and destructive compression test (Arazuri et al., 2007). A destructive compression test (Figure 10.6) was performed to evaluate the behavior of tomatoes during transport. When the container is full, tomatoes placed in the

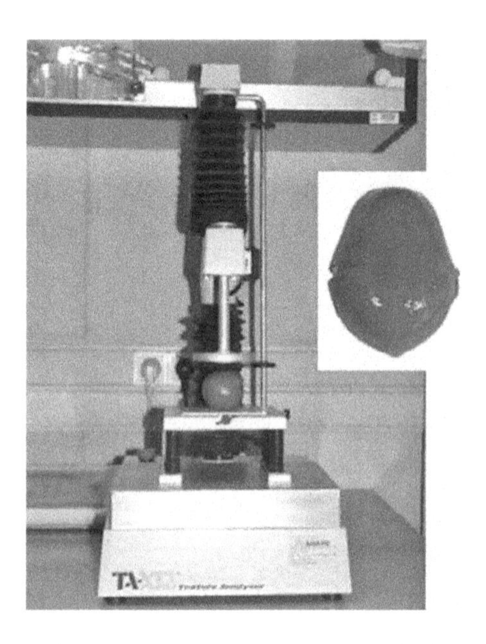

FIGURE 10.6 Destructive compression test and mechanical damage in tomato by compression.

lower and middle portions of the container pars suffer a high compression force due to the weight of the tomatoes above them. Desmet et al. (2004) tested tomato handling chains with the objective of identifying critical points where puncture injury might occur. A PMS-60 instrumented sphere measured impacts as the maximum impact force, the impact duration, and the impact integral, which is equal to the impact force integrated over the impact duration. To relate impact energy to impact force, different levels of impact energy were applied to the instrumented sphere by means of a pendulum. The same levels of impact were performed on tomato samples to determine presence or absence of puncture wounds. From the analysis of chains it was found that the most common deficiencies in the grading lines were wear or lack of cushioning material, lack of deceleration flaps, or bad adjustment of the fall-breaking brushes. The researchers concluded that major improvements with relatively small modifications near the critical points could be obtained by using the PMS-60 instrumented sphere information.

Finally, Salar (2009) evaluated potato harvesters and manipulation lines using different instrumented spheres: IS-100, PTR200 (potato-shaped instrumented device), and Smart Spud. The two last were more similar to potatoes because of their nonspherical shape. The researcher observed differences in the information provided by the devices, but in all cases, critical points were detected by the sensors. As in previously described investigations, samples of potatoes were taken as follows: control samples and samples taken at the detected critical points. Damages were observed and classified depending on their importance. PTR200 data and damage level were well correlated, although the sensitivity of the sensor in low-intensity impacts was lower than with the fruit-shaped electronic sensors. In order to avoid the high standard deviations described by Van Canneyt et al. (2003), repetitive measurements in the same harvest or manipulation time and in two different years were carried out. In addition, the influence of mechanical manipulation on th firmness was analyzed. Texture analyzer TA-XT2 was used and was equipped with compression, puncture, and Warner Bratzler shear probes (Salar et al., 2009).

10.4 TECHNIQUES FOR MEASURING DAMAGE IN FOOD PRODUCTS

The basis of quality assessment is often subjective with attributes such as appearance, smell, texture, and flavor, frequently examined by human inspectors. Consequently, Francis (1980, cited by Brosnan et al., 2004) found that human perception could be easily fooled. The high labor costs, inconsistency, and variability associated with human inspection accentuates the need for objective measurement systems. Recently, automatic inspection systems, mainly based on camera–computer technology have been investigated for the sensory analysis of agricultural and food products (Brosnan et al., 2004).

Mechanical damages have a huge economic cost for producers and wholesalers, and it is necessary to identify these damages as soon as possible to manage products correctly. So, early prediction of damages is one of the main objectives of researchers.

10.4.1 Damage Prediction Models

The special characteristics of agricultural products make it difficult to obtain a general prediction model for damages. Despite this, knowing the properties of one product should make it easier to predict its behavior in an impact or loading, both causes of mechanical damages.

In this sense, Desmet et al. (2003) determined puncture injury susceptibility of tomatoes. Puncture injury of tomatoes is a damage that occurs as a result of impact when the stem of one tomato punctures the skin of another in transit from greenhouse to consumer (Figure 10.7). The punctures induce wound respiration, provide entry sites for decay organisms, and decrease visual appeal. In order to evaluate the tomato skin resistance, a pendulum-induced impact test was carried out. A cylindrical impact probe was chosen with a flat tip and a diameter of an average regular tomato stem (3.7 mm). Due to these impact probe characteristics, the total surface of the tip of the impact probe contacted the tomato during impact. These data provided the necessary information to calculate the susceptibility at a certain level of impact energy as the probe punctures the tomato. Arazuri et al. (2007) evaluated the resistance of the skin of several processing tomato varieties with the aim of knowing the effect of harvest machines. A texture analyzer TA-XT2, equipped with a steel punch of 2 mm maximum diameter, was used. In this case, the object of the analysis was to avoid damage during mechanical harvesting of tomatoes. It was concluded that tomatoes after harvest showed a loss of skin resistance to crack of about 6%.

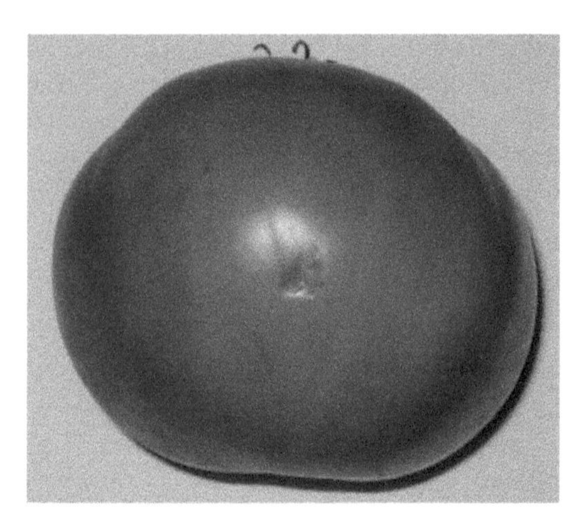

FIGURE 10.7 Puncture injury in tomato skin.

Van Linden et al. (2006) focused their investigation on determination of bruise susceptibility of the tomato. A bruise is a type of subcutaneous damage without rupture of the skin (Mohsenin, 1986) (Figure 10.8). In 2005, Van Linden and De Baerdemaeker found that the bruising mechanism could be a result of both physical injury and the subsequent breakdown of the cell wall components by the action of cell wall–related proteins. Later, these researchers (Van Linden et al., 2006) established one method to determine the bruise susceptibility of tomatoes. An instrumented pendulum was used to produce the necessary impact to develop bruise damage; after two days from the impacts, tomatoes were classified according to the presence or absence of damage. Logistic regression analysis let them develop a bruise damage prediction model using both impact data and characteristics of evaluated tomato varieties. Nevertheless, they concluded that more elaborate logistic models might be

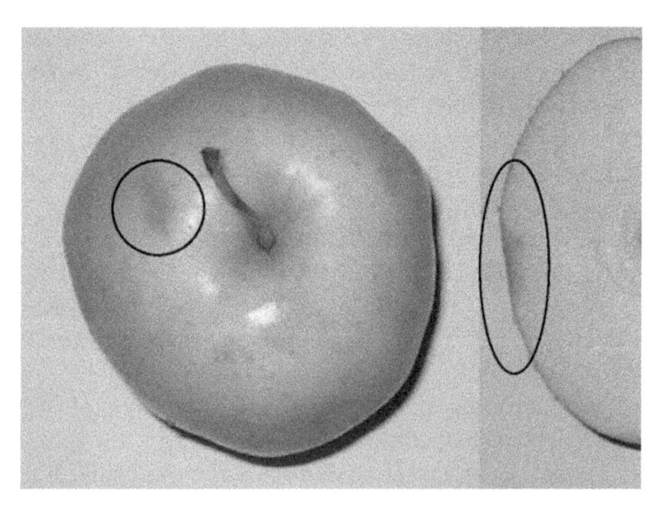

FIGURE 10.8 Bruise damage in apple.

useful to investigate the influence of various fruit properties on the risk of developing bruises after mechanical impact. Simple bruise models should suffice for a quick evaluation of new tomato cultivars regarding their resistance to mechanical impact.

In stone fruits like peaches, nectarines, and plums, researchers are focused on nondestructive tests for bruise evaluation. Valero et al. (2007) studied the direct relationship between nondestructive firmness measurements and relevant fruit texture changes during ripening. In this work, ripening stages and bruising susceptibility were joined.

Most researchers who investigate the effect of impacts and loadings in damage appearance try to develop prediction models. Actually, if the behavior of fruits or vegetables was modeled in different scenarios, the best decision could be made on how to avoid or reduce mechanical damage.

Baritelle et al. (2001) proposed an equation to evaluate the relationship between impact bruising and commodity conditioning. This relationship would estimate bruise threshold as a function of commodity tissue impact properties, Poisson's ratio, and specimen mass and radius of curvature for impacts on a flat, rigid surface. Three factors were found to be most easily controlled: relative turgor, temperature, and strain rate, the latter of which is mainly dependent on handling system design and operation. For example, reducing the relative turgor can reduce tissue elastic modulus (stiffness), which can in turn make a specimen more "self-cushioning," by distributing a given force over a larger area of the specimen's curved surface.

Barreiro et al. (1997) proposed neural bruise prediction models based on the degree of damage in apples and its acceptance or not at market. The prediction relied on European Community (EC) standards. Different models for both quasi-static (compression) and dynamic (impact) loads covering the full commercial ripening period of fruits were developed. The electronic devices used for load sensor calibration were IS-100 and DEA-1, for impact and compression loads, respectively. After a process of variable selection, the bruise prediction models obtained enabled classification of mechanical damages (as acceptable or unacceptable) based on EC standards with errors within the EC tolerance threshold of 10%. The prediction models gather information about bruise susceptibility evolution of fruits in the full commercial ripeness range, information about different loading types, and about fruit physical quality.

Technology advances and computer speed led to the use of high-performance computers, which easily work with large amounts of data. The main problem will be information management. Discrete element analysis and finite element analysis are two examples.

Raji and Favier (2004a) reported that the optimal design and control of many production and postharvest operations require an understanding of the dynamic behavior of agricultural particulates. The resulting mechanical behavior is a complex integrated effect defined by the geometry of the particles, the shape and surface roughness of the containing structure, and the number and strength of the points of interaction of the objects (contacts). Discrete element modeling (DEM), proposed in some research, considers a system as a collection of discrete entities with individual material properties, and calculates the interparticle contact forces, stresses, and particle displacements over a discrete time interval. It involves calculation of the out-of-balance force at very short explicit time intervals, which are so short that any disturbance caused during the time interval is assumed not to propagate further than the immediate neighbors of the particle in question. The process is a cycle with repeated application of Newton's second law of motion integrated to obtain the acceleration, velocity, and displacement. The displacement is used to update the new position during the time interval before the application of the force-displacement law to calculate the new contact force and moment between neighboring particles during the next time interval for a new iteration cycle. In subsequent experiments, a new modeling approach based on DEM was applied to different oilseeds (Raji and Favier, 2004b). The model was validated against experimental data using synthetic spherical particles and canola seeds. It was then used to predict the bulk compression parameters during mechanical oil expression for the three oilseeds. Bed compression was simulated up to the oil point, which is the threshold pressure at which oil emerges from a seed kernel during mechanical seed-oil expression. Similar patterns in the variation of the characteristic parameters were obtained as observed in experimental data. It was concluded that the model was able to predict quite closely the bed strain at the oil point observed in the experiment for each seed type. This suggests that it is a useful tool in the study of mechanical seed-oil expression and other agricultural particulate compression processes as well as provision of data necessary in the design of appropriate machinery.

In contrast to DEM, some authors find finite element analysis as a better system to predict both the behavior of fruit and damage appearance during postharvest manipulation. Dintwa et al. (2008) analyzed the dynamic process of collision among apples and collision of apples with rigid walls. Models were used to investigate the collision of apples in conditions that closely resemble typical practical collision regimes of such fruit during unit operation such as transportation in trucks, sorting operations, or any other handling operations. Information on the quantity of energy loss that can be attributed to the excitation of elastic waves within the body was assessed in isolation to energy dissipation due to the viscoelastic nature of the material. These researchers concluded that that finite element analysis can predict some behaviors of viscoelastic products, but it is necessary to have a more theoretically accurate assessment of the problem.

Finally, the finite element method (FEM) was used to predict bruise damage in watermelon (Sadrnia et al., 2008). One of the main characteristics of the watermelon is the skin thickness, which makes it difficult to detect flesh bruising. It is extremely difficult to measure internal stresses caused by applied compressive forces, and FEM is shown as an alternative tool to model objects of irregular shapes and nonhomogeneous material properties. Watermelons were compressed in longitudinal and transverse directions by parallel plates. The applied forces on models were equal to 10% of breaking force. Two geometric models were developed: One was an axisymmetric (2-D) model while the other one was a three-dimensional (3-D) model. The 2-D model was suitable for analysis of the forces applied by the plates in the longitudinal axis direction of the fruit (blossom–stem). Since the watermelon structure is not axisymmetric in the transverse direction for compression tests, because of rind thickness variation, the 3- model was developed in order to simulate compression in this direction. Different simulations were carried out. In one of them, the modulus of elasticity and the large deformation nonlinearity were considered. After model validation, the researchers concluded that axisymmetric model was suitable for symmetrical loading with two plates at two sides of whole watermelons in the longitudinal direction. And, the 3-D model could be used when the watermelon was loaded with two plates in the transverse direction.

10.4.2 New Technologies for Damage Detection

Until now, the objective has been to describe the mechanism of damages and their prediction from widely used tests such as puncture tests or compression tests. Although these techniques give high-quality information about damages, researchers have focused their investigations on the use of new technologies that are nondestructive and minimal time consumers. Among these technologies are near-infrared spectroscopy (NIRS) and the newest multi- and hyperspectral image technologies.

10.4.2.1 Near-Infrared Technology

Near-infrared (NIR) radiation covers the range of the electromagnetic spectrum between 780 and 2500 nm. In NIR spectroscopy, the product is irradiated with NIR radiation, and the reflected or transmitted radiation is measured. While the radiation penetrates the product, its spectral characteristics change through wavelength-dependent scattering and absorption processes. This change depends on the chemical composition of the product as well as on its light-scattering properties, which are related to its microstructure. Advanced multivariate statistical techniques such as partial least squares regression are then applied to extract the required information from the usually convoluted spectra (Nicolaï et al., 2007).

Xing et al. (2003) applied NIR technology in order to test two types of bruises in apples: bruises produced by impact and by compression. After a canonical discriminant analysis, the researchers obtained different classification models based on spectral data. Although the conclusion was that the accuracy of the models should be improved before incorporating them in grading machines, the researchers found that NIR spectroscopy is a quick alternative for bruise detection.

Van Dijk et al. (2006) evaluated the effect of storage time and temperature on the activities of pectin-degrading enzymes and firmness loss in tomatoes. They observed that the concentration of one of the enzymes, pectin methyl esterase (PME), was below the detection limit for NIR. However, changes produced by PME were reflected in the NIR spectra, so this technique could be applied to evaluate the effect of storage time and temperature.

Following with tomatoes, Hahn (2002) carried out different tests to detect damages produced by fungal infections based on NIR spectra. Fungal spore detection is done regularly by isolation on nutrient agar plates, but it takes a day to determine if the sample is infected. During this period, the infection can increase dramatically causing great losses in containers or storage rooms. Thus, in this research, tomatoes were inoculated with spores and with sterile water to study the feasibility of using optical reflectance for detecting spores. Spectral signatures before and after being inoculated were acquired at the same place to avoid reflectance differences caused by peel color, maturity stage, and sampling place. Fast Fourier transform (FFT) series obtained from the spectral signatures were analyzed by discriminant analysis. As a result, tomatoes infected with *Fusarium oxysporum* were detected with an accuracy of 85.42%.

10.4.2.2 Image Analyses

The potential of computer vision in the food industry has long been recognized. Automated visual inspection is undergoing substantial growth in the food industry because of its cost effectiveness, consistency, superior speed, and accuracy. Traditional visual quality inspection performed by human inspectors has the potential to be replaced by computer vision systems for many tasks (Brosnan et al., 2004).

In general, images are formed by incident light in the visible spectrum falling on a partially reflective, partially absorptive surface, with the scattered photons being gathered up in the camera lens and converted to electrical signals either by vacuum tube or by a charge-coupled device (CCD) (Abdullah, 2008).

Detection of skin defects and damage is the most widely used application of image analysis for the inspection of fruit and vegetables. The presence of external damage is a clear sign of the lack of quality of a product. One difficulty that is common to most of them is to distinguish defective areas of the fruit or vegetable from natural organs like calyxes or stems (Cubero et al., 2010).

Jarimopas and Jaisin (2008) developed an efficient machine vision experimental sorting system for sweet tamarind pods based on image processing techniques. Relevant sorting parameters included shape, size, and defects (cracks). The sorting system involved the use of a CCD camera that was adapted to work with a TV card, microcontrollers, sensors, and a microcomputer. After analyzing images in different situations, sorting performance was deemed to be acceptable according to Thai agricultural and food commodity standards.

In some cases, the color of a defective area of some fruits matches the color of healthy skin from other fruits of the same variety, which makes the task of identifying true defects even more complex (Cubero et al., 2010). In this sense, Leemans and Destain (2004) developed a real-time method for apple classification based on features extracted from defects. After the acquisition of images with CCD cameras, a first segmentation to locate the fruits on the background and a second one to find the possible defects were performed. Once the defects were located, they were characterized by a set of features including color, shape, texture descriptors, as well as the distance of the defects to the nearest calyx or stem end. These data were accumulated for each fruit and summarized. Then, apples were graded using quadratic discriminant analysis obtaining a correct classification rate of 73%. The errors were due to defects that were difficult to segment, such as russet and bruises, or due to some wounds that were located near the stem ends and were probably confused with them.

Most current computer vision systems used in the automatic quality inspection of food are limited to the visible region of the electromagnetic spectrum as they tend to imitate the human eye. However, nonvisible information, such as that provided by near-infrared or ultraviolet regions of the spectrum, can improve the inspection by detecting specific defects or allowing the detection of nonvisible damage (Blasco et al., 2007).

Multi- or hyperspectral cameras permit rapid acquisition of images at many wavelengths. Imaging at fewer than ten wavelengths is generally termed *multispectral*, and more than ten termed *hyperspectral*. The resulting data set can be visualized as a cube with the *X* and *Y* dimensions being the length and width of the image (in pixels) and the *Z* dimension being spectral wavelengths; each data point is an intensity value. Alternatively, the data set could be envisioned as a stack of single-wavelength pictures of the object with as many pictures as the number of wavelengths used. Such imaging provides information about the spatial distribution of constituents near the product's surface (Abbott, 1999).

Near-infrared hyperspectral reflectance imaging was used by Ariana et al. (2006) in order to detect bruises on pickling cucumbers. The researchers selected the wavelength range from 950 to 1650 nm for analysis. After data processing, it was observed that the reflectance of bruised tissue on cucumber fruit was lower than that of normal tissue, and the former increased over time toward that of normal tissue. Therefore, they detected that the wavelength range from 950 to 1350 nm with a bandwidth of 8.8 nm was the most useful in principal component analysis for bruise detection. Finally, the classification accuracies obtained were of 95% at first day and 75% at sixth day after bruising.

Nicolaï et al. (2006) developed a hyperspectral NIR imaging system to identify bitter pit lesions on apples. Bitter pit is a physiological disorder in apples that develops postharvest and produces its decline at the wholesale market or exporter. An indium gallium arsenide near-infrared line scan camera (SU320-1.7RT-V, Sensors Unlimited Inc., Prince, USA) was used for hyperspectral imaging. The optical sensitivity of this camera ranges from 900 to 1700 nm. The researchers successfully validated the calibration and concluded that the system could identify bitter pit injures, even when they were not visible to the naked eye, such as just at harvest.

Bruise detection on apples was the main aim of the research carried out by ElMasry et al. (2008). They investigated the potential of a hyperspectral imaging system for early detection of bruises on different background colors of McIntosh apples. The spectral region was from 400 to 1000 nm and the background colors were green, red, and reddish green. As in other cases, partial least squares (PLS) regression and discriminant analysis were combined in order to obtain as much information as possible from the images. As expected, the efficiency of the developed method was demonstrated and the authors proposed to extend this technique to evaluate damages in other varieties of apple.

Apart from these techniques, magnetic resonance, thermal methods, x-rays, and other techniques can all generate an image (Abdullah, 2008).

Magnetic resonance imaging (MRI) based on the principles of nuclear magnetic resonance (NMR), has achieved general acceptance as a powerful tool for the diagnosis and assessment of clinical conditions following the widespread introduction of medical imaging systems during the 1980s. The ability of MRI to function in a completely noninvasive manner and to encode molecular dynamics through different contrast mechanisms has encouraged development of alternative applications outside the medical domain. Some researchers observed it as a useful tool for pre- and postharvest study and assessment of fruits, vegetables, and other edible commodities (Clark et al., 1997).

Among the possibilities of MRI technology is the detection of mechanical damages. Milczarek et al. (2008) focused their research on assessment of tomato pericarp mechanical damages using multivariate analysis of MRI. As natural biological variation makes image-based quantification of damaged tissue a difficult task, they combined more than one MR image. In order to predict damages, conductivity tests were performed. It was expected that damaged tissue had different conductivity than normal tissue. Once multivariate analysis of the images was carried out, it was proven that this technology was effective for predicting the conductivity score of pericarp tissue in tomatoes, and consequently the mechanical damages.

Other new techniques to evaluate quality based on electromagnetic spectrum characteristics are x-ray imaging and ultraviolet imaging. In the x-ray case, Kotwaliwale et al. (2007) evaluated the presence or absence of defects in pecans. A soft x-ray digital imaging system was used to acquire radiographs. Some pecans with known internal defects and with unknown quality attributes were imaged. After applying contrast stretching or high-frequency emphasis techniques, defects were clearly differentiated by x-ray imaging.

An ultraviolet imaging–based machine vision system was developed by Al-Mallahi et al. (2010) to detect good quality potato tubers on a harvester, in order to remove clods and unwanted potato tubers, especially small tubers. Detection was based on the high degree of ultraviolet reflectance of the tubers compared to that of clods and background. A variable thresholding segmentation method was developed to overcome differences in the lighting conditions and in the water content of clods. The results showed that 98.79% of the clods were detected successfully. Furthermore, the processing time required to segment the objects in each frame was approximately 94 ms, which was enough to work at the normal speed of the conveyor belt of the harvester.

There is a new type of image obtained based on the thermal properties of agricultural products. The *thermograph* is an image processing technique that transforms thermal radiation, recorded by a camera, into a thermographic image or a thermogram. A *thermogram* is a representation of the specific temperature distribution at the object surface (Veraverbeke et al., 2006).

The basic principle of thermal imaging is based on the fact that all materials emit infrared radiation. Thermal imaging systems can detect radiation from the shortwave to long-wave infrared. Typical long-wave infrared systems exhibit maximum sensitivity around room temperature, while mid-wave infrared systems exhibit peak sensitivity at much higher temperatures (Gowen et al., 2010).

Thermography is a very fast measuring technique, which allows measurement of moving objects. It is also a noncontact and nondestructive tool so that no mechanical injury or contamination of the study object can occur during measurement (Veraverbeke et al., 2006). Bruising detection on apples (Varith et al., 2003) and tomatoes (Van Linden et al., 2003) show this technique as a new tool to evaluate the quality of agricultural products.

10.4.2.3 Sonic and Ultrasonic Vibration

Sonic (or acoustic) vibrations encompass the audible frequencies between 20 Hz and ≈ 15kHz; ultrasonic vibrations are above the audible frequency range (>20 kHz). Sonic and ultrasonic waves can be transmitted, reflected, refracted, or diffracted as they interact with the material. Wave propagation velocity, attenuation, and reflection are the important parameters used to evaluate the tissue properties of horticultural commodities. When an object is excited at sonic frequencies, it vibrates. At particular frequencies it will vibrate more vigorously, causing amplitude peaks; such a condition is referred to as *resonance*. Resonant frequencies are related to elasticity, internal friction, or damping, and shape, size, and density. The firmer the flesh, the higher the resonant frequency for products of the same size and shape. The traditional watermelon ripeness test is based on the acoustic principle, where one thumps the melon and listens to the pitch (frequency) of the resonance (Abbott, 1999).

Basing on this principle, Diezma-Iglesias et al. (2004) applied acoustic impulse response techniques to evaluate internal quality in seedless watermelon. These may have a disorder called *hollow heart* produced by alternating between wet and dry soil, and hot and cold temperatures when growing. The researchers developed a device consisting of a microphone, structural elements, and mechanical impact generator. Spectral parameters were examined as potential nondestructive predictors of internal disorders and two frequencies in a specified bandwidth, between 20 and 500 Hz, were the acoustic parameters showing the best ability to detect hollow heart. Therefore, due to the lack of homogeneity in the distribution of disorders inside the fruits, the authors recommended at least five impacts along the surface of the watermelon to improve the results.

Other authors applied this technique to evaluate maturity evolution during storage of apples (De Belie et al., 2000) and tomatoes (Mizrach, 2007). In both cases, recorded data led to good predictions of the quality and proved to be an interesting tool to manage the storage of these products.

10.5 CONCLUSIONS

Physical properties of agricultural products provide information about the susceptibility of these products to mechanical damage during pre- and postharvest processing. This knowledge combined with new technologies will make it possible to indentify critical points of machines in which damage is produced. Thus, sensors for detecting impacts or compressions during agricultural product manipulation could be an interesting solution for damage prevention.

The advance in statistical analysis and the development of new high-speed computers has provided the opportunity to use new technologies such as near-infrared spectroscopy, multi- and hyperspectral imaging analysis, magnetic resonance, and others for obtaining deeper knowledge of the characteristics of agricultural products. In addition, the main objectives of researchers have been focused on nondestructive and rapid techniques for quality evaluation.

Consequently, applying information provided by these systems for the management of agricultural products, quality and acceptability to wholesalers and consumers will be improved.

REFERENCES

Abbot, J.A. 1999. Quality measurement of fruits and vegetables. *Postharvest Biology and Technology.* 15: 207–225.

Abdullah, M.Z. 2008. *Image acquisition systems: Computer vision technology for food quality evaluation.* Maryland Heights, MO: Elsevier.

Al-Mallahi, A., T. Kataoka, H. Okamoto, and Y. Shibata. 2010. Detection of potato tubers using an ultraviolet imaging-based machine vision system. *Biosystems Engineering* 105: 257–265.

Amorim, L., M. Martins, S.A. Lourenço, A.S.D. Gutierrez, F.M. Abreu, and F.P. Gonçalves. 2008. Stone fruit injuries and damage at the wholesale market of São Paulo, Brazil. *Postharvest Biology and Technology* 47: 353–357.

Arana, I., Jarén, C., and S. Arazuri. 2005. Nectarine Woollines Detection by Non-destructive Mechanical Impact. *Biosystems Engineering* 90(1), 37-45.

Arazuri, S., C. Jaren, A.B. Juanena, F. Marinez, and J.J. Perez de Ciriza. 2001. Daños producidos por las cosechadoras de tomate. *Horticultura* 151: 28–35.

Arazuri, S., C. Jarén, J.I. Arana, and J.J. Pérez de Ciriza. 2007. Influence of mechanical harvest on the physical properties of processing tomato (*Licopersicon esculentum* Mill.). *Journal of Food Engineering* 80: 190–198.

Arazuri, S., Arana, I., and C. Jarén. 2010. Evaluation of Mechanical Tomato Harvesting Using Wireless Sensors. *Sensors* 10, 11126-11143.

Ariana, D.P., L. Renfu, and D.E. Guyer. 2006. Near-infrared hyperspectral reflectance imaging for detection of bruises on pickling cucumbers. *Computers and Electronics in Agriculture* 53: 60–70.

Baritelle, A.L., and G.M. Hyde. 2001. Commodity conditioning to reduce impact bruising. *Postharvest Biology and Technology* 21: 331–339.

Barreiro, P., V. Steinmetz, and M. Ruiz-Altisent. 1997. Neural bruise prediction models for fruit handling and machinery evaluation. *Computers and Electronics in Agriculture* 18: 91–103.

Blasco, J., N. Aleixos, J. Gómez, and E. Moltó. 2007. Citrus sorting by identification of the most common defects using multispectral computer vision. *Journal of Food Engineering* 83: 384–393.

Bowden, E.P., and D. Tabor. 1954. *The friction and lubrication of solids.* London: Oxford University Press.

Brosnan, T., and D.W. Sun. 2004. Improving quality inspection of food products by computer vision: A review. *Journal of Food Engineering* 61: 3–16.

Clark, C.J., P.D. Hockings, D.C. Joyce, and R.A. Mazucco. 1997. Application of magnetic resonance imaging to pre- and post-harvest studies of fruits and vegetables. *Postharvest Biology and Technology* 11: 1–21.

Cubero, S., N. Aleixos, E. Moltó, J. Gómez-Sanchis, and J. Blasco. 2010. Advances in machine vision applications for automatic inspection and quality evaluation of fruits and vegetables. *Food and Bioprocess Technology*, doi: 10.1007/s11947-010-0411-8.

De Belie, N., S. Schotte, P. Coucke, and J. De Baerdemaeker. 2000. Development of an automated monitoring device to quantify changes in firmness of apples during storage. *Postharvest Biology and Technology* 18: 1–8.

Desmet, M., J. Lammertyn, N. Scheerlinck, B.E. Verlinden, and B.M. Nicolaï. 2003. Determination of puncture injury susceptibility of tomatoes. *Postharvest Biology and Technology* 27: 293–303.

Desmet, M., V. Van linden, M.L.A.T.M. Hertog, B.E. Verlinden, J. De Baerdemaeker, and B.M. Nicolaï. 2004. Instrumented sphere prediction of tomato stem-puncture injury. *Postharvest Biology and Technology* 34: 81–92.

Diezma-Iglesias, B., M. Ruiz-Altisent, and P. Barreiro. 2004. Detection of internal quality in seedless watermelon by acoustic impulse response. *Biosystems Engineering* 88: 221–230.

Dintwa, E., M. Van Zeebroeck, H. Ramon, and E. Tijskens. 2008. Finite element analysis of the dynamic collision of apple fruit. *Postharvest Biology and Technology* 49: 260–276.

ElMasry, G., N. Wang, C. Vigneault, J. Qiao, and A. ElSayed. 2008. Early detection of apple bruises on different background colors using hyperspectral imaging. *LWT* 41: 337–345.

Garcia-Ramos, F.J., J. Ortiz-Cañavate, and M. Ruiz-Altisent. 2003. Decelerator elements for ramp transfer points in fruit packing lines. *Journal of Food Engineering* 59: 331–337.

García-Ramos, F.J., J. Ortiz-Cañavate, and M. Ruiz-Altisent. 2004. Evaluation and correction of the mechanical aggressiveness of commercial sizers used in stone fruit packing lines. *Journal of Food Engineering* 63: 171–176.

Gowen, A.A., B.K. Tiwari, P.J. Cullen, K. McDonnell, and C.P. O'Donnell. 2010. Applications of thermal imaging in food quality and safety assessment. *Trends in Food Science & Technology* 21: 190–200.

Gunasekaran, S., and M. Mehmet Ak. 2000. Dynamic oscillatory shear testing of foods-selected applications. *Trends in Food Science & Technology* 11: 115–127.

Hahn, F. 2002. Fungal spore detection on tomatoes using spectral Fourier signatures. *Biosystems Engineering* 81: 249–259.

Jarén, C., S. Arazuri, and I. Arana. (2008) Electronic fruits and other sensors. *Chronica Horticulturae* 48 (4): 4–6.

Jarimopas, B., and N. Jaisin. 2008. An experimental machine vision system for sorting sweet tamarind. *Journal of Food Engineering* 89: 291–297.

Kotwaliwale, N., P.R. Weckler, G.H. Brusewitz, G.A. Kranzler, and N.O. Maness. 2007. Non-destructive quality determination of pecans using soft X-rays. *Postharvest Biology and Technology* 45: 372–380.

Leemans, V., and M.F. Destain. 2004. A real-time grading method of apples based on features extracted from defects. *Journal of Food Engineering* 61: 83–89.

Milczarek, R.R., M.E. Saltveit, T.C. Garvey, and M.J. McCarthy. 2008. Assessment of tomato pericarp mechanical damage using multivariate analysis of magnetic resonance images. *Postharvest Biology and Technology* 62: 189–195.

Mizrach, A. 2007. Nondestructive ultrasonic monitoring of tomato quality during shelf-life storage. *Postharvest Biology and Technology* 46: 271–274.

Mohsenin, N.N. 1986. *Physical properties of plant and animal materials: Structure, physical characteristics and mechanical properties*. New York: Gordon and Brach Science Publishers. Inc.

Nicolaï, B.M., K. Beullens, E. Bobelyn, A. Peirs, W. Saeys, K.I. Theron, and J. Lammertyn. 2007. Nondestructive measurement of fruit and vegetable quality by means of NIR spectroscopy: A review. *Postharvest Biology and Technology* 46: 99–118.

Nicolaï, B.M., I. Bulens, J. De Baerdemaeker, B. De Ketelaere, M.L.A.T.M. Hertog, P. Verboven, and J. Lammertyn. (2009) Non-destructive evaluation: Detection of external and internal attributes frequently associated with quality and damage. In *Postharvest handling: A systems approach*, ed. Robert L. Shewfelt and Stanley E. Prussia. Oxford: Academic Press.

Nicolaï, B.M., E. Lötze, A. Peirs, N. Scheerlinck, and K.I. Theron. 2006. Non-destructive measurement of bitter pit in apple fruit using NIR hyperspectral imaging. *Postharvest Biology and Technology* 40: 1–6.

Raji, A.O., and J.F. Favier. 2004a. Model for the deformation in agricultural and food particulate materials under bulk compressive loading using discrete element method. I: Theory, model development and validation. *Journal of Food Engineering* 64: 359–371.

Raji, A.O., and J.F. Favier. 2004b. Model for the deformation in agricultural and food particulate materials under bulk compressive loading using discrete element method. II: Compression of oilseeds. *Journal of Food Engineering* 64: 373–380.

Sadrnia, H., A. Rajabipour, A. Jafari, A. Javadi, Y. Mostofi, J. Kafashan, E. Dintwa, and J. De Baerdemaeker. 2008. Internal bruising prediction in watermelon compression using nonlinear models. *Journal of Food Engineering* 86: 272–280.

Salar, M. 2009. Determinación de daños durante la recolección mecanizada y la manipulación de patata. Doctoral thesis. Universidad Pública de Navarra.

Salar, M., C. Jarén, S. Arazuri, I. Arana, M.J. García, and J. Viscarret. 2009. Modificaciones en las propiedades físicas de las patatas durante la recolección mecánica. *Tierras de Castilla y León* 160: 28–34.

Salvadores, C., C. Jaren, S. Arazuri, A. Juanena, I. Arana, and J.J. Perez de Ciriza. 2001. Damage determination in tomato harvesters using instrumented sphere IS-100. *Proceedings of the 6th International Symposium*. Potsdam (Germany).

Valero, C., C.H. Crisosto, and D. Slaughter. 2007. Relationship between nondestructive firmness measurements and commercially important ripening fruit stages for peaches, nectarines and plums. *Postharvest Biology and Technology* 44: 248–253.

Van Canneyt, T., E. Tijskens, H. Ramon, R. Verschoore, and B. Sonck. 2003. Characterisation of a potato-shaped instrumented sphere. *Biosystems Engineering* 86: 275–285.

Van Dijk, C., C. Boeriu, T.Stolle-Smits, and L.M.M. Tijskens. 2006. The firmness of stored tomatoes (*cv. Tradiro*). 2. Kinetic and Near Infrared models to describe pectin degrading enzymes and firmness loss. *Journal of Food Engineering* 77: 585–593.

Van Linden, V., and J. De Baerdemaeker. 2005. The phenomenon of tomato bruising: Where biomechanics and biochemistry meet. *Acta Horticulturae* 682: 925–932.

Van Linden, V., B. De Ketelaere, M. Desmet, and J. De Baerdemaeker. 2006. Determination of bruise susceptibility of tomato fruit by means of an instrumented pendulum. *Postharvest Biology and Technology* 40: 7–14.

Van Linden, V., R. Vereycken, C. Bravo, H. Ramon, and J. De Baerdemaeker. 2003. Detection technique for tomato bruise damage by thermal imaging. *Acta Horticulturae* 599: 389–394.

Varith, J., G.M. Hyde, A.L. Baritelle, J.K. Fellman, and T. Sattabongkot. 2003. Non-contact bruise detection in apples by thermal imaging. *Innovative Food Science and Emerging Technologies* 4: 211–218.

Veraverbeke, E.A., P. Verboven, J. Lammertyn, P. Cronje, J. De Baerdemaeker, and B.M. Nicolaï. 2006. Thermographic surface quality evaluation of apple. *Journal of Food Engineering* 77: 162–168.

Xing, J., S. Landahl, J. Lammertyn, E. Vrindts, and J. De Baerdemaeker. 2003. Effects of bruise type on discrimination of bruised and non-bruised "Golden Delicious" apples by VIS/NIR spectroscopy. *Postharvest Biology and Technology* 30: 249–258.

11 Measurement of Physical Properties of Fruits Using Image Analysis

José Blasco, Nuria Aleixos, Abelardo Gutierrez, and Enrique Moltó

CONTENTS

11.1 INTRODUCTION

Most of the information that humans obtain from their environment is obtained by vision. The mechanism of human vision depends basically on the eyes, which are sensitive to a narrow part of the electromagnetic spectrum (wavelengths between 400 and 700 nm that are called *visible*), and the brain, which is capable of interpreting the scenes. It is for this reason that most applications to measure external physical properties of fruits are based on this technology.

Light coming from different sources is concentrated by a series of lenses onto a small area inside the eyes, called the retina, where photosensitive cells convert them into neural impulses that are transmitted to the brain via the optic nerve. Similarly, artificial vision systems are composed of a series of lenses and photodetectors (simulating the eye), which turn light into electronic impulses, and an electronic microprocessor to interpret such signals.

Visual information enables humans to estimate important features that are related to the quality of food, such as size, shape, color (which is related to the freshness of the products), the presence of defects (blemishes, rotten parts, etc.), texture, and so forth. Visual information is also one of the first to reach the brain in the process of assessing quality. Moreover, it is a noncontact, nondestructive means of getting such information. For these reasons, automatic devices capable of acquiring and interpreting images are currently a valuable tool for measuring the physical properties of food (Cubero et al., 2010; Sun, 2007 and 2010).

The quality of an image, namely the amount of useful information that it contains, depends on two major factors: the illumination of the scene and the acquisition device (generally a camera).

Cameras use a set of lenses to project the light of the scene onto a photosensor. This photosensor produces an electronic signal which is then digitized so it can be transferred to the memory of a computer system, where it is processed.

In the past, video cameras produced analog signals that were digitized by an electronic card called a *frame grabber.* Recent electronic developments have led to new, compact, high-performance charge-coupled device (CCD) cameras that produce digital images that are immediately available for transfer by standard digital buses, like FireWire, Universal Serial Bus (USB), or gigabit Ethernet (GigE).

Since what we see depends on the reflectance properties of the objects in the scene and the illumination, the design of the illumination system for a machine vision system must take into account the spectral range of interest and the geometrical particularities of the objects to be inspected. Ideally, it has to cover all the spectral range of interest uniformly and illuminate the objects without shadows or specular reflections (often seen as bright spots in the images). However, the arrangement of the light sources and the geometry of the objects mean that some areas are more illuminated than others. This uneven illumination can be corrected during image processing, but it is a computationally time-consuming task.

Directional lighting in a 45° source–camera configuration is used to illuminate flat objects in order to prevent specular reflections toward the camera. In this case, the flat object is illuminated from a light source at 45° and the camera is located vertically (Figure 11.1).

An example of this configuration is described by Sun and Brosnan (2005) using two fluorescent lights in a system for the fully automated quality inspection of pizza bases and sauce spread. A similar configuration with four fluorescent tubes was used by Fernández et al. (2005) to illuminate apple discs in order to control and track their dehydration during frying. A similar system was later used by Pedreschi et al. (2006) to illuminate potato chips in a computer vision system designed to measure the kinetics of color changes at different frying temperatures.

However, when the objects are not flat, as in the case of quasi-spherical fruits or vegetables, it is normally more convenient to use hemispherical diffusers to produce a more uniform scattering of light to illuminate the scene (Figure 11.2). Mateo et al. (2006) used such a method to illuminate samples of meat from tuna fish in order to develop an automated inspection system as part of the search for a means to detect changes in the quality of the fish induced by the method of capture. A hemispheric integrating sphere was used by Riquelme et al. (2008) to inspect olives. Color reveals the quality of olives and was used to detect some skin defects. Since shadows and reflections disturb the correct measurement of this attribute, it was very important to generate homogeneous diffuse lighting.

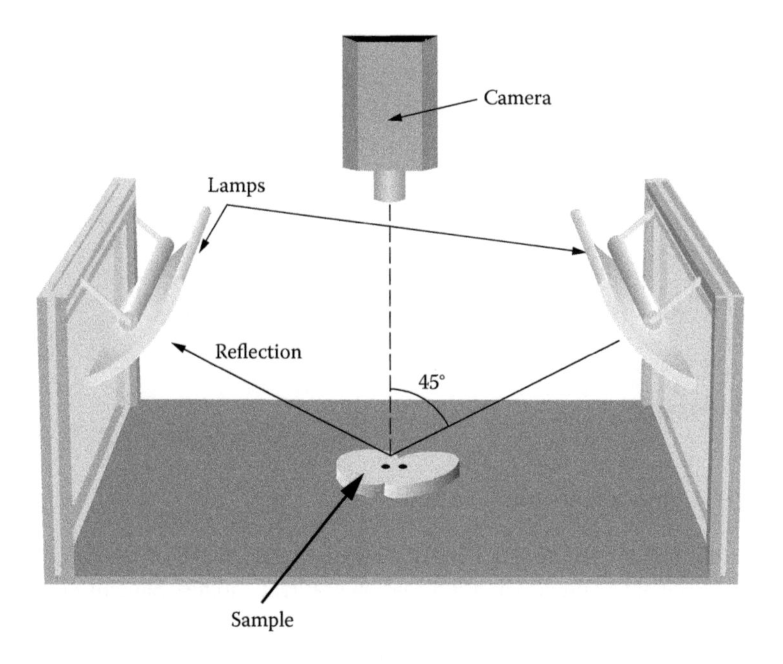

FIGURE 11.1 Example of a 45° arrangement of light source and camera.

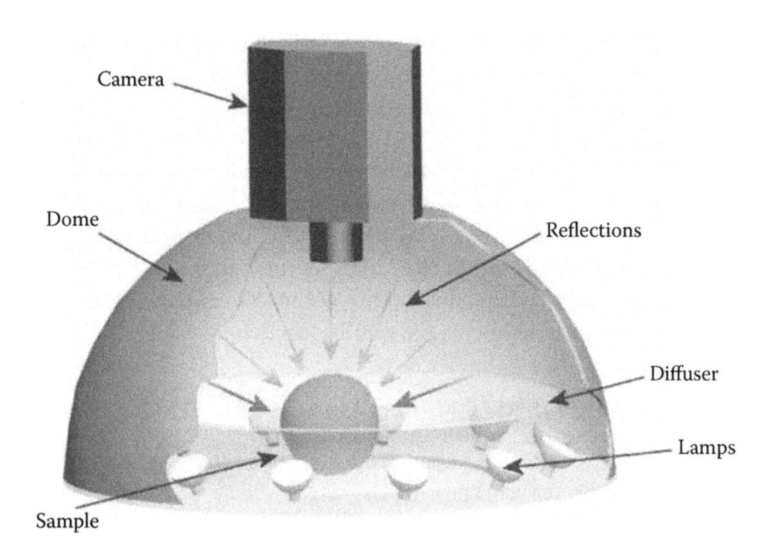

FIGURE 11.2 Example of a diffuse lighting system.

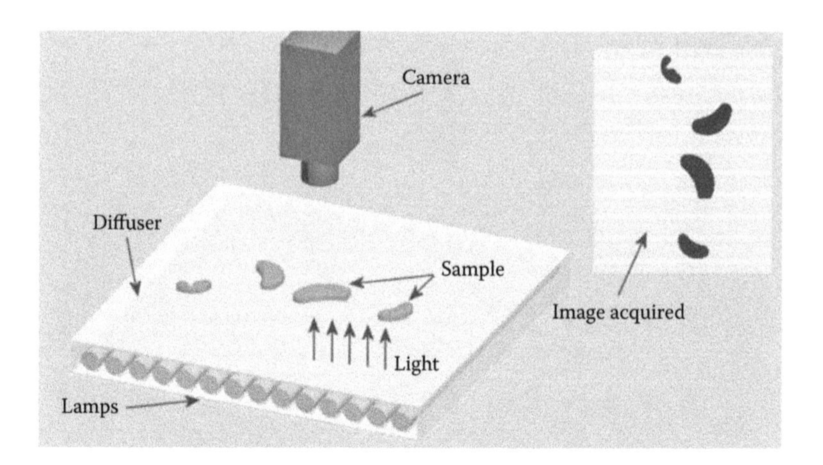

FIGURE 11.3 Example of a backlight model.

In the particular case of spherical objects, the interference of the light with the geometry of the object creates an illumination gradient. Gómez-Sanchis et al. (2008) proposed a methodology for correcting this effect so as to be able to image citrus fruits. They proposed a model for estimating the theoretical height and angle of incidence of the light for each pixel in the fruit image, which could be used to perform mathematical correction of the illumination.

Images with enhanced contrast can be obtained by using backlight illumination (Figure 11.3). These kinds of images are suitable for measuring sizes or shapes because the contour of the objects is strongly enhanced against the background. This method was used by van Eck et al. (1998). In their application, aligned cucumbers appeared as sharply contrasted black shapes against a white background. Backlighting was also used by Menesatti et al. (2008) to estimate the shape of hazelnuts in order to discriminate among different cultivars. In this case, the system used to acquire the images was a flatbed scanner, not a video camera. Costa et al. (2009) estimated the shape of Tarocco oranges by placing the fruits laterally on an illuminated dashboard, thereby increasing the contrast. A metric reference mark was placed by the side of each fruit in order to record its size. Backlighting was also adopted by Blasco et al. (2009a) to perform an in-line automatic inspection of mandarin segments traveling over semitransparent conveyor belts. By illuminating the segments from the background, they appeared clearly silhouetted, thus facilitating the estimation of the size and shape by means of contour analysis. This system also enhanced the possibilities of detecting seeds because segments are semitransparent.

11.2 ESTIMATION OF SIZE USING IMAGE ANALYSIS

The estimation of the size of individual products can be used for quality control by producer companies. Moreover, knowing the size of the products that are put on the market is particularly important in the food industry. In some cases this is because the consumer is more attracted to purchasing larger products and these reach higher market prices. Furthermore, it is often important to package products of a similar size in order to make them more appealing or to optimize their storage. In the case of fresh fruits and vegetables, international quality standards establish commercial categories based on size.

Many image analysis techniques have been described to estimate size and they usually depend on the shape of the product. The features most commonly used in food quality evaluation are area, perimeter, length, and width.

Ni and Gunasekaran (1998) employed a computer vision system to measure the length of cheese shreds. The images were processed by morphological transformation algorithms such as dilation and erosion in order to smooth the image edge contours. Cheese shred lengths were determined from skeletonized images using syntactic methods. Estimation of shred lengths had an error rate below 10%. One of the major problems when measuring the size of objects occurs when the objects are touching or overlapping. Ni and Gunasekaran (2004) developed a method to detect these situations.

Because single images are two-dimensional representations of the scene, sometimes they do not provide enough information for accurate assessment of size; for instance, when objects are neither flat nor perfectly spherical. In such cases, acquisition of several images from different positions may help to estimate size accurately. For instance, Blasco et al. (2003) used four different views of Golden Delicious apples to measure size. They calculated the diameter from the image in which the stem was located nearest to the centroid of the fruit, since international quality standards state that the size of apples has to be measured at their equatorial part. When fruit is not oriented properly, the accuracy of the measurements decreases. Throop et al. (2005) measured the size of apples as they were transported and rotated. Linear and rotational speeds were controlled in order to capture images of one complete revolution of each fruit. They estimated the diameter from the major axis of the apples, their surface, and their volume.

A different approach was taken by Moreda et al. (2007) in an automatic system for in-line determination of the size of nonspherical fruits like tomatoes and kiwifruits. They used an optical ring sensor made up of a large number of equally spaced single infrared emitters and receivers on a circular frame. As the fruit passed through the sensor, it blocked the beams of particular emitters, thus creating shadows over certain receivers. By analyzing these shadows, they estimated the size of the fruit. The authors concluded that the major disadvantage of this method is that the fruit needs to be oriented properly to achieve reliable data, because the fruit swayed to and fro during the measurements.

11.3 SHAPE ESTIMATION

The shape of the objects is also a determining factor in the decision of consumers to select one particular product instead of another. In many cases, products that do not meet quality standards have to be removed.

Simpler shape descriptors such as area, length, width, compactness (ratio between the perimeter and the area), elongation (ratio between the length and the width), and symmetry (similarity in the shape of two halves of an object) can be used to discriminate very different shapes. The aspect ratio (ratio between the major diameter and intermediate diameter) and a particular descriptor the authors called the *ellipsoid ratio* were used by Sadrnia et al. (2007) to estimate the shape of watermelons. Physical characteristics such as mass, volume, dimensions, density, spherical coefficient, and geometric mean diameter were measured. Relations and correlation coefficients obtained among the above characteristics were used to classify fruit shape as either normal or nonstandard. Similarly, simple shape descriptors based on geometrical ratios have been successfully used to classify apples. Such descriptors include the length/width ratio, conicity (distance of the maximum width from a base), squareness (ratio between the product of calyx basin width and distance of the maximum width from the calyx end, and the product of fruit length and fruit width) (Currie et al., 2000). Jarimopas and Jaisin (2008) also classified sweet tamarind pods by shape. A circle centered on the centroid of the pod was drawn and the radial distances from each point on this circle to the closest boundary point were used as shape descriptors.

In many food inspection applications, since objects in the scene are often positioned randomly, shape descriptors should be translation, rotation, and scale invariant (Kim and Kim, 2000). Many shape description methods calculate a vector with features from the point boundaries. This vector is called the *signature* of the object. Signatures are normally standardized so as to be invariant to scaling and rotation (van Otterloo, 1992). Two of the most widely used methods to express signatures are distance between the centroid and boundary points as a function of angles (Titchmarsh, 1967), and the polar or radius signature, used by Tao et al. (1995) to describe the contour of potatoes or by Blasco et al. (2009a) to describe the shape of Satsuma segments. Beyer et al. (2002) fitted a third-order polynomial to the polar signature of sweet cherries, using regression coefficients to describe shape.

Since signatures are functional representations of the contour, if we consider that they are periodic, they can be analyzed using powerful spectral tools like Fourier descriptors derived from processing signatures with the Fourier transform. Estimation of the Fourier transform is very costly in terms of computational resources, and for this reason the fast Fourier transform (FFT) is normally used instead. The FFT is a discrete Fourier transform algorithm that reduces the number of computations needed for N points from $2N^2$ to $2N\log_2 N$ (Bracewell, 1999). Tao et al. (1995) classified potatoes from the first 10 harmonics of the radius signature using the inverse FFT. Currie et al. (2000) also developed Fourier descriptors to classify apples by shape. Principal component analysis was used to select the Fourier descriptors.

Wavelets are mathematical functions that cut data into different frequency components and study each component with a solution that is matched to its scale. They have the advantage over Fourier methods because they can analyze signals with discontinuities and sharp points (Ohm et al., 2000). Moreover, in Fourier analysis the spatial locations of the frequency components of the image are lost. Discrete wavelet descriptors have been proposed to analyze contour signatures, often implemented as a windowing technique, in which a variable spatial and frequency window is employed to reveal the frequency contents at each location on the image. Choudhary et al. (2008) employed discrete wavelet transforms for the classification of different types of grains (wheat, barley, oats, and rye) using 51 morphological, 123 color, and 56 textural features. Iwata et al. (2000) studied the genetic variation of the shape of the root of different varieties of radish in order to analyze the mode of inheritance. They used principal component analysis to select the best Fourier descriptors.

11.4 VOLUME AND MASS ESTIMATION

One of the most cited methods for estimating the volume and surface of agricultural products consists of assuming that the volume can be partially approximated by one or several surfaces of revolution. Thus, the contour of the fruit is extracted from different projections, divided into sections, and then revolved around an axis of inertia. Sabliov et al. (2002) used this method to estimate the volumes of eggs, peaches, lemons, and limes. This work was later enhanced by Wang and Nguang (2007) for eggs, lemons, limes, and tamarillos.

Koc (2007) compared two methods of volume estimation with the volume calculated by water displacement. Assuming that watermelons are ellipsoids, their volume was calculated from the length, major diameter, and minor diameter obtained from two different images. The length and major diameter of each watermelon were measured from the first image, and the minor diameter was measured from an image acquired after rotating the fruit 90° around its longitudinal axis. The second method was similar to the ones described in the previous paragraph (approximation to several surfaces of revolution). This latter method did not show significant statistical differences with volume estimation by water displacement.

Du and Sun (2006) compared two methods to estimate the volumes of ham. One was based on the calculation of three principal dimensions (length, width, and thickness) in order to estimate surface area and volume mathematically. The other divided the ham into a number of sections and assumed surfaces of revolution, and then added the estimated surface area and volume of each section to obtain the total volume. This method produced better results than the previous one.

A simpler approach, based on empirical data, was the one described by Ngouajio et al. (2003), who generated a model to relate bell pepper diameter and length to volume, using the equation of the volume of a sphere as the starting point. They also showed that the method was comparable to measurements performed by water displacement.

Tao et al. (1999) developed an adaptive spherical transform and applied it in an apple sorting system. They used a model to compensate the reflectance intensity gradient on curved objects that can be applied to volume estimation. Similarly, Gómez-Sanchis et al. (2008) created 3D models from 2D images of mandarins. They calculated the centroid of the mandarin and its distance to 16 equidistant points on the contour. Assuming that the height of the fruit is proportional to the average distance (the constant of proportionality being dependent on the variety), they estimated the top point of the fruit. They then traced semi-ellipses connecting this point with the 16 selected points on the contour, thus constructing an interpolation network. From these ellipses, the estimated height of each point in the fruit projection was calculated, thereby composing a volume as shown in Figure 11.4. In the study of Tabatabaeefar et al. (2000) volume estimation is the first step to estimate mass. They proposed different regression models for predicting the mass of oranges from their dimensions and projected areas. They also estimated volume by assuming that fruits were ellipsoids, which offered a better prediction of the mass.

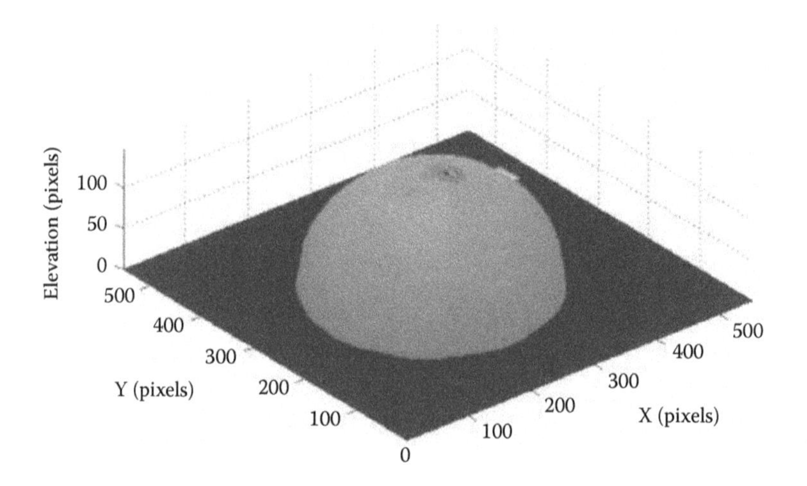

FIGURE 11.4 Reconstruction of a 3-D model of an orange from a 2-D image.

11.5 COLOR ESTIMATION

Color is one of the most important attributes of agrifood products, since consumers associate it with freshness. Producers strive to prevent products with defective colorations from reaching the market, as well as ensure that individual products are packed in batches with a similar color. It is also important to measure how postharvest treatments affect the color of fruits.

The degree of ripeness of some fruits is often estimated by color charts. Colorimeters are electronic devices that express colors in numerical coordinates, CIELAB being the one that is most frequently employed for food. CIELAB coordinates are L* (luminosity), a* (green to red), and b* (blue to yellow). Sometimes standard indexes combine these coordinates in one single ratio that is easier for operators to handle. For instance, the color index in the citrus industry is used to determine the harvesting date or to decide if mandarins should undergo a degreening treatment (Jiménez-Cuesta et al., 1981). Color indexes of tomatoes and their relationship to visual color classification were compared by López Camelo and Gómez, (2004). Different feature indexes using HSI (hue, saturation, intensity) and CIELAB color coordinates were employed.

However, colorimeters are limited to the measurement of small regions of a surface or when the object has a homogeneous color. However, still or video cameras can provide images in which the colors of the pixels are determined individually.

The color of a particular pixel in an image is expressed by three coordinates in a color space. The primary colors red, green, and blue (RGB) are the most widely used in computer vision. When inspected objects have very different colors, sometimes simple ratios between RGB values can discriminate between them, thus saving processing time. For instance, Blasco et al. (2009b) used the R/G ratio to discriminate four categories of pomegranate arils, reaching a success rate similar to the ones obtained by visual inspection.

The mayor drawback of RGB is its dependency on the acquisition device (different devices produce different RGB values for the same pixels in a scene). For this reason, other color spaces, closer to the human perception of color are frequently employed, like HSI. Blasco et al. (2007) compared five color spaces for the identification of external defects in citrus fruits and obtained better results with HSI. Both RGB and HSI were used by Xiaobo et al. (2007) to classify Fuji apples into four color categories. Frequently, individual HSI coordinates provide a simple means of color segmentation. Abdullah et al. (2006) converted RGB into HSI coordinates, and used the H component to classify starfruits into four maturity categories.

However, RGB and HSI are nonuniform color spaces. This means that the same numerical distance between two colors in these spaces may produce distinct differences of human perception depending on the position of such colors in the space. Uniform spaces like CIELAB or HunterLab define distances that produce the same differences in perception regardless of the position of the colors, and for this reason they are very well suited for color comparison. Several works have compared different color spaces, the conclusion being that the most appropriate for measuring fruit color is L*a*b* (Mendoza et al., 2006; Leon et al., 2006).

Simple algorithms based on a single L*a*b* coordinate have been used for the classification of fruits with a characteristic skin color. The a* coordinate was used by Liming and Yanchao (2010) to grade strawberries and the machine vision system achieved a success rate of 89%. Hue angle and chrome are color features derived from the previously mentioned uniform spaces. Kang et al. (2008) quantified the effect of the curvature of the skin on the calculation of hue angle and chrome of mangoes and demonstrated that hue provided a valuable quantitative description of the color and color changes of individual fruit and heterogeneous batches.

11.6 TEXTURE ASSESSMENT

The texture of an image takes into account color and space relationships between neighboring pixels in order to characterize similar regions in an image. Texture has been increasingly used for evaluating and inspecting food attributes (Zheng et al., 2006).

López-García et al. (2010) proposed a multivariate image analysis method that combined color and texture information for the detection of skin defects in four cultivars of citrus. For this purpose, they used the RGB values of each pixel and those of their neighborhood at different scales in a principal components model.

Color co-occurrence matrices (CCMs) and their derived numerical features are a common way to describe texture. A 2D color image can be considered as a set of pixels that carry visual attributes measured in a 3D color space that may have spatial relationships. Different co-occurrence matrices represent these attributes and their relationships. Because co-occurrence matrices are typically large and sparse, various metrics of the matrix are often a more useful set of features (Haralick et al., 1973). Pydipati et al. (2006) used this method to determine whether texture-based HSI color features could be used in conjunction with statistics to identify diseased and normal leaves of citrus trees under laboratory conditions. A similar method was described by Zhao et al. (2009) to discriminate sound skin and five other different types of skin in grapefruit from images acquired under a microscope. These color texture features were also used by Kim et al. (2009) to detect citrus peel diseases. ElMasry et al. (2007) assessed ripeness of strawberries from pseudo-RGB images. Menesatti et al. (2009) used contrast, homogeneity, angular second moment, and correlation features extracted from grey-level co-occurrence matrices (GLCM) to assess the starch content of apples. Visible and near-infrared (NIR) (1000–1700 nm) images were used.

Other approaches use fractal texture features derived from spectral Fourier analysis. For instance, Quevedo et al. (2008) employed them to follow the ripening process of bananas, which is related to the distribution of the senescence spotting of the peel.

11.7 DETECTION OF STEM ENDS, CALYXES, AND EXTERNAL DAMAGES

Detection of superficial damage during the production process is one of the major interests of the food industry. Traditionally, this task is performed by trained workers situated around a conveyor belt that carries the product, and defective products are removed from the conveyor belt manually. This task is tedious and subjective, since different operators have different tolerances. The most widely extended solution to automate this task comes from the use of artificial vision, but the correct detection of all potential cases of damage is still a challenge due to the great diversity of colors, sizes, and textures that can be present.

A particular difficulty hampering the proper identification of true defects of fruits is the confusion with other natural parts of the fruits like the stem or the calyx. Many researchers have used morphological processing for this purpose. For instance, Li et al. (2002) searched for concave areas with a method based on fractal dimensions and neural networks (NN).

Sometimes, different spectral bands can be used to discriminate between stem/calyx and defects. Bennedsen and Peterson (2005) reported a machine vision system for sorting apples by surface defects, including bruises. The system was applied to apples oriented with the stem/calyx axis perpendicular to the camera. Grey-scale images in the visible wavebands were used to verify orientation. Images for detection of defects were acquired using two optical filters at 740 and 950 nm. Six consecutive images were acquired while the fruit was rotated through 360°. Xing et al. (2007) identified several visible and NIR wavelengths that could be used to discriminate the stem end and calyx from sound peel and bruises in apples by means of a hyperspectral system. They used principal components analysis (PCA) to reduce the highly dimensional spectral reflectance data to a few optimal wavebands. Unay and Gosselin (2007) used different visible and NIR interferometric filters to recognize stem ends and calyxes in sound and defective Jonagold apples. NIR filters overcome the problem posed by the fact that this cultivar has a variegated red and yellow surface that makes it more difficult to detect skin damage by using standard color images.

11.8 REAL-TIME INSPECTION OF FRUIT

One of the main interests in the use of machine vision for detecting defects in fruits is the development of automatic systems that can operate in-line and thus assist or replace human inspection. When inspecting agricultural products in-line, special equipment has to be used since they travel at a high speed and cameras must freeze this movement to avoid blurring. Progressive-scan cameras with high shutter speeds decrease this effect by reducing the exposure time of the CCD. As the shutter speed increases, the intensity of the lighting must be increased in order to avoid underexposure.

Several authors have described real-time systems for inspecting different types of fruits. The mechanical implementation of each application depends on the size, shape, and resistance of each particular product. Díaz et al. (2000) designed a system to sort Manzanilla olives into four categories at a rate of 396 olives per second.

Sometimes nonstandard processing hardware is required to achieve real-time specifications. Aleixos et al. (2002) presented a multispectral camera capable of acquiring visible and NIR images from the same scene. They used specific algorithms that were implemented on two processors that worked in parallel. The system was capable of correctly classifying lemons and mandarins at a rate of up to 10 fruits per second, depending on the area of the external defects and the size and color of the fruit.

Fruit sorting systems often require observation of the whole surface of the fruit. To achieve this, sorters rotate the objects under the camera as they advance. Others use several cameras to acquire different views of the product, or mirrors to obtain different views with a single camera (Reese et al., 2009). Leemans and Destain (2004) presented a hierarchical grading method for Jonagold apples. Several images covering the whole surface of the fruits were acquired while they moved on rollers at 1.53 Hz by two cameras, whose acquisition rate was 11.1 images per second. The rotational speed of the rollers was adjusted in such a way that a spherical object having a diameter of 72 mm completed one rotation in exactly four images. After segmentation, blemishes were characterized by 16 features describing their color, shape, texture, and position. Fruits were correctly graded with a 73% success rate.

REFERENCES

Abdullah, M.Z., J. Mohamad-Saleh, A.S. Fathinul-Syahir, and B.M.N. Mohd-Azemi. 2006. Discrimination and classification of fresh-cut starfruits (*Averrhoa carambola* L.) using automated machine vision system. *Journal of Food Engineering* 76:506–523.

Aleixos, N., J. Blasco, F. Navarrón, and E. Moltó. 2002. Multispectral inspection of citrus in real-time using machine vision and digital signal processors. *Computers and Electronics in Agriculture* 33(2):121–137.

Beyer, M., R. Hahn, S. Peschel, M. Harz, and M. Knoche. 2002. Analysing fruit shape in sweet cherry. *Scientia Horticulturae* 96:139–50.

Blasco, J., N. Aleixos, and E. Moltó, E. 2003. Machine vision system for automatic quality grading of fruit. *Biosystems Engineering* 85(4):415–423.

Blasco, J., N. Aleixos, J. Gómez, and E. Moltó. 2007. Citrus sorting by identification of the most common defects using multispectral computer vision. *Journal of Food Engineering* 83(3): 384–393.

Blasco, J., N. Aleixos, S. Cubero, J. Gómez-Sanchis, and E. Moltó. 2009a. Automatic sorting of Satsuma (*Citrus unshiu*) segments using computer vision and morphological features. *Computers and Electronics in Agriculture* 66:1–8.

Blasco, J., S. Cubero, J. Gómez-Sanchis, P. Mira, and E. Moltó. 2009b. Development of a machine for the automatic sorting of pomegranate (*Punica granatum*) arils based on computer vision. *Journal of Food Engineering* 90:27–34.

Bennedsen, B.S., and D.L. Peterson. 2005. Performance of a system for apple surface defect identification in near-infrared images. *Biosystems Engineering* 90(4):419–431.

Bracewell, R. 1999. *The Fourier transform and its applications*, 3rd ed. New York: McGraw-Hill.

Choudhary R., Paliwal J., and Jayas D.S. 2008. Classification of cereal grains using wavelet, morphological, colour, and textural features of non-touching kernel images. *Biosystems Engineering*, 99, 330–337.

Costa, C., P. Menesatti, G. Paglia, F. Pallottino, J. Aguzzi, V. Rimatori, G. Russo, S. Recupero, S., and G.R. Recupero. 2009. Quantitative evaluation of Tarocco sweet orange fruit shape using optoelectronic elliptic Fourier based analysis. *Postharvest Biology and Technology* 54(1):38–47.

Cubero, S., N. Aleixos, E. Moltó, J. Gómez-Sanchis, and J. Blasco. 2010. Advances in machine vision applications for automatic inspection and quality evaluation of fruits and vegetables. *Food and Bioprocess Technology*, doi 10.1007/s11947-010-0411-8:

Currie, A.J., S. Ganeshanandam, D.A. Noiton, D. Garrick, C.J.A. Shelbourne, and N. Orgaguzie. 2000. Quantitative evaluation of apple (*Malus domestica Borkh*) fruit shape by principal component analysis of Fourier descriptors. *Euphytica* 111:219–227.

Díaz, R., G. Faus, M. Blasco, J. Blasco, and E. Moltó. 2000. The application of a fast algorithm for the classification of olives by machine vision. *Food Research International* 33:305–309.

Du, C.-J, and D.-W. Sun. 2006. Estimating the surface area and volume of ellipsoidal ham using computer vision. *Journal of Food Engineering* 73:260–268.

ElMasry, G., N. Wang, A. ElSayed, and M. Ngadi. 2007. Hyperspectral imaging for nondestructive determination of some quality attributes for strawberry. *Journal of Food Engineering* 81:98–107.

Fernández, L., C. Castillero, and J.M. Aguilera. 2005. An application of image analysis to dehydration of apple discs. *Journal of Food Engineering* 67:185–193.

Gómez-Sanchis, J., E. Moltó, G. Camps-Valls, L. Gómez-Chova, N. Aleixos, and J. Blasco. 2008. Automatic correction of the effects of the light source on spherical objects: An application to the analysis of hyperspectral images of citrus fruits. *Journal of Food Engineering* 85(2):191–200.

Haralick, R.M., K. Shanmugam, and I. Dinstein. 1973. Textural features for image classification. *IEEE Transactions on Systems, Man, and Cybernetics SMC* 3(6):610–621.

Iwata, H., S. Niikura, S. Matsuura, Y. Takano, and Y. Ukai. 2000. Diallel analysis of root shape of Japanese radish (*Raphanus sativus* L.) based on elliptic Fourier descriptors. *Breeding Science* 50:73–80.

Jarimopas, B., and N. Jaisin. 2008. An experimental machine vision system for sorting sweet tamarind. *Journal of Food Engineering* 89:291–297.

Jiménez-Cuesta, M., J. Cuquerella, and J.M. Martínez-Jávega. 1981. Determination of a color index for citrus fruit degreening. *Proceedings of the International Society of Citriculture* 2, pp.750–753.

Kang, S.P., A.R. East, and F.J. Trujillo. 2008. Colour vision system evaluation of bicolour fruit: A case study with "B74" mango. *Postharvest Biology and Technology* 49: 77–85.

Kim, D.G., T.F. Burks, J. Qin, and D.M. Bulanon. 2009. Classification of grapefruit peel diseases using color texture feature analysis. *International Journal of Agricultural and Biological Engineering* 2(3): 41–50.

Kim, H.K., and J.D. Kim. 2000. Region-based shape descriptor invariant to rotation, scale and translation. *Signal Processing: Image Communication* 16(1–2): 87–93.

Koc, A.B. 2007. Determination of watermelon volume using ellipsoid approximation and image processing. *Postharvest Biology and Technology* 45:366–371.

Leemans, V., and M.-F. Destain. 2004. A real-time grading method of apples based on features extracted from defects. *Journal of Food Engineering* 6:83–89.

León, K., M. Domingo, F. Pedreschi, and J. León. 2006. Color measurement in L*a*b* units from RGB digital images. *Food Research International* 39:1084–1091.

Li, Q., M. Wang, and W. Gu. 2002. Computer vision based system for apple surface defect detection. *Computers and Electronics in Agriculture* 36:215–223.

Liming, X., and Z. Yanchao. 2010. Automated strawberry grading system based on image processing. *Computers and Electronics in Agriculture* 71(S1):S32–S39.

López Camelo, A.F., and P.A. Gómez. 2004. Comparison of color indexes for tomato ripening *Horticultura Brasileira* 22(3), http://www.scielo.br/scielo.php?script=sci_arttext&pid=S0102-05362004000300006 (accessed September 2010).

López-García, F., G. Andreu-García, J. Blasco, N. Aleixos, and J.M. Valiente. 2010. Automatic detection of skin defects in citrus fruits using a multivariate image analysis approach. *Computers and Electronics in Agriculture* 71:189–197.

Mateo, A., F. Soto, J.A. Villarejo, J. Roca-Dorda, F. De la Gandara, and A. García. 2006. Quality analysis of tuna meat using an automated color inspection system. *Aquacultural Engineering* 35:1–13.

Mendoza, F., P. Dejmek, and J.M. Aguilera. 2006. Calibrated color measurements of agricultural foods using image analysis. *Postharvest Biology and Technology* 41:285–295.

Menesatti, P., C. Costa, G. Paglia, F. Pallottino, S. D'Andrea, V. Rimatori, and J. Aguzzi. 2008. Shape-based methodology for multivariate discrimination among Italian hazelnut cultivars. *Biosystems Engineering* 101:417–424.

Menesatti, P., A. Zanella, S. D'Andrea, C. Costa, G. Paglia, and F. Pallottino. 2009. Supervised multivariate analysis of hyper-spectral NIR images to evaluate the starch index of apples. *Food and Bioprocess Technology* 2:308–314.

Moreda, G.P., J. Ortiz-Cañavate, F.J. García-Ramos, and M. Ruiz-Altinsent. 2007. Effect of orientation on the fruit on-line size determination performed by an optical ring sensor. *Journal of Food Engineering* 81(2):388–398.

Ngouajio, M., W. Kirk, and R. Goldy. 2003. A simple model for rapid and nondestructive estimation of bell pepper fruit volume. *J. Crop Hort. Sci.* 38:509–511.

Ni, H., and S. Gunasekaran. 1998. A computer vision method for determining length of cheese shreds. *Artificial Intelligence Review* 12:27–37.

Ni, H., and S. Gunasekaran. 2004. Image processing algorithm for cheese shred evaluation. *Journal of Food Engineering* 61:37–45.

Ohm, J.-R., F. Bunjamin, W. Liebsch, B. Makai, K. Muller, A. Smolic, and D. Zier. 2000. A set of visual feature descriptors and their combination in a low-level description scheme, *Signal Processing: Image Communication* 16(1):157–179.

Pedreschi. F., J. León, D. Mery, and P. Moyano. 2006. Development of a computer vision system to measure the color of potato chips. *Food Research International* 39:1092–1098.

Pydipati, R., T.F. Burks, and W.S. Lee. 2006. Identification of citrus disease using color texture features and discriminant analysis. *Computers and Electronics in Agriculture* 52:49–59.

Quevedo, R., F. Mendoza, J.M. Aguilera, J. Chanona, and G. Gutiérrez-López. 2008. Determination of senescent spotting in banana (*Musa cavendish*) using fractal texture Fourier image. *Journal of Food Engineering* 84:509–515.

Reese, D., A.M. Lefcourt, M.S. Kim, and Y.M. Lo. 2009. Using parabolic mirrors for complete imaging of apple surfaces *Bioresource Technology* 100:4499–4506.

Riquelme, M.T., P. Barreiro, M. Ruiz-Altisent, and C. Valero. 2008. Olive classification according to external damage using image analysis. *Journal of Food Engineering* 87:371–379.

Sabliov, C.M., D. Boldor, K.M. Keener, and B.E. Farkas. 2002 Image processing method to determine surface area and volume of axi-symmetric agricultural products. International *Journal of Food Properties* 5(3):641–653.

Sadrnia, H., A. Rajabipour, A. Jafary, A. Javadi, and Y. Mostofi. 2007 Classification and analysis of fruit shapes in long type watermelon using image processing. *Journal of Agriculture and Biology* 1:68–70.

Sun, D.-W. 2007. *Computer vision technology for food quality evaluation*. London, UK: Academic Press, Elsevier Science.

Sun, D.-W. 2010. *Hyperspectral imaging for food quality analysis and control*. London, UK: Academic Press, Elsevier Science.

Sun, D.-W., and T. Brosnan. 2005. Pizza quality evaluation using computer vision—part 1. Pizza base and sauce spread. *Journal of Food Engineering* 68:277–287.

Tabatabaeefar, A., A. Vefagh-Nematolahee, and A. Rajabipour. 2000. Modeling of orange mass based on dimensions. *Journal of Agricultural Science and Technology* 2:299–305.

Tao, Y., C.T. Morrow, P.H. Heinemann, and H.J. Sommer. 1995. Fourier-based separation technique for shape grading of potatoes using machine vision. *Transactions of the ASAE* 38(3):949–957.

Tao, Y., and Z. Wen. 1999. An adaptive spherical image transform for high-speed fruit defect detection. *Transactions of the ASAE* 42(1):241–246.

Throop, J.A., D.J. Aneshansley, W.C. Anger, and D.L. Peterson. 2005. Quality evaluation of apples based on surface defects: Development of an automated inspection system. *Postharvest Biology and Technology* 36:281–90.

Titchmarsh, E.C. 1967. *Introduction to the theory of Fourier integrals*. Oxford, UK: Oxford University Press.

Unay, D., and B. Gosselin. 2007. Stem and calyx recognition on Jonagold apples by pattern recognition. *Journal of Food Engineering* 78:597–605.

Van Eck, J.W., G.W.A.M. van der Heijden, and G. Polder. 1998. Accurate measurement of size and shape of cucumber fruits with image analysis. *Journal of Agricultural Engineering Research* 70:335–343.

Van Otterloo, P.J. 1992. *A contour-oriented approach to shape analysis*. Upper Saddle River, NJ: Prentice Hall.

Wang, T.Y., and S.K. Nguang. 2007. Low cost sensor for volume and surface area computation of axi-symmetric agricultural products. *Journal of Food Engineering* 79(3): 870–877.

Xiaobo, Z., Z. Jiewen, and L. Yanxiao. 2007. Apple color grading based on organization feature parameters. *Pattern Recognition Letters* 28:2046–2053.

Xing, J., P. Jancsók, and J. De Baerdemaeker. 2007. Stem-end/calyx identification on apples using contour analysis in multispectral images. *Biosystems Engineering* 96(2):231–237.

Zhao, X., T.F. Burks, J. Qin, and M.A. Ritenour. 2009. Digital microscopic imaging for citrus peel disease classification using color texture features. *Applied Engineering in Agriculture* 25(5):769–776.

Zheng, C., D.-W. Sun, and L. Zheng. 2006. Recent applications of image texture for evaluation of food qualities: A review. *Trends in Food Science & Technology* 17(3):113–128.

12 Novel Measurement Techniques of Physical Properties of Vegetables

J. Ignacio Arana

CONTENTS

12.1 INTRODUCTION

Vegetables are one of the most important foods in the world. Their marketing is very complex because there are great differences among vegetables. In some cases the fruits are marketed but in other cases the products marketed are their leaves, their bulbs, their tubers, their roots, or the whole of the plant. As most vegetables are consumed fresh by people, it is absolutely necessary to ensure sufficient quality for them. Vegetable quality should be characterized as global quality, taking into account all the characteristics of the food and according to the preferences of the consumers. The term *quality* implies the degree of excellence of a product or its suitability for a particular use.

It is possible that the organoleptic characteristics should be the most important to characterize vegetable quality, but the physical characteristics are for consumers to evaluate and many of them can be evaluated before trying the product. For these reasons it is possible to state that physical characteristics are the most important ones for vegetable marketing. Because of the great variability of vegetables and because of the different ways of eating them, fresh or cooked, there are a great number of features characterizing vegetable quality. It is possible to highlight geometric characteristics as size and shape, dynamic characteristics as density, optical as color, and textural as firmness or mealiness. In addition, some of the physical properties of vegetables affect their ability to be harvested mechanically or are related to their maturity and can be used to select the moment for collecting them and also affect their conservation potential. Some of these characteristics are easily characterized using objective methods, but others have been usually evaluated using sensory methods.

Novel marketing techniques make it necessary to properly characterize vegetables, to improve their global quality, and to ensure that all vegetables marketed have a minimum quality level. Thus, it is necessary to develop quick, inexpensive, objective, and nondestructive methods to be installed in the vegetable selection chains. Novel measurement techniques are necessary to evaluate parameters characterizing physical properties of vegetables.

There are different characteristics determining the quality of different vegetables. Tomato quality can be characterized by flavor, firmness, deformability, mealiness, color, acidity, size, or shape. The most important features for peppers can be color, size, shape, and piquancy. Potato and onion quality can be diminished during the conservation process and quality reduction can be characterized by weight loss, damages, and sprouting.

12.2 MEASURING PHYSICAL CHARACTERISTICS OF VEGETABLES ON-LINE

The physical characteristics affecting vegetable quality are geometric and dynamic features. Among the geometric characteristics it is possible to emphasize size and shape, and among the dynamic features, weight. There are some parameters characterizing these features such as volume, width, length, height, diameter, and specific gravity or density.

All parameters related to vegetable quality should be measured accurately enough and the devices used to measure them should be installed on-line to evaluate vegetable quality and to sort the vegetables in real time.

12.2.1 Size and Shape

For determination of the quality of agricultural produce, size and shape are necessary because they are primary parameters. The major problem is that agricultural products have large variability in size and shape.

Fruit size can be characterized by means of physical parameters such as volume, weight, diameter, circumference, projected area, or any combination of these (Peleg, 1985). The fruit volume can be used to index ripeness to predict optimum harvest time (Hahn and Sanchez, 2000) or to predict yield (Mitchell, 1986). Fruit size determination allows the sorting of fruits into size groups, its classification into batches of uniform size, and it allows pattern packing. Grading of fruits into size groups is often necessary in the food industry to meet the requirements of some processing machines or to assign process differentials of large and small produce. Peeling machines in artichoke canning factories require correctly sized vegetables to work.

Sizers can be classified into two categories, those based on weight and those based on dimensions, the latter referring to maximum diameter, volume, axes, projected area, and perimeter measurements. The United States and the European Community marketing standards have ruled that the size is defined by the fruit diameter or the maximum diameter in the equatorial section, respectively. However, packers can use weight sizers for vegetables when there is a relationship between diameter and weight or volume.

12.2.2 Volume

Geometric characteristics of vegetables are usually measured by means of calipers. These methods can be accurate enough, but they are slow and costly. In most of the cases, precise measurement of the geometric parameters is not essential, but it is necessary to sort vegetables according to their size in the selection chain. The measure on-line of the size has to be nondestructive, quick, inexpensive, and accurate enough. Some sorting devices are based on weight, and there are devices based on volume measurement that usually involve computer imaging.

The reference method for measuring the volume of vegetables is by water displacement based on Archimedes' principle. Mosehnin (1970) developed a methodology based on this principle. In situations where fruit growth continues, it is preferable to estimate fruit volume by air displacement using an air pycnometer. This is because submerging fruits in water can make them more susceptible to subsequent fungal attack. Air pycnometers are based on the ideal gas (Boyle-Mariotte) law. The main disadvantage of these instruments is the long measuring time. Iraguen et al. (2006) designed a portable air pycnometer to measure the volume of grape clusters attached to vine plants. This method can be used to measure most vegetables.

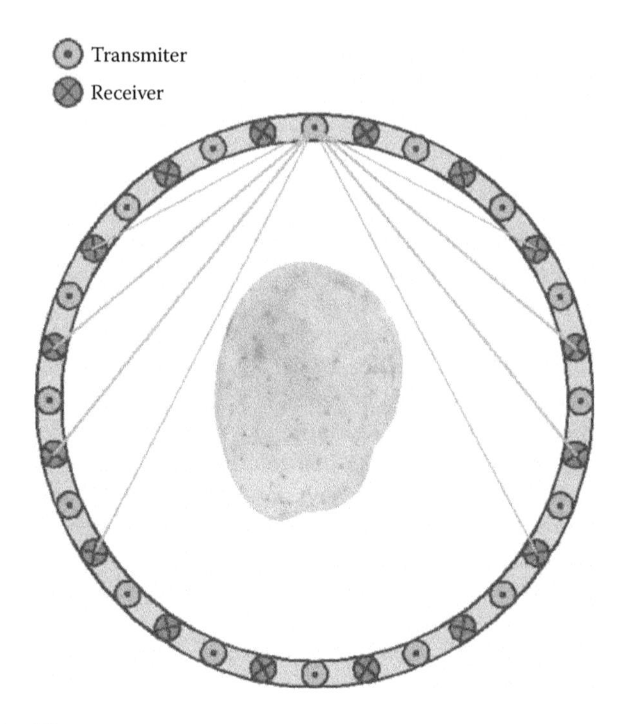

FIGURE 12.1 Ring sensor.

Recently, novel techniques to measure vegetable size have been developed. These techniques are nondestructive and can be used in sorting chains.

Gall (1997) developed an optical ring sensor system to measure fruit and vegetable size. The ring sensor consists of a number of single emitters and receivers arranged alternately and spaced equally on the circumference of a circle (ring). The emitters and receivers have a Lambertian angular response. The emitters are switched in sequence round the ring. From one activated transmitter, the signal to each optical receiver can be likened to a snapshot. It is a light-blocking-based system. When the inner space of the ring is clear, every receiver detects the light emitted by the activated transmitter owing to the Lambertian response of all the transducers, but when an object is introduced in the ring, a shadow zone appears, corresponding to the receivers being obscured from the activated transmitters, as shown in Figure 12.1. The two closest noninterrupted paths, which are cords in the ring circle, can be used to describe the curvature of the object. After a scanning revolution of the ring when there is a static object within the scanned area, a basic picture showing all cords, including the cross section of the object, is obtained. The information is coded in an array Z whose size in the two-dimensional case is equal to the number of transmitters. The ring sensor can be used to measure volume, length, position of the largest slice of an object, the major and the minor axes of the largest and the middle slice, the circumferences in the crosswise section, and mass if the density is known or density if the mass is known. It can be used in plant breeding in the vegetable industry. Gall et al. (1998) used this ring sensor to measure the size of potatoes. The ring circle

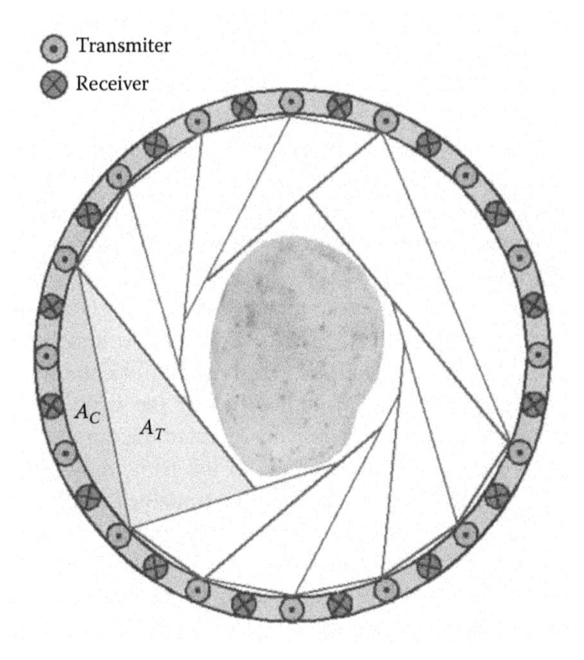

FIGURE 12.2 Segmentation of the ring sensor into triangles and segments of a circle.

was segmented into definite elements: triangles and segments of a circle as shown in Figure 12.2. The cross-sectional area of the potato is computed as the difference between the area of the circle encompassed by the ring and the sum of the triangles and segments of the circle that comprise the gap between the object's contour and the ring (Equation [12.1])

$$A = \pi r^2 - \Sigma (A_T + A_C) \tag{12.1}$$

where A is the area of the fruit cross-section, r is the ring radius, A_T is the area of a triangle, and A_C is the area of the segment.

The volume of the potatoes can be measured by extending the algorithms into 3D space. The ring sensor was used to measure the cross-section area, the volume, and the length of the major axes of the potatoes.

A practical test was made using a ring consisting of 64 transmitters and 64 receivers built by Argus Electronic (1997). The rings diameter is 175 mm with a scanning rate of 400 revolutions per second. The test was performed on 12 different potatoes passing through the ring at a constant speed of 1.0 m s^{-1}. Each potato was measured 50 times using different positions in the ring.

The accuracy of the ring depends on the number of transmitters and receivers and the size of the object. The error in determination of the diameter and the area decreases as the diameter of the cross section increases. The error in determination of the volume decreases as the volume of the potatoes and the number of transmitters and receivers increase. The accuracy of the ring sensor is better than 6% for potatoes having a volume larger than 70 cm^3 and 5% for potatoes having a volume larger than 100 cm^3. The ring sensor is able to scan 3.6 tons per hour. The current mechanical size-grading systems for potatoes can process 10 tons per hour but with an accuracy of only 10–30%.

Moreda (2004) used an optical ring sensor system to estimate the volume of cucumbers and zucchinis. The measures were achieved at high speed, up to 2 m/s. The system generates an enveloping helix used to estimate the fruit's length, volume, and major and minor axes of two particular cross sections. Fruit volume is calculated by summing the individual products of the cross-section areas multiplied by the helix pitch. Moreda obtained a root mean squared error (RMSE) of 26 ml (root mean square error of prediction [RMSPE] = 5.8%) for volume estimation of zucchini. Moreda et al. (2007) analyzed the effect of fruit orientation on the precision of measurements of tomato volume obtained using the optical ring sensor. They concluded that random orientation negatively affected the precision of volume measurements owing to the swinging movement of the fruit when crossing the optical ring sensor. Using a feeder longer than 2 m, the tomatoes would arrive completely stabilized to the ring sensor and the errors would decrease. The ring sensor method is the most versatile of the systems to measure size based on measuring the volume of the gap between the fruit and an outer casing because it can measure fruit axes in addition to volume. This method requires oriented fruits for axis measurements but not for volume measurements. In the case of zucchini, and owing to their length, correct orientation is necessary to accurately measure volume, but this correct orientation is easy to achieve using angled belts.

Iwamoto and Chuma (1981) described an optoelectronic system based on the blocking of light that measures the horizontal width of fruits in the direction of movement. It consists of a conveyor and a couple of transmitters and receivers placed at opposite sides of the conveyor where the fruits are placed. When a fruit passes through the system, the light path is blocked. For a certain conveyor speed, the fruit width is proportional to the duration of the light blocking. This system can be used for oblate fruits such as conventional beefsteak tomatoes because they have a clear resting stability position when tossed in a horizontal position on a horizontal surface, and because the equatorial diameter determines size, according to most marketing standards. However, this system cannot measure fruit height.

Hahn and Sanchez (2000) measured the volume of carrots using integrated imaging areas and orthonormal imaging. In the first case, the volume algorithm added the integration areas from the 106 different images taken of the fruit. The volume from each slice was the area of the slice multiplied by the cylinder length and the volume of the carrot was the sum of the 106 carrot slices. In the second case, a simpler algorithm especially designed for carrot volume evaluation used only two orthogonal images. The entire volume can be obtained after adding n parallel discs, where n is the number of columns present along the other orthonormal image (main carrot length axis). Each disc size varied according to its respective column height. The real volume of carrots was obtained using a system based on Archimedes' principle and compared with the volume predicted using the two different algorithms. Correlation coefficients of 0.985 and 0.986 were achieved between the real and predicted volume values using the area-integrating and the orthonormal algorithms, respectively. The time required for predicting the volume with the area-integrating algorithm was 2.5 and 3 s for carrots 25 and 35 cm long, respectively. The orthonormal algorithm specially designed for carrot measurements worked well with only two images, allowing a quicker measurement of volume.

Novel measuring techniques based on two-dimensional machine vision have been used on vegetables. Sarkar and Wolfe (1985) developed algorithms using digital image analysis and pattern recognition techniques for orientation of fresh market tomatoes and classification based on size, shape, color, and surface defects. For shape and size assessment, they selected the perimeter of the tomato profile image because they had found that perimeter and area correlated equally well with the mean equatorial diameter, and that perimeter measurements were computationally less expensive than area measurements. They classified the tomatoes in three categories according to size with an overall error rate of 5.6%. The tomatoes were classified as normal or abnormal according to their shape, and the error was 2.1%. The mean gray level of the tomato was used as a measure of color. The tomatoes were classified as green and red and the error was of 0.7%. The orientation of the tomatoes was determined using the ratio of maximum to mean gray value within the centralized region of the tomatoes. As the stem scars are usually lighter than tomato skin, the ratio is higher for stem ends than for blossom ends. The error in the orientation detection was 3.1%. Blossom-end defects were detected by defining a global ray level threshold to discriminate between defective and normal skin pixels. This threshold was found at the valley of the bimodal cumulative histogram formed by adding the histograms of defective and nondefective blossom-end images. The error in the detection of the blossom-end defects was 7.6%. Stem-end defects were detected by defining a threshold in the gradient image after enhancing the contrast of the raw image using a gradient operator. The error in the detection of the stem-end defects was 6.3%. Therefore, digital image algorithms are effective in classifying tomatoes based on size, shape, and color, for determining its orientation, and for detecting its defects.

The fruit size is mainly characterized by the equatorial or transverse diameter and the polar diameter or length. The reference method to measure these diameters is the caliper. Owing to the odd shapes of horticultural produce, there are several possible equatorial and polar diameters on the same fruit. Marchant (1990), using dimensional analysis, stated that for a solid of fixed shape, any size measure is proportional to volume. He stated that the volume of potatoes was equivalent to the quotient of the square of the projected area and the length of the potatoes. He assumed that tubers had a constant density and developed a system to estimate the weight of potatoes based on this proportionality. The system uses a roller conveyor without singulation across the line. The potato weight was estimated from 12 images that were captured as the tubers rolled along the conveyor. Marchant's system achieved an RMSE for the estimated weight of 14.2 g (Marchant et al. 1989) equivalent to an RMSPE of 7% of the mean tuber weight, which was 200 g. Because they assumed that the density is constant, the relative error for the estimation of the volume of potatoes was 7%. The system by Marchant has a throughput of up to 40 tubers/s and has been tested on-line. Forbes and Tattersfield (1999) developed another system to estimate potato volume. A machine vision algorithm estimated fruit volume from 2D digital images. The method included a neural network, is rotationally invariant, and does not require conveyor mechanisms to align the potatoes. It can be mounted in independent lines. The RMSPE achieved by them for estimating the volume of potatoes, using four images, was 5.3% (Forbes, 2000). The system by Marchant has a higher throughput than the one by Forbes and Tattersfield. It is apparently cheaper than the Forbes system because it only uses one camera, whereas the Forbes system uses four cameras. Teneze et al. (1989) developed an image analysis system to identify French beans and to eliminate as many undesirable objects as possible during bean collection.

Hryniewicz et al. (2005) developed a contactless system for 3D surface modeling. It uses two Internet cameras with a slide projector as an independent light source. The methodology is based on the assumption that most of the fruits have circular cross sections and therefore only applies to such fruits. Circular cross sections were assembled to reconstruct the surface of carrots and tomatoes. The software calculates the middles and radiuses in 3D, based on rectified images. The system can also be used for volume and mass calculation. The carrot dimensions obtained by them, using this device, were similar to reality. The device accuracy was checked with 100 apples and 70 tomatoes with known max diameters, which were classified with 100% success, according to the 5% fruit max diameter tolerance defined by European Commission Regulations (2001). The 5% device tolerance is sufficient for the setup to be implemented as the grading device. The system could be improved by obtaining stereo pair images from different stations or viewpoints. The device developed by them could be a universal tool for collecting information about vegetable quality during the growing season.

Lee et al. (2006) used the volume intersection (VI) method to measure surface area and volume of fresh produce having irregular shapes, such as tomatoes. This is a nondestructive method that studies the fruit shape from silhouettes. Two-dimensional silhouettes are obtained by taking pictures of the fruit from different directions by rotating the fruit at a fixed angular interval. The silhouettes and the viewpoints form cones and the intersection of these cones form a region that surrounds the fruit. By working with several silhouettes taken from different viewpoints, the region will approximate the shape of the fruit. The region obtained reconstructs the 3D shape of the fruit owing to virtual small cubes called *cubic pixels*. These authors did not use several cameras, but placed tomatoes on a turntable. The turntable rotated at a fixed angular interval and a fixed camera generated the silhouettes from different directions. A mathematical model was developed to calculate surface area and volume, and each silhouette was treated as a cross section of the tomato taken at a specific angular position. The accuracy of the method was assessed by comparing the obtained results with the ones obtained using a reference method based on Archimedes' principle. The system provides a measurement calibration accuracy adjustment by using objects with known surface area and volume. The precision of the method depended on the angular interval of the imaging system as well as on the concavities of the fruit examined. For apples, they obtained an RMSE of 6 ml and a RMSPE of 1.9%, but apples have fewer concavities than tomatoes. For tomatoes, the average surface area measurement and volume difference stayed within 2% and 4%, respectively. The method is easy and cheap because it uses only one camera. Compared with the traditional methods, it is fast and accurate, although it requires some time for turning the turntable.

The systems based on 2D machine vision have the advantage that they can classify vegetables according to surface color and external defects in addition to size and shape. A limitation of the applicability of this technique to horticultural produce inspection could be the appearance of reflections that tend to saturate charge-coupled devices (CCD) and complementary metal-oxide semiconductor (CMOS) sensors of video cameras. This limitation could be eliminated by using indirect lighting, and where necessary, by mounting filters in the optics of the video camera. Another option would be the utilization of charge injection device (CID) sensors in cameras, which do not saturate as easily as CCD or CMOS sensors (Moreda et al., 2009).

Hatou et al. (1996) used an active 3D machine technique to measure tomato size. They used a laser range scanner to obtain measures of the tomatoes as they traveled on a conveyor belt beneath the range finder of the scanner. An ideal tomato shape was built from these measures, which was compared with the reference shape and the differences between the two shapes were used for tomato classification. The classification system includes neural networks and expert systems and the grading results were similar to those achieved by human inspectors. The grading procedure contains two steps: first the rough classification, including 3 categories (good, normal, bad) using the neural networks, and second the elaborate classification, involving 2 grades in the category "good," using expert systems. However, the method was more time consuming than the manual classification because it took 5 seconds to classify each tomato. The systems using active 3D machine vision technology are easier to apply to the industrial process because of the simplified lighting and camera adjustments utilized by 3D systems for taking images of the fruits.

Kanali et al. (1998) developed an artificial retina machine vision system with 49 photo sensors. In this system there is a direct transfer of the acquired information to the processing algorithm and no digital image analysis is required. This is an advantage because this analysis often requires specialized and expensive software to correctly process the images. The artificial retina uses photo sensors that are cheaper than the ones used in CCD or CMOS sensors equipped with video cameras. However, the artificial retina can calculate fruit volume but not fruit diameter (Gall et al., 1998). They used the artificial retina to classify oranges and eggplants obtaining overall size classification rates of 74% for eggplants. For oranges, they obtained a very accurate classification (99%) probably due to the high size variability of the orange sample. Although it is a desktop system, the technique can be used for automated inspection of vegetables.

Hoffman et al. (2005) sorted potato tubers according to their starch content, which is intensely related to their density. They calculated the density as the ratio between the weight and the volume. The weight was determined dynamically with a belt balance and the volume was measured dynamically using an optoelectronic device. The device consisted of a ring-shaped basic body with four cameras. The tubers fall individually through the basic body, while the camera system records their geometric dimensions in three axes and the surface of the tubers. A processor calculates the volume of the tuber with elliptic integrals from the data supplied by the camera. At the start of the installation, the tubers are transferred from a feeding conveyor moving at low speed to a belt with a higher conveying speed. As a result of the increase in speed on changing belts, the product is separated, ensuring that there is always only one tuber lying on the 300-mm-long weighing belt. After the mass has been determined, the tubers are then fed to the camera system to determine the volume. A processor calculates the density from the mass and volume data and issues an appropriate control command to the ejector to pass the tubers onto the appropriate classification belt. It was possible to classify the tubers in three categories according to their starch content, with a class width of 6%, calculated from its density, because of the high repeatability of the tuber volume determination and the high precision achieved in the mass measurement device with an error essentially below 1%. The only large-scale technical procedure can be substituted for this technique for nondestructive classification of potatoes on the basis of their starch content, which places the potatoes in salt baths or similar solutions and dispersions. The system is able to determine the volume of 20–25 tubers per second, but only 4–5 tubers can be weighed by the belt balance. Therefore, it is necessary to install several weighing lanes in parallel to feed the volume measurement system.

Al-Mallahi et al. (2010) developed a machine vision system to detect good quality potato tubers on a harvester, in order to remove clods and unwanted potato tubers, especially small tubers.

Electronic-based techniques to estimate the size of foods have been used mainly on fruits, but they have been also used on vegetables. The on-line determination of diameters, perimeter, projected area, volume, and density of vegetables having irregular shapes has been investigated in recent years. Today, systems based on 2D machine vision and electronic weighing are most commonly used to measure the size and volume of vegetables and also to sort them, but it is reasonable to expect that 3D imaging systems will be increasingly used to complement the 2D systems.

12.2.3 Other Geometric Parameters

Volume is not sufficient to characterize the geometry of many vegetables, and the characterization of their geometric quality needs to use shape parameters such as length, width, or flexure. Cao and Nagata (1997) developed a method to judge the shape of sweet peppers using only a single side view image of the samples to present the shape properties. In addition, defects or damages can severely reduce the quality of vegetables and the affected fruits should be eliminated on line.

Hahn (2002) developed a sizer system for jalapeño chilies based on the blocking of a plane or sheet of laser light. Fruits have to be placed on a conveyor belt correctly oriented with their polar axes perpendicular to the line. The system was designed with two 20-cm-wide belt conveyors. The first band ran at a speed of 30 m min^{-1}. Chilies cannot roll because of their peduncles, but when they fall from the first band to the second band, they bounce. The baby suckers of the second band stop the chilies without causing impact damages. The second band ran at a speed twice that of the first one, and separated chilies for individual sensing and sorting. The second band includes the sensors. Two sensors were used, one detected chili necrosis optically and the other measured chili width.

The first detector was a radiometer, which activated a piston for removing chilies showing necrosis. One hundred healthy chilies and 100 showing necrosis were tested using a spectrophotometer in the 500–850 nm range. Relative reflectance showed spectral differences at the visible band. Necrotic tissue absorbed all the visible radiation and its reflectance peak at 720 nm was twice that of the healthy tissue. The radiometer was built with two silicon photodetectors and the signal obtained from both of them was converted to digital and sent to a microcontroller, which activated a piston whenever the ratio between the 550 nm value and the 720 nm was lower than 0.1. The accuracy of necrosis detection was 96.3%.

The chili-width sensor was located close to the end of the band and detected both vertical and horizontal chili widths. After passing through the radiometer, the chilies passed between a laser light generator and a bar of photodetectors, both of which were arranged vertically. The photodetector bar scanned the laser stripe every 20 ms. Each scan showed the number of photodetectors that did not receive the laser stripes and these data were recorded by a microcontroller. The fruit height or the vertical width of each chili corresponds to the maximum value recorded by the microcontroller. The horizontal width measurements were obtained by counting the number of blocked laser stripes as the chili advanced on the conveyor belt. This system can also measure the projected area of the fruit by integrating the different height measures as the fruit passes. Chilies were classified in three categories according to their width with an accuracy of 92% when the chilies were manually oriented and 87% when the system works in fully automatic mode. In addition, first-class chilies, which are the most important from the economic point of view, were sorted with success rates of 92.3%. The belt ran at a speed of 1 m/s, allowing sorting up to 15 fruits per second per lane. It is possible to increase sorting accuracy using a machine vision including cameras to view the top and the lateral chili sides and measure chili horizontal and vertical widths. This method is adequate for prolate fruits, such as eggplants. This is because the height measured by the system corresponds to the maximum diameter due to the stability position.

Okayama et al. (2006) developed a machine vision method for the objective and rapid classification of the shape of bell (sweet) peppers. Based on the local grading standards, bell peppers are classified in two grades based on external color, shape, and incidence of bruises or disease. Significant features to represent shape properties were selected and a classification test was conducted using these features. These features were calculated from three images of each sample. The results by a neural network classifier showed that the classification accuracy went up following an increase in the number of side views from 2 to 18. A rate of 91% of correctly classified peppers was obtained using 4 views and 92% using 8 views. When only 2 side views are used, the best angle between cameras is 90°. Nam-Hong et al. (2007) developed an automatic sorting system for green peppers using machine vision. The system included a feeding and individuation mechanism, an image inspection and processing system, and a discharging system. Using an on-line grading algorithm developed with Visual C/C++, the green peppers could be graded into four classes (large, medium, small sizes, and high curved shape), based on the measurement of two geometric parameters (length and flexure). The first derivative of thickness profile was used as a parameter to remove the stem area of a segmented image of each pepper. At a sorting speed of 0.45 m/s, the grading accuracies for large, medium, and small peppers were 86.0, 81.3, and 90.6 %, respectively. The system provided the performance capability of 121 kg/h, which is about five times higher than that obtained with conventional

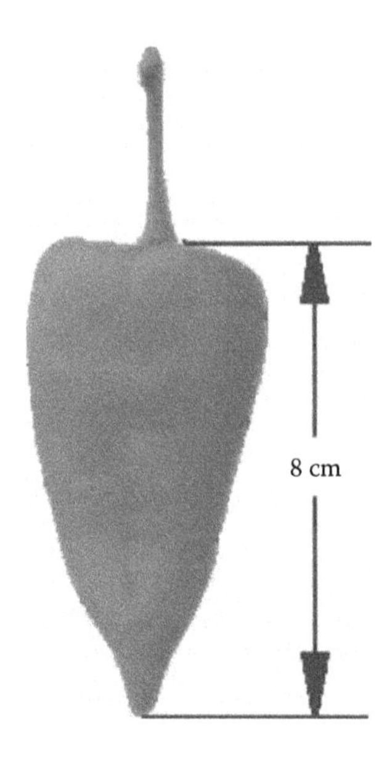

FIGURE 12.3 High-quality piquillo pepper.

manual sorting operations. The prototype system was economically feasible for sorting green peppers by showing a cost reduction of about 40%.

The shape of peppers is a cultivar feature and the quality of peppers belonging to a certain cultivar or specific denomination depends on their accordance with a certain shape determined by certain values of geometric parameters. Jarén et al. (2004) developed a procedure for sorting peppers, belonging to the piquillo cultivar, using image and color analyses. Piquillo peppers were classified into three quality categories according to the concordance between their physical parameters and the standards defined in the specific denomination. These parameters are related to size, shape, and texture of the peppers. The piquillo pepper of Lodosa is small, about 8 cm in length, with a flat triangular shape with its tip ending in a point, a bright red color, meaty, compacted, and consistent. Figure 12.3 shows a high-quality piquillo pepper.

The authors concluded that it is possible to correctly classify 98% of the peppers using color and image analyses and all peppers classified as belonging to the first category were really included in this category. Thus, it is possible to use the image and color analyses to develop a fast, reliable method based on piquillo pepper features, mainly geometric shape and bright red color. Figure 12.4 shows the graphic representation of the discriminate analysis used to classify piquillo peppers.

Van de Vooren et al. (1992) employed an image analysis technique that distinguished 80% of the mushroom cultivars used in the trial based on area, shape factor, and eccentricity of the cap and the shape factor of the gills. Color, shape, stem cut, and cap veil opening of mushrooms were quantified using image analysis in order to inspect and grade the mushrooms using an automated system. The shape criterion was determined from two-dimensional moments of inertia and misclassification based on it ranged from 14 to 24%.

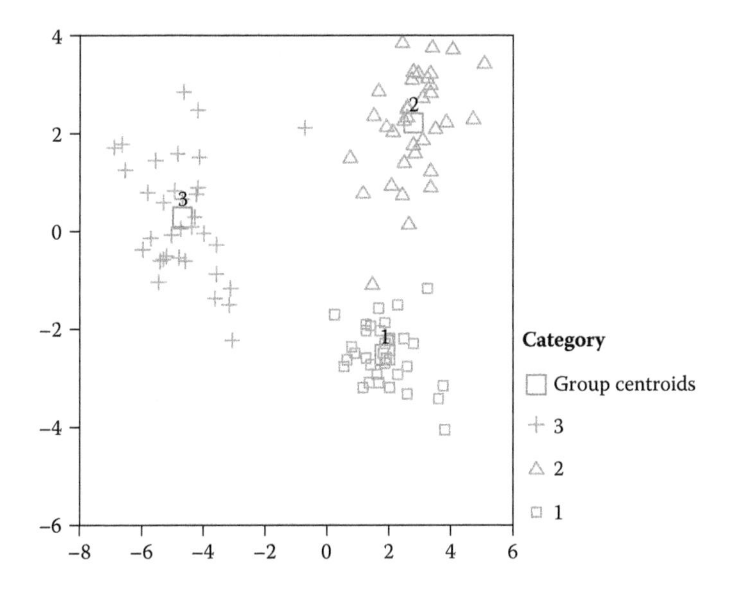

FIGURE 12.4 Graphic from the discriminate analysis used to classify piquillo peppers in three categories.

12.2.4 Weight

The measurement of weight is used for sorting vegetables because this parameter is closely related to volume and size and it is possible to measure it on-line. There are mechanical and electronic weight sizers; the former have a measurement point in each ejection point whereas the latter have a measurement point per lane. Francés et al. (2000), using an electronic weight sizer, processed a 150-g weight pattern piece 100 times at a speed of 16 fruits per second, obtaining a coefficient of variation (CV) of 0.34%. Weight can be measured indirectly using dimensional measurements such as projected area using a model or equation (Jahns et al., 2001; Varghese et al., 1991; Jarimopas et al., 1991; Davenel et al., 1988). Quiao et al. (2005) used hyperspectral imaging to measure pepper weight. The weight of fruits is also related to yield. Quiao et al. (2004) developed the Mobile Fruit Grading Robot (MGFR), proposed by Shibusawa and Kondo (2002), for sweet peppers. The prototype robot consisted of a manipulator, an end effector, a machine vision system, and a mobile mechanism. The robot could acquire 5 fruit images from 4 sides and the top while its manipulator transported the fruit received from the operator. A fruit mass prediction method was developed using 5 images. Results indicated that the fruit mass was successfully estimated with projection areas. The robot was used for mapping yield and quality of pepper crops.

Weight loss during the conservation process is not usually measured. It is a parameter that should be measured continuously because this information is necessary to improve the conservation of vegetables such as potatoes and onions. This is because weight loss is produced by water loss, which the quality of these vegetables. When weight loss occurs, storage conditions should be modified by humidifying or

FIGURE 12.5 The Mooij Weight Watcher on the potatoes.

renewing air through a humidifier or fans, respectively. A device to measure weight loss in potatoes and onions has been developed by Mooij. The Mooij Weight Watcher continuously measures and registers the weight loss of a sample bag of potatoes or onions during storage. One Mooij Weight Watchers is positioned in the pressure chamber and another one is placed on top of the pile, as shown in Figure 12.5. In this way it is possible to take measures in both places and to detect the time and the importance of the weight loss happening in the stored potatoes. The Mooij Weight Watcher is directly connected to the Orion storage control computer and the information is available on the Orion controller as well on the PC. It is a novel technique that results in more consciousness of the store manager and it leads to improvements of the storage system and finally a better quality of the crop during unloading the store.

12.2.5 Specific Gravity or Density

Tuber specific gravity is a measure of potato internal quality and it is used by the potato processing industry for assessing crop acceptability. There is a relationship between specific gravity and starch content (Hoffman et al., 2004). Starch content affects the taste, method of cooking, and processing of potato tubers. Potato specific gravity can be measured by means of a hydrometer by comparing the potato weight in air and in water and by using saline solutions. Freve and Dube (1987) developed a weighing system to simplify and accelerate the determination of specific gravity and to measure different grade sizes of potato samples using the same electronic scale. Specific gravity can be measured as the quotient between weight, which can be measured on-line by means of charge cells, and volume, which can be measured by means of computer image. Kakimoto and Izumoto (1995) estimated the specific gravity of hardwoods with image processing.

Mora and Schimleck (2009) determined the specific gravity of green *Pinus taeda* samples by near-infrared (NIR) spectroscopy. Komiyama et al. (2007) developed a nondestructive method for determination of starch content measuring it by specific gravity using visible and NIR spectroscopy. Potatoes were placed on a conveyor system and a photometric sensor was continuously applied. They concluded that the visible and NIR transmittance method is effective for accurate and rapid selection of potato tubers based on starch content measured by specific gravity, and thus could be used in packinghouses.

12.2.6 Vegetable Powder

Some vegetables, such as tomato, onion, or coffee, can be marketed as powders. In this case there are two important physical parameters: particle size and stickiness.

Singh et al. (2005) analyzed particle size of blanched and unblanched samples of tomato powders using a particle size analyzer based on the fact that the smaller the particle size, the larger the diffraction angle. Particle size analyzer systems consist of three basic units: optical units, dry powder feeder, and a computer system. The result obtained is the size distribution of equivalent spherical particles. The tomato powder sample falls into a feeder by compressed air. The sample then passes through the tubing and is fired through the air cell fitted to the Mastersizer, where it is measured and then collected by vacuum-extracting system.

The importance of the sticky behavior of amorphous food powders has been recognized over many decades in the food industry due to its influence on process, handling abilities, and quality of the powders. Stickiness is a major problem affecting product quality and yield during processing operations. It can be described in terms of cohesion and adhesion (Boonyai et al2004). The technique for measuring stickiness can be divided into direct and indirect. Lazar et al. (1956) used a conventional method that is propeller driven to measure the sticky point temperature of spray-dried tomato powder. Jaya and Das (2004) used this device to determine the sticky point temperature of instant coffee, tomato soup powder, and mango powder. There are other direct and indirect different to measure stickiness. Among them, the glass transition temperature method is one of the most used for food powders. Bahdra et al. (2009) developed a new technology to measure the sticky point temperature of coffee powder using a rheometer.

12.3 NOVEL TECHNIQUES TO CHARACTERIZE VEGETABLE TEXTURE

For most of vegetables, texture seems to be a primary quality attribute together with flavor and appearance. Only humans can perceive, analyze, and interpret most of the textural sensations. This is why it is difficult to introduce nonsensory methods to measure textural parameters, but since sensory analysis is subjective, expensive, time consuming, and requires training, new instrumental methods are being proposed continuously.

FIGURE 12.6 Measure of asparagus fibrosity using the texture analyzer.

The reference traditional methods for texture assessment of vegetables are sensory evaluation by means of expert tasters or consumers and destructive texture mechanical tests. Sensory methods are subjective, costly, time consuming, and require a large number of trained panelists. The reference destructive method to measure firmness, highly related to fruit quality, is the Magness-Taylor test, which measures the maximum force for a cylindrical plunger that penetrates into the flesh of the vegetable. Besides penetration tests, (confined) compression, tensile, and bending tests have been described in the literature with regard to the measurement of mechanical properties of vegetables. Adegoroye et al. (1988) related the mechanical properties of tomatoes with the test parameters from Magness-Taylor. They related epicarp strength to force (*N*) at bioyield; locular resistance to the residual force following bioyield; firmness to the maximum slope of the force–deformation curve; compliance with the deformation per unit epicarp strength and toughness with respect to the total energy consumed during the test, equivalent to the area under the force–deformation curve. Texture analyzers have been widely used to measure texture parameters of vegetables. Jarén et al. (2006) used a Stable Microsystem TA-TX2 Texture Analyser (Stable Micro Systems Ltd., Surrey, UK) to measure asparagus fibrosity, Arazuri et al. (2007) measured tomato firmness and tomato skin resistance, and Lemmens et al. (2009) measured carrot firmness using the same texture analyzer. Figure 12.6 shows the measure of asparagus fibrosity using the texture analyzer. Taherian and Ramaswany (2009) used a Universal Testing Machine (Lloyd Model LRX-2500 N; Lloyd Instruments Ltd., Fareham, Hans, UK) to measure the texture of turnip and beet roots. They defined the root vegetable firmness as the force–deformation ratio (N(mm), the stiffness as the stress-to-strain ratio (modulus of deformation, N/mm^2) and the springiness (elasticity) as the percentage of recoverability of input energy (percentage ratio of the area under the force–deformation curve during relaxation in relation to the total area).

Novel techniques have been developed for nondestructive sensing of vegetable texture. These techniques are based on vibration measurements, impact response, microdeformation measurements, ultrasonic wave propagation, nuclear magnetic resonance (NMR) spectroscopy and imaging, x-ray, time-resolved diffuse reflectance spectroscopy (TDRS), and NIR spectroscopy.

12.3.1 Techniques Based on Vibration Measurements

In vibration-based measurements, the fruit is struck with a small hammer and the sound produced is recorded using a microphone. A computer calculates the first resonance frequency of the time domain signal by means of a fast Fourier transform. The higher the resonance frequency, the firmer the fruit.

Kojima et al. (1991) determined tomato firmness using a resonance frequency–based method. They demonstrated that the minimum stress–relaxation time decreased as the tomatoes matured. They predicted tomato maturity using this parameter more accurately than using traditional techniques based on color. Felföldi et al. (1996) used an acoustic response method to evaluate tomato firmness, using the stiffness factor as a firmness parameter and concluding that it is appropriate to evaluate tomato softening. Muramatsu et al. (1996) measured firmness and characterized tomato maturity by measuring the sound amplitude propagated through the tomatoes. Sound frequencies are generated on one side of the tomato, and the sound wavelength propagated within the tomato is determined on the opposite side. Mature vegetables propagate shorter wavelengths and consequently more sound waves than immature ones. The difference in wave count was prominent above 500 Hz.

De Katelaere et al. (2006) assessed tomato firmness using a commercial desktop unit (AFS, AWETA, Nootdorp, The Netherlands). It uses an electromagnetic-driven probe to excite the fruit, a small microphone to capture the vibration of it, and a weight cell to weigh the fruit. The tomatoes where tested at four equidistant places around its bottom floral end in order to fit into the specimen holder. The only two parameters needed to obtain the firmness index S are the resonant frequency of the elliptical mode and the mass of the fruit (Cooke, 1972). The AFS method works correctly for firm tomatoes but is inaccurate for soft tomatoes.

The firmness index was developed by Cooke and Rand (1973) as shown in Equation (12.2)

$$S = f_R^{\,2} m^{2/3} \tag{12.2}$$

where f_R is the resonant frequency of the first elliptical mode in Hz and m is the mass of the object (kg).

Nicolaï et al. (2005) reported that there is a good relationship between this index and the elastic properties of the tomatoes. However, if the fruit shape is far from spherical, it is necessary to use an adapted firmness index that also includes a measure of shape S, as shown in Equation (12.3) (e.g., the length–diameter ratio)

$$F = f_R^{\,2} m^{2/3}/(aS + b) \qquad (12.\ 3)$$

where a and b are constants.

Sakurai et al. (2005) evaluated the texture of cucumbers by an acoustic measurement of crispness. Vibration signals were transformed to spectral data and a new index to evaluate the texture, named the sharpness index, was calculated from spectral data. Taniwaki et al. (2006) developed a method for quantifying the texture of blanched bunching onions, which are unsuitable for human sensory evaluation. They used a piezoelectric sensor and a knife-edge probe. The signal amplitude density was computed by summing up the amplitude of the texture signals and divided by the data length; this value was used as a texture index. Analyzing the texture signals in the lowest frequency band (0–50Hz) revealed the leaf structure of the blanched bunching onion tissue. They developed a device that enables direct measurement of food texture and has been applied to the green onion texture measurement. Taniwaki and Sakurai (2008) developed a new texture index (TI) to evaluate the texture of cabbages and the energy density, which was determined by the integration of squared amplitudes of texture signals multiplied by a factor of a frequency band. This TI enabled evaluation of acoustical signals in the high-frequency region (>1000 Hz) more sensitively than the previously used index amplitude density.

Van Zeebroeck at al. (2007) used acoustic stiffness as an objective measurement of tomato ripeness and measured the energy of the impact of a pendulum with the tomatoes and its radius of curvature to predict its bruise susceptibility.

12.3.2 Techniques Based on Impact Response

Determination of vegetable firmness based on impact force measurements has been widely applied. A steel sphere is dropped from a predetermined height onto the vegetable. In other versions, the impact is lateral and the force acting on the sphere is held by a spring controlled by a solenoid.

Chen et al. (1993) applied this technique to determine the firmness of intact radishes (*Raphaus sativus L.*) and pumpkins (*Cucurbita pepo L.*). The impact forces of the sample struck by the hammer of a rigid sphere were studied as a means for nondestructive determination of firmness. The correlation coefficients between the impact characteristic K, and the flesh firmness of pumpkins and Sakurajima radishes measured by a penetrometer were 0.93 and 0.92, respectively; hence, it is possible to determine firmness using an impact test.

De Katelaere et al. (2006) assessed tomato firmness using a low-mass impact sensor (SIQ-FT). It is a benchtop version (Sinclair iQ^{TM} firmness tester) provided by Sinclair International Ltd. (Norwich, England). The sensor hits the fruit by air pressure and captures the impact signal. The tomatoes were tested at four equidistant locations along its equator. The firmness is calculated according to the impact signal as a dynamic measure of fruit tissue spring constant and can be expressed by Equation (12.4) reported by Schmulevich et al. (2003):

$$SIQ-FT = C\left(\frac{P_{\max}}{\int p(t)dt}\right)^2 \tag{12.4}$$

where *SIQ – FT* is an internal quality value, *C* is a system constant, P_{max} is the peak amplitude of the impact response, and *p*(*t*) the impact response as a function of time.

They reported that the results obtained from the impact and the acoustic tests indicate that SIQ-FT would be the preferred choice for soft tomatoes due to its lower uncertainty. They also concluded that the results obtained using both devices (SIQ-FT and AFS) correlate well when considering a large firmness variance. Both instruments are capable of sensing firmness loss as induced by storage. The time might have come to consider those nondestructive techniques as new standards for fruit firmness evaluation replacing the older destructive standard.

12.3.3 Techniques Based on Microdeformation Measurements

Nondestructive microdeformation sensors have been developed to measure very small deformations produced on fruits during compression. A sphere penetrates into the skin of the fruit and the fruit surface deformation is measured using a laser displacement sensor. A different noncontact type of force–deformation sensor uses an air blast to deform the fruit surface (Prussia et al., 1994). Fruit firmness is related to the local deformation. Felföldi et al. (1996) used a hand-held penetrometer (MGA 1091) to measure the compressive stress with a very small mechanically limited deformation of 0.15 mm. A controlling microcomputer (data logger) calculated the coefficient of elasticity, which is the ratio of the compressive stress to deformation, expressed in kPa/mm. They concluded that the coefficient of elasticity is an appropriate characteristic to evaluate the softening of the tomato.

Hertog et al. (2004) evaluated tomato firmness using a noninvasive compression force. The force readings were taken by recording the maximum force required to compress the fruit for 2 mm using two horizontal parallel plates with the tomato placed on its side. As the compression deformation was fixed at 2 mm, the maximum compression force registered was linearly related to the slope of the force–deformation curve resembling the Young's modulus. The pretest speed was 5 mm s^{-1}, the test speed was 1 mm s^{-1}, and the trigger force was 1 *N*.

12.3.4 Techniques Based on Ultrasonic Wave Propagation

Ultrasonic wave propagation–based methods have been used to measure wood texture and recently to measure vegetable texture.

Ha et al. (1991) detected defective potatoes using the pulse transmission method. They found that the attenuation in defective potatoes was much higher than in sound ones. Hansen et al. (1992) used ultrasound methods to distinguish between uninfested and infested potatoes. Cheng and Haugh (1994) detected hollow hearths in potatoes using ultrasound on the basis of ultrasonic power transmitted through the whole potato tuber. They found that the wave form of transmitted ultrasonic signals through a hollow heart potato differed from that through a sound potato. Verlinden et al. (2004) evaluated tomato texture changes related to chilling injury by acoustic firmness measurements and ultrasonic wave propagation. Mizrach (2007) used a nondestructive ultrasonic method to monitor the physiochemical changes in firmness of greenhouse tomatoes during their shelf life. The acoustic wave attenuation in the fruit tissue was measured with a continuous-touch ultrasonic system by means of ultrasonic probes in contact with the peel. The attenuation was found to be linearly related to the firmness of the fruit during 8 storage days, which suggests that this method might be used for nondestructive firmness monitoring of tomatoes during their shelf life.

12.3.5 NMR Techniques

Nuclear magnetic resonance techniques have been used for on-line measurements of the oil content of avocados. NMR devices are very expensive and their use is so far limited to research applications.

McCarthy and Kauten (1990) used magnetic resonance imaging (RMI) to differentiate ripe tomatoes from green tomatoes. Ishida et al. (2000) used RMI to monitor changes during ripening of products such as tomatoes and cucumbers. Thybo et al. (2004) used MRI images of potatoes to detect nonvisible internal bruises produced by mechanical impacts and spraying symptoms caused by virus. Wang and Wang (1992) used NMR imaging techniques for detecting changes in the internal structure of zucchini squash (*Cucurbita pepo L.*, Ambassador) during exposure to chilling temperatures. Duce et al. (1992) used NMR imaging to identify fresh and frozen courgettes. Sequi et al. (2007) used MRI spectroscopy to unambiguously assess the place of origin and the cultivar of fresh cherry tomatoes. They were able to distinguish protected geographical indication (PGI) status cherry tomatoes from non-PGI ones, as well as cv. Naomi from cv. Shiren samples. The method determines informative physical and morphological parameters, transverse relaxation times, and thicknesses, respectively, which can be combined into four empirical equations; two of them are used to determine the cultivated variety and the other two to assess the place of origin. This approach has successfully recognized the cultivated variety of approximately 90% of the analyzed samples and the geographical origin of approximately 80% of the investigated cherry tomatoes.

12.3.6 Techniques Based on X-Ray Imaging

The properties of reflection, absorption, and transmission of x-rays are used to measure vegetable quality. X-ray imaging provides a cross-sectional view of the interior of an object and has been used to detect internal defects in fresh vegetables. X-rays pass through the material and are absorbed along the way according to the density and composition of the matter they encounter. There are two ways in which x-ray imaging can be performed: computed tomography (CT) and line scanning. Computed tomography scanning provides a more accurate view of the interior of the object than line scanning but line scanning is more applicable for an on-line inspection in a packing house.

Nylund and Lutz (1950) used x-rays to detect hollow hearth in potato. Thai et al. (1991) used x-ray (CT) imaging to detect tomato fruit maturity and Thai et al. (1997) used this technique to detect insect infestation in sweet potato roots. X-ray CT has been used to detect internal disorders of vegetables like hollow heart in potato. This disorder affects the density and water content of the internal tissue. X-ray imaging has been commercially used to inspect hollow hearth in potatoes and the overall quality of lettuce (Dull, 1986).

12.3.7 Techniques Based on Chlorophyll Fluorescence

Fluorescence is the phenomenon by which absorbed radiant energy is in part immediately re-emitted in form of light. Chlorophyll fluorescence has been used as a rapid and nondestructive method for quality measurement of vegetables. Lurie et al. (1994) used pulse-amplitude modulated (PAM) fluorometry to measure chilling injury nondestructively before tissue damage is visible in green peppers. Toivonen (1992) reported that chlorophyll fluorescence is a reliable, rapid, nondestructive indicator of broccoli quality during modified-atmosphere packaging, and it could be used to determine if broccoli has developed off odors without opening the bag and disrupting the package atmosphere. Lin and Jolliffe (2000) measured chlorophyll fluorescence of greenhouse cucumber to monitor senescence, temperature stress, and desiccation during storage and used fluorescence parameters as quality indicators. De Ell and Toivonen (2006) proposed chlorophyll fluorescence as an indicator of physiological changes related to color and texture in cold-stored broccoli after transfer to room temperature. Other features have been determined using chlorophyll fluorescence such as the frost sensitivity of potatoes and their tolerance to heat and cold (Hetherington et al., 1983) or the stress tolerance and stress-induced injury in cucumbers and cabbage leaves (Smillie and Hetherington, 1983).

12.3.8 Techniques Based on TDRS

Time-resolved diffuse reflectance spectroscopy has been developed in the biomedical field and allows the complete nondestructive characterization of biological tissues. A portable prototype was constructed for measuring quality parameters including texture of fruits and vegetables like tomatoes (Nicolaï and Verlinden, 2003).

12.3.9 Techniques Based on NIR Spectroscopy

Since the late 1980s, NIR spectroscopy has been used to measure vegetable texture. It is related to chemical composition of vegetables, which is related to their texture parameters.

Brach et al. (1982) used this technique to evaluate lettuce maturity with a coefficient of correlation of 0.96. Isakson and Kjolstad (1990) used NIR transmission and reflectance spectroscopy to measure vining pea maturity and quality parameters, achieving a coefficient of determination of 0.88. This method could be substituted for the traditional technique, which is the tenderometer test. Chalukova et al. (2000) developed mathematical models for calibration and prediction of some indices of maturity of green peas. They established that the indexes of maturity can be determined by NIR with values of R and RMSEP of 0.964 and 2.85 finometer degrees of firmness, respectively. In the prediction of five sensory attributes and the degree of maturity, R values varied from 0.733 to 0.896. The results obtained are the basis for the development of a portable NIR analyzer suitable for field application. Kjolstad et al. (2006) used NIR reflectance to predict the values of hardness and mealiness of frozen green peas. They developed models to predict the values of these textural variables from the acquired NIR spectra and from the tenderometer readings, being the reference measure method the sensory analysis. The model developed using variables from NIR spectra achieved a higher accuracy than the one achieved by the model developed from the tenderometer readings. This study suggests that NIR analysis could be a useful tool in instrumentally assessing the quality of frozen peas.

NIR spectroscopy has been used to predict the fiber content of green asparagus, intensely related to the fibrosity (Garrido et al. 2000), which is the most important feature to characterize asparagus quality. Jarén et al. (2006) used NIR spectroscopy to determine the harvest time necessary to achieve higher-quality asparagus. They found quality differences, related to fibrosity, among asparagus collected at different times and demonstrated that the base of the asparagus turned out to be the best part to use in order to establish the harvest date.

Van Dijk et al. (2006) tried to predict the firmness values of tomatoes using NIR spectroscopy. They reported that their calibration models could not predict the firmness of the individual tomatoes, although they could predict the firmness loss of a homogeneous batch of tomatoes. Shao et al. (2007) developed a model to predict tomato firmness using visible and near-infrared spectroscopy defined by the maximum resistance exerted by tomatoes during a compression test (F_c) and during the puncture test (F_p). The standard error of prediction was 16.017 N for F_c, and 1.18 N for F_p. The coefficient of correlation was 0.81 for F_c, and 0.83 for F_p.

12.3.10 Other Techniques to Measure Textural Properties

Tu et al. (2000) used laser-scattering imaging to study tomato fruit quality. They compared the results obtained using this novel technique with the ones obtained using the traditional compression test and the acoustic technique and developed a second-order polynomial equation to express the relation between the stiffness factor and the size of the scattering image (determination coefficient $R^2 = 0.62$). These systems can be used to evaluate tomato maturity. Ariana et al. (2006) used NIR hyperspectral imaging to detect bruises on pickling cucumbers. They developed a system consisting of an imaging spectrograph attached to an InGaAs camera with line-light fiber bundles as an illumination source and a specially designed sample roller for positioning the fruit for imaging.

The effect of external loads on vegetables has been investigated using finite element structural and modal analysis. Vibration response under impact loading has been studied for tomatoes (Langenakens et al., 1997). Mechanistic models have been developed to explicitly incorporate tissue features. A three-dimensional extension was developed by Gao et al. (1989, 1990) and Gao and Pitt (1991) and fitted to potato tissue. Ariana et al. (2006) used NIR spectroscopy to detect bruises on pickling cucumbers.

12.3.11 Surface Texture

The surface texture of vegetables can be characterized by either sensory panel tests or physical instrument tests. Physical techniques used for surface characterization are categorized into two groups: surface contacting and nonsurface contacting (Chen, 2007). The surface contacting techniques are the tribometer, surface force apparatus, contact profilometry, atomic force microscopy, and friction force microscopy. The nonsurface contacting techniques are the gloss meter, fiber-optic reflectometer, angle-resolved light scattering, surface glistening points method, and confocal laser scanning microscopy. The most frequently used parameters to characterize surface texture are root-mean square roughness, average roughness, peak to valley height, and average slope of surface asperities. Only a few surface-related characteristics can be quantified using a physical method (e.g., roughness/smoothness, glossiness, grittiness, moistness). Montouto-Grana et al. (2002) proposed the use of skin color and surface roughness as key to assessing parameters of raw potatoes. Lu et al. (1999) used the surface glistening point method to examine the surfaces of tomatoes and found that the root-mean-square roughness of tomato was 20 μm, and that tomatoes were slightly rougher than apples. Quevedo and Aguilera (2004) characterized the surfaces of tomatoes and potatoes using graphics relating the glistening density and the incident angle of illumination and found that a larger negative slope would suggest a smoother surface. Goula et al. (2007) used a contact probe test to characterize the surface stickiness of tomato pulp. The sample is brought in contact with the probe at a fixed speed of about 50 mm/min, and when the drop surface makes a good contact with the probe, it is withdrawn at the same speed. The instantaneous tensile force curve was recorded during the probe withdrawal, from which the maximum tensile force and other parameters were cross-examined against images of bonding, debonding, and failure of the material.

12.4 NOVEL TECHNIQUES TO MEASURE OTHER PHYSICAL PARAMETERS RELATED TO VEGETABLE QUALITY

Although most of the novel techniques used to measure the physical parameters of vegetables are related to their physical characteristics and texture, there are some novel techniques applied to measure optical, electrical, thermal, or acoustic parameters that are useful in characterizing vegetable quality.

12.4.1 Optical Parameters

Optical properties of vegetables were widely used in classifying them. Traditionally, color was measured using colorimeters, with results expressed in the RGB and CIELAB coordinates. Color can be predicted using NIR technology; it is not a breakthrough, although it may reduce the number of devices needed in a selection chain or a packing line. Cozzolino et al. (2004) used a diode array spectrophotometer (400–1100 nm) to predict the color of grapes.

Qin and Lu (2008) used a spatially resolved steady-state hyperspectral imaging technique combined with inverse algorithms developed based on the diffusion theory model for measuring the absorption and scattering properties of cucumbers, zucchini squash, and tomatoes over the spectral range of 500–1000 nm. Using this noncontact technique, they identified the ripeness stages of tomatoes. In Chapter 4, novel techniques for measuring gloss in eggplants have been included.

12.4.2 Electrical Parameters

The dielectric behavior of vegetables is important to characterize their ability to be cooked using microwaves. The most important parameters characterizing dielectric properties of vegetables are the dielectric constant and the loss factor. The resonant cavity method is the traditional method used to measure these parameters. Because those interested in the dielectric properties of agricultural materials are not often properly equipped for measuring these parameters, some effort has been devoted to the development of models for estimating the dielectric properties. Sharma and Prasad (2002) developed predictive models to obtain the dielectric constant and loss factor of garlic as functions of moisture content and temperature, which could be used to predict microwave heating patterns of garlic as continuous functions of changing moisture content during drying.

12.4.3 Thermal Parameters

Thermal conductivity and thermal diffusivity of vegetables are important parameters in industrial processes at high and low temperatures. Wang and Brennan (1992) developed a line-source system to measure the thermal conductivity of potato in the of 40°–70°C temperature range. Van Gelder (1998) used a thermistor-based method to measure the thermal conductivity of potato and tomato paste at high temperatures. Abhayawik et al. (2002) used a CT meter (TELEPH, Meylan, France) to measure the thermal conductivity and diffusivity and the volumic heat capacity of ground onion powder. The range of the power that could be provided was 0–2.5 W and the accuracy was 5%. The apparatus measures the response of the product to a thermal shock with the help of a flat probe composed of an annular heat source that heats the product and a thermocouple sensor placed in the center of the ring to measure the product temperature. A known heat flux is transmitted through the product during a selected heating period using the annular heat source and the sensor is used to follow the evolution of the product temperature.

Thermal diffusivity is associated with the speed of propagation of heat from the solid body during changes of temperature with time. The higher the thermal diffusivity, the faster the propagation of heat in the medium. Gordon and Thorne (1990) determined the thermal diffusivity of vegetables during cooling in a medium at constant temperature. They used two methods. The first one, called the slope method, calculates thermal diffusivity from the slope of the cooling curve of the food center, while the second method, called the lag method, uses the maximum temperature difference between the center of the food and a point halfway from the center to the surface. Dincer (1996) developed an analytical model for determining the thermal diffusivity of cucumber subjected to cooling. Morikawa (2002) used a high-speed infrared (IR) camera to develop a new type of two-dimensional thermal analysis of the freezing process in onion epidermal cells. He developed a multiple measurement system able to measure thermal conductivity, thermal diffusivity, and to observe latent heat on the phase transition and the distribution of thermal diffusivity in the complex materials.

12.5 CONCLUSIONS

The ideal method covering all requirements of current and future applications in production, storage, and marketing of vegetables has not yet emerged. As sensors often measure only a single quality property, combined techniques will have to be optimized to measure the overall quality of vegetables. Future advances in vegetable quality measurement are likely to come from developing a better understanding of consumer expectations. Consumer decisions and satisfaction are determined by how vegetable quality is perceived. Measurement of consumer perception is essential, and it is more important to measure the vegetable properties that determine consumer acceptance than the properties that are easiest to measure.

Today, physical characteristics of vegetables must be evaluated to determine their quality, and novel techniques have been developed to measure physical parameters accurately, quickly, and inexpensively. Most of these techniques can measure on-line and some of them can measure continuously. Using these parameters, it is possible to characterize not only size but also shape, which is very important when evaluating the quality of many vegetables that are characterized by a particular shape. Image analysis is an important tool to measure most of these physical parameters.

As consumers not only take into account size and shape but also textural features, it is necessary to evaluate the texture of vegetables. Novel techniques to measure textural parameters have been developed in recent years. They are based on vibration measurements, impact response, micro-deformation measurements, ultrasonic wave propagation, nuclear magnetic resonance, x-ray imaging, chlorophyll fluorescence, and spectroscopy, and have been widely applied to vegetables.

Using novel techniques to measure parameters that characterize the size, shape, and texture of vegetables, it is possible to evaluate their global quality on-line, with sufficient accuracy and without excessive cost.

REFERENCES

Abhayawick. L., J.C. Laguerre, V. Tauzin, and A. Duquenoy. 2002. *Journal of Food Engineering* 55:253–262.

Adegoroye, A.S., P.A. Jolliffe, and M.A. Tung. 1988. Instrumental measures of texture of tomato fruit and their correlation with propectine content: Effect of fruit size and stages of ripeness. *Journal of Food Science Technology* 25(2):72–74.

Al-Mallahi, A., T. Kataoka, H. Okamoto, and Y. Shibata. 2010. Detection of potato tubers using an ultraviolet imaging-based machine vision system. *Biosystems Engineering* 105(2):257–265.

Arazuri, S., C. Jarén, J.I. Arana, and J.J. Perez De Ciriza. 2007. Influence of mechanical harvest on the physical properties of processing tomato (*Lycopersicon esculentum Mill.*). *Journal of Food Engineering* 80:190–198.

Ariana D.P., R. Lu, and D.E. Guyer. 2006. Near-infrared hyperspectral reflectance imaging for detection of bruises on pickling cucumbers. *Computers and Electronics in Agriculture* 53:60–70.

Bhadra, R., K. Muthukumarappan, and K.A. Rosentrater. 2009. Measurement of sticky point temperature of coffee powder with a rheometer. Paper presented at the ASABE/CSBE North Central Intersectional Meeting.

Boonyai, P., B. Bhandari, and T. Howes. 2004. Stickiness measurement techniques for food powders: A review. *Powder Technology* 145:34–46.

Brach, E.I., C.T. Phan, G. Pouskinshky, J.J. Jasmin, and C.B. Aubé. 1982. Lettuce maturity detection in the visible (380–720 nm) infrared (680–750 nm) and near infrared (800–1.850 nm) wavelength band. *Agronomie* 2:685–694.

Cao, Q., and M. Nagata. 1997. Study on grade judgment of fruit vegetables using machine vision (part 3): Judgment for shape of green pepper using neural network. *Journal of Society of High Technology in Agriculture* 9(1):49–59.

Chalukova, R., M. Krivosiev, V. Kalinov, and C. Scotter. 2000. Determination of green pea maturity by meassurement of whole pea transmittance in the NIR region. *Lebensmittel-Wissenschaft und Technologie* 33(7):489–498.

Chen, J. 2007. Surface texture of foods: Perception and characterization. *Critical Reviews in Food Science and Nutrition* 47:583–598.

Chen, J.Y., M. Miyazato, E. Ishiguro, and N. Namba. 1993. Determination of firmness in intact pumpkins and Skurajima radishes by impact forces. *Journal of the Japanese Society for Horticultural Sciences* 61:951–956.

Cheng, I., and G. Haugh. 1994. Detecting hollow heart in potatoes using ultrasounds. *Transactions of the ASABE* 37(1): 217–222.

Cooke, J.R. 1972. An interpretation of the resonant behaviour of intact fruits and vegetables. *Transactions of the ASAE* 15:1075–1080.

Cooke, J.R., and R.H. Rand. 1973. A mathematical study of resonance in intact fruits and vegetables using a 3-media elastic sphere model. *Journal of Agricultural Science* 18:141–157.

Cozzolino, D., M.B. Esler, R.G. Dambers, et al. 2004. Prediction of colour and pH in grapes using a diode array spectrophotometer (400–1100 nm). *Journal of Near Infrared Spectroscopy* 12:105–111.

Davenel, A., Ch. Guizard, T. Labarre, and F. Sevila. 1988. Automatic detection of surface defects on fruit by using a vision system. *Journal of Agricultural Engineering Research* 41(1):1–9.

De Ell, J.R., and P.M.A. Toivonen. 2006. Chlorophyll fluorescence as an indicator of physiological changes in cold-stored broccoli after transfer to room temperature. *Journal of Food Science* 64(3):501–503.

De Ketelaere, B., M.S. Howard, L. Crezee, et al. 2006. Postharvest firmness changes as measured by acoustic and low-mass impact devices: A comparison of techniques. *Postharvest Biology and Technology* 41:275–284.

Dincer, I. 1996. Determination of thermal diffusivities of cylindrical bodies being cooled. *International Communications in Heat and Mass Transfer* 23:713–720.

Duce, S.L., T.A. Carpenter, and L.D. Halla. 1992. Nuclear magnetic resonance imaging of fresh and frozen courgettes. *Journal of Food Engineering* 16(3):165–172.

Dull, G.G. 1986. Non-destructive evaluation of quality of stored fruits and vegetables. *Journal of Food Technology* 5:106–110.

European Community (EC). 2001. Commission Regulation No. 1619/2001 of 6 August 2001:2–3.

Felföldi, J., A. Fekete, and M. Gilinger. 1996. Firmness-based assessment of tomato shelf-life. Paper presented at AgEng, Madrid, Paper No. 96F-012.

Forbes, K. 2000. Volume estimation of fruit from digital profile images. Master's thesis. Cape Town, South Africa: University of Cape Town, Department of Electrical Engineering, http://www.dip.ee.uct.ac.za/(kforbes/Publications/ msckaf.pdf. (accessed September 30, 2007).

Forbes, K.A., and G.M. Tattersfield. 1999. Estimating fruit volume from digital images.

Fifth Africon Conference in Africa: AfriCon, vol. 1. IEEE, Los Alamitos, CA, 107–112.
Francés, J.V., Calpe, J., Soria, E. et al. 2000. Application of ARMA modeling to the improvement of weight estimations in fruit sorting and grading machinery. International Conference on Acoustics, Speech and Signal Processing 6:3666–3669. IEEE, Los Alamitos, Cal., USA.
Freve, A., and Y. Dube. 1987. Tackle-Scal-Uter: A system with computer, electronic scale and tackle for measuring specific gravity and other weights of potato. *American Potato Journal* 64(8):439–440.
Gall, H. 1997. A ring sensor system using a modified polar coordinate system to describe the shape of irregular objects. *Measurement Science and Technology* 8(11):1228–1235.
Gall, H., A. Muir, J. Fleming, R. Pohlmann, L. Göcke, and W. Hossack. 1998. A ring sensor system for the determination of volume and axis measurements of irregular objects. *Measurement Science and Technology* 9(11):1809–1820.
Gao, Q., and R.E. Pitt. 1991. Mechanics of pharenchyma tissue based on cell orientation and microstructure. *Transactions of the ASAE* 34:232–238.
Gao, Q., R.E. Pitt, and J.A. Bartsch. 1989. Elastic-plastic constitutive reactions of the cell walls of apple and tomato parenchyma. *Journal of Rheology* 33:233–356.
Gao, Q., R.E. Pitt, and A. Ruina. 1990. A mechanics model of compression of cells with finite initial contact area. *Biorheology* 27:225–240.
Garrido, A., M.T. Sánchez, and M.D. Pérez. 2000. Application of NIR spectroscopy to plant material analysis. (Aplicación de la espectroscopía NIR al analísis de productos de origen vegetal.) *Alimentaria* 57:57–62.
Gordon, C., and S. Thorne. 1990. Determination of the thermal diffusivity of foods from temperature measurements during cooling. *Journal of Food Engineering* 11(2):133–145.
Goula, M.A., and T.D. Karapantsios. 2007. Characterization of tomato pulp stickiness during spray drying using a contact probe method. *Drying Technology* 25:591–598.
Ha, K., H. Kanai, N. Chubachi, and K. Kamicura. 1991. Basic study on nondestructive evaluation of potatoes using ultrasound. *Journal of Applied Physics Part I* 30:80–82.
Hahn, F. 2002. Automatic jalapeño chili grading by width. *Biosystems Engineering* 83(4):433–440.
Hahn, F., and S. Sanchez. 2000. Carrot volume evaluation using imaging algorithms. *Journal of Agricultural Engineering Research* 75(3):243–249.
Hansen, J.D., C.L. Emerson, and D.A. Signorotti. 1992. Visual detection of sweet-potato weevil by noninvasive methods. *Florida Entomologist* 75:369–375.
Hatou, K., T. Morimoto, J. De Jager, and Y. Hashimoto. 1996. Measurement and recognition of 3-D body in intelligent plant factory. *Abstracts of the International Conference on Agricultural Engineering* (AgEng) 2:861–862. Paper No. 96F-027. Madrid.
Hertog, M.L.A.T.M., R. Ben-Arie, E. Róth, and B. Nicolăi. 2004. Humidity and temperature effects on invasive and non-invasive firmness measures. Postharvest *Biology and technology* 33:79–91.
Hetherington, S.E., R.M. Smillie, P. Malagamba, and Z. Huamán. 1983. Heat tolerance and cold tolerance of cultivated potatoes measured by the chlorophyll-fluorescence method. *Planta* 159(2):119–124.
Hoffmann, T., C. Fürll, and J. Ludwig. 2004. A system for the on-line starch determination at potato tubers. In *Proceedings of the International Conference on Agricultural Engineering (AgEng).* Technologisch Instituut vzw, CD-ROM.
Hoffmann, T., G. Wormans, C. Fürll, and J. Poller. 2005. A system for determining starch in potatoes on line, http://vddb-dt.library.lt/fedora/get/LT-eLABa-0001:J.04~2005~ISSN_1392-1134.V_37.N_2.PG_34-43/DS.002.1.01.ARTIC (accessed August 15, 2008).

Hryniewicz, M., I. Sotome, J. Anthonis, H. Ramon, and J. De Baerdemaeker. 2005. 3-D surface modeling with stereovision. In *Proceedings of the 3rd International Symposium on Applications of Modeling as an Innovative Technology in the Agri-Food Chain*, ed. M.L.A.T.M. Hertog and B.M. Nicolaï. MODEL-IT 2005. *ISHS Acta Horticulturae* 674.

Iraguen, V., A. Guesalaga, and E. Agosin. 2006. A portable non-destructive volume meter for wine grape clusters. *Measurement Science and Technology* 17(12):N92–N96.

Isakson, T., and L. Kjolstad. 1990. Prediction of sensory quality of peas by NIR and NIR transmittance. In *The Proceedings of the 2nd International NIRS Conference*, ed. M. Iwamoto and S. Kawano. Tokyo: Korin Publishing Co.

Ishida, N., M. Koizumi, H. Ogawa, and H. Kano. 2000. Micro-MRI in food science. *Journal of the Japanese Society for Food Science Technology* 47(6):407–423.

Iwamoto, M., and Y. Chuma. 1981. Recent studies on development in automated citrus packinghouse facility in Japan. *Proceedings of the International Society of Citriculture* 2:831–834.

Jahns, G., H.M. Nielsen, and W. Paul. 2001. Measuring image analysis attributes and modeling of fuzzy consumer aspects for tomato quality grading. *Computers and Electronics in Agriculture* 31(1):17–29.

Jarén, C., I. Arana, J.M. Senosiain, and S. Arazuri. 2004. Clasificación por color y análisis de imagen de pimiento del piquillo (Color and image analysis classification of Piquillo peppers). VII Colour National Congress. Pamplona (Spain).

Jarén, C., S. Arazuri, M.J. García, P. Arnal, and J.I. Arana. 2006. White asparagus harvest date discrimination using NIRS technology. *International Journal of Infrared and Millimeter Waves* 27(3):391–401.

Jarimopas, B., P. Siriratchatapong, T. Chaiyaboonyathanit, and S. Niemhom. 1991. Image-processed mango sizing machine. *Kasetsart Journal* 25(5):131–139.

Jaya, S., and H. Das. 2004. Glass transition and sticky point temperatures and stability/mobility diagram of fruit powders. *Food and Bioprocess Technology* 2(1):89–95.

Jun, Qiao, A. Sasao, S. Shibusawa, N. Kondo, and E. Morimoto. 2004. Mobile fruit grading robot (Part 1): Development of a robotic system for grading sweet peppers. *Journal of the Japanese Society of Agricultural Machinery* 66(2):113–122.

Jun, Qiao, A. Sasao, S. Shibusawa, N. Kondo, and E. Morimoto. 2005. Mapping yield and quality using the mobile fruit grading robot. *Biosystems Engineering* 90(2):135–142.

Jun Qiao, Ning Wang, Michael O. Ngadi, and Singh Baljinder. 2005. Water content and weight estimation for potatoes using hyperspectral imaging. Paper No. 053126, ASAE Annual Meeting.

Kanali, C., H. Murase, and N. Honami. 1998. Three-dimensional shape recognition using a charge-simulation method to process primary image features. *Journal of Agricultural Engineering Research* 70(2):195–208.

Kakimoto, T., and Y. Izumoto. 1995. Estimation of specific gravity of hardwoods with image processing. Memoirs of Osaka Kyoiku University. III, *Natural Science and Applied Science* 43(2):211–217.

Kjolstad, L., T. Isakson, and H.J. Rosenfeld. 2006. Prediction of sensory quality by near infrared reflectance analysis of frozen and freeze dried green peas (*Pisum sativum*). *Journal of the Science of Food and Agriculture* 51(2):247–260.

Kojima, K., N. Sakurai, S. Kuraisi, R. Yamamoto, and J.J. Nevins. 1991. Novel technique for measuring tissue firmness within tomato (*Lycopersicon esculentum Mill*) fruit. *Plant Physiology* 96:545–550.

Komiyama, S., J. Kato, H. Honda, and K. Matsushima. 2007. Development of sorting system based on potato starch content using visible and near-infrared spectroscopy. *Journal of the Japanese Society for Food Science and Technology* 54(6):304–309.

Langenakens, J., X. Vandewalle, and J. De Baerdemaeker. 1997. Influence of global shape and internal structure of tomatoes on the resonant frequency. *Journal of Agricultural Engineering Research* 66:41–49.

Lazar, M.E., A.H. Brown, G.S. Smith, F.F. Wong, and F.E. Lindquist. 1956. Experimental production of tomato powder by spray drying, *Food Technol*ogy 10:129–134.

Lee, D.J., X. Xu, J. Eifert, and P. Zhan. 2006. Area and volume measurements of objects with irregular shapes using multiple silhouettes. *Optical Engineering* 45 (2):027202–11.

Lemmens, L., E. Tibäck, C. Svelander, et al. 2009. Thermal pretreatments of carrot pieces using different heating techniques: Effects on quality related aspects. *Innovative Food Science and Emerging Technologies* 10:522–529.

Lin, W.C., and P.A. Jolliffe. 2000. Chlorophyll fluorescence of long English cucumber affected by storage conditions. XXV International Horticultural Congress Part 7: Quality of Horticultural Products. *Acta Horticulturae* 517:449–456.

Lu, R., J.J. Koenderink, and A.M.L. Kappers. 1999. Surface roughness from highlight structure. *Applied Optics* 38:2886–2894.

Lurie, S., P. Ronen, and S. Meier. 1994. Determining chilling injury using non-destructive pulse amplitude (PAM) fluorometry. *Journal of the American Society for Horticultural Science* 119(1):59–62.

Marchant, J.A. 1990. A mechatronic approach to produce grading. Proceedings of the Institution of Mechanical Engineers, International Conference. *Mechatronics: Designing Intelligent Machines* 159–164.

Marchant, J.A., C.M. Onyango, and E. Elipe. 1989. Weight and dimensional measurements on potatoes at high speed using image analysis. In *Agrotique 89: Proceedings of the Second International Conference, Teknea*, ed. J.P. Sagaspe and A. Villeger, 41–52.

McCarthy, M.J., and R.J. Kauten. 1990. Magnetic resonance imaging applications in food research. *Trends in Food Science & Technology* 1:134–139.

Mitchell, P.D. 1986. Pear fruit growth and the use of diameter to estimate fruit volume and weight. *HortScience* 21(4):1003–1005.

Mizrach, A. 2007. Non-destructive ultrasonic monitoring of tomato quality during shef-life storage. *Postharvest Biology and Technology* 46(3):271–274.

Mohsenin, N.N. 1970. Physical characteristics. In *Physical properties of plant and animal materials,* vol. I, 51–87. New York: Gordon & Breach Science Publishers Inc.

Montouto-Grana, M., E. Fernández Fernández, M.L. Vázquez-Oderiz, and M.A. Romero-Rodriguez. 2002. Development of a sensory profile for the specific denomination Galician potato. *Food Quality and Preference* 13:99–106.

Mora, C.R., and L.R. Schimleck,. 2009. Determination of specific gravity of green *Pinus taeda* samples by near infrared spectroscopy: Comparison of pre-processing methods using multivariate figures of merit. *Wood Science and Technology* 43(5/6):441–456.

Moreda, G.P., 2004. Diseño y evaluación de un sistema para la determinación en línea del tamaño de frutas y hortalizas mediante la utilización de un anillo óptico (Design and assessment of a system for on-line size determination of fruits and vegetables, using an optical ring sensor). Ph.D. dissertation. Universidad Politécnica de Madrid, Spain.

Moreda, G.P., J. Ortiz-Cañavate, F.J. García-Ramos, and M. Ruiz-Altisent. 2007. Effect of orientation on the fruit on-line size determination performed by an optical ring sensor. *Journal of Food Engineering* 81(2):388–398.

Moreda, G.P., J. Ortiz-Cañavate, F.J. García Ramos, and M. Ruiz-Altisent. 2009. Non-destructive technologies for fruit and vegetable size determination: A review. *Journal of Food Engineering* 92:119–136.

Morikawa, J. 2002. Measurement of thermal property by high speed & micro scale IR camera: 2. Development for multiple measurement system. *Termophys Prop* 23:217–219.

Muramatsu, N., N. Sakurai, R. Yamamoto, and J.J. Nevins. 1996. Nondestructive acoustic measurement of firmness for nectarines, apricots, plums and tomatoes. *HortScience* 31:1199–1202.

Nam-Hong Cho, Dong-Il Chang, Soo-Hee Lee, Hak-jin Kim, and Lee Young-Hee. 2007. Development of automatic sorting system for green pepper using machine vision. In *ASAE Annual Meeting* 076106. St. Joseph, MI: American Society of Agricultural and Biological Engineers.

Nicolaï, B., J. Lammertyn, E.A. Veraberbeke, et al. 2005. Non-destructive techniques for measuring quality of fruit and vegetables. *Acta Horticulturae* 682:1333–1339.

Nicolaï, B., and B.E. Verlinden. 2003. Textures assessment of perishable products. *Acta Horticulturae* 600:513–519.

Nylund, R.E., and J.M. Lutz. 1950. Separation of hollow heart potato tubers by means of size grading, specific gravity, and x-ray examination. *American Journal of Potato Research* 27:214–222.

Okayama, T., J. Quiao, H. Tanaka, N. Komdo, and S. Sakae. 2006. Classification of shape of bell pepper by machine vision system. *Agricultural Information Research* 15(2):113–122.

Peleg, K. 1985. Sorting operations. In *Produce handling, packaging and distribution*, 53–87. Westport, CN: AVI Publishing Co.

Prussia, S.E., J.J. Astleford, Y.C. Humg, and R. Hewlett. 1994. Non-destructive firmness measuring device. US Patent No. 5,372,030.

Qin, J., and R. Lu. 2008. Measurement of the optical properties of fruits and vegetables using spatially resolved hyperspectral diffuse reflectance imaging technique. *Postharvest Biology and Technology* 49:355–365.

Quevedo, R., and J.M. Aguilera. 2004. Characterization of food surface roughness using the glistening points method. *Journal of Food Engineering* 65:1–7.

Sakurai, N., S. Iwatami, S. Terasaki, and R. Yamamoto. 2005. Texture evaluation of cucumber by a new acoustic vibration method. *Journal of the Japanese Society for Horticultural Science* 74:31–15.

Sarkar, N., and R.R. Wolfe. 1985. Feature extraction techniques for sorting tomatoes by computer vision. *Transactions of the ASAE* 28(3):970–979.

Sequi, P., M.T. Dell'Abate, and M. Valentini. 2007. Identification of cherry tomatoes growth origin by means of magnetic resonance imaging. *Journal of the Science of Food and Agriculture* 87(1):127–132.

Shao, Y., Y. He, A.H. Gómez, A.G. Pereir, Z. Qiu, and Y. Zhanh. 2007. Visible/NIR spectrometric technique for nondestructive assessment of tomato "Heatwave" (*Licopersicum esculentum*) quality characteristics. *Journal of Food Engineering* 81(4): 672–678.

Sharma, G.P., and S. Prasad. 2002. Dielectric properties of garlic (*Allium Sativum L.*) at 2450 MHz as function of temperature and moisture content. *Journal of Food Engineering* 52:343–348.

Shibusawa, S., and N. Kondo. 2002. Grading system for agro-products. Patent No. 2003-103864. Japan.

Shmulevich, I. N. Galili, and M.S. Howarth. 2003. Nondestructive dynamic testing of apples for firmness evaluation. *Postharvest Biology and Technology* 29:287–299.

Singh, R. S. Mangaraj, and S.D. Kulkarni. 2005. Particle-size analysis of tomato powder. *Journal of Food Processing and Preservation* 30:87–98.

Smillie, R.M., and S.E. Hetherington. 1983. Stress tolerance and stress-induced injury in crop plants measured by chlorophyll fluorescence in vivo. *Plant Physiology* 72:1043–1050.

Taherian, A.R., and H.S. Ramaswany. 2009. Kinetic considerations of texture softening in heat treated root vegetables. *International Journal of Food Properties* 12:114–128.

Taniwaki, M., and N. Sakurai. 2008. Texture measurement of cabbages using an acoustical vibration method. *Postaharvest Biology and Technology* 50(2–3):176–181.

Taniwaki, M., H. Takanori, and N. Sakurai. 2006. Development of a method for quantifying food texture using blanched bunching onions. *Journal of the Japanese Society for Horticultural Science* 75(5):410–414.

Teneze, B., P. Baylou, and P. Riboulet. 1989. Presentation of the image analysis part of an automatic programme for French beans developed by AGROTEC (Presentation de la partie d´un programme "Tri automatique de haricots verts" elabore par AGROTEC). In *Agrotique 89: Proceedings of the Second International Conference*, ed. J.P. Sagaspe and A. Villeger, 53–87.

Thai, C.N., J.N. Pease, and E.W. Tollner. 1991. Determination of green tomato maturity from x-ray computed tomography imaging. Proceedings of the 19th Symposium on Automated Agriculture for the 21st Century, 134–143. St. Joseph, MI: ASAE.

Thai, C.N., E.W. Tollner, K. Morito, and S.J. Kays. 1997. VIS-NIR and x-ray characterization of sweet potato weevil larvae development and subsequent damage in infested roots. In *Proceedings: Sensor for nondestructive testing. Measuring the quality of fresh fruits and vegetables*, 361–368. February 18–21, Orlando, Florida. USA: North Regional Agricultural Engineering Services, Ithaca, NY.

Thybo, A.K., S.N. Jespersen, P.E. Laerke, and H.J. Stødkilde., 2004. Nondestructive detection of internal bruise and sprain disease symptoms in potatoes using magnetic resonance imaging. *Magnetic Resonance Imaging* 222(9):1311–1317.

Toivonen, P.M.A. 1992. Chlorophyll fluorescence as a non-destructive indicator of freshness in harvested broccoli. *HortScience* 27:1010–1015.

Tu, K., P. Jansók, B. Nicolaï, and J. De Baerdemaeker. 2000. Use of a laser-scattering imaging to study tomato-fruit quality in relation to acoustic and compression measurements. *International Journal of Food Science and Technology* 35:503–510.

Van de Vooren, J.G., G. Polder, and G.W.A.M. Van der Heijden. 1992. Identification of mushroom cultivars using image analysis. *Transactions of the ASABE* 35(1): 347–350.

Van Dijk, C., C. Boeriu, F. Meter, T. Stolle-Smits, and L.M.M. Tijskens. 2006. The firmness of stores tomatoes (cv. Tradiro). 1. Kinetic and near infrared models to describe firmness and moisture loss. *Journal of Food Engineering* 77(3):575–584.

Van Gelder, M. F. 1998. A thermistor based method for measurement of thermal conductivity and thermal diffusivity of moist food materials at high temperatures. PhD Thesis. Faculty of the Virginia Polytechnic Institute and State University, Blacksburg, Virginia.

Van Zeebroeck, M., V. Van Linden, P. Darius, B. de Katelaere, H. Ramon, and L.M.M. Tijskens. 2007. The effect of fruit properties on the cruise susceptibility of tomatoes. *Postharvest Biology and Technology* 45:168–175.

Varghese, Z., C.T. Morrow, P.H. Heinemann, H. Joseph Sommer III, Y. Tao, and R.M. Crassweller. 1991. Automated inspection of golden delicious apples using color computer vision. ASAE Paper No. 917002. St. Joseph, MI: ASAE.

Verlinden, B.E., V. De Smedt, and B. Nicolai. 2004. Evaluation of ultrasonic wave propagation to measure chilling injury in tomatoes. *Postharvest Biology and Technology* 32(1):109–113.

Wang, C.Y., and P.C. Wang. 1989. Nondestructive detection of core breakdown in Bartlett pears with nuclear magnetic resonance imaging. *Hortscience* 24(1):106–109.

Wang, N., and J.G. Brennan. 1992. Thermal conductivity of potato as a function of moisture content. *Journal of Food Engineering* 17:153–160.

Wang, C.Y., and P.C. Wang. 1992. Differences in nuclear magnetic resonance in images between chilled and non-chilled zucchini squash. *Enviromental and Experimental Botany* 32(3): 213–219.

13 Physical Properties of Cereal Products

Measurement Techniques and Applications

Iñigo Arozarena, Asunción Iguaz, Maria José Noriega, Gloria Bobo, and Paloma Vírseda

CONTENTS

13.1 INTRODUCTION

In this chapter new methods to characterize cereals and their products are described in order to determine the effects of manufacturing processes and storage conditions on the thermal, mechanical, and structural properties of grains, cereal flours, and their products.

The methods to determine physical properties in cereals have been classified in three different cases depending on raw matter state: before, during, and after production; and the current techniques in image analysis, light and electron microscopy, and NMR spectroscopy used to analyze the microstructure of cereal products. This chapter also discusses the methods used to optimize processing parameters and formulations to produce end products with desirable sensory and textural properties, the shelf life of cereal products, and the relationships between the sensory and physical characteristics of cereal foods.

13.2 PHYSICAL PROPERTIES OF GRAIN

13.2.1 Mechanical Properties

When talking about mechanical properties of grains, it is necessary to distinguish between bulk grain mechanical properties and mechanical properties of individual kernels as they have quite different implications and applications.

The physical handling of bulk grains, in storage and transport, is an important issue in cereal technology. The bulk grain will behave like granular materials. The flow properties of such materials are strongly influenced by the variability of grain size, the porosity between grains, and the elastic properties of the individual whole grains (Anderssen and Haraszi 2009). Important mechanical properties of bulk grain are modulus of elasticity, Poisson ratio, internal friction angle, and apparent cohesion. The measurement of these properties lets the engineer design and optimize storage systems and processing plants of grains (Molenda and Stasiak 2002; Stasiak et al., 2007).

On the other hand, the mechanical properties of individual kernels are more related to quality parameters characteristic of the different cereals. Between the individual kernel mechanical properties that have been studied, we find the modulus of elasticity and the maximum compressive force or the average breaking force. Thus, the relationship between kernel hardness, a quality parameter used for grading cereals like wheat, and some mechanical properties like the maximum compressive stress has been proven (Osborne and Anderssen 2003). As for rice, the average breaking force measured with a bending test has been shown to be correlated to the percentage of fissured kernels at the end of any storage period after drying (Nguyen and Kunze 1984).

13.2.1.1 Measurement of Mechanical Properties of Bulk Grain

The design of equipment for primary silos for granular agricultural material requires knowledge of the properties of stored materials that influence the loads of the structure. Traditional methods of silo design took into account some properties of stored material such as specific weight, internal friction angle, and the grain-to-wall friction coefficient. Recently, numerical methods have been proposed for silo design. The use of this technique implies the consideration of additional material properties such as elasticity modulus or Poisson's ratio (Moya et al., 2002).

There are a number of testing methods for bulk grain property evaluation, and many of them have been adapted from methods used in soil mechanics. For example, Poisson's ratio relates to the strain of a material in the lateral and longitudinal directions. Usual methods for its measurement are triaxial and biaxial tests. However, other methods are available as the one considered by the European Standards (European Committee for Standarization, 1995) or lambdameters proposed by Kwade et al. (1994).

The Young's modulus of elasticity is the ratio of stress to strain when an elastic solid material, like a bulk grain sample, is compressed. The determination of the modulus of elasticity can be carried out using a uniaxial compression tester like an oedometer (Bauer 1992) or the experimental set for the uniaxial compression test proposed by Molenda and Stasiak (2002). A triaxial test can also be performed to determine this parameter.

Acoustic methods have also been proposed for the determination of elastic parameters of bulk cereal grains. Ultrasound velocity measurement has been used to evaluate textural properties of raw materials and products on-line during technological processes in the food industry (Gan et al., 2002). Stasiak et al. (2007) developed an experimental setup, similar to the one shown in Figure 13.1, which generated and recorded acoustic shear waves. With the aid of a digital oscilloscope, the time of propagation of an acoustic shear wave through the sample of grain was measured and the modulus of elasticity was calculated from Equation (13.1) (Timoshenko and Goodier 1970):

$$E = 2\rho V_s^2(1 + \nu) \tag{13.1}$$

where E is the modulus of elasticity (MPa), ρ is the bulk density (kg m^{-3}), V_s is the speed of acoustic shear waves (m s^{-1}), and ν is Poisson's ratio.

Experiments with wheat, barley, corn, oats, and triticale were carried out. The modulus of elasticity was found to increase with a buildup in hydrostatic pressure and decrease with increasing moisture content.

The coefficient of friction, along with bulk density and pressure ratio, are parameters commonly used to calculate loads exerted by grain on storage structures. It depends on the type of grain, its bulk density and moisture content, and roughness

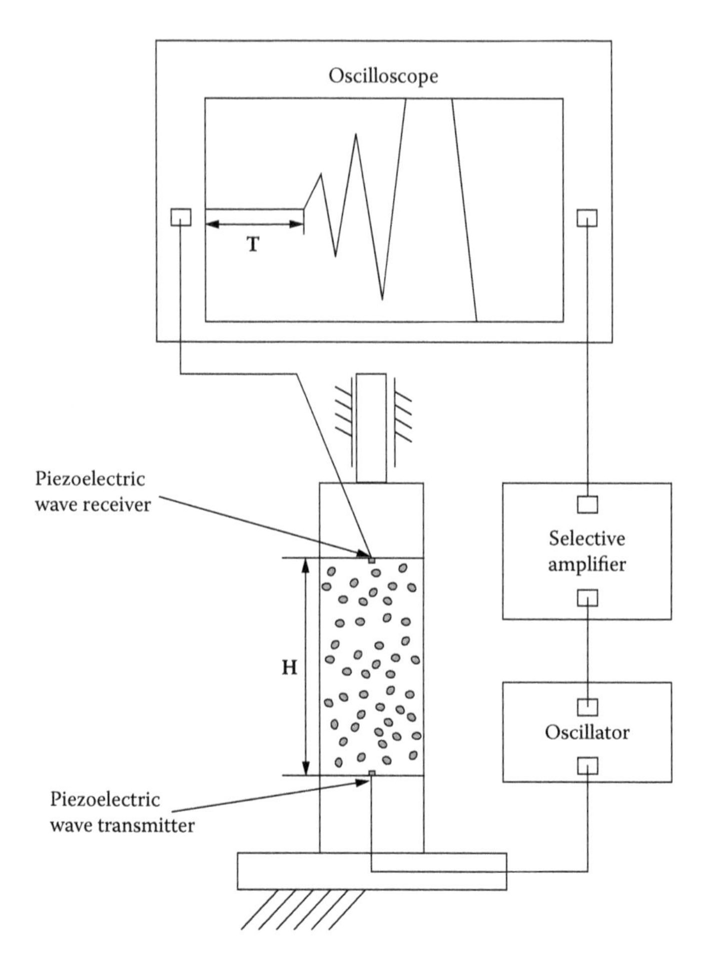

FIGURE 13.1 Experimental setup for generating and recording acoustic shear waves in cylindrical hydrostatic chamber. (Reprinted with permission from Stasiak, M., M. Molenda, and J. Horabik, J. 2007. Determination of modulus of elasticity of cereals and rapeseeds using acoustic method. *Journal of Food Engineering* 82: 51–57.)

of the wall surface (Rusinek and Molenda 2007). The direct shear test has been frequently used to measure frictional properties of granular materials because of its simplicity and versatility. This test creates a shear zone within the sample by the relative movement of one cell with respect to the other.

13.2.1.2 Measurement of Individual Kernel Mechanical Properties

Milling is a process of applying mechanical loads to grain kernels. Therefore, milling quality of any kind of cereal is directly related to mechanical properties of individual kernels. The study of the relation of force to stresses and strains in an individual wheat kernel may be useful in understanding how a wheat kernel is fractured when subjected to forces intended to reduce particle size. The mechanical properties of

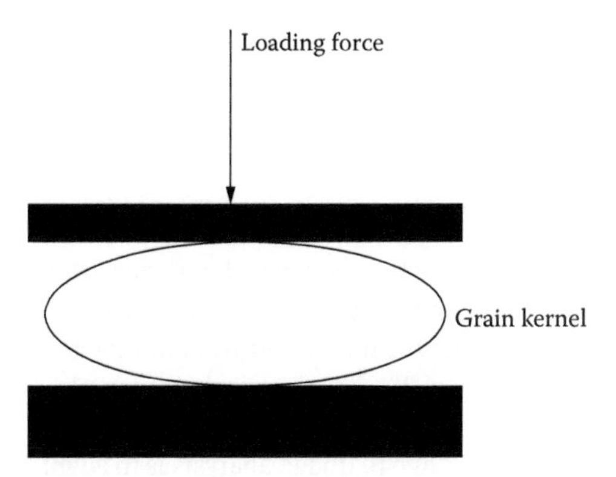

FIGURE 13.2 Schematic of a uniaxial compression test for a single grain kernel. Typical force–deformation curve.

rice kernels are also crucial for understanding the fissuring problem and to optimize drying and milling operations.

The methods used to study mechanical properties of individual grain kernels are the compression test and bending test.

Uniaxial compression tests on single grains are routinely conducted using fully automatic test equipment such as the Instron Universal Testing apparatus. For the standard method (ASAE, 2009), single intact cereal grains are placed between two parallel plates of the lower and upper heads of the compression-testing instrument (Figure 13.2). A constant loading rate of 1.25 mm/min ± 50% is recommended for seeds and grain compression tests. The complete force–deformation curve is recorded through the point of rupture. From the force–deformation plotting and using adequate equations, values of several mechanical parameters, such as modulus of elasticity and maximum compressive stress or bioyield stress/strain, can be calculated. Kang et al. (1995) determined the mechanical properties of five different wheat classes and found that moisture content had a large influence on the modulus of elasticity and bioyield stress and strain values. They proposed using the modulus of deformability at a moisture content of 18% (d.b.) to separate soft and hard wheat classes. Similar conclusions can be drawn from the study of Sayyah and Minaei (2004). The relation between other elastic properties with quality parameters like wheat hardness has also been studied. Osborne et al. (2007) found that the maximum compressive strength measured in wheat kernels using a uniaxial compression test was correlated with the endosperm strength determined from the crush–response profiles that are obtained in the single-kernel characterization system (SKCS).

As for bending tests, the three-point bending test is frequently used for measuring elastic properties of grains. In this test a single grain rests on two supports and a compressing bar moves down between the two supports, bending the grain until it snaps. Deformation is linearly related to applied load up to the break point. From this curve, several mechanical parameters such as as bending strength, modulus of elasticity, or deformation at breakage can be determined. Lu and Siebenmorgen (1995) tried to correlate head rice yield (HRY) with selected mechanical properties obtained from compression and bending tests. They found that compression tests did not accurately reflect HRY, but it was possible to predict HRYs from bending tests because this parameter was closely related to the percentage of rough rice that failed in bending under a certain critical breaking force.

13.2.2 Geometrical Properties: Shape and Size by Image Analysis

In recent years there have been attempts to develop nondestructive, noninvasive sensors for assessing composition and quality of food products. Various sensors such as the charge-coupled device (CCD) camera, ultrasound, magnetic resonance imaging (MRI), computed tomography (CT), and electrical tomography (ET) are used widely to obtain images of food products.

In general, the biggest challenge in image analysis is to adapt existing applications of imaging technology to a specific problem. This requires that we recognize the problem and determine if it can be solved by image processing, know and understand the capabilities and limitations of image processing, and use the previously discussed information on fundamental techniques of algorithms and its implementation.

In some cases the images we acquire are of 3-D objects, such as a dispersion of starch granules or cereal grains for size measurement. These pictures may be taken with a macro camera or a scanning electron microscope (SEM), depending on the magnification required. If care taken in dispersing the particles on a contrasting surface so that small particles do not hide behind large ones, there should be no difficulty in interpreting the results.

Compared with other features such as color and texture, shape is easier to measure using image processing techniques. Frequently, the objects of one class can be distinguished from the others by their shapes, which are physical dimensional measurements that characterize the appearance of an object. Shape features can be measured independently and by combining size measurements.

Among the most common applications of image analysis in the grain sector includes the identification and classification of seeds through their color, size, shape, and texture characteristics.

Sakai et al. (1996) analyzed the effects of rice varieties and polishing methods on the shape of brown rice and polished rice by image processing to investigate the possibility of separating the rice varieties by their differences in shape. Area, perimeter, maximum length, maximum width, compactness, and elongation were measured.

Four parameters were measured:

- *Area:* This is the projected area of a particle on the 2-D plane.
- *Perimeter:* This is the length of the perimeter of a particle's projected image on the 2-D plane.
- *Maximum length:* This is defined as the maximum length of a straight line connected by two points on the perimeter of a particle.
- *Maximum width:* This is defined as the maximum length of a straight line perpendicular to the line of maximum length.

Two shape factors were defined and derived from the dimensions as follows:

- *Compactness:* This is the term defined by Equation (13.2) (takes the value of 1 for a circle):

$$C = \frac{4\pi S}{R^2} \tag{13.2}$$

- where C is compactness (dimensionless), S is the projected area of the sample (mm^2), and R is the perimeter of the sample (mm).
- *Elongation:* This is the ratio of maximum length divided by maximum width.
- *Particle mass:* The mass of each group of 100 particles is measured on an electric balance (minimum reading = 1 mg) after image processing. Particle mass is used for the calculation of the polishing percentage.
- *Polishing percentage:* This is calculated from the differences in particle mass before and after polishing.

The first problem is to select the measuring method (whether two-dimensional [2-D] or three-dimensional [3-D]). The authors have previously reported 3-D analytical results by image processing for soybean seed (Sakai and Yonekawa 1992), but this method is inadequate for a rice particle because of its light permeability. The 2-D measurement has some problems, such as reduced information on a particle, but has merit because the measuring system is simple and many samples can be treated together. Therefore, the 2-D measurement was selected for this study.

The maximum length, maximum width, and elongation were different from the traditional dimensions such as length and width. Separating the rice varieties was possible at a probability level of 95.45% with combined dimensions and shape factors or with single ones.

Shape variation, based on grain morphology, was quantified in 15 Indian wheat varieties by digital image analysis using custom-built software (Shouche et al. 2001). Gray images of the grains of different varieties were captured in a crease-down position using an HP scan jet IICX/T scanner in transparency mode. The software rotates each grain in the captured image for normalization of orientation. These rotated images were used for further analysis. Geometric features such as area, perimeter, compactness, major and minor axis length and their ratios, slenderness, and spread were computed on the binary image. Five other shape factors were derived from these basic geometric features. Moment analysis for calculating standard, central, normalized central, and invariant moments of grains was done using the gray images of grains. The data for each parameter for every grain in the image as well as the mean, standard deviation (S.D.), and standard error (S.E.) for that parameter were stored for further use. The wheat varieties used in this study showed differences in geometric and shape-related parameters. It was concluded that Euclidean distance calculated on the basis of these differences could serve as a basis of distinguishing between samples and for their identification.

More recently, Zapotoczny et al. (2008) applied image analysis for varietal classification of Polish spring barley. The objective of this study was to determine the utility of morphological features for classifying individual kernels of five varieties of barley. Morphological features are based on measurements of linear dimensions (i.e., minimal and maximal length, width, perimeter, and convex perimeter). On the basis of the measurement of basic parameters, shape factors such as aspect ratio, area ratio, circularity, eccentricity, elongation, slenderness, compactness, and corrugation were calculated. The measurements of morphological features also include calculation of the center of gravity and geometrical momentum. Furthermore, coefficients based on the measurement of the distance of the center of gravity from the object perimeter were calculated. In summary, each barley kernel was described using 74 morphological features. The selection was carried out using three methods based on Fisher's coefficient, probability of error, and average correlation coefficient and mutual information.

With the same purpose of inspection and evaluation, Igathinathane et al. (2009) developed a machine vision ImageJ plug-in for Java for orthogonal length and width determination of singulated particles from digital images. A flatbed scanner obtained the digital images of particulate samples. The pixel-march method, which compared pixel colors to determine object boundaries for dimensional measurements, utilized only the ImageJ fitted-ellipse centroid coordinates and major axis inclination. The pixel-march started from centroid objects and proceeded along the fitted ellipse's major and minor axes for boundary identification. Actual dimensions of selected reference particles measured using digital calipers validated the plug-in. The plug-in was applied to measure orthogonal dimensions of eight types of food grains. The plug-in has overall accuracy greater than 96.6%, computation speed of 254 ± 125 particles/s, handles all shapes and particle orientations, makes repeatable measurements, and is economical. Applications of the developed plug-in may include routine laboratory dimensional measurements, physical dimensional characteristics, size-based grading, and sieve analysis simulation for particle size distribution.

13.2.3 Color by Image Analysis

Color is used extensively for grain measurement. Thus, Neuman et al. (1989) developed a color digital image processing workstation to instrumentally evaluate the color of individual cereal grains and other objects identified within digital images. Digital color images were acquired by video digitization of the RGB (red, green, blue) signals produced by a Saticon-type video camera. A low-cost microcomputer frame grabber system was used to control digitization, to perform image segmentation, and to extract color features. Color attributes of cultivars belonging to different wheat classes were examined. In general, significant differences were discerned between varieties of different class (e.g., amber durum and hard red) and some varieties within the same class (e.g., hard red spring varieties Neepawa and Columbus). On the other hand, Casady et al. (1992) developed a trainable algorithm on a color machine vision system for inspection of soybean seed quality. The variables used for classification were color chromaticity coordinates and seed sphericity. Ahmad et al. (1999) developed an RGB color feature–based multivariate decision model to discriminate between asymptomatic and symptomatic soybean seeds for inspection and grading, which comprises six color features including averages, minimums, and variances for RGB pixel values. Ruan et al. (2001) developed an automatic system to determine the weight percentage of scabby wheat kernels, based on color features of scabby kernels captured by machine vision.

On the other hand, Liu et al. (2005) developed an image analysis algorithm based on color and morphological features for identifying different varieties of rice seeds. Seven color features and fourteen morphological features were used for discriminate analysis. A two-layer neural network model was developed for classification. A CCD color camera was used to record images of each rice seed variety. Thresholding is the approach used for image segmentation. Threshold value is generated according the results of the histogram analysis.

13.2.4 Texture by Image Analysis

In image analysis, texture is an attribute representing the spatial arrangement of the grey levels of the pixels in a region. The texture of a segmented area is an important feature for area description, which quantifies some characteristic of the grey-level variation within the object. Among the texture analysis methods for food quality evaluation, most approaches are statistical including the pixel-value run length method. A digital image analysis algorithm was developed to facilitate classification of cereal grains using textural features of individual grains (Majumdar and Jayas 2000). The textural features of individual kernels were extracted from different colors and color band combinations of images. There were 25 textural features used in the discriminant analysis; that is, 10 grey-level co-occurrence matrix features, 12 grey-level run length matrix features, and three grey-level features.

Nevertheless, we can also find in the literature other applications of image analysis, for example, to measure the volume of the grain or to identify damaged kernels. Thus, the work of Ávila et al. (2002) presented in the XIV International Graphic Engineering Congress (Santander, Spain), explored obtaining a three-dimensional computerized reconstruction of a cereal grain from images, which permits measuring the volume of each grain and different parts in it, such as the starch and the embryo. Although it is possible with relative ease to measure the three main dimensions of the grain (length, width, and thickness), it is difficult to estimate their volume from these measurements. The analytical technique used was the single-kernel characterization system (SKCS), which allows for the measurement of individual properties of each grain such as diameter, hardness, density, and so on. To do this, the following activities were carried out: a suitable technique to embed the grain was developed, serial sections were acquired from the embeddings, the section images were digitized, a computerized three-dimensional reconstruction of the kernel was created, and the starch and embryo volume of the model was measured (Figure 13.3).

Steenoek et al. (2001) implemented a computer vision system for corn kernel damage evaluation. Major categories of corn damage in the Midwestern US grain market were blue-eye mold damage and germ damage. The officially sampled Federal Grain Inspection Service (FGIS) provided 720 corn kernels. Inspectors classified these kernels into blue-eye mold, germ-damaged, and sound kernels at an 88% agreement rate. A color vision system and lighting chamber were developed to capture replicate images from each sample kernel. Images were segmented via input of RGB values into a neural network trained to recognize color patterns of blue-eye mold,

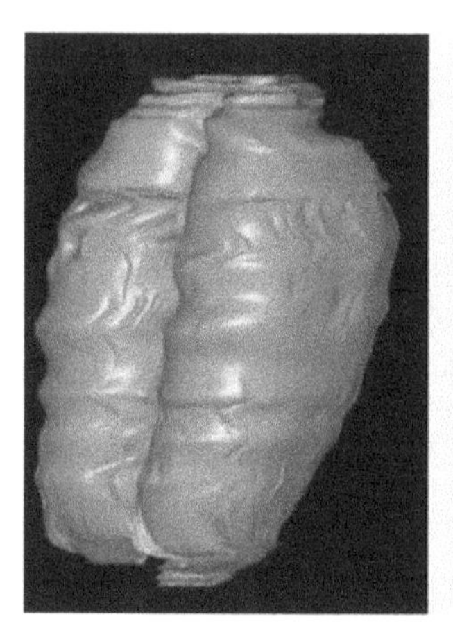
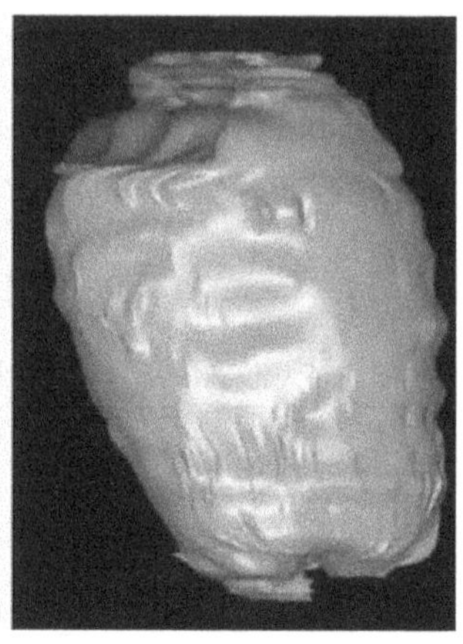

FIGURE 13.3 Complete reconstruction of the grain volume with Microstation Modeler software. The image on the left is the anterior view and the image on the right is the rear view. (Reprinted with permission from Fernando Fadon Salazar Univesidad de Cantabria [Spain], dated: March 16, 2011.)

germ-damaged, sound germ, shadows in sound germ, hard starch, and soft starch. Morphological features (area and number of occurrences) from each of these color group areas were input in a genetic-based probabilistic neural network for computer vision image classification of kernels into blue-eye mold, germ-damaged, and sound categories. Correct classification by the network on unseen images was 78%, 94%, and 93%, respectively. Correct classification for sound and damaged categories on unseen images was 92% and 93%, respectively.

In the cereal industry, the proportion of vitreous durum kernels in a sample is an important grading attribute in assessing the quality of durum wheat. The current standard method of determining wheat vitreousness is performed by visual inspection, which can be tedious and subjective. Thus, Xie et al. (2004) evaluated an automated machine vision inspection system to detect wheat vitreousness using reflectance and transmittance images. Two subclasses of durum wheat were investigated in this study: hard vitreous and of amber color (HVAC) and not hard vitreous and of amber color (NHVAC). A total of 4,907 kernels in the calibration set and 4,407 kernels in the validation set were imaged using a Cervitec 1625 grain inspection system. Classification models were developed with stepwise discriminant analysis and an artificial neural network (ANN). A discriminant model correctly classified 94.9% of the HVAC and 91.0% of the NHVAC in the calibration set, and 92.4% of the HVAC and 92.7% of the NHVAC in the validation set. The classification results using the ANN were not as good as with the discriminant methods, but the ANN only used features from reflectance images. Among all the kernels, mottled kernels were the most difficult to classify. Both reflectance and transmittance images were helpful in classification. In conclusion, the Cervitec 1625 automated vision-based wheat quality inspection system may provide the grain industry with a rapid, objective, and accurate method to determine the vitreousness of durum wheat.

13.2.5 Microstructure by Scanning Electron Microscopy

Recent developments in the microscopy field have changed our understanding of food structures and the types of information we can now expect to glean about them. Electron and light microscopy are the two most commonly used tools for studying food microstructure (James 2009).

A scanning electron microscope uses a finely focused beam of electrons to reveal the detailed surface characteristics of a specimen and provide information relating to its three-dimensional structure.

Ogawa et al. (2003) and Dang and Copeland (2004) used scanning electron microscopy (SEM) to study morphological differences between rice grains of different varieties. Ogawa et al. (2003) compared histological and morphological features in raw and cooked whole grain sections of milled, medium grain California rice using fluorescence and SEM. Milled, raw grains contain fine cracks throughout the endosperm and it is observed that once the grains are cooked, the cracks are wider and more defined. The cracks seem to serve as microchannels for water migration into the grain during cooking. The combination of fluorescence and SEM allowed the authors to determine that dense regions with minimal starch gelatinization in cooked grains were evidently areas with low water penetration. The voided regions were areas of high water penetration. For these reasons, the cracks or absence of them create the unequal uptake of water into the grain during cooking.

Dang and Copeland (2004) used environmental scanning electron microscopy (ESEM) to explore the morphological differences between uncooked and parboiled rice grains. They observed that the sizes of endosperm cells varied between varieties but were not related to the size of starch granules or to the amylose and amylopectin content of the starch. Differences between varieties were also noted in the fracture patterns of grains. The tendency of rice grains to crack (under sun or during milling) is an important aspect of grain quality that has been related to defects in the grain. The interface between the cell wall and the cytoplasm inside endosperm cells is considered to be an area of weakness that makes grains more susceptible to cracking.

Noda et al. (2004), working with wheat grains, evaluated changes in the structure of partially digested starch produced by sprout-induced α-amylase. SEM was used to visualize the surface structure of the starch granules. Wheat starch samples at a significantly delayed harvest were used because they are associated with high activity of sprout-induced α-amylase (Ichinose et. al., 2001). Relatively small changes in surface structure of starch granules were visualized by SEM.

The structure of the vitreous endosperm of raw and popped grains of popcorn maize and sorghum were examined by SEM. Parker et al. (1999), in researching new ready-to-use or ready-to-eat products and for diversified food and allied uses of the foxtail millet, suggested that popping-induced wall fragmentation improves the accessibility of the protein and starch reserves of the endosperm to digestive enzymes. Ushakumari et al. (2004) corroborated that traditional (popping and flaking) as well as contemporary methods (roller drying and extrusion cooking) of cereal processing could be successfully applied to foxtail millet to prepare ready-to-eat products. The changes in the starch granular structure caused by heat treatment were examined by SEM and it could be established that the degree of starch gelatinization was highest in the case of roller-dried millet followed by popped, flaked, and extruded products. Puffed cereals are commonly used as ready-to-eat breakfast foods or as ingredients in snack formulations. The changes associated with the gun puffing process were evaluated for six different grains (common wheat, emmer wheat, rye, barley, rice, and buckwheat). Mariotti et al. (2006) demonstrated that the effect of the puffing treatment is strongly influenced by the morphology and composition of the kernel. Besides, it is well known that moisture content and moisture transfer between bread components are significant factors contributing to bread staling (Gray and Bemiller 2003); thus, the use of low amounts of puffed flours in bread dough could slow the staling phenomenon so that the product would stay softer for a longer period. Significant changes in starch structure and physical properties induced by puffing modify the water-holding capacity of the product. This fact suggested that flours from puffed cereals could be successfully used to control water migration in baked products.

Changes in starch microstructure in soft and hard wheat grains after cooking in a pressure cooker was investigated by Srikaeo et al. (2006), who used SEM and a light microscope (LM) in conjunction with image analysis.

A novel processing technology was developed by Das et al. (2008) to polish rice in a more selective way with the help of xylanase and cellulase enzymes. They used SEM to evaluate the enzyme-degraded surface structure of the rice grain, which allowed easy uptake of water through the bran layers and led to quicker gelatinization of the rice on cooking.

Wijngaard et al. (2007) used SEM and confocal laser scanning microscopy (CLSM) to study how the malting process affected the structures of buckwheat and barley. SEM proved that buckwheat starch is degraded by both pitting and surface erosion. A concentric sphere structure was visible when buckwheat starch granules were partly broken down.

Another interesting use of SEM was presented by Mills et al. (2005). Combining the specificity of antibodies with topographical information from electron microscopy provided information on the distribution of components within wheat grains. Frozen, fractured grains are particularly useful for immunolabeling studies because the aleurone layer, and outer (subaleurone) and inner endosperm can be clearly identified on the fracture face. This technique could develop a more precise biophysical description of the grain and hence predict its milling and baking properties more effectively.

Another example of the use of SEM to evaluate how *Fusarium*-damaged kernels (FDK) are affected (Jackowiak et al., 2005). Examination of the FDK fraction confirmed localization of *Fusarium* hyphae on the surface and inside the tissues of kernels. In addition, endosperm from *Fusarium*-infected kernels revealed the presence of fungal hyphae in the endosperm and a partial or complete lack of the protein matrix, damage to large and small starch granules caused by enzymes, disappearance of small starch granules as the colonization progressed, and complete disappearance of the starchy endosperm under severe infection.

Jaisut et al. (2008) developed a drying process that consists of high-temperature fluidized bed drying, tempering, and ventilation to reduce the glycemic index of brown fragrant rice. The glycemic index is related to the risk of Type 2 diabetes. SEM revealed that during processing, starch granules lost their polygonal shape and partial gelatinization of rice starch took place. During drying and tempering steps, the bonds between amylose and amylopectin molecules in starch granules were relaxed at their gelatinization temperatures, 68–78°C, and leaching of some amylase from starch granules led to partial gelatinization (Sander 1996).

It is possible to investigate ancient cereal cooking practices as Valamoti et al. (2008) did in the microstructure of preserved starch in charred ground cereal remains recovered from prehistoric sites in Greece and Bulgaria. SEM demonstrated that, under some conditions, distinctive cooked starch structure survives the charring process. The morphology of starch granules change predictably according to the processing that they have undergone. Thus, detection of specific starch granule forms can be used to interpret ancient food preparation techniques. This type of study has focused on desiccated cereal foods, where preservation at the macroscopic and microstructural levels is often excellent.

13.3 HYGROSCOPIC PROPERTIES

13.3.1 Equilibrium Moisture Content

In preserving hygroscopic materials, like cereals, the state of water plays a crucial role. The quality of stored grain depends not only on its moisture content, but also on moisture migration during storage. During storage, moisture migration can occur by natural convection of the interstitial air or by water infiltration due to structure leakage. Typically, convection problems occur at the top or bottom of bins where warm moist air may come in contact with cool headspace and ducting surface and condense. Grain will exchange water and interstitial air until the vapour pressure of water in the grain matches that of interstitial air. At this point, the moisture content of the grain will be the equilibrium moisture content (EMC). During the storage of cereals, it is of vital importance to know the equilibrium moisture content of grain because together with temperature, it will determine the environmental conditions of the interstitial storage air, which in turn is the primary factor for favorable or unfavorable conditions for insect or mold development.

EMC can be expressed as a function of relative humidity and temperature of air using the so-called moisture sorption isotherms. These sorption isotherms are graphical representations, for different temperatures, of the relation between the relative humidity of the air and the equilibrium moisture content of grain in equilibrium with this air. Many methods are available for determining water sorption isotherms. These methods are (a) gravimetic, (b) manometric, and (c) hygrometric. The *gravimetric method* involves the measurement of mass changes that can occur and be measured both continuously and discontinuously in dynamic and static systems. The *manometric method* involves sensitive manometers to measure vapor pressure of water in equilibrium with the sample at a given moisture content. The *hygrometric method* measures the equilibrium relative humidity of the air in contact with a food material at given moisture content. The hygrometers used in these methods can be dew point hygrometers that detect the condensation of cooling water vapor or electronic hygrometers that measure the change of conductance or capacitance of hygro sensors (Basu et al., 2006).

Once equilibrium conditions are determined for a given temperature, data can be fitted to different expressions that later can be used to predict equilibrium moisture content of grain if the air temperature and relative humidity are known.

The most common technique to determine the equilibrium moisture content is a static gravimetric method that was standardized and recommended by the Water Activity Group of the European COST 90 project (Wolf et al., 1985). The method uses thermostated jars filled at the bottom with supersaturated salt solutions that maintain the desired air relative humidity. The sample is held at each relative humidity until the weight stops changing and then the sample moisture content is calculated. This method has been extensively used for the determination of sorption isotherms of grains like rough rice (San Martín et al. 2001; Iguaz and Vírseda 2007), barley (Basunia and Abe 2005), or corn (Samapundo et al., 2007). Although the salt-saturated method provides accurate results, it has many associated disadvantages. The time required to reach equilibrium can be large (up to a month). Because of its manual nature, the method is exposed to different error sources like measurement errors derived from the removal of a sample from the storage container for periodic weighing or analytical balance inaccuracies. The long equilibration periods enhance the risk of mould formations or other sample contaminations.

To avoid the drawbacks of the saturated salt solution method, new automated water sorption instruments have been developed. These instruments can conveniently and precisely control both relative humidity and temperature, providing a faster and more robust method for the determination of sorption isotherms of different materials including foods and grains.

Besides these new automated devices, there is another research work focused on the design and testing of different technologies that enable the determination of equilibrium moisture content of grain stored in bins by using temperature and relative humidity sensors.

13.3.1.1 Automated Water Sorption Systems

These instruments, including the Dynamic Vapor Sorption (DVS) instrument from Surface Measurement Systems Ltd., the Aquadyne DVS from Quantachrome Instruments, the IGAsorp from Hiden Isochema, the Cisorp Water Sorption Analyser from C.I. Electronics Ltd., or the Q5000SA from TA Instruments, measure adsorption and desorption isotherms of water vapor accurately with minimal operator involvement. A scheme of automated water sorption equipment is shown in Figure 13.4. With this new technique, a sample is subjected to varying conditions of humidity and temperature by using mass flow controllers, one for dry air and the other for air saturated with vapor. These mass flow controllers adjust the amounts of saturated and dry gas to obtain humidity from 0 to 98%. The weight of samples is constantly monitored and recorded using highly sensitive and stable digital microbalance instrument (0.1 μg weighting resolution). These instruments include a separate temperature-controlled zone where the balance is located, which ensures long-term stability in weight measurements. Humidity and temperature

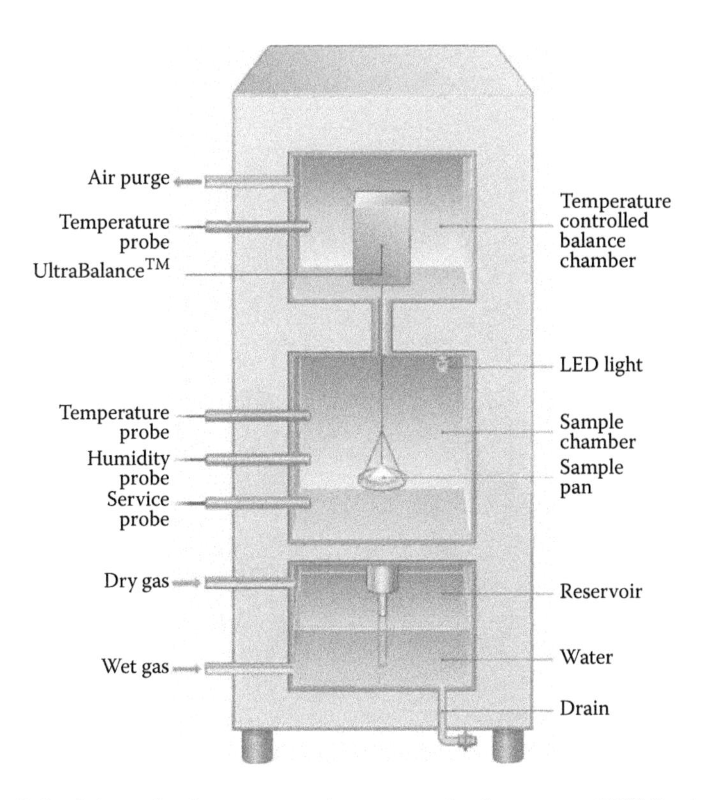

FIGURE 13.4 Schematic of an automated water sorption instrument (DVS Intrinsic from SMS). (Courtesy of Surface Measurement Systems, Ltd.)

sensors are located in the sample chamber, providing continuous indications of the experiment performance.

It is convenient to keep in mind that the true equilibrium between a sample and the humid air surrounding it requires an infinitely long period of time to be achieved. All of these automated water sorption instruments use an apparent equilibrium when weight measurements stop changing by a tolerable level. Increasing the tolerable weight change will speed up the process but may compromise equilibrium moisture content measurement.

13.3.1.2 Equilibrium Moisture Content Measurement of Grain Stored in Bins

Accurate measurement of moisture content of grain is very important in postharvest handling and processing operations such as drying and storage. When the moisture content of grain in storage is maintained below a critical level, the development of microorganisms, insects, and molds can be retarded and chemical activity can be kept stable. Equilibrium moisture content (EMC) of grain stored in a bin can be measured in samples obtained using sampling probes and analyzing them externally, but this is tedious and does not provide measurement in real time.

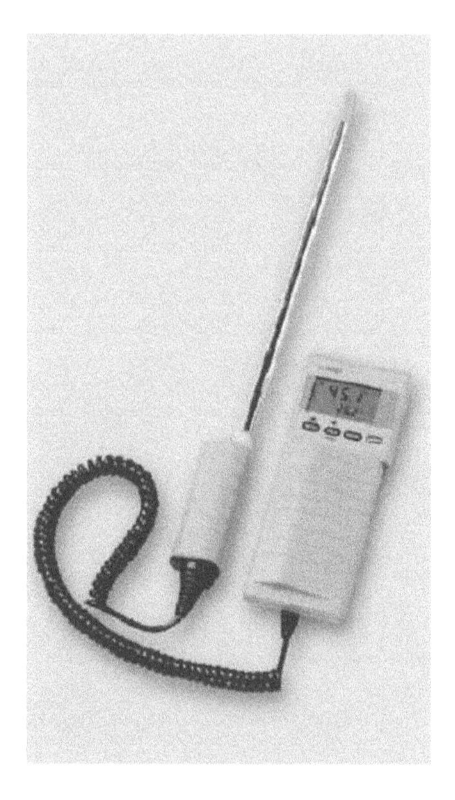

FIGURE 13.5 Relative humidity and temperature sensors that can be used for grain equilibrium moisture content prediction (Model HPM46/HMI41 from VAISALA). (Courtesy of VAISALA.)

The possibility of measuring in-bin moisture content of grain with inexpensive and reliable relative humidity (RH) and temperature sensors, such as the one shown in Figure 13.5, would be desirable for proper grain storage management. The equilibrium relative humidity (ERH) technique consists of the use of sorption isotherms for determining the equilibrium moisture content of a mass of grain by measuring the equilibrium relative humidity and temperature of the interstitial air within it (Young, 1991).

According to Young (1991), some aspects of this technique must be considered: (1) accurate relationships among moisture content, relative humidity, and temperature should be established; (2) grain equilibration with the environment needs to be established when the measurement is made; (3) at high relative humidity levels, a small error in relative humidity measurement causes a large error in grain moisture content; and (4) measurement of relative humidity is subject to considerable error.

A serious handicap of using RH and temperature measurements for quick EMC determination is the slow response time of the sensors. Chen (2001) was the first to investigate whether the use of this technique could be feasible for rapid moisture content control of bulk-stored grain. He determined that a measurement time of 10 minutes was required for the RH and temperature sensors he used (with a stagnant air environment) to equilibrate the grain environment for accurate measurement. By calibrating the RH sensors with saturated salt solutions, the EMC of grain was predicted to within an accuracy of 1% with respect to the moisture content determined by the standard oven method for RH values below 85%.

Uddin et al. (2006) evaluated the contributions of RH and temperature sensor error in predicting the EMC of grain. Using a *stagnant air environment* protected sensor they concluded that temperature error effect was negligible on EMC prediction compared to RH error. They evaluated the RH sensor error contribution on EMC prediction of about ±0.65% for RH between 20% and 70%, regardless of which isotherm equation (Modified Henderson, Chung-Pfost and Oswin) was used. For RH levels above 70% the EMC prediction error increased substantially.

Amstrong and Weiting (2008) presented a new prototype RH and temperature sensor that uses a thinner filter element to protect the sensor and a forced air flow (0.6 m^3/h) inside the sensor to improve the response time of the instrument. The difference between the temperature of the probe prior to insertion into the grain and the grain temperature was proved to have a significant effect on the response and measurement time of the sensor, but in all cases the response time ranged from 4 to 5 minutes.

13.3.2 Water Diffusivity

Postharvest processing of cereals includes several operations such as drying, tempering, or storage. All of these operations involve internal moisture and heat transfer. To design optimum strategies for drying, tempering, and storage operations, a good understanding of these transfer processes is necessary. Many mathematical models have been proposed for the study of grain drying (Iguaz et al. 2004a; Nishiyama et al. 2006), postdrying tempering (Yang et al., 2002), and storage (Iguaz et al., 2004b, 2004c; Aregba and Nadeau 2007). These models assume that moisture transfer to the kernel surface is facilitated by diffusion of water, in the form of liquid or vapor, and were derived using a number of assumptions to simplify these models for computation (Gruwel et al., 2008). For example, all these models assume the initial moisture distribution within the grain kernel to be uniform or a constant water diffusivity coefficient for the whole cereal grain. Recent studies with magnetic resonance imaging have shown that water distribution in seeds is nonuniform prior to and during drying (Ghosh et al., 2006). On the other hand, Ghosh et al. (2007) showed that variations in water diffusivity coefficients in the different grain components can significantly reduce the accuracy of grain drying and storage models.

Traditionally, the water diffusivity coefficient of grains have been determined as a bulk value obtained by the gravimetric method from drying kinetic experiments with a simplified geometry (Iguaz et al., 2003). Novel techniques like magnetic resonance imaging (MRI) allow for the observation of moisture movement and distribution inside intact kernels in a nondestructive and noninvasive way. It also can be used to explain the moisture pattern during operations like drying (Ghosh et al., 2006) and to measure the moisture diffusivity values of different grain components (Gruwel et al., 2008).

13.3.2.1 Magnetic Resonance Image Analysis

Magnetic resonance imaging uses radio waves and powerful magnets to generate images of tissue. A strong magnetic field partially aligns the hydrogen atoms of water molecules in the tissue. A radio wave then disturbs the built-up magnetization, and radio waves are in turn emitted as the magnetization returns to its starting location. These radio waves are detected and used to construct an image (Ghosh et al., 2006).

The acquired MR images are gray-scale representations of the number of protons in the water-containing parts of cereal kernels, which in turn represent water distribution. The brightness of the image is directly proportional to the proton density and hence the moisture content.

Ghosh (2007) used a magnetic resonance imaging (MRI) experimental setup to study the actual physical phenomena that occur during the drying of a wheat kernel, similar to the one shown in Figure 13.6. A 11.7 T (500 MHz) Magnex (Magnex Scientific Ltd., Yarnton, UK) super-conducting vertical-bore magnet equipped with a Magnex SGRAD 123/72/S 72 mm self-shielded, water-cooled, gradient set capable of producing a maximum gradient strength of 550 mT m^{-1} was used for MRI

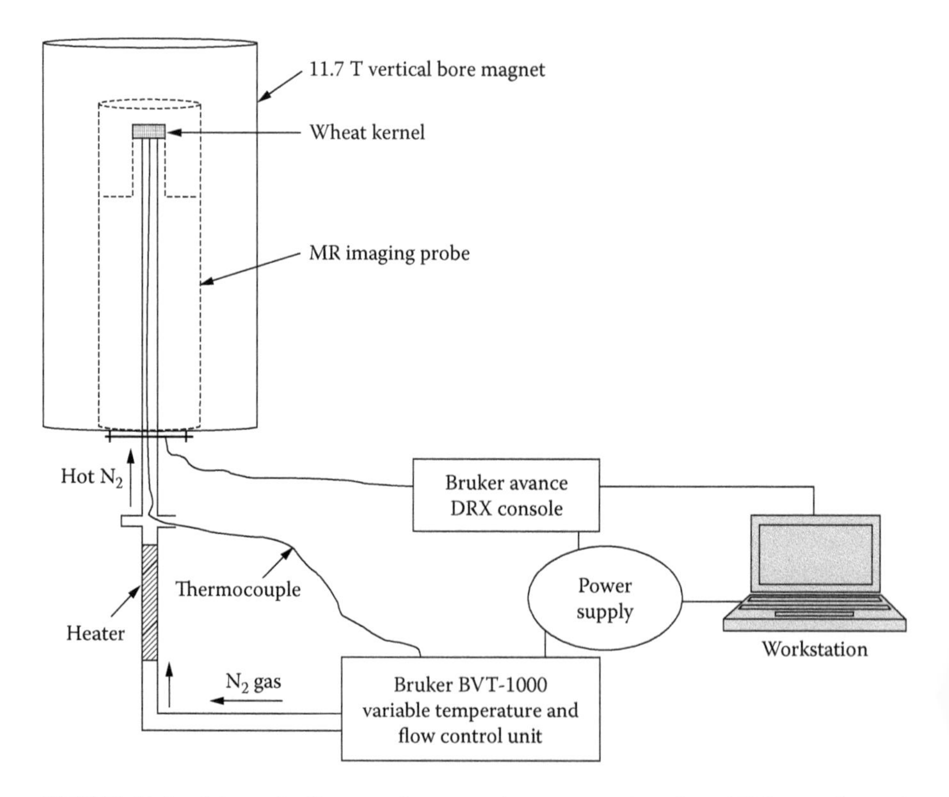

FIGURE 13.6 Schematic diagram of a magnetic resonance imaging (MRI) experimental setup with the dryer assembly. (Reprinted with permission from Ghosh, P.K., D.S. Jayas, M.L.H. Gruwel, and N.D.G. White. 2007. A magnetic resonance imaging study of wheat drying kinetics. *Biosystems Engineering* 97: 189–190.)

experiments. A custom-made MR imaging probe was designed and constructed to specifically fit a single wheat grain. For grain drying, dry nitrogen gas at 30°, 40°, and 50°C was supplied using an in-line variable temperature and flow controller unit. The MR images show the variation in the internal distribution of water prior to and during drying. Water was concentrated mainly in the embryo region and the embryo signal intensity remained high, even after 4 hours of drying at all three temperatures. Using a calibration curve of the MRI intensity of kernels with a known moisture content, drying kinetic curves for the different components of the wheat kernel can be elaborated from MR images. A calibration curve is needed because the magnetic resonance signal intensity is not always proportional to the actual moisture content at different locations of an image.

Diffusion-weighted imaging (DWI) is an application of MRI that targets the random molecular mobility of water in biological tissues and measures the diffusivity values (also known as the apparent diffusion coefficient or ADC) of individual biological components (Thomas et al., 2000). In DWI, instead of the homogeneous magnetic field used in traditional MRI, the homogeneity is varied linearly by a pulsed field gradient. DWI has been successfully used by several researchers to measure moisture diffusivity in cereal grains such as wheat (Gruwel et al., 2008) or corn (Bačić et al., 1992). Ghosh et al. (2009) used DWI to obtain diffusion coefficients of water inside barley kernels obtained values of $2.2 \cdot 10^{-5}$ mm^2/s and $1.0 \cdot 10^{-5}$ mm^2/s for embryo and endosperm, respectively.

13.3.3 Grain Specific Weight

Grain specific weight is usually used as an indicator of grain quality in commercial trading. It measures bulk density of grain and it is defined as the weight of grains needed to fill a container of known volume under specific conditions. In the United States, specific weight is measured in pounds of grain per volumetric bushel (2150 in^3) and in Europe as kilograms per hectoliter. Although there are some instruments on the market that can perform test weight determinations automatically, it is more usual that test weight is determined manually. The apparatus traditionally used for specific weight determinations consists of a hopper equipped with a slide gate supported above a container of known volume. Grain for testing is placed in the hopper. When the gate is opened, grain drops into the container, fills it, and flows over the sides. The operator strikes off the top of the container with a leveling stick and weighs it. These apparatuses need to meet some specifications according to the official inspection regulations of each country.

Reference methods for grain bulk density determination are often tedious, time consuming, and very often limited to the testing of a few samples. That is why some research has been focused on developing other indirect methods that, although not being accepted for official trading purposes, can be used to estimate bulk density with reasonably accurate results.

13.3.3.1 Image Analysis and Neural Network

Öste et al. (1996) proposed a device to estimate quality parameters of cereal kernels, bulk density among them, by producing images of the cereal kernels. It is an automatic and nondestructive method that is carried out quickly and easily (a sample of 100 g takes about 4 min to analyze). The instrument consists of an endless belt provided with transverse grooves where the kernels are properly arranged. A video camera is located above the belt. At one signal the belt stops and the video camera takes an image of the kernels on the belt. The images of the kernels are taken in common visible light. The video camera must take images of the whole sample of kernels, which cover the belt without overlapping and without interspaces. That means that each kernel passing under the video camera will appear in exactly one image and each image will include a plurality of kernels. Once the video camera has taken the kernel pictures, a computer processes the images. Different geometric parameters, such as surface area, length, and width, as well as color parameters, such as values of the intensity of the red, green, and blue color components, are determined. Moreover, the number of kernels included in the analyzed sample can also be determined. The device includes a scale so the thousand kernel weight can easily be calculated. Mean value and standard deviations of each geometric and color parameter are then combined and used as input signals to a neural network. The best results, with a correlation of 0.89 between actual bulk density and bulk density measured with the novel device, were obtained when the following parameters were used as input signals to the neural network: mean and standard deviation of length values, mean and standard deviation of width values, mean and standard deviation of surface values, total sum of the lengths of the sample per unit of weight of the sample, total sum of the widths of the sample per unit of weight of the sample, total sum of the surface areas of the sample per unit of weight of the sample, and thousand kernel weight.

13.3.3.2 Dielectric Properties

Microwave techniques are of growing interest to industrial process monitoring as the methods are noninvasive, nondestructive, penetrative, sensitive, accurate, reliable, easily calibrated, and lead to simple handling and servicing (Krasweski, 1996). Microwave sensor performance is based on the measurement of the dielectric properties of a material, which provides a basis for developing methods of indirect real-time determination of its physical properties.

The electrical properties of materials, known as dielectric properties, are of critical importance in understanding the interaction of microwave electromagnetic energy with those materials as they determine the absorption of microwave energy. The more usual dielectric property is the relative complex permittivity defined by Equation (13.3):

$$\varepsilon = \varepsilon' - j\varepsilon'' \tag{13.3}$$

where the real part ε', or dielectric constant, represents the ability of a material to store the electric field energy and the imaginary part ε'', or dielectric loss factors reflects the ability of a material to dissipate electric energy in the form of heat.

The dielectric properties of most materials vary with several properties. In hygroscopic materials such as foods, the most relevant factor is moisture content. In granular or particulate materials, such as cereals, the bulk density of the air–grain mixture is another important factor that influences dielectric properties. Other factors that affect dielectric properties of materials are the applied alternating electric field, the temperature or the composition, and the structure of the material (Nelson, 1981).

The relationship between dielectric properties and the physical properties of foods and other materials constitute the basis for the development of methods and sensors for the indirect determination of physical properties. Trabelsi et al. (2006) measured the relative complex permittivity of several cereal grains of different moisture content, bulk density, and temperature over a broad frequency range with a pair of horn-lens antennas providing a focused beam and proposed explicit analytical relationships for indirect grain physical property determination. As shown in Figure 13.7, they found that both ε' and ε'' increased linearly with bulk density of samples of wheat, soybeans, and corn at a given moisture content and temperature, which was expected as an increase in bulk density results in an increase of the volumetric mass of water.

They used the complex-plane representation for investigating the variations of ε' as a function of ε''. They plotted the normalized dielectric constant ε'/ρ against the normalized dielectric loss factor (ε''/ρ) and they found that, for each cereal, data corresponding to samples of different moisture content and temperature fell along the same straight line, indicating that moisture content and temperature have interchangeable effects on the dielectric properties. Based on this approach, they presented Equations (13.4), (13.5), and (13.6) for predicting bulk density of each cereal as a function of its dielectric properties and independent of moisture content and temperature.

For wheat (M = 10.4–18% w.b., T = 13–37°C):

$$\rho = \frac{0.4395 \cdot \varepsilon' - \varepsilon''}{1.0981} \tag{13.4}$$

For soybeans (M = 9.7–18.7% w.b., T = 10–38°C):

$$\rho = \frac{0.5729 \cdot \varepsilon' - \varepsilon''}{1.5918} \tag{13.5}$$

For corn (M = 10.7–20.1% w.b., T = –12–23°C):

$$\rho = \frac{0.5086 \cdot \varepsilon' - \varepsilon''}{1.2974} \tag{13.6}$$

The standard error of calibration and relative error for each cereal indicated that the method is as accurate, if not better, than other commonly used methods. Among the advantages of this method, the authors underline the possibility of determining bulk density of a sample without knowing its moisture content or temperature, its potential for real-time on-line sensing of bulk density of grain under dynamic conditions, and its safety of use (Trabelsi et al., 2001).

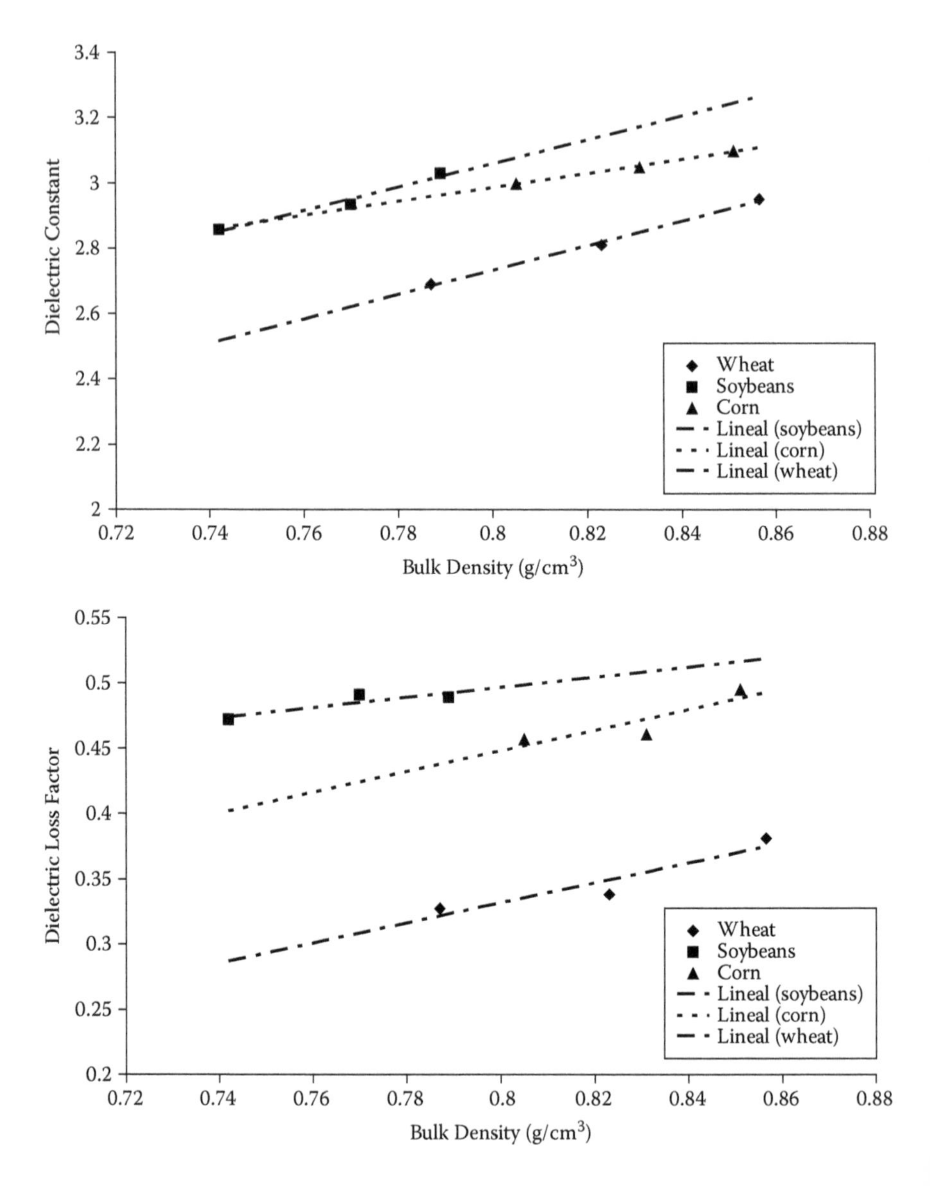

FIGURE 13.7 Dielectric constant (a) and dielectric loss factor (b) as a function of bulk density for wheat (M^1 = 13.9% w.b., T^2 = 24°C), soybeans (M = 14.4% w.b., T = 25°C) and corn (M = 14.0% w.b., T = 23°C) at 7.0 GHz. ([1] Moisture content, [2]Temperature)

13.4 PHYSICAL PROPERTIES IN FLOURS: RHEOLOGICAL DOUGH PROPERTIES

In this field, the most important properties of cereal flour dough are the rheological properties, which determine performance during the various stages in the transformation process and have a great influence on the final quality of cereal-derived products.

There is a wide range of physical test methods designed to measure the dough response to the application of different kinds of deformations or strains. These methods can be divided into two very distinct groups: the techniques that obtain fundamental rheological properties and those that give descriptive empirical measurements of rheological properties.

13.4.1 Empirical Rheological Tests

Empirical rheological measurements must be obtained by specific instruments used under specific procedures that include the way the dough is obtained, the geometry and size of the test sample, and the deformation conditions. Therefore, these measurements are absolutely descriptive and dependent on the type of instrument and test performance condition, which means that properties measured by a particular instrument cannot be directly compared with those obtained with a different instrument, or even with the same instrument if the analysis procedure is not exactly the same.

These instrument devices have been designed by a few companies in several countries including Brabender Instruments in Germany (Farinograph, Extensigraph, Amylograph, Viscograph), Chopin in France (Alveograph, Consistograph, Mixolab), Newport Scientific in Australia (DoughLAB, Rapid Visco Analyzer), and TMCO National Manufacturing in the United States (Mixograph), among others. Most of these instruments are robust and easy to use by personnel who do not require much technical training in order to provide relatively accurate and reproducible results. From recorded data, a wide array of different parameters can be obtained. They are denoted with descriptive terms such as dough development time, stability, extensibility, resistance, work of deformation, or viscosity (apparent), which provide practical information in the evaluation of quality and performance during processing of cereal raw materials and products. Some of these measurements have been approved by different international organizations (e.g., AACC, ICC, International Organization for Standardization [ISO]) and are included as standards for the wheat and flour quality evaluation in the national food legislation of various countries (Table 13.1). All these facts explain why empirical descriptive physical tests are commonly employed by the cereals industry as part of their quality control and research and development (R&D) activities.

Nevertheless, a revision of the scientific literature shows that these methods have been and are extensively used by many researchers in multiple fields of cereal science and technology (Table 13.2). Probably the most extended application of empirical physical tests is the characterization of end-user quality of different wheat genotypes (Bordes et al., 2008; van Bockstaele et al., 2008), including new breeding lines obtained by genetic engineering (Dumur et al., 2010). There also are many references that treat the evaluation of how the addition of different ingredients is suitable to modify the physical behavior of dough and the quality of end products. These include enzymes (Collar et al., 2005; Pham Van et al., 2007), nonstarch polysaccharides and hydrocolloids (Collar et al., 2005; Angionoli and Collar, 2008; Yalcin and Basman, 2008; Sudha and Venkateswara Rao, 2009), oxidants (Junqueira et al., 2007), blends of flour

TABLE 13.1
Main Empirical Rheological Tests

Instrument (Manufacturer)	Matrix	Variable Measured	Test Parameters	Reference Methods
		Mixing Properties		
Farinograph (Brabender)	Wheat	Torque	Water absorption Development time Stability Degree of softening	ICC 115/1 AACC 54-21.01 ISO 5530-1
Mixograph (TMCO National Manufacturing)	Wheat	Torque	Similar to farinograph	AACC 54-40.01
Consistograph (Chopin)	Wheat	Pressure	Stability	AACC 54-50.01
		Deformation Behavior		
Extensigraph (Brabender)	Wheat	Resistance to (uniaxial) extension	Tenacity (P) Extensibility (E) Ratio R/E Energy (area under the curve)	ICC 114/1 AACC 54-10.01 ISO 5530-2
Alveograph (Chopin)	Wheat	Pressure: resistance to (biaxial) extension	Tenacity (P) Extensibility (L) Ratio P/L Work of deformation (area under the curve)	ICC 121 AACC 54-30.02 ISO 5530-4
		Pasting Properties		
Amylograph (Brabender)	Wheat	Apparent viscosity	Amilase activity Gelatinization properties	ICC 126/1 AACC 22-10.01 AACC 22-12.01 ISO 7973 AACC 61-01.01 (Rice)
Viscograph (Brabender)	Starch	Apparent viscosity	Gelatinization properties	ICC 169 AACC 61-01 ISO 7973
Falling number instrument (Perten)	Grain and malted cereals	Time (apparent viscosity)	Amilase activity	ICC 107/1 AACC 56-81.03 ISO/DIS 3093
Rapid Visco-analyser (Newport-Perten)	Grain flours and starches	Apparent viscosity	Amilase activity Gelatinization properties	ICC 161 AACC 22-08.01 ICC 162 AACC 76-21.01 (wheat and rye) AACC 76-22.01 (oat) AACC 61-02.01 (rice)
Mixolab (Chopin)	Wheat	Torque	Mixing properties Gelatinization properties	ICC 173 AACC 54-60.01

TABLE 13.2
Some Approximate Data about the Presence of References to Various Empirical Rheological Test Instruments in the Scientific Literature[a]

Instrument	Total n° of References	Distribution of References over Time (%) 1945–1970	1971–1980	1981–1990	1991–2000	2001–2010	1st Reference Year
Farinograph	1500	5	23	17	17	38	1948
Mixograph	600	1	9	13	26	51	1956
Extensigraph	560	7	24	15	15	39	1945
Alveograph	420	4	21	17	18	40	1952
Amylograph	720	6	25	29	20	20	1945
RVA	700	—	—	1	16	83	1987
Mixolab	40	—	—	—	—	100	2007

[a] Data obtain from searches in the ISI Web of Knowledge (October 2010). Instrument names were used as topic terms of the search. Document types different than "article" were excluded.

with waxy flours, and starches (Gianibelli et al., 2005), among other compounds and additives.

In the following paragraphs, a brief description of the main empirical physical test is presented. These tests are commonly classified in three groups, which can be associated with the different phases in the baking process: (1) the mixing properties to evaluate the behavior of dough during and after its development, (2) the instruments that measure the resistance of dough to extension, and (3) the pasting properties that provide information about the changes in physical properties of dough in conditions that resemble those that take place during the baking process.

13.4.1.1 Mixing Properties

Recording dough mixers are designed to measure the resistance to mixing (consistency) over time, from the moment at which water is added to flour until the dough acquires its maximum development, and subsequently, when it is subjected to overmixing.

The farinograph is the most widely used recording mixer. The instrument records the torque (consistency, in farinograph units [FUs]) as a measure of the opposite deflection of the motor housing caused by the resistance of the dough against the blades in the mixer. The farinograms showed an initial ascending part, until the consistency reaches a maximum value when the dough is fully developed, followed by a second phase of decline due to overmixing and progressive breakdown of the dough structure. From the curve, three main characteristics can be obtained: development time, stability, and degree of softening. Prior to doing the farinograph test, it is necessary to determine the water absorption by flour, defined as the amount of water required for dough to reach a maximum consistency of 500 FU. A strong flour is characterized by high water absorption and a long mixing time required to became a completely developed dough, which maintains its consistency for a long period of time (high stability), until it shows a moderate degree of softening. The standard farinograph mixer has two rotating Z-shaped blades, which submit the suspension of flour (300 g) and water to a defined mechanical stress at constant temperature. It is possible to attach other mixing tools to this mixer, which allows one to work with small sample sizes (50 g and 10 g), with other materials like rye doughs and sponge butters, or to impart intensive mixing and high shear to the dough (resistograph mixer).

The mixograph is another recording mixer that is widely used, particularly in the United States. The interpretation of mixograms resembled that of farinograms. This instrument submits the dough to a more intense mixing action than the farinograph, leading to shorter test times. There are several models for sample sizes of 35 g, 10 g, and 2 g.

In the consistograph, the resistance to mixing is recorded by monitoring the pressure on one side of the mixing bowl. A first test is usually made in which the water absorption at a target pressure is calculated on the basis of the maximum peak pressure reached when dough is obtained with an amount of water based on the initial moisture content of the flour. In a subsequent test performed at the hydration level previously determined, physical properties of the wheat flour dough are determined. Finally, it must be noted that the alveo-consistograph joins the alveograph and consitograph utilities together in a single piece of equipment.

The most recent dough recorder mixer is the doughLAB, which is designed for small sample sizes (4 g), and can work at high and variable speeds and at different temperatures (controlled in an external bath). It can be used for the evaluation of flour from soft, hard, and durum wheat, and other grains. With specific conditions, test time can be reduced to 10 minutes, and data can be correlated with standard methods.

13.4.1.2 Extensional Measurements

Once the dough is formed, the second phase to evaluate baking quality is extensional measurements. It is concerned with the behavior of dough against extension forces, in order to simulate the expansion of gas during fermentation and early baking stages. There are several instruments designed to measure the resistance of dough to extension: the extensograph (uniaxial extension) and the alveograph (multiaxial extension) are the most widely used.

In the case of the extensograph, the dough is previously developed in a farinograph until it reaches a 500 FU consistency (AACC Method 54-10) or for a fixed 5-minute period (ICC Standard 114). After that, the dough is divided into two test pieces. Each cylindrically shaped piece is transversally stretched until it breaks. The extensogram represents the exerted force versus stretching length (time). The main empirical parameters calculated from extensograms are: resistances to extension at 5 cm and maximum, extensibility or distance from the beginning of test to the dough rupture, the ratios of resistances to extensibility, and the area below the curve, which is proportional to the energy required to stretch the dough. Usually, for each flour sample, the extensograph test is performed at three different proving times (45 min, 90 min, and 135 min). Besides the application of the standard

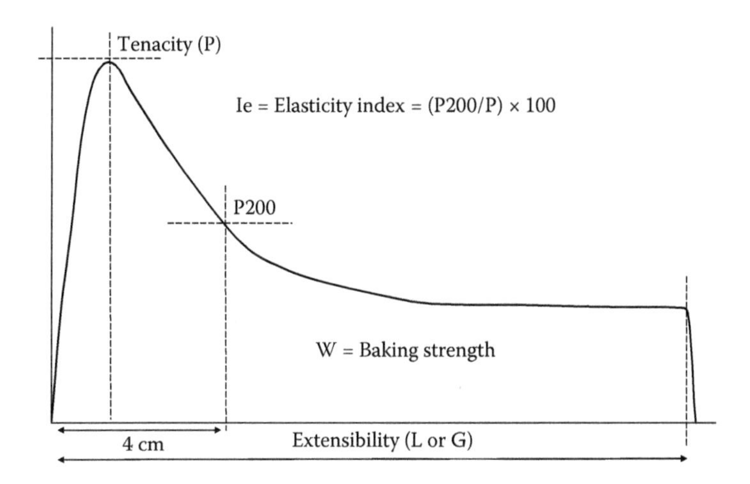

FIGURE 13.8 Alveogram.

methods, the extensograph is very useful to evaluate the influence of flour additives on the dough characteristics.

The alveograph is probably the most empirical rheological instrument used in Europe. The alveographic test is an official standard method to evaluate rheological properties of dough in the EU countries (European Standard CEN/TS 15731). While in the extensograph, the dough is stretched in one direction; in the alveograph a three-dimensional extension is produced by blowing air below a circular piece of dough. As a result, the dough expands into a bubble, which seems to simulate well the expansion of the dough gas cells during fermentation and rising while baking. The alveograms are obtained by monitoring the air pressure under the bubble until it breaks. The interpretation of the main parameters obtained from the alveogram (Figure 13.8) is similar to that of the extensogram: tenacity (P) or maximum pressure, extensibility (E) or length of the curve, baking strength (W) calculated from the area under the curve, and configuration ratio of the curve (P/L). Table 13.3 lists the alveographic characteristics of several different flours produced by the Spanish milling industry and the end products for which these flours are recommended.

13.4.1.3 Pasting Properties

During baking, physical properties of dough change drastically, partially due to the gelatinizing of starch and its hydrolysis by amylolytic enzymes. These changes can be simulated by several instruments that work in a similar fashion: A suspension of flour and water is controlled, heated, and its apparent viscosity is continuously measured. During heating, two opposite phenomena take place: The gelatinization process causes the progressive rise of the suspension viscosity, and gelatinized starch becomes highly susceptible to amylolytic enzymes, particularly to α-amilase, which is much more thermostable than β-amilase. Hydrolysis of the starch diminishes the suspension viscosity. Therefore the maximum viscosity can be used to estimate the amylolitic activity of the flour. These variables are inversely related; the higher the amylolitic activity, the lower is the maximum viscosity.

TABLE 13.3
Alveographic Parameters of Several Flours and Recommended End Products

Flour	P/L	W (kJ × 10^{-4})	General Description	Recommended End Products
Flour 1	0.20–0.30	85–110	Very soft flour, very extensible	Cookies, biscuits, soft cakes Industrial bread (short fermentation)
Flour 2	0.30–0.40	130–150	Soft-Medium force flour Very extensible	Cakes Industrial bread
Flour 3	0.45–0.55	180–210	Medium force flour with good stretching properties	Artisanal bread (long fermentation) Controlled fermentation
Flour 4	0.60–0.70	200–220	Strong flours	Artisanal long fermentation and high volume breads
Flour 5	0.65–0.85	220–240	High tenacity Medium stretching properties	
Flour 6	0.7–0.90	310–350	Very strong flours	Hamburger
Flour 7	0.9–1.20	350–380	Very high tenacity Little stretching properties	Panned bread and toasted bread Frozen bread and yeast-leavened dough Puff products Improver flour for blending

The amylograph was the first equipment developed for the analysis of pasting properties of flours. The standard method (AACC 22-10.01) for the measurement of α-amylase activity in wheat flour is widely used. It fixed the temperature increase of the flour and water suspension at 1.5°C/min (Figure 13.8). The maximum viscosity is called the *amylograph value*, *peak viscosity*, or *malt index*. Flour with a normal enzymatic activity usually has an apparent viscosity of 400–600 AU (amylograph units). Values below and above this limit will indicate strong and weak amylolitic capacities, respectively. This standard procedure must be modified according to method AACC 22-12.01 in the case of the analysis of wheat flour supplemented with fungal α-amilase.

The Rapid Visco Analyzer (RVA) and the viscograph are able to measure the rheological behavior of starch and products containing starch of any origin: cereals (Hagenimana and Ding, 2005), tubers (Svegmark et al., 2002), and legumes (Liu and Eskin, 1998). In these instruments, sample viscosity is recorded during a temperature program that commonly includes three phases: heating, holding, and cooling. The temperature profile—heating and cooling rates (between 0.5 and 3°C/min in the case of the viscograph, and until 14°C/min in the RVA), holding time, and temperature—can be modified according to the application requirements. From curves, different parameters can be obtained: temperature at which gelatinization starts and finishes, maximum or peak viscosity during heating, trough or minimum viscosity during the holding phase at maximum temperature, breakdown or difference between peak viscosity and trough, and final viscosity at the end of cooling. The setback values—setback viscosity minus peak and/or trough viscosities—are indicative of the retrogradation tendency of starch (Karim et al., 2000). It is outstanding that RVA is the rheological descriptive measurement instrument probably most cited in the scientific literature (Table 13.3) in the last decade. It has been widely used, not only in the study of flours and starches of any origin, but also in the analysis of hydrocolloids, dairy products, protein isolates, snacks, ready-to-eat cereals, and feed products, among other food products and ingredients.

Finally, the falling number (FN) must be cited. It is defined as the time in seconds required to stir and then allow the stirrer to fall a measured distance through a hot flour aqueous solution that is undergoing liquefaction (AACC method 56.81.03). As peak viscosity in the amylograph test, FN is inversely correlated with α-amylase activity. Because of its simplicity, this test is widely employed for grain and flour quality control in many milling, malting, baking, and pasta industries.

13.4.1.4 The Mixolab

The Chopin Mixolab is the last available piece of laboratory equipment of the empirical rheological testing device family. The main innovation in this instrument is that it allows the characterization, in a single test, of both the mixing and the pasting properties of dough. Flour is placed in the bowl and water required for optimum consistency is added. The instrument measures the torque (expressed in Nm) produced by the passage of the ingredients between the two kneading arms. The temperature of the bowls is controlled and programmed. Usually, the typical temperature profile shows five phases: mixing is made at a constant temperature (i.e., 30°C), follow by heating, holding at maximum temperature (i.e., 90°C), cooling, and holding at the end temperature (i.e., 50°C). Rosell et al. (2010) detailed the physical changes of dough in the Mixolab test, distinguishing six stages (Figure 13.9). The two initial stages allow for the evaluation of the mixing properties of the dough. Torque rapidly rises until the dough is fully developed (point C1). Then, a period of almost constant torque occurs that determines the dough stability. Overmixing causes a decrease of the torque, due to the weakening of the protein network. In the third stage, when heating is applied, protein destabilization and unfolding start, and an acceleration of the torque declines until a minimum value (C2) is observed. This is followed by a drastic increase (stage 4: C2–C3) in the torque in which the viscosity increment caused by starch gelatinization has the major role. When mechanical stress and temperature constraints lead to the physical breakdown of the starch granules, viscosity and torque are reduced (stage 5: C3–C4), which can be related to the cooking stability of starch. Stage 6 (C4–C6) corresponds to the final rise of the torque associated with the retrogradation and reordering of the gelatinized starch molecules that take place when temperature diminishes. Rosell el al. (2010) compared Mixolab mixing and pasting properties of fiber-enriched bread doughs with those obtained with the farinograph and Rapid Visco Analyzer, respectively, and found a broad range of significant correlations. Good correlations were also obtained between Mixolab and the alveo-consitograph (Ozturk et al., 2008), alveograph (Koksel et al., 2009), and extensograph results (Cato and Mills, 2008). With regard to pasting properties, it must be noted that while in traditional devices, such as the amylograph, viscograph,

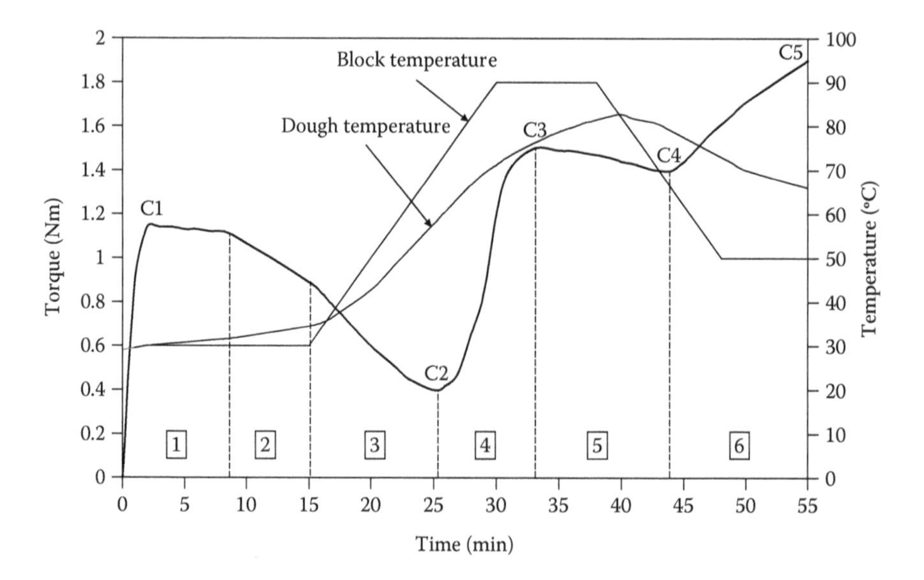

FIGURE 13.9 A typical mixolab curve.

and Rapid Visco Analyzer, tests are performed in flour slurries, in the Mixolab, torque measures are made in a dough system with limited water content (Rossel et al., 2007). This suggests that a better simulation of the modifications undergone during real baking should be obtained in Mixolab than in the traditional pasting test devices. Therefore, this new device seems to be a promising tool for the rheological assessment of bread dough. Two years after its introduction, the use of the Mixolab has been approved as a new AACC International Method (54-60.01).

13.4.2 Fundamental Rheological Measurements

Although empirical rheological tests are very useful for providing practical information for the cereal industries, they are not sufficient for determining the fundamental behavior of dough processing and baking quality (Yihu and Qiang, 2007). This limitation must be overcome by the determination of fundamental rheological measurements that, by definition, should be tractable and universal, and independent of size, shape, and instrumental conditions.

Dobraszcyk and Morgenstern (2003) give a complete and detailed description of the principles, applications, advantages, and limitations of the fundamental rheological measurements used in dough testing. They distinguished three main types of tests: dynamic oscillation, creep and relaxation, and extensional flow measurements.

According to these authors, the most preferred are the dynamic oscillation tests, which provide information about the viscous and elastic components of dough, measuring its response to the application, over time, of sinusoidally oscillating stress or strain. The unique viscoelastic behavior of wheat dough is primary caused by the presence of gluten proteins. Dynamic oscillating testing has been applied to study mechanical properties of gluten previously extracted from dough. This approach has become successful for examining the molecular basis of dough and gluten properties (Yihu and Qiang, 2007). Nevertheless, conventional shear rheological tests are usually made under small deformation conditions that are far from those observed during actual mixing, fermentation, and baking, and therefore, they are not appropriate in predicting end-use quality. This criticism can be extended to most of the methods that have been used to measure the fundamental rheological properties of dough in extension (Dobraszcyk and Morgenstern, 2003).

Relaxation measurements have been useful not only to study the structure and properties of gluten proteins, but have also been found to be significantly correlated with baking quality parameters and empirical rheological measurements (Van Bockstaele et al., 2008).

13.5 PHYSICAL PROPERTIES IN CEREALS PRODUCTS

13.5.1 Texture in Bakery Products

In the mid-1960s, texture profile analysis (TPA) was created at General Foods as an imitative test that purports to provide standardized values of food texture. Great advances in the techniques and equipment have been achieved in the last 30 years. As a result, a convergence between instrumental physical measurement and techniques of sensory science are set.

TPA has been used at length to study the texture changes in dough and bakery products (bread, muffins, sponge cake, or noodles) made of different cereals such as wheat, rice, or oats. However, apart from TPA, there are other tests such as tensile deformation (Gujral et al., 2008), extrusion test in batter (Sciarini et al., 2010), and cut test in bread (Giannou and Tzia, 2007). Lostie et al. (2002) also investigated textural evolution based on photography and image analysis to evaluate pore distribution. Most of these tests are compared with results obtained from sensory evaluation.

The technological quality of bread has been investigated with respect to dough texture properties (springiness, hardness, cohesiveness, adhesiveness, chewiness, etc. Armero and Collar (1997) studied the effect of ingredients, the bread-making process, and antistaling additives. Giannou and Tzia (2007) developed research to determine the effects on textural behavior and quality of frozen dough bread and to establish a prediction of final product characteristics.

There is an interest on the effect of frozen storage time in bread that is partially baked (Bárcenas and Rosell 2006); fully baked and frozen, or partially baked and frozen (Curic et al., 2008); or its effect in products such as partially baked chapati (Gujral et al., 2008).

TAP was also utilized to evaluate texture attributes of batter and bread quality in a gluten-free product (Miñarro et al., 2010; Sciarini et al., 2010).

There are some studies with muffins where texture properties are evaluated; for example, the addition of alternative ingredients to improve nutritional, textural, and sensorial parameters, and the quality of the product (Baixauli et al., 2007; Jisha et al., 2010).

Katagiri and Kitabatake (2010) studied the rheological properties of traditional Japanese wheat noodles where a compression test was used to compare hand-stretched process with machine-made noodles. In the work of Nakamura et al. (2010), a continuous progressive compression (CPC) method was put into practice in high-quality and bio-functional wheat and rice bread and noodles.

13.5.2 Microstructure by Scanning Electron Microscopy

SEM has the particular advantage of providing great depth of field. These features turn this kind of microscopy into a useful tool to evaluate microstructures in food and ingredients due to comparative information obtained on the gross size and shape of the particles, and details of the internal microstructure can be determined. The form and interaction of individual components, including crystalline inclusions and air bubbles, can be examined; for example, in different types of bakery products. It is possible to reveal the form of the starch, in particular the intactness of starch granules together with the levels of association of the protein strands. This can often be related to the way in which the product is likely to break down in the mouth (Webb and Holgate, 2003).

Kàlab et al. (1995) mentioned that imaging techniques can be used to help evaluate changes produced by processing in terms of morphology and composition. Most foods are of biological origin, but are processed to varying degrees, sometimes to such an extent that their biological origin is not readily apparent, for example, grain versus bread or milk versus cheese. Visual changes due to processing (e.g., milling of grain and gelatinization of starch) are the result of changes at the microscopic and molecular levels.

SEM is a relatively straightforward technique in its preparation of dry materials, and for this reason it is widely used for the characterization of powders (flour, sugars, tea, or coffee) (Webb and Holgate, 2003). A new kind of SEM has been developed to satisfy research needs. Conventional SEM works with dried samples and operates at high vacuum. There is another alternative in which the sample is frozen below –80°C called cryo-SEM. However, there are delicate hydrated or volatile samples often present in food-systems, which are not vacuum tolerant. For this reason, low-vacuum SEM (LV-SEM) or variable-pressure SEM (VP-SEM) and a saturated water vapor environment called environmental SEM (ESEM) were developed (Kaláb et al., 1995; Webb and Holgate, 2003).

13.5.2.1 Bakery Products

To determine the nature of pore spaces in baked breads, various heating modes such as microwave-infrared (MIR), microwave-jet (MJET) impingement, and jet (JET) impingement are used. Datta et al. (2007) combined several novel and old techniques (liquid extrusion porosimetry, scanned image analysis, pycnometry, the volume displacement method, and SEM) to characterize in terms of total porosity, fraction of closed, blind and flow-through pores, and pore size distributions. One technique cannot cover the large range of pore sizes, total porosity, and flow-through versus closed pores. A significant fraction of the pores were found to be closed. Breads baked in JET had the highest total porosity followed by MJET and MIR. According to SEM analysis, breads baked in JET ovens looked quite different than the ones baked in other ovens. It was also presented that the crust region of the bread had smaller pores, which were close to each other compared to the crumb region.

Also investigating how temperature affects bakery products, Sanchez-Pardo et al. (2008) utilized imaging, light microscopy, and SEM to compare the microstructure of crumbs from pound cakes baked in a microwave or conventional oven. SEM was used to observe and compare, in 3-D, the starch granules and protein matrix in crumb cake. The conclusion was that conventionally baked products had a greater amount of protein matrix throughout; the matrix structure of the crumb was comparable between microwave and conventionally baked pound cakes. Turabi et al. (2010) stated the fact that SEM has been widely used to study qualitative changes that occur during baking. However, there is no study in which quantitative information was obtained from SEM images of bakery products. Therefore, Turbai et al.'s research was designed to obtain quantitative and qualitative information on the macro- and microstructure of gluten-free rice cakes containing xanthan and xanthan-guar gum that were baked in conventional or IR-MW combination baking. First, more porous cakes were obtained when a xanthan and xanthan-guar gum blend was used. Besides, Turabi et al. (2010) found, unlike Sanchez-Pardo et al. (2008), that the baking method in this case could affect porosity of the cakes and formation of pores having a larger area. The microstructure of baked cakes was also different; that is, there were more deformed starches in conventionally baked cakes than in the cakes baked in an IR-MW combination oven. They reached the conclusion that quantitative analysis of SEM micrographs was possible.

On the opposite side of the thermometer, that is to say, in processes of refrigeration and freezing, some studies also exist in which SEM has been used. For example, Yi and Kerr (2009) compared the effects of freezing temperature and rate as well as storage temperature and time on the quality of frozen dough. Yeasted bread dough was analyzed by cryo-SEM, which showed that dough stored at –30°C and –35°C had the least damaged gluten network. The main conclusion was that faster freezing and lower storage temperatures promote less damage to the gluten network, and thus help retain the elastic properties of the dough. On the other hand, relatively lower freezing rates and storage temperatures promote yeast viability and gassing power. In addition, prolonged storage times lead to detrimental changes in both gluten structure and yeast viability.

A study was carried out by Ben Aissa et al. (2010) to better understand the impact of the baking time on the contraction of the crumb during chilling after baking and during freezing. SEM pictures showed for longer baking times, the starch granules were fully gelatinized and no ghosts of starch granules were visible. The magnitude of the contraction was thus associated with the degree of baking and with the degree of starch granule destructuration. Results showed that a longer baking time resulted in a lower contraction of the bread.

Microencapsulated high-fat powders in bread and its effect on the structure compared with commercial bread with partially hydrated vegetable fat was studied by O'Brien et al. (2000). SEM showed differences in structure between the standard bread, bread with commercial vegetable fat, and breads containing microencapsulated fat.

Another study considered the effect of an ingredient to improve the quality of the bread, which was developed by Bárcenas and Rosell (2005). In this case, the effect of adding hydroxypropyl methylcellulose (HPMC) in a basic bread formulation was analyzed in terms of microstructure, as a bread improver and antistaling agent. The microstructure was analyzed by cryo-SEM. The gas cell wall of the control showed a strong connection among all the components, forming a complex structure with numerous cavities. In opposition, the gas cell walls of the crumb containing HPMC showed a smooth structure with fewer cavities; the underneath components were not easily envisaged. The microstructure analysis revealed the possible interaction between the HPMC and the bread constituents, which could partially explain the antistaling effect of this hydrocolloid. The results confirm the ability of HPMC to improve fresh bread quality and for delay staling.

Cryo-SEM was also used to evaluate the effect of an automatic dosing unit compared to a manual dosing unit on the rheological, microstructural, and textural properties of an aerated batter for preparing a bakery product. The batters' microstructure revealed many changes that could be related to the dosing step (Baixauli et al., 2007).

Saleem et al. (2005) determined the physical and mechanical properties required to develop a numerical solution procedure capable of predicting the stress state of a semisweet biscuit during the postbaking process, because rich tea biscuit varieties can sometimes develop cracks up to a few hours after baking. In the electron micrographs of the cross section they observed that the commercial biscuits had a more open structure than pilot-scale biscuits, consisting of gaps between the material layers. This implies that the mobility of water will be highest in commercial biscuits due to their highly porous structure.

Indrani et al. (2010) evaluated the effect of replacing wheat flour with a 5%, 10%, 15%, and 20% multigrain mix (MGM) (soya bean, oats, fenugreek seeds, flaxseed, and sesame seeds) on rheological and bread-making characteristics of wheat flour. SEM evaluation of wheat flour dough showed that the protein matrix formed a smooth, enveloping, veil-like network stretched over the starch granules. The results showed that bread with improved quality characteristics and perceptible multigrain taste could be produced by adding 15% MGM and a combination of additives.

There is another interesting study investigating the hypothesis that contamination was present in ingested food. Gatti et al. (2009) developed, within a European project called Nanopathology, a new diagnostic tool for determining the presence of inorganic materials in wheat bread and wheat biscuits from different countries. Inorganic, micro-, and nanoscaled contaminants were analyzed by means of ESEM. The results indicate that 40% of the samples analyzed contained foreign bodies such as ceramic and metallic debris of probable environmental or industrial origin.

13.5.2.2 Other Products and Technologies

13.5.2.2.1 Spaghetti and Noodles

Heneen and Brismar (2003) studied the structure of cooked spaghetti containing durum and bread wheat using LV-SEM. This research achieved characterization of the different regions of spaghetti in unprecedented detail. The structural features provided novel information on the changes accompanying starch gelatinization. The changes were expressed in the swelling of starch granules and the appearance of voids around them, changes in the internal structure of starch granules, fusions between neighboring granules with discontinuities in the protein matrix, and final starch deformation, segmentation, and possible retrogradation close to the surface.

The effect of fermentation of whole polished rice grains on the physical properties of rice flour and the rheological characteristics of rice noodles were investigated by Lu et al. (2005). SEM showed slight superficial corrosion in fermented rice starch granules. The fermented samples had very shallow pits but the control samples had no pits. The presence of the pits in the starch granules seems to indicate some breakdown of the starch.

13.5.2.2.2 High Hydrostatic Pressure

Celiac disease requires research about the use grains as alternatives to wheat or rye. Hüttner et al. (2009) investigated the effects of high hydrostatic pressure (HP) on oat batters (10 min at 200, 300, 350, 400, or 500 MPa). SEM and bright field microscopy were used to observe the effect of high HP on microstructure. The results indicated that high HP had a significant effect on oat batter microstructure, and both starch and proteins were affected. In addition, batter viscosity and elasticity were significantly improved, and changes became more evident at high pressures (500 MPa), where SEM revealed that the starch granules of treated batters became swollen and slightly disintegrated. Overall, the majority of oat starch granules retained their granular structure, but significant changes in their surface appearance were visible. Hüttner et al. (2010) described the use of high HP as a tool to improve the bread-making performance of oat flour by adding treated samples (200, 350, and 500 MPa) to an oat-bread recipe, replacing 10%, 20%, or 40% of untreated oat flour. Bread analysis revealed significantly improved bread volume, and upon addition of 10% oat batter treated at 200 MPa, the staling rate was reduced in all breads containing oat batter treated at 200 MPa. The improved bread-making performance was attributed to the weakening of the protein structure, moisture redistribution, and possibly changed interactions between proteins and starch nce. Oat batters treated at pressures greater than or equal to 350 MPa produced deteriorated bread quality. This technique might also be beneficial for the production of other freshly baked gluten-free breads, which are predominantly starch based and therefore characterized by fast staling.

Other research alternatives as with respect to gluten replacement for gluten-free products, Vallons et al. (2010) investigated the application of high-pressure processing of sorghum batters (200 to 600 MPa at 20°C) in the production of sorghum breads, and the microstructure was investigated using SEM. At pressures over 300 MPa, the batter consistency increased and treatment with pressures greater than or equal to 400 MPa resulted in microstructural changes. The starch granules became swollen and deformed, which was also observed by Hüttner et al. (2010). However, at 600 MPa, their granular structure remained intact. Furthermore, freeze-dried sorghum batters treated at 200 MPa (weakest batter) and at 600 MPa (strongest batter) were added to a sorghum bread recipe, replacing 2% and 10% of untreated sorghum flour. The results showed that breads containing 2% or 10% at 200 MPa were not significantly different from the control bread. However, the quality of breads containing 2% at 600 MPa delayed staling.

A study of the effect of high HP in wheat batters was performed by Bárcenas et al. (2010). They developed analyses in microbiological, color, mechanical, and texture surface parameters. SEM was utilized and suggested that proteins were affected when subjected to pressure levels higher than 50 MPa, but starch modification required higher pressure levels, as Vallons (2010) and Hüttner (2010) observed in oat and sorghum. High HP–treated yeasted doughs led to wheat breads with a different appearance and technological characteristics. This study suggests that novel textured cereal-based products treated with high HP in the 50–200 MPa range could be obtained.

REFERENCES

AACC (American Association of Cereal Chemists). *Approved Methods of Analysis*, 11th ed. (online), http://www.aaccnet.org/ApprovedMethods/.

AACC Method 56-81.03. Determination of Falling Number. AACC International Approved Methods. 11th edition (online): http://www.aaccnet.org/ApprovedMethods

AACC Method 54-60.01. Determination of Rheological Behavior as a Function of Mixing and Temperature Increase in Wheat Flour and Whole Wheat Meal. AACC International Approved Methods. 11th edition (online): http://www.aaccnet.org/ApprovedMethods

AACC method 22-10.01. Measurement of alpha-Amylase Activity with the Amylograph. AACC International Approved Methods. 11th edition (online): http://www.aaccnet.org/ApprovedMethods

Ahmad, I.S., J.F. Reid, M.R. Paulsen, and J.B. Sinclair. (1999). Colour classifier for symptomatic soybean seeds using image processing. *Plant Disease* 83: 320–327.

Anderssen, R.S., and R. Haraazi. 2009. Characterizing and exploiting the rheology of wheat hardness. *European Food Research and Technology* 229: 159–174.

Angioloni, A., and C. Collar. 2008. Functional response of diluted dough matrixes in high-fibre systems: A viscometric and rheological approach. *Food Research International* 41: 803–812.

Aregba, A.W., and J.P. Nadeau. 2007. Comparison of two non-equilibrium models for static grain deep-bed drying by numerical simulation. *Journal of Food Engineering* 78: 1174–1187.

Armero, E., and C. Collar (1997). Texture properties of formulated wheat doughs: Relationships with dough and bread technological quality. *European Food Research and Technology* 204(2): 136–145.

Armstrong, P.R., and M. Weiting. 2008. Design and testing of an instrument to measure equilibrium moisture content of grain. *Transactions of the ASABE* 24(5): 617–624.

ASAE Standards. 2009. S368.4 (R2008). Compression Test of food materials of convex shape. St. Joseph, Mich: ASABE.

Ávila, M., J.M. Gomis, and S. Pandiella. (2002). Reconstrucción tridimensional computerizada de granos de cereales. XIV Congreso Internacional de Ingeniería Gráfica, Santander, España.

Bačić, G., R. Srejić, G. Lahajnar, I. Zupanči, and S. Ratković, S. 1992. Water and lipids in maize seed embryos: A proton NMR relaxation and diffusion study. *Seed Science and Technology* 20:233–240.

Baixauli, R., T. Sanz, et al. (2007). Influence of the dosing process on the rheological and microstructural properties of a bakery product. *Food Hydrocolloids* 21(2): 230–236.

Bárcenas, M.E., R. Altamirano-Fortoul, et al. (2010). Effect of high pressure processing on wheat dough and bread characteristics. *LWT - Food Science and Technology* 43(1): 12–19.

Bárcenas, M.E., and C.M. Rosell (2005). Effect of HPMC addition on the microstructure, quality and aging of wheat bread. *Food Hydrocolloids* 19(6): 1037–1043.

Bárcenas, M.E., and C.M. Rosell (2006). Effect of frozen storage time on the bread crumb and aging of par-baked bread. *Food Chemistry* 95(3): 438–445.

Basu, S., U.S. Shivhare, and A.S. Mujumdar. 2006. Models for sorption isotherms for foods: A review. *Drying Technology* 24: 917–930.

Basunia, M.A., and T. Abe. 2005. Adsorption isotherms of barley at low and high temperatures. *Journal of Food Engineering* 66: 129–136.

Bauer, E. 1992. Zum mechanischen Verhalten granularer Stoffe unter vorweigend oedometrischer Beanspruchung. In Gutehus, G., Natan, O. (eds): Veröffentl. Inst. F. Bodenmechanik und Felsmechanik, Universität Karlsruhe, No. 130.

Ben Aissa, M.F., J.Y. Monteau, et al. (2010). Volume change of bread and bread crumb during cooling, chilling and freezing, and the impact of baking. *Journal of Cereal Science* 51(1): 115–119.

Bordes, J., G. Branlard, F.X. Oury, G. Charmet, and F. Balfourier. 2008. Agronomic characteristics, grain quality and flour rheology of 372 bread wheats in a worldwide core collection. *Journal of Cereal Science* 48: 569–579.

Casady, W.W., M.R. Paulsen, J.F. Reid, and J.B. Sinclair. (1992). A trainable algorithm for inspection of soybean seed quality. *Transactions of the ASAE* 35: 2027–2034.

Cato, L., and C. Mills. 2008. Evaluation of the Mixolab for assessment of flour quality. *Food Australia* 60: 577–581.

Chen, C. 2001. Moisture measurement of grain using humidity sensors. *Transactions of the ASAE* 44(5): 1241–1245.

Collar, C., and C. Bollain. 2005. Relationships between dough functional indicators during breadmaking steps in formulated samples. *European Food Research and Technology* 220: 372–379.

Curic, D., D. Novotni, et al. (2008). Design of a quality index for the objective evaluation of bread quality: Application to wheat breads using selected bake off technology for bread making. *Food Research International* 41(7): 714–719.

Dang, J.M.C., and L. Copeland (2004). Studies of the fracture surface of rice grains using environmental scanning electron microscopy. *Journal of the Science of Food and Agriculture* 84(7): 707–713.

Das, M., S. Gupta, et al. (2008). Enzymatic polishing of rice: A new processing technology. *LWT – Food Science and Technology* 41(10): 2079–2084.

Datta, A.K., S. Sahin, et al. (2007). Porous media characterization of breads baked using novel heating modes. *Journal of Food Engineering* 79(1): 106–116.

Dobraszczyk, B.J., and M.P. Morgenstern. 2003. Rheology and the breadmaking process. *Journal of Cereal Science* 38:229–245.

Dumur, J., J. Jahier, M. Dardevet, H. Chiron, A.M. Tanguy, and G. Branlard. 2010. Effects of the replacement of Glu-A1 by Glu-D1 locus on agronomic performance and bread-making quality of the hexaploid wheat cv. Courtot. *Journal of Cereal Science* 51:175–181

Eccles, C.D., P.T. Callaghan, and C.F. Jenner. 1988. Measurement of the self-diffusion coefficient of water as a function of position in the wheat grain using nuclear magnetic resonance. *Biophysical Journal* 53: 77–81.

European Committee for Standarization (CEN). 1995. Eurocode 1 (ENV 1991-4:1995). Basis and design and actions on structures– Part 4: Actions in silos and tanks. Brussels. Belgium

European Standard CEN/TS 15731:2008. Cereals and cereal products – Common wheat (Triticum aestivum L.) – Determination of alveograph properties of dough at adapted hydration from commercial or test flours and test milling methodology

Gan, T.H., D.A. Hutchins, and D.R. Billson. 2002. Preliminary studies of a novel air-coupled ultrasonic inspection system for food containers. *Journal of Food Engineering* 53: 313–323.

Gatti, A.M., D. Tossini, et al. (2009). Investigation of the presence of inorganic micro- and nanosized contaminants in bread and biscuits by environmental scanning electron microscopy. *Critical Reviews in Food Science and Nutrition* 49(3): 275–282.

Ghosh, P.K., D.S. Jayas, M.L.H. Gruwel, and N.D.G. White. 2006. Magnetic resonance imaging studies to determine the moisture removal patterns in wheat during drying. *Canadian Biosystems Engineering* 48(7): 713–718.

Ghosh, P.K., D.S. Jayas, M.L.H. Gruwel, and N.D.G. White. 2007. A magnetic resonance imaging study of wheat drying kinetics. *Biosystems Engineering* 97: 189–190.

Ghosh, P.K., D.S. Jayas, and M.L.H.Gruwel. 2009. Measurement of water diffusivity in barley components using diffusion weighted imaging and validation with a drying model. *Drying Technology* 27: 382–392.

Gianibelli, M.C., M.J. Sissons, and I. Batey. 2005. Effect of source and proportion of waxy starches on pasta cooking quality. *Cereal Chemistry* 82: 321–327.

Giannou, V., and C. Tzia (2007). Frozen dough bread: Quality and textural behavior during prolonged storage. Prediction of final product characteristics. *Journal of Food Engineering* 79(3): 929–934.

Gray, J.A., and J.N. Bemiller (2003). Bread staling: Molecular basis and control. *Comprehensive Reviews in Food Science and Food Safety* 2(1): 1–21.

Gruwell, M.L.H., Ghosh, P.K, Latta, P., and D.S. Jayas. 2008. On the diffusion constant of water in wheat. *Journal of Agricultural and Food Chemistry* 56, 59–62.

Gujral, H.S., G.S. Singh, et al. (2008). Extending shelf life of chapatti by partial baking and frozen storage. *Journal of Food Engineering* 89(4): 466–471.

Hagenimana, A., and X.L. Ding. 2005. A comparative study of pasting and hydration properties of native rice starches and their mixtures. *Cereal Chemistry* 82: 70–76.

Heneen, W.K., and K. Brismar (2003). Structure of cooked spaghetti of durum and bread wheats. *Starch/Staerke* 55(12): 546–557.

Hüttner, E.K., F.D. Bello, et al. (2010). Fundamental study on the effect of hydrostatic pressure treatment on the bread-making performance of oat flour. *European Food Research and Technology* 230(6): 827–835.

Ichinose, Y., K. Takata, et al. (2001). Effects of increase in α-amylase and endo-protease activities during germination on the breadmaking quality of wheat. *Food Science and Technology Research* **7**(3): 214–219.

Igathinathane, C., L.O. Pordesimo, and W.D. Batchelor. (2009). Major orthogonal dimensions measurement of food grains by machine vision using ImageJ. *Food Research International* 42: 76–84.

Iguaz, A., A. Esnoz, C. Arrqoui, and P. Vírseda. 2004a. Modelizaciόn y simulaciόn del secado de arroz en cáscara en secaderos de flujo mixto. In *Proceedings of the III Spanish Congress on Food Engineering* CESIA 2004. Pamplona (Spain).

Iguaz, A., C. Arroqui, A. Esnoz, and P. Vírseda. 2004b. Modelling and validation of heat transfer in stored rough rice without aeration. *Biosystems Engineering* 88(4): 429–439.

Iguaz, A., C. Arroqui, A. Esnoz, and P. Vírseda. 2004c. Modelling and validation of heat transfer in stored rough rice with aeration. *Biosystems Engineering* 89(1): 69–77.

Iguaz, A., M.B. San Martín, J.I. Maté, T. Fernández, and P. Vírseda. 2003. Modelling effective diffusivity of rough rice (*Lido cultivar*) at low drying temperatures. *Journal of Food Engineering* 59(2&3): 253–258.

Iguaz, A., and P. Vírseda. 2007. Moisture desorption isotherms of rough rice at high temperatures. *Journal of Food Engineering* 79(3): 794–802.

Indrani, D., C. Soumya, et al. (2010). Multigrain bread its dough rheology, microstructure, quality and nutritional characteristics. *Journal of Texture Studies* 41(3): 302–319.

Jackowiak, H., D. Packa, et al. (2005). Scanning electron microscopy of *Fusarium* damaged kernels of spring wheat. *International Journal of Food Microbiology* 98(2): 113–123.

Jaisut, D., S. Prachayawarakorn, et al. (2008). Effects of drying temperature and tempering time on starch digestibility of brown fragrant rice. *Journal of Food Engineering* 86(2): 251–258.

James, B. (2009). Advances in "wet" electron microscopy techniques and their application to the study of food structure. *Trends in Food Science and Technology* 20(3–4): 114–124.

Jisha, S., G. Padmaja, et al. (2010). Nutritional and textural studies on dietary fiber-enriched muffins and biscuits from cassava-based composite flours. *Journal of Food Quality* 33(SUPPL. 1): 79–99.

Junqueira, R.M., I.A. Castro, J.A.G. Areas, A.C.C. Silva, M.B.S. Scholz, S. Mendes, and K.G. Oliveira. 2007. Application of response surface methodology for the optimization of oxidants in wheat flour. *Food Chemistry* 101: 131–139.

Kaláb, M., P. Allan-Wojtas, et al. (1995). Microscopy and other imaging techniques in food structure analysis. *Trends in Food Science & Technology* 6(6): 177–186.

Kang, Y.S., C.K. Spillman, J.L. Steele, and D.S. Chung. 1995. Mechanical properties of wheat. *Transactions of the ASAE* 38(2): 573–478.

Karim, A.A., M.H. Norziah, and C.C. Seow. 2000. Methods for the study of starch retrogradation. *Food Chemistry* 71: 9–36.

Katagiri, M., and N. Kitabatake (2010). Rheological properties of somen noodles: A traditional Japanese wheat product. *Journal of Food Science* 75(1).

Koksel, H., K. Kahraman, T. Sanal, D.S. Ozay, and A. Dubat. 2009. Potential utilization of Mixolab for quality evaluation of bread wheat genotypes. *Cereal Chemistry* 86: 522–526.

Kraszewski, A.W. 1996. *Microwave aquametry*. New York: IEEE Press.

Kwade, A., D. Schulze, and J. Schwedes. 1994. Determination of the stress ratio in uniaxial compression tests. *Powder Handling and Processing* 6(1): 61–65.

Liu, Z., F. Cheng, Y. Ying, and X. Rao. (2005). Identification of rice seed varieties using neural network. *Journal of Zhejiang University SCIENCE* 6B(11): 1095–1100.

Liu, H., and N.A.M. Eskin. 1998. Interactions of native and acetylated pea starch with yellow mustard mucilage, locust beam and gelatine. *Food Hydrocolloids* 12: 37–41.

Lostie, M., R. Peczalski, et al. (2002). Study of sponge cake batter baking process. Part I: Experimental data. *Journal of Food Engineering* 51(2): 131–137.

Lu, R., and T.J. Siebenmorgen. 1995. Correlation of head rice yield to selected physical and mechanical properties of rice kernels. 1995. *Transactions of the ASAE* 38(3): 889–894.

Lu, Z.H., L.T. Li, et al. (2005). The effects of natural fermentation on the physical properties of rice flour and the rheological characteristics of rice noodles. *International Journal of Food Science and Technology* 40(9): 985–992.

Majumdar, S., and D.S. Jayas. (2000). Classification of cereal grains using machine vision: IV. Combined morphology, color, and texture models. *Transactions of the ASAE* 43: 1689–1694.

Mariotti, M., C. Alamprese, et al. (2006). Effect of puffing on ultrastructure and physical characteristics of cereal grains and flours. *Journal of Cereal Science* 43(1): 47–56.

Mills, E. N. C., M. L. Parker, et al. (2005). Chemical imaging: The distribution of ions and molecules in developing and mature wheat grain. *Journal of Cereal Science* 41(2): 193–201.

Miñarro, B., I. Normahomed, et al. (2010). Influence of unicellular protein on gluten-free bread characteristics. *European Food Research and Technology* 231(2): 171–179.

Molenda, M., and M. Stasiak. 2002. Determination of the elastic constant of cereal grains in a uniaxial compression test. *International Agrophysics* 16:61–65.

Moya, M., F. Ayuga, M. Guaita, and P. Aguado, P. 2002. Mechanical properties of granular agricultural materials. *Transactions of the ASAE* 45(5): 1569–1577.

Nakamura, S., H. Satoh, et al. (2010). Palatable and bio-functional wheat/rice products developed from pre-germinated brown rice of super-hard cultivar EM10. *Bioscience, Biotechnology and Biochemistry* 74(6): 1164–1172.

Nelson, S.O. 1981. Review of factors influencing the dielectric properties of cereal grains. *Cereal Chemistry* 58(6):487–492.

Neuman, M.R., H.D. Sapirstein, E. Shwedyk, and W. Buchuk. (1989). Wheat grain colour analysis by digital image processing. I. Methodology. *Journal of Cereal Science* 10(3): 175–182.

Nguyen, C.N., and O.R. Kunze. 1984. Fissures related to post-drying treatments in rough rice. *Cereal Chemistry* 61(1): 63–68.

Nishiyama, Y., W. Cao, and B. Li. 2006. Grain intermittent drying characteristics analysed by a simplified model. *Journal of Food Engineering* 76(3): 272–279.

Noda, T., S. Takigawa, et al. (2004). The physicochemical properties of partially digested starch from sprouted wheat grain. *Carbohydrate Polymers* 56(3): 271–277.

O'Brien, C.M., H. Grau, et al. (2000). Functionality of microencapsulated high-fat powders in wheat bread. *European Food Research and Technology* 212(1): 64–69.

Ogawa, Y., G.M. Glenn, et al. (2003). Histological structures of cooked rice grain. *Journal of Agricultural and Food Chemistry* 51(24): 7019–7023.

Osborne, B.G., and R.S. Anderssen. 2003. Single-kernel characterization principles and applications. *Cereal Chemistry* 80: 613–622.

Osborne, B.G., R.J. Henry, and M.D. Southan. 2007. Assessment of commercial milling potential of hard what by measurement of the rheological properties of whole grain. *Journal of Cereal Science* 45: 122–127.

Öste, R. P. Egelberg, C. Peterson, E. Svenson, and O. Monsoon. 1999. Methods and devices for automatic assessment of corn. U.S. Patent No. 5,898,792.

Ozturk, S., K. Kahraman, B. Tiftik, and H. Koksel. 2008. Predicting the cookie quality of flours by using Mixolab. *European Food Research and Technology* 227:1549–1554.

Parker, M.L., A. Grant, et al. (1999). Effects of popping on the endosperm cell walls of sorghum and maize. *Journal of Cereal Science* 30(3): 209–216.

Pham Van, H., T. Maeda, M. Fujita, N. Morita. 2007. Dough properties and breadmaking qualities of whole waxy wheat flour and effects of additional enzymes. *Journal of the Science of Food and Agriculture* 87: 2538–2543.

Rosell, C.M., C. Collar, and M. Haros. 2007. Assessment of hydrocolloid effects on the thermo-mechanical properties of wheat using the Mixolab. *Food Hydrocolloids* 21: 452–462.

Rosell, C.M., E. Santos, and C. Collar. 2010. Physical characterization of fiber-enriched doughs by dual mixing and temperature constraints using the Mixolab. *European Food Research and Technology* 231: 535–544.

Ruan, R., N. Shu, L.Q. Luo, C. Xia, P. Chen, R. Jones, W. Wilcke, and R.V. Morey. (2001). Estimation of weight percentage of scabby wheat kernels using an automatic machine vision and neural network based system. *Transactions of the ASAE* 44: 983–988.

Rusinek, R., and M. Molenda. 2007. Static and kinetic friction of rapeseed. 2007. *Research in Agricultural Engineering* 53(1): 14–19.

Sakai, N., and S. Yonekawa. (1992). Three-dimensional image analysis of the shape of soybean seed. *Journal of Food Engineering* 1: 221–234.

Sakai, N., S. Yonekawa, and A. Matsuzaki. (1996). Two-dimensional image analysis of the shape of rice and its application to separating varieties. *Journal of Food Engineering* 21: 397–407.

Saleem, Q., R. D. Wildman, et al. (2005). Material properties of semi-sweet biscuits for finite element modelling of biscuit cracking. *Journal of Food Engineering* 68(1): 19–32.

Samapundo, S., F. Devliehere, B. De Meulanaer, A. Atukwase, Y. Lamboni, and J.M. Debevere. 2007. Sorption isotherms and isosteric heats of sorption of whole yellow dent corn. *Journal of Food Engineering* 79: 168–175.

Sander, J.P.M. 1996. Starch manufacturing in the world. Advanced Post Academic Course on Tapioca Starch Technology. AIT Center, Bangkok, Thailand 22-26 January and 19-23 February.

San Martín, M.B., J.I. Maté, T. Fernández, and P. Vírseda. 2001. Modelling adsorption equilibrium moisture characteristics of rough rice. *Drying Technology* 19(3&4): 681–690.

Sánchez-Pardo, M. E., A. Ortiz-Moreno, et al. (2008). Comparison of crumb microstructure from pound cakes baked in a microwave or conventional oven. *LWT – Food Science and Technology* 41(4): 620–627.

Sayyah, A.H.A., and S. Mineaei. 2004. Behavior of wheat kernels under quasi-static loading and its relation to grain hardness. *Journal of Agricultural Science and Technology* 6: 11–19.

Sciarini, L.S., P.D. Ribotta, et al. (2010). Influence of gluten-free flours and their mixtures on batter properties and bread quality. *Food and Bioprocess Technology* 3(4): 577–585.

Shouche, S.P., R. Rastogi, S.G. Bhagwat, and J. Krishna. (2001). Shape analysis of grains of Indian wheat varieties. *Computers and Electronics in Agriculture* 33: 55–76.

Srikaeo, K., J.E. Furst, et al. (2006). Microstructural changes of starch in cooked wheat grains as affected by cooking temperatures and times. *LWT – Food Science and Technology* 39(5): 528–533.

Stasiak, M., M. Molenda, and J. Horabik, J. 2007. Determination of modulus of elasticity of cereals and rapeseeds using acoustic method. *Journal of Food Engineering* 82: 51–57.

Steenoek, L.W., M.K. Misra, C.R. Hurburgh Jr., and C.J. Bern. (2001). Implementing a computer vision system for corn kernel damage evaluation. *Applied Engineering in Agriculture* 17(2): 235–240.

Sudha, M.L., and G. Venkateswara Rao. 2009. Influence of hydroxypropyl methylcellulose on the rheological and microstructural characteristics of whole wheat flour dough and quality of puri. *Journal of Texture Studies* 40: 172–191.

Svegmark, K., K. Helmersson, G. Nilsson, P.O. Nilsson, R. Andersson, and E. Svensson. 2002. Comparison of potato amylopectin starches and potato starches: Influence of year and variety. *Carbohydrate Polymers* 47: 331–340.

Thomas, D.L., M.F. Lythgoe, G.S. Pell, F. Calamante, and R.J. Ordidge. 2000. The measurement of diffusion and perfusion in biological systems using magnetic resonance imaging. *Physics in Medicine and Biology* 45: R97–R138.

Timoshenko, S.P., and J.N. Goodier. 1970. *Theory of elasticity*. McGrow-Hill Tokyo.

Trabelsi, S., A.W. Kraszewski, and S.O. Nelson. 2001 Determining bulk density of granular materials from microwave measurements of their dielectric properties. *Proceedings of the 18th IEEE Instrumentation and Measurement Technology Conference* (1–3): 1887–1892.

Trabelsi, S., and S.O. Nelson. 2006. Nondestructive sensing of physical properties of granular materials by microwave permittivity measurement. *IEEE Transactions on Instrumentation and Measurement* 55(3): 953–963.

Turabi, E., G. Sumnu, et al. (2010). Quantitative analysis of macro and micro-structure of gluten-free rice cakes containing different types of gums baked in different ovens. *Food Hydrocolloids* 24(8): 755–762.

Uddin, M.S., P.R. Armstrong, and N. Zhang. 2006. Accuracy of grain moisture content prediction using temperature and relative humidity sensors. *Transactions of the ASABE* 22(2): 267–273.

Ushakumari, S.R., S. Latha, et al. (2004). The functional properties of popped, flaked, extruded and roller-dried foxtail millet (*Setaria italica*). *International Journal of Food Science and Technology* 39(9): 907–915.

Valamoti, S.M., D. Samuel, et al. (2008). Prehistoric cereal foods from Greece and Bulgaria: Investigation of starch microstructure in experimental and archaeological charred remains. *Vegetation History and Archaeobotany* 17(SUPPL. 1), 265–276.

Vallons, K.J.R., L.A.M. Ryan, et al. (2010). High pressure-treated sorghum flour as a functional ingredient in the production of sorghum bread. *European Food Research and Technology* 231(5): 711–717.

Van Bockstaele, F., I. De Leyn, M. Eeckhout, and K. Dewettinck. 2008. Rheological properties of wheat flour dough and the relationship with bread volume. I. Creep-recovery measurements. *Cereal Chemistry*: 85(6): 753–761.

Webb, J., and J.H. Holgate (2003). Microscopy: Scanning electron microscopy. In *Encyclopedia of food sciences and nutrition*, ed. C. Benjamin, 3922–3928. Oxford, UK: Academic Press.

Wijngaard, H.H., S. Renzetti, et al. (2007). Microstructure of buckwheat and barley during malting observed by confocal scanning laser microscopy and scanning electron microscopy. *Journal of the Institute of Brewing* 113(1): 34–41.

Wolf, W., W.E.L. Spiess, and G. Jung. 1985. Standarization of isotherm measurements. In *Properties of water in foods*, ed. D. Simatos and J.L. Multon, 661–679. The Netherlands: Martinus Nijhoff.

Xie, F., T. Perason, F. Dowell, and N. Zang. (2004). Detecting vitreous wheat kernels using reflectance and transmittance image analysis. *Cereal Chemistry* 81(5): 594–597.

Yalcin, S., and A. Basman. 2008. Quality characteristics of corn noodles containing gelatinized starch, transglutaminase and gum. *Journal of Food Quality* 31: 465–479.

Yang W., C. Jia, T.J. Siebenmorgen, T.A. Howell, and A.G. Cnossen. 2002. Intra-kernel moisture responses of rice to drying and tempering treatments by finite element simulation. *Transactions of the ASAE* 45(4): 1037–1044.

Yi, J.H., and W.L. Kerr (2009). Combined effects of freezing rate, storage temperature and time on bread dough and baking properties. *LWT – Food Science and Technology* 42(9): 1474–1483.

Yihu, S., and Z. Qiang. 2007. Dynamic rheological properties of wheat flour dough and proteins. *Trends in Food Science & Technology* 18:132–138.

Young, J.H. 1991. Moisture. In *Instrumentation and Measurement for Environmental Science,* 3rd ed., ed. Z.A. Henry, G.C. Zoerb, and G.S. Birth. St Joseph, MI: ASAE.

Zapotocznya, P., M. Zielinskaa, and Z. Nitab. (2008). Application of image analysis for the varietal classification of barley: Morphological features. *Journal of Cereal Science* 48: 104–110.

14 Physical Properties of Meat and Meat Products

Measurement Techniques and Applications

Kizkitza Insausti, Maria José Beriain, and Maria Victoria Sarriés

CONTENTS

14.1 INTRODUCTION

The use of nondestructive physical measurements for the rapid characterization of the composition of raw materials and end products is essential in order to improve the quality of meat industry products.

Classification systems are marketing instruments that try to predict the commercial value of meat depending on differences that determine the economic value, such as carcass yield or the quantity of edible meat, or the sensory quality expected by consumers. Most classification systems in the world are based on carcass quality categories or commercial cuts. In these systems, visual parameters are assessed because they are easily applied in the slaughter process and these parameters are interesting for livestock producers, retailers, and butchers. However, meat quality is determined by properties that are perceived by the senses: color, texture, juiciness, and flavor. These properties of the carcass are not easily assessed visually and they are related to human perception, so they can change depending of geographic, commercial, and psychological aspects. Therefore, developing a method for predicting meat quality characteristics of a live animal or at the slaughterhouse could help the meat industry to ensure beef quality.

In addition, falling meat consumption in recent years has led the meat industry to develop brands that offer a product with a differentiated level of quality, thereby enhancing consumer satisfaction. Different methods based on physical properties of meat have therefore been studied with the purpose of achieving this goal.

Some of the technologies that are used nowadays for measuring physical parameters affecting meat quality, characterization, and classification are image analysis, ultrasound, color reflectance, hyperstectral imaging, computer tomography, and so on. These techniques are applied not only to meat and meat products, but also to the carcass and to the live animal because meat characterization and classification does not start in the steak. The whole production chain must be taken into account. The present chapter will therefore focus on the application of different technologies in order to objectively measure physical parameters of meat that can be used to characterize and classify the quality of meat and meat products, namely, marbling, tenderness, and color, which are the main characteristics that determine meat quality. These techniques can be applied throughout the meat production chain, that is, from the farm to the final meat product. This is why this chapter includes novel measurement techniques on the live animal, on the carcass, on meat, and on meat products.

14.2 NOVEL MEASUREMENT TECHNIQUES ON LIVE ANIMALS

Predicting an animal's readiness for slaughter or for selection with superior carcass traits will help producers with valuable information on the time needed on the feedlot to obtain meat for a certain market and will also help to identify important parameters for breeding programs.

14.2.1 Ultrasound Scanning

Ultrasound is used in many fields, typically to penetrate a medium and to measure the reflection signature or supply focused energy. The reflection signature can reveal details about the inner structure of the medium. The most well-known application of ultrasound is its use in sonography to produce pictures that allow visualization of subcutaneous body structures such as muscles.

FIGURE 14.1 Transducer designed specifically for ultrasound scanning in cattle.

Ultrasound has been used for over forty years to determine body composition in live animals (Stouffer et al., 1959). Ultrasound offers the ability to rapidly and economically estimate carcass characteristics of live animals. Generally, most researchers have found ultrasound to estimate carcass back fat thickness with an acceptable degree of accuracy (Brethour, 1992). Results for the longissimus muscle area have been less conclusive (Smith et al., 1992). The development of a transducer designed specifically for cattle use that would allow one to obtain an image of the entire longissimus muscle area, has resulted in improved accuracy of these traits (Figures 14.1 and 14.2) (Herring et al., 1994). The image obtained by this technique gives information on the muscle area and also on the ultrasound grey level, which is related to the amount of marbling in the scanned muscle.

Ultrasound has been, in general, identified as the technology with the greatest potential for beef because it can be applied to the live animal (Whittaker et al., 1992;

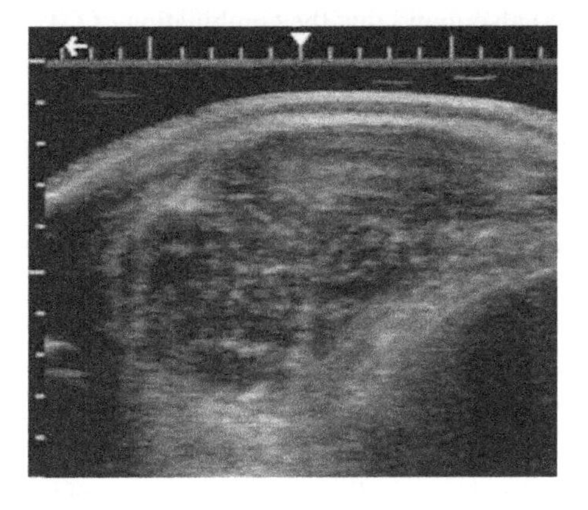

FIGURE 14.2 Ultrasound scan where the longissimus dorsi muscle is shown at the 12th rib.

Herring et al., 1994; Williams et al., 1997; Wall et al., 2004). However, there is a lack of recent literature comparing the accuracy of real-time ultrasound measurements taken at ages other than shortly before harvest (Wall et al., 2004).

Whittaker et al. (1992) suggested the idea that live animal scanning is better related to carcass and meat traits than carcass scanning because changes occur in muscle during slaughter and aging. This suggests that when this technology is applied to live animals, better correlation coefficients between ultrasonic readings and grading systems can be estimated. This technology can thus be used by producers in order to better manage the time on the feedlot depending on the desired carcass quality.

Another application of ultrasound has been to study the relationship between the fatty acid profile and ultrasound scans. Huerta-Leidenz et al. (1993) showed a linear relationship between the average of three ultrasound measurements of subcutaneous fat in live animals (at the rib-lumbar area over the 12th–13th ribs) and monounsaturated/saturated fatty acid ratio (MUFA/SFA ratio).

14.2.2 Others

Computer tomography (CT) measurement of areas and average densities of fat, muscle, and bone in cross-sectional scans through the body of a live animal can be used to predict total weights of each tissue, that is, the retail yield of the animal. Sheep and pigs have been successfully CT scanned in vivo for the last few decades, and the resulting data on carcass composition have been used in commercial breeding programs, at least for sheep (Jopson et al., 2004). Recent research in different sheep breeds has also shown strong negative genetic and phenotypic correlations between intramuscular fat and CT muscle density (Karamichou et al., 2006; Navajas et al., 2006). Since fat has a higher density than muscle fiber, a higher concentration of intramuscular fat will reduce the overall lean tissue density. Taste panel scores for flavor, juiciness, and overall palatability have also shown strong negative correlations with CT muscle density, although no genetic association with toughness has been identified (Karamichou et al., 2006).

Lambe et al. (2008) also found that the combination of CT and live animal video image analysis could be employed to effect genetic improvement of carcass quality traits and intramuscular fat content within breeding programs for different sheep breeds, but it would be of little use as selection criteria to improve shear force or ultimate pH.

Surface electromyography (SEMG) records electrical signals in active muscle fibers through the skin surface. SEMG recordings have been related to muscle fiber composition and muscle fiber diameters in different species, which have important effects on meat quality (Ravn et al., 2008). A work in pigs and lambs has also shown associations of SEMG with the glycogen content of muscle fiber, postmortem rate of pH decline, and shear force measurements of meat. There is, therefore, potential to predict and classify animals depending on meat quality and meat eating quality in livestock using SEMG analysis.

14.3 NOVEL MEASUREMENT TECHNIQUES ON THE CARCASS

The price of beef carcasses depends on the SEUROP carcass classification system (European Community, 1991, 2006). This system implements a scale for grading the conformation and another scale for grading the fat covering of the carcass as a basis for grading carcass quality.

The conformation and fat cover scores are furnished by slaughterhouse personnel who have been suitably trained in the grading of beef carcasses with the aid of the photographic patterns employed in the SEUROP system. This classification system is, however, sometimes criticized because of the level of subjectivity involved, which can cause classifications to vary between slaughterhouses and even between graders working at the same slaughterhouse (Fisher, 2007).

There are other grading systems, such as those from the US Department of Agriculture (USDA) and Meat Standards Australia (MSA), which include meat quality related to parameters in their classification schemes. But even these grading systems are trying to complement the classification with the objective measurements of different carcass physical properties. Therefore, new grading tools have been developed in an effort to achieve more objective classifications, for instance, ultrasound, infrared or electromagnetic systems, and image analysis (Oliver et al., 2010). All these techniques try to characterize and classify carcass quality and value in relation to the parameters that represent the real value of the meat, which is the consumed product obtained from the carcass, but always measuring physical properties that can be related to meat eating quality.

14.3.1 Image Analysis

Image analysis involves taking images of a carcass, then using software to extract data, such as linear measurements, volumes, angles, curvatures, and colors, which are used to predict the conformation and fat class. They can also predict more objective indicators of carcass value such as percentage hindquarter, yield of high-priced cuts and saleable yield.

The image analysis system is based on the digitalization of the images that have been taken with a video that has an optic objective adapted to it (photographic, microscopic, etc.). Digitalization converts the saved image in a spot matrix, which is identified by a computer depending on position coordinates, lightness, and color (Swatland, 1995).

The images to be analyzed can be taken with a holding frame or without stopping the carcass at the slaughter line. The first type of system has been more frequently used in Europe, while the second was developed in Australia. Images can be taken from the whole carcass (Figure 14.3) but also on the rib cut (Figure 14.4). The latter aims to quantify the longissimus dorsi muscle area as well as the amount and distribution of fat flecks, which is directly related to the palatability of meat. With both measurements, an accurate classification of carcasses according to the final meat eating quality could be achieved. Data from all systems can be linked to improve predictions that are made using regression analysis.

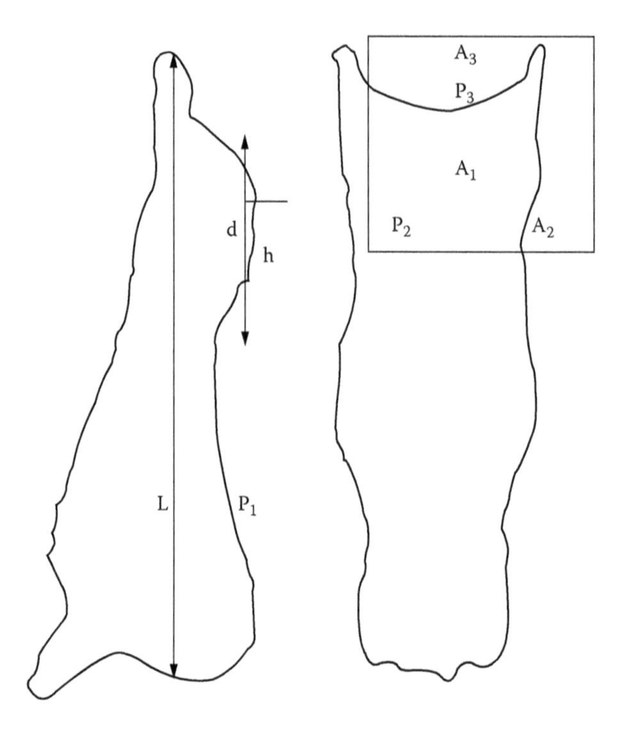

FIGURE 14.3 Measurements taken on a bovine carcass (lateral and dorsal) related to conformation. (From Mendizabal, J. A., A. Legarra, and A. Purroy. 2005. Relación entre la nota de conformación y diferentes medidas morfométricas realizadas mediante análisis de imagen en canales bovinas de diferentes conformaciones. *Comunicaciones de las XI jornadas sobre producción animal. Información Técnica Económica Agraria* 26(II):765–757. Zaragoza, España.)

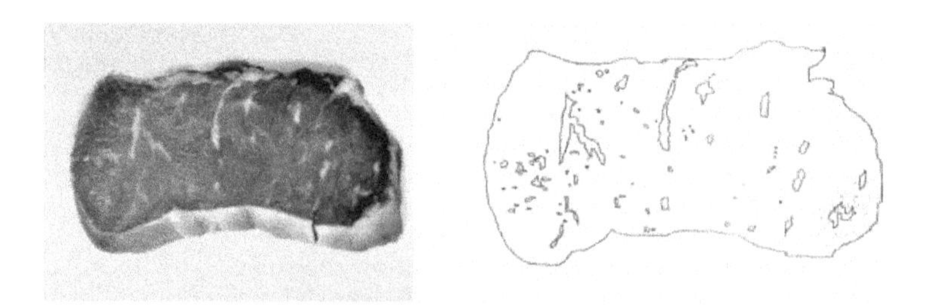

FIGURE 14.4 Marbling measured by image analysis (From Goñi M.V., M. J. Beriain, G. Indurain, et al. 2007. Predicting longissimus dorsi characteristics in beef based on early postmortem colour measurements. *Meat Science* 76:38–45.)

Image analysis has been approved by the European Commission as an official method for pig carcass classification (European Community, 2008, 2009) and it is in fact used to grade pig carcasses at a number of slaughterhouses in Europe. Implementing these procedures at slaughterhouses is harder in the case of cattle, in which the range of genotypes and production systems gives rise to high variability among carcasses.

In this sense, Mendizabal et al. (2005) applied this technology to a wide range of beef carcasses classified from S to P for conformation (European carcass classification system) in cattle from different productive purposes and breeds. In a second assay, image analysis was applied to carcasses obtained from pure breed beef-purpose yearling bulls. When comparing results from both assays, it could be stated that image analysis applied at the slaughterhouse explained around 90% of conformation variability in pure breed young bulls. However, the explained variability was 69% in the first assay. These results provide evidence that image analysis offers good perspectives for classifying carcasses, but also, that these new technologies have to be adapted to each production system.

Different video image analysis (VIA) systems have therefore been developed, which differ in the number of cameras and photographs taken, the speed of classification, if no holding frame is used, the lighting system, the application for hot carcass or cold carcass grading, and so on. Some of the most important systems are BCC-2, which was developed by SFK Technology in Denmark; VBS 2000, which was developed by E + V in Germany; MAC, which was developed by Normaclass in France; VIAScan, which was developed by Meat and Livestock Australia; and CVS, which that was developed by Lacombe University in Canada. Further information on these systems and on their applications and accuracies are reported in Cannell et al. (1999, 2002), Belk et al. (2000), Vote et al. (2003), and Wyle et al. (2003).

Nevertheless, it is important to note that the VIA systems have to be prepared and calibrated using reference scores determined by one or more human classifiers. Thus, any inaccuracy and inconsistency in the reference scores is included in the measured error of the VIA systems. It is therefore important to make sure that the best possible reference is used when systems are being calibrated or assessed. This may involve using a panel of experienced classifiers rather than a single classifier.

A common feature of all the trials of VIA systems is that conformation class is predicted with greater accuracy and repeatability than fat class. This may reflect the fact that human classifiers find it more difficult to accurately determine fat class than to determine conformation class.

In this sense, in Japan, beef marbling is considered one of the most important factors that determine beef eating quality, and highly marbled beef is usually consumed (Zembayashi and Lunt, 1995; Dubeski et al., 1997). In Japan, the ratio of marbling area to longissimus dorsi area (MA/LDA) was around 8% for the Japanese Black and 5% for the Japanese Shorthorn (Kuchida et al., 1992). In the United States, Gerrard et al. (1996) reported 5.28% of intramuscular fat for a sample representative of the US market. These data differ from the ones observed in Spanish local cattle breeds, which showed MA/LDA ratios between 0.9% for Asturiana and 2.9% for Avileña breeds in young bulls slaughtered at 300 kg live weight (Goñi et al., 1999). Thus, the MA/LDA ratio measured by image analysis could estimate the eating quality of beef determined by its intramuscular fat content.

14.3.2 Ultrasounds

As already mentioned (Section 14.2.1), ultrasound has a great potential for carcass classification when applied to live animals, but it can also be used on the carcass when preharvest measurements cannot be taken. With regard to this, Indurain et al. (2006) found a significant and positive correlation between marbling and the grey level of ultrasonic image scanning at the 12th rib ($r = 0.63$; $p < 0.001$). In addition, there were specific fatty acids that were correlated to the ultrasound grey level (Indurain et al., 2006). In intramuscular fat, C18:1 ($r = 3.37$, $p < 0.05$) and monounsaturated fatty acids ($r = 0.36$; $p < 0.05$) percentages were positively correlated to the whiteness of the ultrasonic image scanned at the 12th rib. However, it had a negative correlation coefficient with C18:2n-6 ($r = -0.51$, $p < 0.01$) and 18:3n-6 ($r = -0.37$, $p < 0.05$).

Thus, ultrasound measurements on the carcass can be used to sort carcasses as a function of longissimus muscle area and marbling, but also depending on the fatty acid composition of the intramuscular fat meat.

14.3.3 Color Reflectance

The Commission Internationale de l'Éclairage (CIE, 1976) has defined different representation systems in order to specify color in an objective way. In these systems, color is determined by three coordinates but the recommended one for meat is the CIE L*a*b* system because it shows the higher correlation with visual assessments (Hunt et al., 1991). Spectrophotocolorimeters (Figure 14.5) and colorimeters (Figure 14.6) are the instruments for measuring color. They emit a light shaft and then receive the wave length coming from the objects. The obtained coordinates are lightness (L*), redness (a*), and yellowness (b*). The chroma (C*) and hue (H) can also be calculated as a function of a* and b*.

Most research has mainly been focused on the use of color reflectance measured on the carcass to estimate tenderness and color stability after aging. This still remains a challenge, but it could be of great interest for the meat industry to be able to categorize carcasses depending on the tenderness and color stability that meat will reach after aging. These two are major quality parameters in meat quality but they are very difficult to predict due to the heterogeneity of meat.

Regarding the estimation of tenderness, Jeremiah et al. (1991), Wulf et al. (1997), and Tatum et al. (1999) reported that objective measurements of longissimus dorsi (LD) muscle color were related closely enough to beef tenderness to allow carcasses to be sorted into carcass groupings of different palatability. Wulf and Page (2000) also found that when introducing L* and b* in multiple regression equations to predict longissimus dorsi shear force and palatability, the predicted R^2 was higher than considering only marbling. The measurement of the longissimus dorsi muscle color would therefore be a noninvasive method and a potentially useful predictor of beef tenderness.

Goñi et al. (2007) found that carcass surface muscle color on the day of slaughter (45 minutes after slaughter) would have a less significant relationship with tenderness of beef after aging 7 and 14 days than color measurements taken 24 hours postmortem. However, the regression analysis showed that beef texture at 7 days can

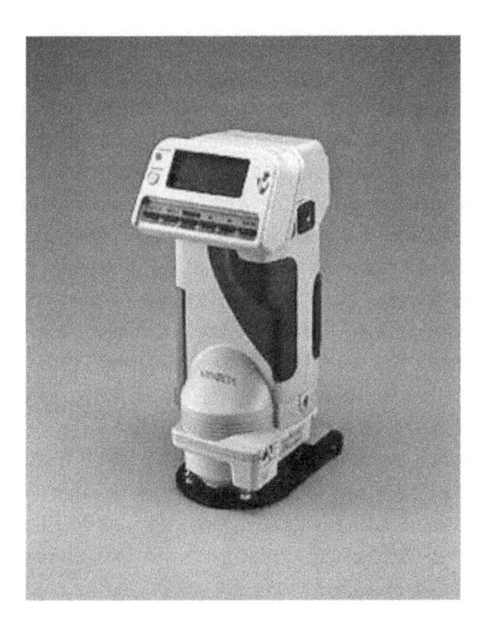

FIGURE 14.5 Spectrophotocolorimeter.

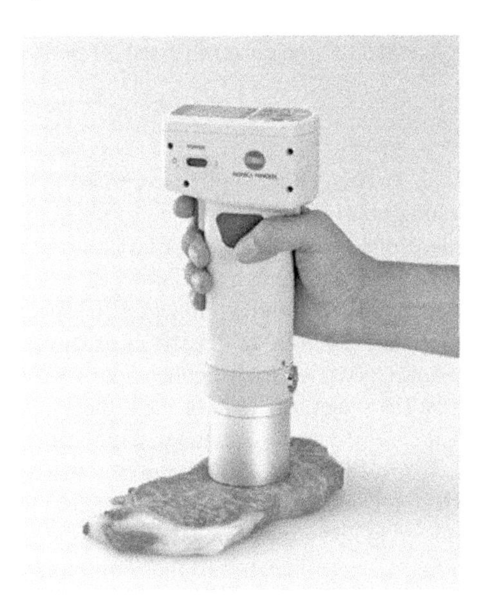

FIGURE 14.6 Colorimeter.

be predicted quite accurately just 45 minutes after slaughter. This fact could be very important for the meat industry in Spain because most beef quality labels recommend that beef meat should be aged 7 days.

Color stability is regulated by many interrelated factors (oxygen availability, enzymatic systems, mitochondrial activity, muscle integrity, lipid oxidation, microbial spoilage, etc.). In beef from young bulls, these factors provide meat with an acceptable color for consumers after 1 hour of blooming. On the contrary, after 48 hours display, a high percentage of beef samples, measured by reflectance with a spectrophotocolorimeter, might be rejected by consumers due to discoloration problems, which will lead to an important economic loss (Insausti et al., 2006). Due to the relationship between metmyoglobin (MMb) and the acceptability of meat by consumers, the MMb percentage was used to segregate between acceptable and unacceptable beef color. MMb percentage was also used to estimate which samples are more likely bound to undergo discoloration, which is of great interest for the meat industry because it is necessary that carcass classification can be able to predict, at the slaughterhouse, the sensory quality of a product that will be consumed some days after slaughter. In this sense, the best regression equation to predict color stability (expressed as MMb% after 48 h display) in beef aged 7 days included a* measured on the carcass 45 minutes postmortem ($R^2 = 0.38$, $p < 0.001$, Beriain et al., 2009).

Finally, color is an important quality characteristic and is a major determinant in pricing veal carcasses (Hulsegge et al., 2001). In The Netherlands, veal carcasses are, in addition to classification for conformation and fatness (SEUROP-system), assigned to different color classes by visually matching the color of the muscle rectus abdominis with fixed colors from a 10-point color scale, developed by the Institute for Animal Science and Health (ID Lelystad) (Hulsegge et al., 1996). The Dutch veal industry is interested in an on-line instrument method to determine the color of veal that produces results equivalent to the currently used 10-point color scale.

14.3.4 Others

Other technologies that have been either applied to carcass grading or have potential for this application include near-infrared reflectance, total body electrical conductivity (TOBEC), bioelectrical impedance analysis (BIA), magnetic inductance technology (MIT), and x-ray computer tomography (CT).

Visible and near-infrared (VIS/NIR) reflectance spectroscopy technology appeared to have the greatest potential to classify beef carcasses according to tenderness (Shackelford et al., 2005). An early application of VIS/NIR tenderness classification was focused on US Select carcasses because these carcasses are discounted relative to US Choice carcasses, even though the vast majority of US Select carcasses are acceptably tender. Shackelford et al. (2005) reported that VIS.NIR spectroscopy was effective in identifying carcasses that produced tender meat after aging. In the validation data set reported in that study, carcasses predicted to be tender had higher sensory panel tenderness ratings compared to those not predicted to be tender (5.6 versus 5.1, respectively). Furthermore, only 5.5% of carcasses predicted to be tender had slice shear force values greater than 25 kg compared to 30.1% of those not predicted to be tender. Future research objectives related to tenderness prediction include the application of this technology to carcasses of all quality grades, muscles other than the longissimus, and other species (Oliver et al., 2010).

Berg et al. (1994) showed that TOBEC could predict the fat and lean content of lamb carcasses with sufficient accuracy for grading applications, but the technology is not applicable to beef carcass grading due to size limitations. However, scanning individual cuts with TOBEC can give accurate predictions of carcass wholesale value. BIA has also been shown to be a useful predictor of beef carcass composition (Zollinger et al., 2010). It is feasible that BIA could be used to augment VIA data because BIA measures the internal rather than the external fat, which is assessed by VIA. There are practical difficulties, however, related to the placement of electrodes on the carcass. MIT is similar in principle to TOBEC, but whereas the latter generates a field inside a large coil through which the carcass passes, with MIT the carcass passes through the field generated between pairs of coils. By using a series of coil pairs, it should be possible to gain information about the distribution of the lean tissues as well as the total amount of lean in the carcass.

X-ray CT scanning facilitates the measurement of areas and average densities of fat, muscle, and bone in cross-sectional scans through the body of an animal. CT is based on the attenuation of x-rays through tissues and objects depending on their different densities. These differences are reflected in different CT values, which appear in the CT image in a grey scale. In this way, CT provides valuable information on total carcass composition in a fast way, and is a lower-cost alternative to manual dissection (Kongsro et al., 2009, Navajas et al., 2010, Jopson et al., 2004). Since commercially available CT scanning machines have been designed for use in human medicine, only livestock species small enough to fit through the circular gantry can undergo whole-body scanning.

14.4 NOVEL MEASUREMENT TECHNIQUES ON MEAT

Many investigators have attempted to apply technologies such as Tendertec (Belk et al., 2001), connective tissue probe (Swatland et al., 1998), elastography (Berg et al.,1999), near-infrared spectroscopy (Park et al., 1998), ultrasound (Park et al., 1994), image analysis (Li et al., 2001), lean color attributes (Wulf et al., 1997), image analysis traits using prototype BeefCam® modules (Belk et al., 2000), and a combination of colorimetric, marbling, and hump height traits (Wulf and Page, 2000) for the purpose of classifying meat quality. In this regard, it must be taken into account that the main meat quality parameters are color (when purchasing), tenderness (while eating), and marbling (which determines characteristic flavor and palatability).

14.4.1 Color

The color of muscle foods is critically appraised by consumers and it is often their basis for product selection or rejection. The color of uncooked meat and meat products is usually described as pink or red, but colors range from nearly white to dark red (Hunt et al., 1991).

There are many factors that may modify meat color. Thus, meat color depends on antemortem factors (breed, sex, age, type of muscular fiber, handling, feeding, stress previous to slaughter, and amount of pigments), factors at slaughter (development of rigor mortis and chilling, final pH, electrical stimulation, or bleeding) and postmortem factors (aging, packaging, storing conditions, microbial contamination, or lipid oxidation) (Faustman and Cassens, 1990; Renerre, 1999).

Myoglobin may be present in three different chemical states: it has a purple color when it is in the reduced state; it oxygenates in the presence of oxygen, changing meat color to brilliant red, but it can oxidize to metmyoglobin (MMb) changing meat to a brownish color (Rosset and Roussel-Ciquard, 1978). The proportion of the different states of myoglobin at the meat surface changes with storage conditions and the atmosphere around the meat and may determine the shelf life of meat. Oxygen and status of the enzymatic systems of myoglobin oxidation and reduction are the key factors that determine the balance among the different statuses of myoglobin, and in this way, beef color and consumer acceptability.

After slaughter and during storage, the rate of MMb accumulation on the surface of meat increases, causing a discoloration. MacDougall (1982) reported that consumer discrimination begins at 20% MMb, and at that moment, half of consumers would refuse the meat (Hood and Riordan, 1973). A brownish color is evident at 40% MMb, and when meat achieves 50% MMb, 100% of consumers refuse the meat (Van den Oord and Wesdorp, 1971).

Discoloration of meat cuts can result in a large monetary loss to retailers as well as to the other segments of the industry. The strategy for maximizing acceptable color (and color shelf life) must involve delaying pigment oxidation and/or enhancing reduction of oxidized pigments (Renerre, 1999).

Color is traditionally assessed by photometric or spectrometric methods, using external (surface spectrophotometers or colorimeters) or internal (optic probes) measurements (Eikelenboom et al., 1992).

As already mentioned (Section 14.3.3), there are different systems to measure color, but in meat the most used coordinates are L^* (lightness), a^* (redness) and, b^* (yellowness). If a spectrophotocolorimeter is used, $L^*a^*b^*$ coordinates are recorded along with the reflectance at different wavelengths. The advantage of this is that the percentage of the different myoglobin forms at the meat surface can be calculated, as described by Hunt et al. (1991). Thus, the usefulness of this measuring technique applied directly on the meat cut is that it can help to follow meat discoloration during retail display, and eventually estimate the shelf life of meat under different packaging conditions. In this regard, Insausti et al. (1999) found that color degradation could be predicted by means of metmyoglobin (calculated by reflectance data) and b^* value in beef samples stored under vacuum and modified atmosphere for 15 days.

14.4.2 Texture

A number of techniques have been proposed to measure meat toughness using mechanical measurements. However, according to Tornberg (1996), there has been no real progress in this field since the 1930s, when the well-known Warner-Bratzler shear device was conceived. Nevertheless, other instrumental measurements, such as compression, tensile, penetrometry, and bite tests, are also used (Lepetit and Culioli, 1994). However, most rheological studies of meat quality use the Warner-Bratzler shear force (WBSF) device in cooked meat and compression tests in raw meat. The equipment types reported in the literature to carry out these tests are the Instron (Figure 14.7) and the Texture Analyzer TA.XT (Figure 14.8). Different attachments or devices can be used in these devices in order to perform rheological tests.

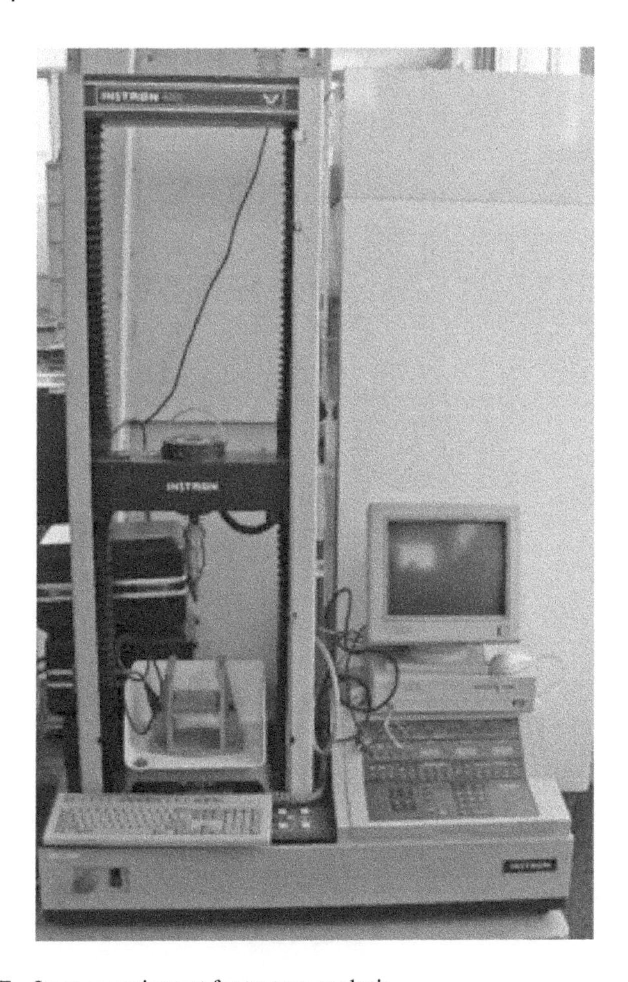

FIGURE 14.7 Instron equipment for texture analysis.

The WBSF device (Figure 14.9) measures the strength (kg or Newton) that is needed to cut through a piece of cooked meat so that the higher shear values are directly related to harder meat. To do so, the steak is previously cooked and left to cool for 2 to 24 hours in the refrigerator. The meat is then cut into pieces (1 cm × 1 cm × 3.5 cm) in parallel to the muscle fibers. It is recommended to repeat the shear test in eight pieces of each steak.

Texture of raw meat is analyzed using a modified compression device (Figure 14.10, Campo et al., 2000) that avoids transversal elongation of the sample. The stress is usually assessed at the maximum rate of compression and at 20% and 80% of maximum compression. These values are related to the amount and type of connective tissue of the muscle.

Comparing these two measurement devices, Campo et al. (2000) observed that the WBSF test is not as good in assessing meat texture evolution through aging as the modified compression device using raw meat. These authors explained this difference in relation to the influence that heating has on muscle protein textural

FIGURE 14.8 Texture Analyser TA.XT.

FIGURE 14.9 Warner Bratzler shearing device.

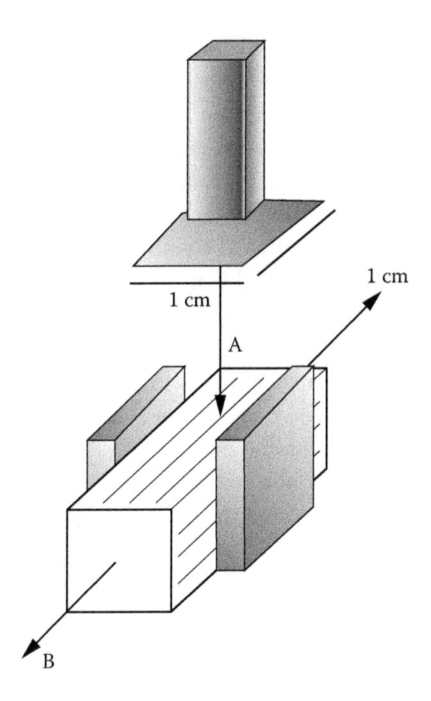

FIGURE 14.10 Compression device.

characteristics. They also reported that low compression rates, in raw meat, are clearly influenced by aging in the development of meat tenderness, and high compression rates demonstrate the influence of connective tissue composition.

As already mentioned (Section 14.4), tenderness is the most important sensory attribute affecting consumer acceptability of beef (Robbins et al., 2003; Feuz et al., 2004) and is also extremely variable. Aging is the practice of holding meat at low temperatures to improve tenderness; however, aging does not occur uniformly in all animals. There are species differences, genetic differences within species, and sex differences that occur during aging (Kuber et al., 2004) as well as large variations between individuals. Because aging is primarily the result of enzymatic activity, the amounts and activities of proteases and the condition and isotypic composition of substrate proteins all presumably influence the potential degree of tenderization (Ahn et al., 2003; Shimada and Takahashi 2003). Rapidly growing animals have higher levels of activity of muscle protease because of the need for rapid turnover, and would be expected to age to a greater degree. In contrast, mature animals, with low growth and lower protein turnover, would be expected to age relatively little (Huff-Lonergan et al., 1995).

Because of the inconsistencies in measured tenderness changes during aging, Novakofski and Brewer (2006) attempted to look at the relationship between aging and initial tenderness, irrespective of animal age, genetics, or nutritional status. They observed that initially tough steaks will benefit from aging while very tender steaks may be adversely affected.

The expected effect of aging on all meat is to increase tenderness. Therefore, they were particularly surprised that some steaks became tougher during aging. Evaluating this phenomenon in more detail when steaks were grouped using initial tenderness or initial shear as a basis for grouping, clear aging patterns emerged. This implies that the benefit of aging is a function of initial tenderness or shear. The implication of this work is that sorting by tenderness would allow preservation of the highest quality and identification of steaks with the greatest potential benefit from aging. The solution is a sensing method that can predict the best aging duration for each piece of meat. Such sensing needs to be very fast, nondestructive, inexpensive, hygienic, and safe (Clerjon and Damez, 2007).

Nevertheless, there are other techniques to measure meat tenderness. In this sense, a technical jump was made by Hildrum et al. (1994) who showed that sensory hardness and tenderness can be predicted by NIR spectroscopy (reflection mode). This technique is rapid and nondestructive when it is applied on meat cuts. Bowling et al. (2009) confirmed the potential of NIR spectroscopy for predicting toughness.

Park et al. (1994) tentatively measured beef tenderness, juiciness, and flavor using ultrasonic spectral analysis. Accuracy of their prediction models was not adequate for use as a tool in practice, but they considered that this approach had potential for nondestructive sensory attribute measurements. One advantage of ultrasonic methods is that they can be applied on the live animal or the whole carcass.

Jeremiah et al. (1991), Wulf et al. (1997), Tatum et al. (1999), and Goñi et al. (2007) reported that objective measurements of beef longissimus dorsi (LD) muscle color were related strongly enough to beef tenderness, so carcasses could be sorted according to LD muscle color in order to group carcasses that differ in palatability. This technology may result in a noninvasive, useful predictor of beef tenderness.

Other technologies may also have potential as predictors of beef tenderness, such as image analysis traits using the prototype BeefCam (Belk et al., 2000; Canell et al., 2002) under commercial processing conditions with a simplified method of shear force determination (Shackelford et al., 1999), texture features computed from muscle images (Li et al., 1999), and a combination of colorimeter, marbling, and hump height traits (Wulf and Page, 2000).

Clerjon and Damez (2007) also reported the importance of monitoring changes in muscle structure during the aging of bovine meat. They presented a polarimetric microwave method consisting of free space and contact reflection coefficient measurements using a horn antenna and reflectance probes. This method is based on the measurement of dielectric properties of tissues parallel and perpendicular to muscle fibers. The final objective of this research was to provide sensors that can be used on-line to detect, with an acceptable error, insufficiently aged meat.

Finally, Ranasinghesagara et al. (2010) recently investigated the potential of a novel optical reflectance imaging method to predict beef tenderness. In this research, they observed that analyzing 2-dimensional reflectance images of meat surfaces provide valuable information regarding the physical characteristics of meat that are responsible for beef tenderness.

14.4.3 Others

There are other important meat quality parameters apart from color and texture. Some of them are marbling, drip loss, and palatability. There is also a wide variety of equipment under study with the aim of trying to predict these quality parameters as a function of physical properties of meat. With respect to this, computer vision systems offer a promising alternative to human expert grading. Some advantages of computer vision systems is that they are noninvasive, objective, consistent, and able to render rapid estimations of meat palatability (Jackman et al., 2009a).

Many digital imaging systems have been developed to process images of raw beef in order to create predictive models for meat palatability (Jackman et al., 2010a). Among these studies, the method of Jackman et al. (2009d) has been the most successful. In this method, palatability indicators of longissimus dorsi muscle marbling, color, and surface texture were compared to the palatability assessment results. These authors have published a lot of research works in past years dealing with the application of surface texture features to predict beef tenderness, color, marbling, and palatability: Zheng et al. (2007), Jackman et al. (2008), Jackman et al. (2009b, 2009c, 2009d), and Jackman et al. (2010a, 2010b).

Frisullo et al. (2010) applied the x-ray microtomography technique to quantify intramuscular fat and also to study fat distribution in different breeds and commercial meat joints in order to provide a more accurate description of the fat microstructure and meat quality.

Finally, hyperspectral imaging systems were used in pork to extract spectral and spatial characteristics simultaneously to determine quality attributes such as drip loss, pH, color, and marbling (Qiao et al., 2007a, 2007b).

14.5 NOVEL MEASUREMENT TECHNIQUES ON MEAT PRODUCTS

Meat products can be obtained using entire muscles or meat pieces that are marinated, injected, or cured, or they can be formed by meat emulsions. In the latter, traditional meat products contain up to 30% fat. The fat plays an important role in meat product processing, stabilizing meat emulsions, reducing cooking loss, improving water-holding capacity, and providing flavor, juiciness, and desirable mouthfeel (Choi et al., 2010). Thus, any modification in the traditional formulation will have a direct effect on the physical properties of meat products. The use of accurate measurement techniques is therefore of great importance regarding the characterization and classification of these emerging new meat products.

14.5.1 Optical Properties

Cassens et al. (1995) stated the importance of establishing a reference method for the assessment of meat color so that results can be compared. Brewer et al. (2001) stated that in a standard comparison of data generated using different instruments, illuminants and/or muscles may not be valid, and that a preselection of the best combination of muscle, color measurement, and instrument/illuminant is necessary for optimum results. Ansorena et al. (1997) therefore evaluated the color of Chorizo de

Pamplona, a Spanish dry fermented sausage. They concluded that the Hunter system was better for separating samples according to redness while the CIE system was better for separating samples according to yellowness.

14.5.2 Texture and Technological Properties

Regarding the texture of meat products, shear value, compression, and viscosity are of great interest because new trends in meat products are focused on the production of low-fat and low-salt products and/or products that substitute vegetable fat for animal fat. In the latter, the new source of fat has to be emulsified before fabrication, and any change in the traditional formulation of meat products will have an effect on its texture.

Shear value or tenderness can be measured with a WBSF device as described in Section 14.4.2.

Regarding compression, texture profile analysis (TPA) is usually applied where samples are compressed to 50% of their original length with a compression device (see Section IV.B). Then, texture variables from the force and area measurements can be calculated: hardness (Hd) = peak force (N) required for first compression; cohesiveness (Ch) = ratio (dimensionless) of active work done under the second compression curve to that done under the first compression curve; springiness (Sp) = distance (mm) of sample recovery after the first compression; gumminess (Gm) = $Hd \times Ch$ (N); and chewiness (Cw) = $Hd \times Ch \times Sp$ ($N \times$ mm).

Another important texture property of meat products is viscosity. *Viscosity* is a measure of the resistance of a fluid that is being deformed by either shear or tensile stress. Therefore, it describes the internal resistance to flow and may be thought of as a measure of fluid friction. This parameter is calculated with a viscometer (Figure 14.11) and it is of great importance in meat emulsions where replacement or reduction of some compounds has been made. In this sense, Choi et al. (2009, 2010) reported that substituting various vegetable oil and dietary fiber for fat will increase the viscosity of low-fat meat batters. They also found correlations between emulsion viscosity and emulsion stability because high viscosity emulsions are not easily broken.

There are also some technological aspects that depend on physical properties that are of great economic importance. Because texture properties change with variations in the formulation, technological changes occur in emulsion capacity and stability, cooking yield, and water and fat retention. The estimation and measurement of all these parameters can help the industry to optimize processing techniques and obtain the maximum economic benefit. Some of the newest technologies that are applied in this field are response surface and image processing methodologies (Velioglu et al., 2010) in combination with mathematical models.

14.5.3 Thermal Properties

Traditionally, the meat industry has used animal fat in the formulation of cooked meat products. Today, consumers are sensitive to saturated fat consumption because of the health implications of this type of fat. However, the production of cooked meat

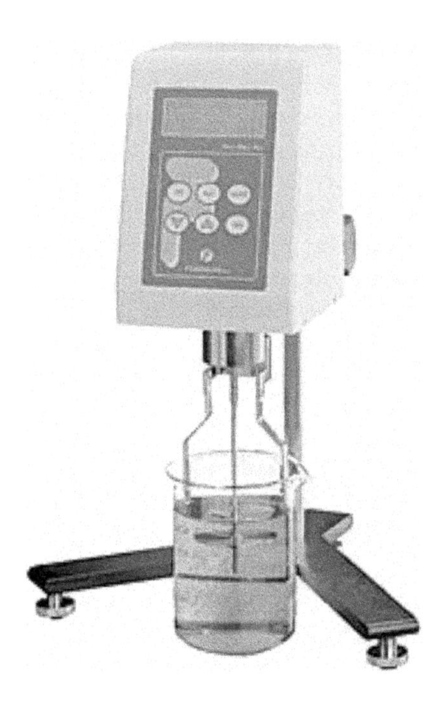

FIGURE 14.11 Viscosimeter.

products with more health-friendly vegetable oil would have an effect on formulation and on the elaboration process, especially on the heat treatment that the product undergoes during cooking (Dieguez et al., 2010). Regarding this, thermal conductivity and specific heat capacity both vary with chemical composition, that is, with the ingredients of the meat emulsion. Thermal conductivity of a foodstuff depends greatly on its porosity, structure, and chemical composition (mainly associated with the properties of air, water, proteins, and fat present in the food). The heat capacity of a food material depends mainly on its water content, but it has been shown to be affected by other components, such as proteins and fat. In addition, thermal conductivity and heat capacity oscillate as a function of temperature (Dieguez et al., 2010).

Dieguez et al. (2010) used the finite element method to calculate the heat transfer processes that take place during cooking of meat emulsions formulated with different amounts of pork fat or olive oil. To do so, experimental data acquired in the pilot plant were used to validate the theoretical model obtained by computer simulation. The results indicated that the computer simulation predicts to a good approximation the heat transfer process when the fat source and amount are changed at the meat emulsion formulation. This could be of importance when new meat emulsions are developed now that food innovation is a key goal for the meat industry.

14.5.4 The Particular Case of Ham

When talking about meat products, there is a wide variety of emulsified products, as already mentioned. However, entire muscles or meat cuts can also be marinated, injected, or cured giving a variety of hams that have different quality characteristics, and thus, different physical properties that must be considered.

Hams differ in quality primarily due to the number of muscle pieces forming the ham. The highest-quality hams are cut from a single muscle with minimal or no brine injection to increase yield. The more muscles that are required to make up the ham, the more brine is needed to form the ham and produce an acceptable yield. Hams requiring a lot of brine are thus of low quality. However, low-quality hams are cheaper to produce, so they are economically viable. Various qualities of ham can look quite similar, which makes the job of expert graders or factory managers difficult. By introducing a computer vision system to observe the ham, finer details can be examined under consistent and objective conditions, allowing effective discrimination between various classes or qualities of ham (Jackman et al., 2010c).

The term *ham* implies the thigh and rump of pork, cut from the haunch of a pig or boar, which may be fresh, dry cured (country hams), or wet-cured (city hams), and then boiled or smoked (Iqbal et al., 2010). Hams consist of at least 20.5% protein (not counting fat portions) without added water. However, the term can be legally applied to products like turkey hams, if the meat is taken from the breast or leg of this bird. A variety of hams made from countless products are now available, which are cured and smoked by different methods. Presliced hams, also known as picnic hams, are considered inexpensive substitutes for regular hams. The production and commercialization of thin ham slices made particularly from pork, turkey, and other poultry is now going through its most significant increase as demanded by the industry and consumers for catering, delicatessen, and ingredient usage. Accordingly, quality inspection based on the identification of objective methods and the quality features of these products become increasingly important to presliced ham product manufacturers (Iqbal et al., 2010).

New measurement techniques applied to hams are used to predict carcass composition. In this sense, Jia et al. (2010) indicated that video image analysis of ham cross-section slices combined with back fat depth at the 10th rib can be used for accurate estimation for total carcass lean or fat composition.

In ham products, the fat-connective tissue size distribution (FSD) represents a fundamental physical property used for quality assessment purposes. FSD is related to sensory properties such as texture, taste, and visual appearance; therefore, the accurate representation of this microstructural property is needed for objective quality characterizations and predictions (Mendoza et al., 2009b). Thus, for objective characterization, image analysis techniques need to take into account the high variability in texture appearance.

Thus, the visual texture of pork ham slices reveals a great deal of information about the different qualities and the perceived image roughness and texture. Recent studies have demonstrated relationships between textural appearance and other fractal measurements such as the directional fractal dimensions based on the variogram of the intensity image in different color scales (Mendoza et al., 2009a) and lacunarity analysis related to pores/defects and fat-connective tissue (Valous et al., 2009). Nevertheless, complete spectra for characterizing and modeling FSDs with high precision represent a relevant approach to get invariant texture descriptors or textural signatures of pork ham surfaces. To do so, ham images are captured using color-calibrated image acquisition systems. They are then processed using software in order to obtain directional fractal dimensions, which are the measurements of the image roughness degree along a certain spatial direction (Mendoza et al., 2009a).

Suitable descriptors for the complex texture appearance of meat products, such as presliced hams, could provide objective information about the characteristic uniformity or dissimilarity in texture between regions on the same ham slice and also between different production batches. Moreover, these texture descriptors could be related to the product specifications of the manufacturer for chemical composition, physical properties, and sensory attributes.

When talking about ham, there is the particular case of Iberian dry-cured hams that has to be mentioned.

Iberian ham is one of the most expensive luxury food products produced in Europe. It is obtained from pigs left free to graze on grass and acorns. Depending on the feeding regime during the final growing period, hams are classified in three commercial categories that can be priced from 25 to 125 euros per kilo (Pérez-Marín et al., 2009).

Quality control programs to determine and guarantee animal feeding regimes are based on on-farm inspection of the animals themselves and analysis of fatty acid composition in subcutaneous fat by gas chromatography, because the feeding regime of the animal is directly related with the type and amount of the fatty acid in the resulting profile.

The high cost and time-consuming nature of farm inspections and analysis of melted fat by gas chromatography prevents the application of these quality control programs to all the animals produced. Therefore, the Iberian pork industry needs fast, accurate, and low-cost methods for the quality control of animals and their expensive luxury products.

For Iberian pig fat, several tests have highlighted the possibilities of NIR spectroscopy technology as a tool for classifying Iberian pig carcasses into commercial categories on the basis of feeding regime. Thus, the incorporation of NIR spectroscopy technology at various points in the Iberian pork production process would provide a spectrum unique to each live animal and/or carcass, which would be of enormous value for traceability and the monitoring of quality specifications.

Another important quality trait of Iberian dry-cured hams is oiliness and brightness. These qualities are highly dependent on the fat solid–liquid ratio at a given temperature. Because of this, estimating the fat solid–liquid ratio at the different temperatures of analysis could be of interest in estimating the quality of hams. Thus, it is of great interest to determine the thermal behavior of this type of dry-cured ham (Niñoles et al., 2010). The characterization of the melting properties can be carried out by differential scanning calorimetry (DSC), but also by ultrasonic measurement, which is a nondestructive, fast, inexpensive, and reliable technique as reported by Niñoles et al. (2010). These data can also be related to sensory attributes in Iberian dry-cured hams.

14.6 CONCLUSIONS

It is not easy to talk about quality of meat and meat products because there is a wide range of factors that have an effect on it. These factors include the whole production chain, from farm to fork; thus, there are a lot of processes, practices, and people involved in obtaining a high-quality final product. In this context, it is easy to infer that is hard to make a measurement of meat quality in an objective and effective way.

Falling consumption of meat and meat products in recent years has forced the meat industry to develop two main strategies. On one hand, objective measurement techniques are being implemented along the production chain in order to ensure meat quality, traceability, homogeneity, and safety. On the other hand, food innovation is being carried out with the purpose of offering more healthy meat products to consumers.

As discussed, there are a lot of physical properties that can be measured to characterize, classify, and estimate meat quality. So far, the main quality parameters studied in meat and meat products are color, texture, and marbling, and the main technologies applied are ultrasound, image analysis, and color reflectance measurements. Currently, there are lots of new technologies being researched to obtain useful tools for the meat industry. The general basis for all of them is to try and measure meat structure and fat distribution, which are responsible for color, tenderness, and palatability.

REFERENCES

Ahn, D.H., K. Shimada, and K. Takahashi. 2003. Relationship between weakening of Z-disks and liberation of phospholipids during postmortem aging of pork and beef. *J. Food Sci.* 68(1):94–98.

Ansorena, D., M. P. De Peña, I. Astiasarán, and J. Bello. 1997. Colour evaluation of Chorizo de Pamplona, a Spanish dry fermented sausage: Comparison between the CIE L* a* b*and the Hunter Lab Systems with Illuminants D65 and C. *Meat Science* 46:313–318.

Belk, K. E., M. H. George, J. D. Tatum, et al. 2001. Evaluation of the Tendertec beef grading instrument to *y* predict the tenderness of steaks from beef carcasses. *Journal of Animal Science* 79 (3):688–697.

Belk, K. E., J. A. Scanga, A. M. Wyle, et al. 2000. The use of video image analysis and instrumentation to predict beef palatability. *Proc. Recip. Meat Conf.* 53:10–15.

Berg, E. P., J. C. Forrest, D. L. Thomas, et al. 1994. Electromagnetic scanning to predict lamb carcass composition. *Journal of Animal Science* 72:1728–1736.

Berg, E. P., F. Kallel, F. Hussain, et al. 1999. The use of elastography to measure quality characteristics of pork semimembranous muscle. *Meat Science* 53 (1):31–35.

Beriain, M. J., M. V. Goñi, G. Indurain, et al. 2009. Predicting longissimus dorsi myoglobin oxidation in aged beef based on early post-mortem colour measurements on the carcass as a colour stability index. *Meat Science* 81:439–445.

Bowling, M. B., D. J. Vote, K. E. Belk, et al. 2009. Using reflectance spectroscopy to predict beef tenderness. *Meat Science* 82:1–5.

Brethour, J. R. 1992. The repeatability and accuracy of ultrasound in measuring back fat thickness. *Journal of Animal Science* 70:1039–1044.

Brewer, M. S., L. G. Zhu, B. Bidner, et al. 2001. Measuring pork color: Effects of bloom time, muscle, pH and relationship to instrumental parameters. *Meat Science* 57:176–196.

Campo, M. M., P. Santolaria, C. Sañudo, et al. 2000. Assessment of breed type and ageing time effects on beef meat quality using two different texture devices. *Meat Science* 55:371–378.

Cannell, R.C., K. E. Belk, J. D. Tatum, et al. 2002. Online evaluation of a commercial video image analysis system (computer vision system) to predict beef carcass red meat yield and for augmenting the assignment of USDA yield grades. *Journal of Animal Science* 80:1195–1201.

Cannell, R.C., J. D. Tatum, K. E. Belk, et al. 1999. Dual-component video image analysis system (VIASCAN) as a predictor of beef carcass red meat yield percentage and for augmenting application of USDA yield grades. *Journal of Animal Science* 77:2942–2950.

Cassens, R. G., D. Demeyer, G. Eikenlenboom, et al. 1995. Recommendations of reference methods for assessment of meat colour. *In Proceedings 41st International Congress of Meat Science and Technology* 2:410–411.

Choi, Y. S., J. H. Choi, D. J. Han, et al. 2009. Characteristics of low-fat meat emulsion systems with pork fat replaced by vegetable oils and rice bran fiber. *Meat Science* 82(2):266–271.

Choi, Yun-Sang, Ji-Hun Choi, Doo-Jeong Han, et al. 2010. Optimization of replacing pork back fat with grape seed oil and rice bran fiber for reduced-fat meat emulsion systems. *Meat Science* 84:212–218.

CIE (Commission Internationale de l'Éclairage). 1976. Official recommendations on uniform colour spaces. Colour difference equations and metric colour terms, Suppl. No. 2. CIE Publication No. 15 *Colourimetry*. Paris.

Clerjon, S., and J. L. Damez. 2007. Microwave sensing for meat and fish structure evaluation. *Measurement Science and Technology* 18:1038–1045.

Dieguez, P.M., M. J. Beriain, K. Insausti, et al. 2010. Thermal analysis of meat emulsion cooking process by computer simulation and experimental measurement. *International Journal of Food Engineering* 6(1): Article 8.

Dubeski, P.L., D. M. Jones, J. L. Aalhus, et al. 1997. Canadian, American, and Japanese carcass grades of heifers fed to heavy weights to enhance marbling. *Can. J. Anim. Sci.* 77:393–402.

Eikelenboom, G., A. H. Hovingbolink, and B. Hulsegge. 1992. Evaluation of invasive instruments for assessment of veal color at time of classification. *Meat Science* 31(3):343–349.

European Community. 1991. Council Regulation No. 1026/91 of 22 April 1991 amending regulation (EEC) No. 1208/81 determining the Community scale for the classification of carcasses of adult bovine animals. *Official Journal of the European Communities* 106.

European Community. 2006. Council Regulation No. 1183/2006 of 24 July 2006 concerning the Community scale for the classification of carcasses of adult bovine animals (codified version). *Official Journal of the European Communities* 214.

European Community. 2008. Commission Decision of 28 July 2008 amending Decision 2006/784/EC as regards the authorization of a method for grading pig carcasses in France (notified under document number C(2008) 3803. *Official Journal of the European Union* 221.

European Community. 2009. Commission Decision of 19 December 2008 authorizing methods for grading pig carcasses in Spain (notified under document number C(2008) 8477. *Official Journal of the European Union* 6.

Faustman, C., and R. G. Cassens. 1990. The biochemical basis for discoloration in fresh meat: A review. *Journal of Muscle Food* 1:217–1243.

Feuz, D. M., W. J. Umberger, C. R. Calkins, et al. 2004. U.S. consumers' willingness to pay for flavor and tenderness in steaks as determined with an experimental auction. *J. Ag. Resource Econ.* 29(3):501–516.

Fisher, A. 2007. Beef carcass classification in the EU: An historical perspective. In *Evaluation of carcass and meat quality in cattle and sheep*. EAAP Publication 123:19-30. The Netherlands: Wageningen Academic Publishers.

Frisullo, P., O. Mario, J. Laverse, et al. 2010. Assessment of intramuscular fat level and distribution in beef muscles using x-ray microcomputed tomography. *Meat Science* 85:250–255.

Gerrard, D. E., X. Gao, and J. Tan. 1996. Beef marbling and color score determination by image analysis. *Journal of Food Science* 61:145–147.

Goñi M.V., M. J. Beriain, G. Indurain, et al. 2007. Predicting longissimus dorsi characteristics in beef based on early post-mortem colour measurements. *Meat Science* 76:38–45.

Goñi MV, Mendizabal JA, Beriain MJ, Alberti P, Arana A, Eguinoa P, Purroy A, Indurain Insausti K. 1999. Marbrure de la viande de veaux de sept races à viande espagnoles determinée par analyse d'image Procedings of the 6th Rencontres Recherches Ruminants, pp 278. December 1–2, 1999. Paris, France.

Herring, W.O., D. C. Miller, J. K. Bertrand, et al. 1994. Evaluation of machine, technician and interpreter effects on ultrasonic measures of backfat and longissimus muscle area in beef cattle. *Journal of Animal Science* 72:2216–2226.

Hildrum, K. I., B. N. Nilsen, M. Mielnik, et al. 1994. Prediction of sensory characteristics of beef by near-infrared spectroscopy. *Meat Science* 38(1):67–80.

Hood, D. E., and E. B. Riordan. 1973. Discolouration in pre-packaged beef: measurement by reflectance spectrophotometry and shopper discrimination. *Journal of Food Technology* 8:333–334.

Huerta-Leidenz, N. O., H. R. Cross, J. W. Savell, et al. 1993. Comparison of the fatty acid composition of subcutaneous adipose tissue from mature Brahman and Hereford cow. *Journal of Animal Science* 71:625–630.

Huff-Lonergan, E., F. C. Parrish Jr., and R. M. Robson. 1995. Effects of postmortem aging time, animal age, and sex on degradation of titin and nebulin in bovine longissimus muscle. *Journal Animal Science* 73(4):1064–1073.

Hulsegge, B., G. Eikelenboom, A. H. Hoving-Bolink, et al. 1996. *Ontwikkeling van kleurenstandaarden voor de kleurclassificatie van runderkarkassen en karkassen van rose en traditioneel gemeste kalveren.* ID-DLO Report No. 96.030.

Hulsegge, B., B. Engel, W. Buist, et al. 2001. Instrumental colour classification of veal carcasses. *Meat Science* 57:191–195.

Hunt, M. C., J. C. Acton, R. C. Benedict, et al. 1991. Guidelines for meat color evaluation. In *Proceedings, 44th Reciprocating Meat Conference of the American Meat Science Association*, 3–14. Champaign, IL: American Meat Science Association.

Indurain, G., M. J. Beriain, M. V. Goñi, et al. 2006. Composition and estimation of intramuscular and subcutaneous fatty acid composition in Spanish young bulls. *Meat Science* 73:326–334.

Insausti, K., M. J. Beriain, A. Purroy, et al. 1999. Colour stability of beef from Spanish native cattle breeds stored under vacuum and modified atmosphere. *Meat Science* 53:241–249.

Insausti, K., G. Indurain, M. V. Goñi, et al. 2006. Colour stability of beef stored under vacuum packaging and its relationship to carcass fatness. *Proceedings of 52st ICoMST*, 525–536. Dublin, Ireland.

Iqbal, Abdullah, Nektarios, A. Valous, Fernando Mendoza, et al. 2010. Classification of presliced pork and Turkey ham qualities based on image colour and textural features and their relationships with consumer responses. *Meat Science* 84:455–465.

Jackman, P., Da-Wen Sun, Cheng-Jin Du, et al. 2008. Prediction of beef eating quality from colour, marbling and wavelet texture features. *Meat Science* 80:1273–1281.

Jackman, P., Da-Wen Sun, Du C.J., and Paul Allen. 2009a. Prediction of beef eating qualities from colour, marbling and wavelength surface texture features using homogeneous carcass treatment. *Pattern Recognition* 42, 751–763.

Jackman, P., Da-Wen Sun, and Paul Allen. 2009b. Comparison of various wavelet texture features to predict beef palatability. *Meat Science* 83:82–87.

Jackman, P., Da-Wen Sun, and Paul Allen. 2009c. Automatic segmentation of beef longissimus dorsi muscle and marbling by an adaptable algorithm. *Meat Science* 83:187–194.

Jackman, P., Da-Wen Sun, and Paul Allen. 2009d. Comparison of the predictive power of beef surface wavelet texture features at high and low magnification. *Meat Science* 82:353–356.

Jackman, P., Da-Wen Sun, and Paul Allen. 2010. Prediction of beef palatability from colour, marbling and surface texture features of *longissimus dorsi*. *Journal of Food Engineering* 96:151–165.

Jackman, P., Da-Wen Sun, Paul Allen, et al. 2010. Correlation of consumer assessment of longissimus dorsi beef palatability with image colour, marbling and surface texture features. *Meat Science* 84:564–568.

Jackman, P., Da-Wen Sun, Paul Allen, et al. 2010. Identification of important image features for pork and turkey ham classification using colour and wavelet texture features and genetic selection. *Meat Science* 84:711–717.

Jeremiah, L. E., A. K. W. Tong, and L. L. Gibson. 1991. The usefulness of muscle color and pH for segregating beef carcasses into tenderness groups. *Meat Science* 30:97–114.

Jia, Jiancheng, A. P. Schinckel, J. C. Forrest, et al. 2010. Prediction of lean and fat composition in swine carcasses from ham area measurements with image analysis. *Meat Science* 85:240–244.

Jopson, N. B., P. Amer, and J. C. McEwan. 2004. Comparison of two-stage selection breeding programmes for terminal sire sheep. *Proceedings of the NZ Society of Animal Production* 64:212–216.

Karamichou, E., R. I. Richardson, G. R. Nute, et al. 2006. Genetic analysis of carcass composition, as assessed by x-ray computer tomography, and meat quality traits in Scottish Blackface sheep. *Animal Science* 82:151–162.

Kongsro, J., M. Roe, K. Kvaal, et al. 2009. Prediction of fat, muscle and value in Norwegian lamb carcasses using EUROP classification, carcass shape and length measurements, visible light reflectance and computer tomography (CT). *Meat Science* 81:102–107.

Kuber, P. S., J. R. Busboom, E. Huff-Lonergan, et al. 2004. Effects of biological type and dietary fat treatment on factors associated with tenderness: I. Measurements on beef longissimus muscle. *Journal Animal Science* 82(3):770–778.

Kuchida, K., K. Yamaki, T. Yamagishi, et al. 1992. Evaluation of meat quality in Japanese beef cattle by computer image analysis. *Anim. Sci. Technol.* 63:121–128.

Lambe, N. R., E. A. Navajas, C. P. Schofield, et al. 2008. The use of various live animal measurements to predict carcass and meat quality in two divergent lamb breeds. *Meat Science* 80:1138–1149.

Lepetit, J., and J. Culioli. 1994. Mechanical properties of meat. *Meat Science* 36:203–237.

Li, J., J. Tan, F. A. Martz, et al. 1999. Image texture features as indicators of beef tenderness. *Meat Science* 53:17–22.

Li, J., J. Tan, and P. Shatadal. 2001. Classification of tough and tender beef by image texture analysis. *Meat Science* 57(4):341–346.

MacDougall, D. B. 1982. Changes on the colour and opacity of meat. *Food Chemistry* 9:75–88.

Mendizabal, J. A., A. Legarra, and A. Purroy. 2005. Relación entre la nota de conformación y diferentes medidas morfométricas realizadas mediante análisis de imagen en canales bovinas de diferentes conformaciones. *Comunicaciones de las XI jornadas sobre producción animal. Información Técnica Económica Agraria* 26(II):765–757. Zaragoza, España.

Mendoza, Fernando, Nektarios A. Valous, Paul Allen, et al. 2009a. Analysis and classification of commercial ham slice images using directional fractal dimension features. *Meat Science* 81:313–320.

Mendoza, F., N. A. Valous, Da-Wen Sun, et al. 2009b. Characterization of fat-connective tissue size distribution in pre-sliced pork hams using multifractal analysis. *Meat Science* 83:713–722.

Navajas, E. A., C. A. Glasbey, K. A. McLean, et al. 2006. In vivo measurements of muscle volume by automatic image analysis of spiral computed tomography. *Animal Science* 82:545-553.

Navajas, E. A., C. A. Glasbey, A.V. Fisher, et al. 2010. Assessing beef carcass tissue weights using computed tomography spirals of primal cuts. *Meat Science* 84:30–38.

Niñoles, L., A. Mulet, S. Ventanas, et al. 2010. Ultrasonic assessment of the melting behaviour in fat from Iberian dry-cured hams. *Meat Science* 85:26–32.

Novakofski, J., and S. Brewer. 2006. The paradox of toughening during the aging of tender steaks. *Journal of Food Science* 71(6):473–479.

Oliver, A., J. A. Mendizabal, G. Ripoll, et al. 2010. Predicting meat yields and commercial meat cuts from carcasses of young bulls of Spanish breeds by the SEUROP method and an image analysis system. *Meat Science* 84:628–633.

Park, B., Y. R. Chen, W. R. Hruschka, et al. 1998. Near-infrared reflectance analysis for predicting beef *longissimus* tenderness. *Journal of Animal Science* 76(8):2115–2120.

Park, B., A. D. Whittaker, R. K. Miller, et al. 1994. Ultrasonic spectral-analysis for beef sensory attributes. *Journal of Food Science* 59(4):697–701.

Pérez-Marín D., E. De Pedro Sanz, J. E. Guerrero-Ginel, et al. 2009. A feasibility study on the use of near-infrared spectroscopy for prediction of the fatty acid profile in live Iberian pigs and carcasses. *Meat Science* 83:627–633.

Qiao, J., N. Wang, M. O. Ngadi, et al. 2007b. Prediction of drip-loss, pH, and color for pork using a hyperspectral imaging technique. *Meat Science* 76:1–8.

Qiao, J., Michael Ngadi, et al. 2007a. Pork quality and marbling level assessment using a hyperspectral imaging system. *Journal of Food Engineering* 83:10–16.

Ranasinghesagara, J., T. M. Nath, S. J. Wells, et al. 2010. Imaging optical diffuse reflectance in beef muscles for tenderness prediction. *Meat Science* 84:413–421.

Ravn, L. S., N. K. Andersen, M. A. Rasmussen, et al. 2008. De electricitatis catholici musculari: Concerning the electrical properties of muscles, with emphasis on meat quality. *Meat Science* 80:423–430.

Renerre, M. 1999. Review: Biochemical basis of fresh meat color. In *Proceedings of the 45th International Congress of Meat Science and Technology*, 344–352. Yokohama, Japan.

Robbins, K., J. Jensen, K. J. Ryan, et al. 2003.Consumer attitudes towards beef and acceptability of enhanced beef. *Meat Science* 65(2):721–729.

Rosset, R., and N. Roussel-Ciquard. 1978. Storage of fresh meat. *Industries Alimentaires et Agricoles* 95(4):357–369.

Shackelford, S.D., T. L. Wheeler, and M. Koohmaraie. 1999. Tenderness classification of beef: II. Design and analysis of a system to measure beef longissimus shear force under commercial processing conditions. *Journal of Animal Science* 77(6):1474–1481.

Shackelford, S. D., T. L. Wheeler, and M. Koohmaraie. 2005. On-line classification of US Select beef carcasses for longissimus tenderness using visible and near-infrared reflectance spectroscopy. *Meat Science* 69:409–415.

Shimada, K., and K. Takahashi. 2003. Relationship between fragmentation of myofibrils and liberation of phospholipids from Z-disks induced by calcium ions at 0.1 mM: Mechanism of tenderization of pork and beef during postmortem aging. *Journal Food Science* 68(9):2623–2629.

Smith, M. T., J. W. Oltjen, H. G. Dolezal, et al. 1992. Evaluation of ultrasound for prediction of carcass fat thickness and longissimus muscle area in feedlot steers. *Journal of Animal Science* 70:29–37.

Stouffer, J. R., M. V. Wellentine, and G. H. Wellington. 1959. Ultrasonic measurement of fat thickness and loin eye area on live cattle and hogs. *Journal of Animal Science* 18, 1483 (Abstract).

Swatland, H. J. 1995. Video image analysis. In *On-line evaluation of meat*, 271–290. Lancaster, PA: Technomic Publishing Company.

Swatland, H. J., J. C. Brooks, and M. F. Miller. 1998. Possibilities for predicting taste and tenderness of broiled beef steaks using an optical-electromechanical probe. *Meat Science* 50(1):1–12.

Tatum, J.D., K. E. Belk, M. H. George, et al. 1999. Identification of quality management practices to reduce the incidence of retail beef tenderness problems: Development and evaluation of a prototype quality system to produce tender beef. *Journal of Animal Science* 77:2112–2118.

Tornberg, E. 1996. Biophysical aspects of meat tenderness. Conference Information: 42nd International Congress of Meat Science and Technology—Meat for the Consumer. *Meat Science* 43:175–191.

Valous, N. A., F. Mendoza, D. Sun, et al. 2009. Texture appearance characterization of pre-sliced pork ham images using fractal metrics: Fourier analysis dimension and lacunarity. *Food Research International* 42(3):353–362.

Van den Oord, A. H., and J. J. Wesdorp. 1971. Analysis of pigments in intact beef samples. *Journal of Food Technology* 6:1–8.

Velioğlu. Hasan Murat, Serap Durakl Velioğlu, Ismail Hakk Boyaci, et al. 2010. Investigating the effects of ingredient levels on physical quality properties of cooked hamburger patties using response surface methodology and image processing technology. *Meat Science* 84:477–483.

Vote, D. J., K. E. Belk, J. D. Tatum, et al. 2003. Online prediction of beef tenderness using a computer vision system equipped with a BeefCam module. *Journal of Animal Science* 81:457–465.

Wall, P. B., G. H. Rouse, D. E. Wilson, et al. 2004. Use of ultrasound to predict body composition changes in steers at 100 and 65 days before slaughter. *Journal of Animal Science* 82:1621–1623.

Whittaker, A. D., B. Park, B. R. Thane, et al. 1992. Principles of ultrasound and measurement of intramuscular fat. *Journal of Animal Science* 70:942–952.

Williams, R. E., J. K. Bertrand, S. E. Williams, et al. 1997. Biceps femoris and rump fat as additional ultrasound measurements for predicting retail product and trimmable fat in beef carcasses. *Journal of Animal Science* 75:7–13.

Wulf, D. M., and J. K. Page. 2000. Using measurements of muscle color, pH, and electrical impedance to augment the current USDA beef quality grading standards and improve the accuracy and precision of sorting carcasses into palatability groups. *Journal of Animal Science* 78:2595–2607.

Wulf, D. M., S. F. O'Connor, J. D. Tatum, et al. 1997. Using objective measures of color to predict beef longissimus tenderness. *Journal of Animal Science* 75:685–692.

Wyle, A. M., D. J. Vote, D. L. Roeber, et al. 2003. Effectiveness of the SmartMV prototype BeefCam System to sort beef carcasses into expected palatability groups. *Journal of Animal Science* 81:441–448.

Zembayashi, M., and D. K. Lunt. 1995. Distribution of intramuscular lipid throughout M. longissimus thoracis and lumbarum in Japanese Black, Japanese Shorthorn, Holstein and Japanese Black crossbreds. *Meat Science* 40:211–216.

Zheng Chaoxin, Da-Wen Sun, and Liyun Zheng. 2007. A new region-primitive method for classification of colour meat image texture based on size, orientation, and contrast. *Meat Science* 76:620–627.

Zollinger B. L., R. L. Farrow, T. E. Lawrence, et al. 2010. Prediction of beef carcass salable yield and trimmable fat using bioelectrical impedance analysis. *Meat Science* 84:449–454.

15 Physical Properties of Dairy Products

Lourdes Sánchez and Maria Dolores Pérez

CONTENTS

15.1 INTRODUCTION

Milk is a complex biological secretion that contains various components in different physical states. Whey proteins are present in milk in colloidal solution, while caseins that form a supramolecular structure, the micelles, are in colloidal suspension. Milk fat is present as an emulsion in an aqueous protein phase, wherein the fat structured as globules, in which the lipids are arranged as layers. Fat globules are covered by a membrane that originates from the mammary cell in the secretion process. Therefore, milk components are present in different physical states and also show different particle sizes, as shown in Table 15.1.

The relationships among milk components determine the physical characteristics of liquid milk. Furthermore, the technological processes that are applied to milk to create different dairy products result in considerable changes in the physical properties of milk and in the relations among their components. The quality of milk products can be evaluated by determining some of its physical properties. Measurement of the physical properties of milk and dairy products is important in designing dairy equipment, estimating the concentration of certain components, or assessing the extent of a chemical or physical change. The great advantage of physical measurements for such purposes is their speed and simplicity, as well as their potential to be automated.

Physical properties of liquid milk are measured by traditional methods that are included in the control of milk at reception. Some of these properties are routinely controlled because they can give indirect information about alterations or adulterations of milk. The main measured properties are the following: the specific density, which ranges from 1.029 to 1.039 (15°C), and abnormal values may indicate the addition of water or other substances to milk; the freezing point, which is between −0.53 and −0.55°C, is also suitable for detecting the addition of water; the pH, which ranges from 6.5 to 6.75; the refractive index (n_D^{20}) from 1.3410 to 1.3480, and the specific conductivity, which is 4–5.5 × 10^{-3} $ohm^{-1}cm^{-1}$ (25°C) and may indicate the presence of mastitis in cows (Belitz and Grosch, 2009; Hamann and Zecconi, 1998). The main advances in techniques applied to liquid milk have been made in relation to composition analysis by applying spectroscopic techniques, and especially since

TABLE 15.1
Physical State and Particle Size Distribution in Milk

Compartment	Size, Diameter (nm)	Type of Particles
Emulsion	2,000–6,000	Fat globules
Colloidal dispersion	50–300	Casein-calcium phosphate
	4–6	Whey proteins
True solution	0.5	Lactose, salts, other substances

Source: Chandan, R. C. 2006. Milk composition, physical and processing characteristics. In *Manufacturing yogurt and fermented milks*, ed. R.C. Chandan, C.H. White, A. Kilara, and Y.H. Hui, 17–40. Ames, IA: Blackwell Publishing.

near-infrared spectroscopy (NIRS) has been developed to determine milk constituents, such as fat, protein, and lactose, and applied in on-line analysis of milk quality, even in the farm (Tsenkova et al., 1999; 2000). The advantages of including NIRS are mainly its high-speed analysis and the fact that it is a nondestructive measurement.

In this chapter, we will consider the physical properties of the main milk products. Taking into consideration that the characteristics and quality attributes of each product are very specific, we will describe the methods applied of measuring their physical properties independently.

15.2 MEASUREMENT OF PHYSICAL PROPERTIES OF POWDERED MILK

Milk powder is composed of a matrix of amorphous lactose in which proteins, fat, and air vacuoles are dispersed. The chemical composition of whole milk powder is, on average, 26.5% protein, 27.4% fat, 37.7% lactose, 5.7% minerals, and 2.7% water, and that of skimmed milk powder is 38.2% protein, 0.9% fat, 49.6% lactose, 8.2% minerals, and 3.0% water (Thomas et al., 2004).

Spray drying is the principal process used to produce milk powders. Spray drying involves atomization of concentrated milk into a drying medium, resulting in an extremely rapid evaporation of water. Atomization results from the dispersion of concentrated milk using either a spinning disk atomizer or a series of high-pressure nozzles. This is done inside a large drying chamber in a flow of hot air. The feed travels through the dryer until it reaches the desired level of water content in the product (Vega and Roos, 2006). Roller drying is also used to obtain milk powder, specifically for the chocolate industry, since the higher temperature applied in this process results in the Maillard reaction and then in desirable cooked flavors (Thomas et al., 2004).

Milk powders can be produced with a variety of different physical properties. The measurement of physical properties is important because they can help to define the powder and because they intrinsically affect its behavior during processing and storage (Teunou et al., 1999).

15.2.1 Particle Size and Density

The average sizes and size distribution of particles in milk powders are usually measured by static light scattering using a Malvern Mastersizer with a laser beam at a wavelength of 633 nm with a 300-mm RF lens. The instrument uses the refractive index of the dispersed phase and its absorption (Keogh et al., 2003; Gaiani et al., 2006; Ilari and Mekkaoui, 2005; Fitzpatrick et al., 2004a, 2004b, 2008).

The apparent density, which takes into account the space filled by air in spray-dried powders, may be expressed as bulk or packed density. *Bulk density* is the solid weight per volume unit of the powder, whereas *packed density* includes the air trapped within particles but not the void volume between particles (Thomas et al., 2004). Bulk density is determined by weighing a tared cylinder filled with a fixed volume of powder; packed or tapped density is determined by weighing the same cylinder after tapping several times on a rubber mat to achieve constant volume (Mistry and Pulgar, 1996; Fitzpatrick et al., 2004 a,b). Bulk and packed densities can be also measured using a Powder Tester (Hosokawa Micron, Osaka, Japan) (Gaiani et al., 2006).

The particle density is calculated by measuring the air-free volume of a known weight of powder using a Micromeritis pycnometer (Malvern Instruments, Malvern, UK). The pycnometer determines the density and volume of milk powder by measuring the pressure change of a gas (helium or nitrogen) in a calibrated volume (Keogh et al., 2003; Fitzpatrick et al., 2004a, 2004b). Using these data and the bulk density of the powder, the vacuole volume and interstitial air can be calculated.

15.2.2 Flow Properties

Powder flowability is the ability of a powder to flow as individual particles. It is considered one the most important physical properties of milk powder and is directly influenced by two properties, cohesion, and compressibility (Thomas et al., 2004). Flow properties of milk powder are important in handling and processing operations, such as flow from hoppers and silos, transportation, mixing, compression, and packaging. One of the major industrial powder problems is obtaining reliable and consistent flow out of hoppers and feeders without excessive spillage and dust generation (Teunou et al., 1999; Ilari, 2002; Fitzpatrick et al., 2004a, 2004b, 2007).

Therefore, measurement of flow properties has been an important issue in milk powder industry and several techniques have been developed to measure it.

Jenike (1964) pioneered the application of shear cell techniques for measuring powder flow properties. In this technique, the cell is loaded with the powder, and the powder is compacted to a known bulk density (consolidation stress). A normal stress, inferior or equal to the consolidation stress, is applied to the sample. A shear stress is then applied in an orthogonal direction until the compacted powder fractures (Figure 15.1). The measured flow properties used in this methodology are the flow function, the effective angle of internal friction, and the angle of wall friction. The flow function is a plot of the unconfined yield stress of the powder versus major consolidating stress, and represents the strength developed within a powder when consolidated, which must be overcome to make the powder flow. The flow index, defined as the inverse slope of the flow function, is used to classify powder flowability with higher values representing easier flow.

Fitzpatrick et al., (2004a, 2007) used an annular shear cell to measure the flow function and effective angle of internal friction of milk powders. Besides, these authors used a Jenike shear cell to determine the angle of friction of milk powders, whereby the cylindrical base of the cell was replaced by a flat of stainless steel with a rough surface (Teunou et al., 1999; Fitzpatrick et al., 2004a, 2004b, 2007).

The flowability of milk powders has also been determined by measuring the angle of repose (a static measure of relative flowability) using simple equipment (Figure 15.2) (Kim et al., 2005). In this technique, a certain amount of milk powder is placed in the top box of the equipment with the trap door closed. The trap door is then opened allowing the powder to flow downward and to form a heap. This method allows for the measurement of the drained angle of repose using horizontal still

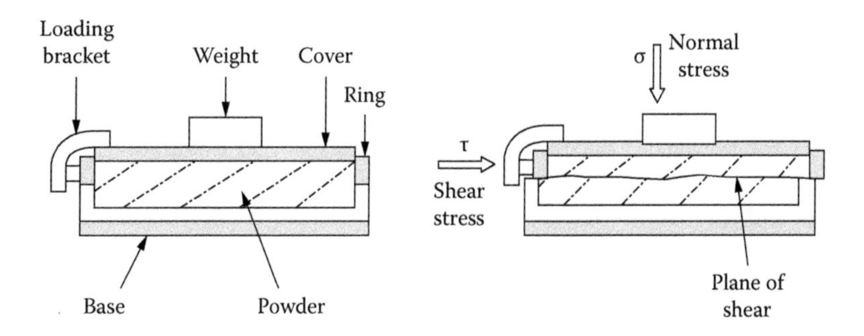

FIGURE 15.1 Jenicke shear cell used to determine powder flowability. Powder is first placed in the cell and compacted under a known consolidation stress. A normal stress (σ) is applied to the sample. A shear stress (τ) is then applied in an orthogonal direction until the compact fracture. A yield locus is obtained for different (until the consolidation stress). The experiment is carried out for different consolidation stresses. (From Thomas, M.E.C., J.L. Scher, S. Desobry-Banon, and S. Desobry. 2004. Milk powders ageing: Effect on physical and functional properties. *Critical Reviews in Food Science and Nutrition*, 44: 297–322.)

photographs and a protractor. More free-flowing powders tend to have lower drained angles of repose.

Flow properties can also be determined using a commercial powder tester (Micrometrics Laboratory, Hosokawa, Japan) that has been recommended for dairy powders. It evaluates four physical properties: the angle of repose, angle of spatula, compressibility, and cohesion, as described by Carr (1965). These four physical properties were assessed in arriving at a flowability index with a theoretical value from 1 to 100 (Ilari, 2002; Ilari and Mekkaoui., 2005).

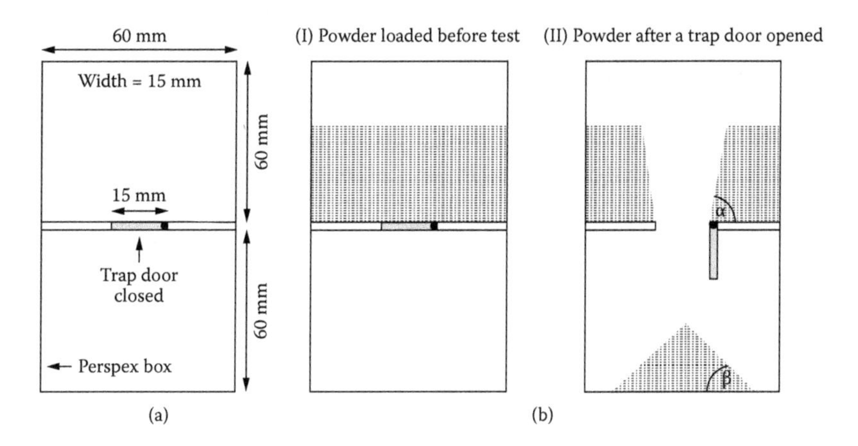

FIGURE 15.2 Schematic representation of the flowability test equipment (A) and the flowability test (B). α: drained angle of repose, β: poured angle of repose. (From Kim, E.H.J., X.D. Chen, and D. Pearce. 2005. Effect of surface composition on the flowability of industrial spray-dried dairy powders. *Colloids and Surfaces B: Biointerfaces* 46: 182–187. Reprinted with permission from Elsevier, License number 2637830789973, dated March 28, 2011.)

15.2.3 Stickiness and Caking Properties

Stickiness is a major problem encountered during drying and handling of dried dairy products because it greatly affects quality and yield during processing operations. It causes the powder to become much more cohesive and eventually caking, and also causes a powder to adhere more to a surface. These changes alter the handling behavior and appearance of the powders (Boonyai et al., 2004, Thomas et al., 2004; Vega and Roos, 2006).

Lactose present in spray-dried dairy powders is in its amorphous state and it is the main component that has influence on the sticking and caking behavior of dairy powders. When an amorphous component is given sufficient conditions of temperature and water content, it can mobilize as a high-viscosity flow, which can make it sticky and lead to caking. Furthermore, molecular mobility enables amorphous lactose to crystallize over time. Crystallization will only take place if the powder temperature is greater than its glass transition temperature (*Tg*), whereby the molecules have sufficient mobility to initiate crystallization. *Tg* is usually well above the storage temperature for most dry powders. However, lactose in its amorphous state is very hygroscopic and will readily adsorb water from ambient air. This increase in water will cause a significant reduction in *Tg*. At a temperature above the *Tg* (for a given water content), the powder particles become sticky, and this is often described as the sticky point temperature. It is generally accepted that the sticky point of lactose is about 10–20°C above the onset *Tg* (Roos, 2002; Schuck et al., 2007).

The problem of stickiness in dairy powders has led to significant costs in the industry, which justifies the need for research to develop measurement techniques and approaches for predicting powder caking problems in advance, so that strategies can be implemented to prevent them from being realized in practice. Various direct and indirect techniques using several instrumentations have been developed for characterization of the stickiness behavior of food powders, which were reviewed by Boonyai et al. (2004). Direct techniques involve measuring shear force, viscosity, cohesion, and adhesion of the milk powder sample as it changes from a free-flowing state to a sticky state as a function of moisture and/or temperature. Indirect techniques are based on the determination of the glass transition temperature, and results obtained can be indirectly correlated to the stickiness.

Direct conventional techniques involve mechanical movement, shearing, or compression of the powder sample, whereas direct pneumatic techniques involve the use of an air stream with predetermined temperature and humidity to blow or suspend the powder particles. The results generated from these methods can be directly expressed in terms of sticky-point temperature and/or adhesiveness and cohesiveness (Boonyai et al., 2004).

Propeller-driven methods are the oldest of all the stickiness measurement techniques used. It consists of a test tube containing a sample of known moisture content, submerged in a heating medium. An impeller embedded in the sample is either turned manually (Chuy and Labuza, 1994) or electrically driven (Hennings et al., 2001; Ozkan et al., 2002). With increasing temperature, the particle surface becomes less viscous allowing stickiness at the contacting surfaces. This produces an increase in force required to turn the propeller; consequently the stirring force increases sharply and the temperature at this point is referred to as the *sticky-point temperature*. Henning et al., (2001) developed a mechanically driven device to measure the sticky point of milk powders by measuring the output of electric resistance from the stirrer. The sticky-point temperature was taken when the voltage increased sharply and the stirrer stopped.

The ampule method has also been used to measure the surface caking temperature of infant milk powders (Chuy and Labuza, 1994). In this technique, milk powder samples of known moisture content contained in a glass ampule are heated in a bath. During heating, the ampule is tapped sharply on a hard surface to separate the powder. The surface caking temperature is taken as the temperature at which the powder failed to separate into fine particles, appearing as clumps.

Rennie et al. (1999) used an unconfined yield test to measure the cohesion of milk powders that provides quantitative information regarding the strength of caked powders. Previously, consolidated plugs were prepared by placing a certain amount of powder into the consolidation blank and vibrating the die to pack the powder in the die. Once the powder was compacted, the consolidation plunger was placed on the powder in the die. When testing, the plunger was removed from the die and the consolidation die was dismantled, yielding a cylindrical plug of powder. The cylinder was axially loaded by using a loading frame that gradually increased the load on the plug. The load at which the powder failed was defined as the *unconfined yield stress*.

Ozkam et al. (2002) developed a viscometer technique to measure stickiness of milk powders based on the measurement of the torque required to turn a propeller inserted in the milk powder sample (Figure 15.3). The bath temperature and the torque required to rotate a homemade L-shaped spindle (propeller or stirrer) inserted into milk powders were recorded. The sticky temperature was determined as the point where a sharp increase in torque occurred. These authors also used a

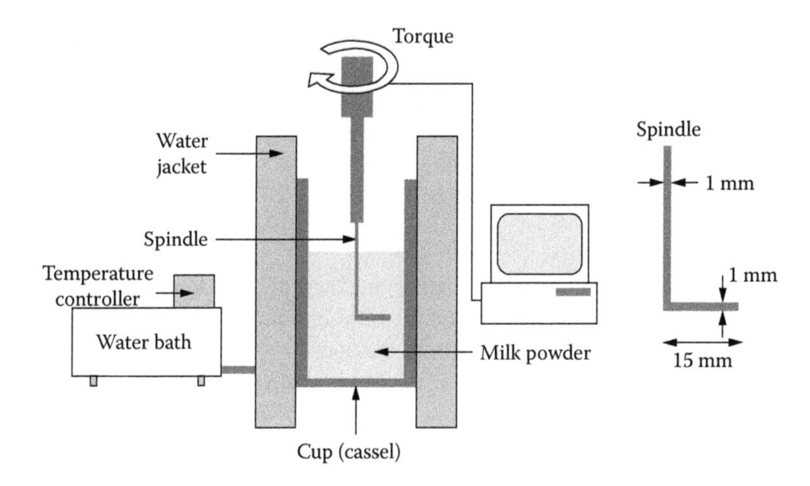

FIGURE 15.3 Schematic illustration of the viscometer technique (From Ozkan, N., N. Walisinghe, and X.D. Chen. 2002. Characterization of stickiness and cake formation in whole and skim milk powders. *Journal of Food Engineering* 55: 293–303. Reprinted with permission from Elsevier, License number 26377791197007, dated March 28, 2011.)

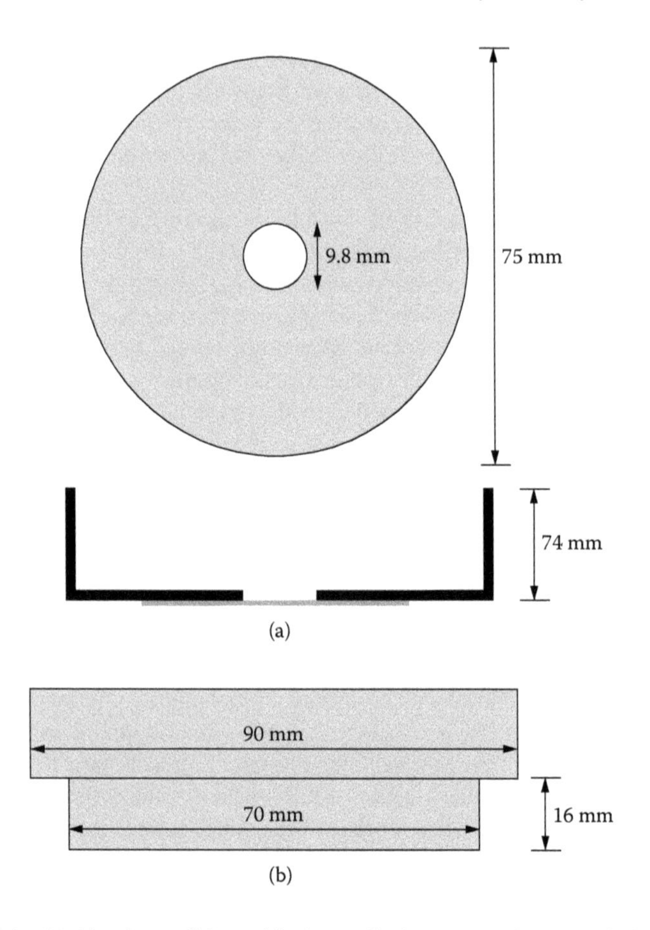

FIGURE 15.4 (a) Aluminum dish used in force–displacement cake strength determination (with central hole and foil cover underneath to prevent powder falling through); (b) powder surface leveling tool. (From Fitzpatrick, J.J., E. O'Callaghan, and J. O'Flynn. 2008. Application of a novel cake strength tester for investigating caking of skim milk powder. *Food and Bioproducts Processing* 86: 198–203. Reprinted with permission from Elsevier, License number 2637790845929, dated March 28, 2011.)

penetration test with a cylindrical indenter attached to an Instrom testing machine to record the force required to penetrate the powder compact and found good agreement between the results obtained by both methods.

A new easy-to-use empirical tester for investigating caking of skimmed milk powder was developed by Fitzpatrick et al., (2007, 2008). It consisted of placing a certain amount of powder into a cylindrical aluminum dish with a circular hole in the center as illustrated in (Figure 15.4). Aluminum foil was taped to the bottom of the dish to cover the hole, preventing the powder from falling through. The powder was spread out across the dish and then exposed to a condition that caused the powder to cake. Following exposure, the foil was removed and the dish was centered below a rod attached to a JJ tester. The rod was then moved downward through the caked powder at a constant speed and the force versus displacement was measured. The peak force was used as an index of cake strength.

Patterson et al. (2007) used the particle-gun method to study the stickiness of skim milk powders. This apparatus reproduces the sort of conditions that are present in commercial spray drier ducts and cyclones. The particle-gun rig consisted of a controlled humidity and temperature air supply system and a particle feeding system that "fires" the particles onto a deposition plate at high velocity. Particles were manually introduced slowly at the top of the tube using a venturi funnel arrangement where they were accelerated before being impacted against a stainless steel disc that was placed under the particle-gun tip. The fraction of the feed particles that stick to the plate was recorded as an index of the stickiness of the powder under the conditions used. The gun method was further modified to include a vibratory feeder and to enclose the feed area within a plastic hood (Murti et al., 2009).

Murti et al. (2010) used a fluidized bed rig to determine the stickiness of milk powders. The technique involved the increase of humidity of the fluidizing air maintaining constant the temperature. The sticky point, determined by visual observation, was defined as the humidity value that caused the bed to lose fluidization; that is, the bed collapse could not be refluidized by a sudden vibration.

A cyclone test device was used by Intipunya et al. (2009) for characterizing the stickiness of milk powder. The device mainly consisted of an air heater, a spraying chamber, and a cyclone chamber. The cyclonic air stream generates a rotary motion of the powder particles at the bottom of the cyclone chamber. Stickiness was observed within 1–2 minutes; the particles stuck to each other and some attached to the chamber wall.

As indicated previously, the measurement of the glass transition temperature is an indirect method to determine the stickiness of milk powder. Differential scanning calorimetry (DSC) has been used widely to determine the *Tg* of milk powders (Chuy and Labuza, 1994; Sherstha et al., 2007; Fitzpatrick et al., 2007; Intipunya et al., 2009; Boonyai et al., 2007; Hogan et al., 2010). Using this technique, *Tg* is characterized by a measurable discontinuity in specific heat capacity. The *Tg* values can be estimated from the heat scanned curve at the onset point or the midpoint of the curve. The glass transition occurs over a temperature range that is often a relatively narrow range of 10 to 20°C for amorphous lactose.

A thermal mechanical compression test (TMCT) was used by Boonyai et al. (2007) to perform glass–rubber transition analysis of milk powders. They employed a thermally controlled sample cell that was attached to a texture analyzer. The thermal compression test involves the application of a compression force on a thin layer of powder sample in a thermally controlled sample cell attached to a texture analyzer equipped with a cylindrical probe. During heat scanning, the measured force decreases slightly until the glass–rubber transition temperature (*Tgr*) is reached, resulting in a rapid drop in the force. Then, the *Tgr* can be simply determined by performing linear regression on the TMCT data.

Hogan et al. (2010) further described a rheological technique for determining glass–rubber transition of spray-dried dairy powders. Glass–rubber transition was determined, using a laboratory rheometer, by two methods involving compression of a powder sample between a steel parallel plate and a Peltier plate. In one method, the gap between plates was measured under a constant force; in the other method, normal force was measured at a constant gap. *Tgr* was identified by a significant increase in the rate of change in gap distance or normal force during heating.

15.2.4 Solubility Properties

The solubility of milk powder is an essential attribute of quality. It depends on different dissolution steps in water (wettability, sinkability, and dispersibility). *Wettability* is the penetration of the liquid into the pores of the powder particles. *Sinkability* is defined by the fall of powder below the surface of the liquid. *Dispersibility* is the ability to disperse in single particles throughout the water. Solubility involves the separation of particles in water with low-energy stirring (Thomas et al., 2004). Depending on powder properties, each step could take more or less time in the solubilization process. Several attempts have been made to determine these steps in milk powders, resulting in different techniques designed to characterize them.

The wettability of milk powders has been tested using static and dynamic wetting tests (Kim et al., 2002; Freudig et al., 1999). In static tests, a constant amount of powder is placed on a slide covering a water reservoir. By pulling the slide away, the powder layer was brought into contact with the water. The wetting time, which is necessary for the submersion of the last powder particle, is measured. In dynamic tests, the powder is fed to the surface of a large, slowly agitated vessel via a pipe that equalizes and guides the particle flow. The feed rate is increased until layer formation is observed. This method measures the maximum powder rate per surface area of layer formation (Freudig et al., 1999).

A rehydration method was developed by Gaiani et al. (2006). The method involved dispersing milk powder in a stirred vessel equipped with a turbidity sensor under standardized conditions. The changes of turbidity occurring during powder rehydration highlighted several stages. These stages include particle wetting, and then swelling as the water penetrates into the powder bed, followed by a slow dispersion of the particles. This is reflected by a quick increase of turbidity immediately after the addition of milk powder until a maximum, which is related to the powder wetting. Next to this stage, turbidity is stabilized indicating the end of rehydration.

The dispersibility of milk powder is usually determined from the amount of undispersed material collected on a filter after reconstitution in water under set conditions. Dispersibility is inversely correlated to the weight percentage of residue remaining on the mesh (IDF method, 1979).

However, as dissolution steps are difficult to study independently, solubility appears the more reliable criterion to evaluate milk powder behavior toward water (Thomas et al., 2004). All methods used to determine solubility are based on the same principle. Powder is first dispersed into water under controlled conditions. After centrifugation, the quantity of insoluble material is determined in the sediment. Generally, the solubility of milk powders is determined from the volume in milliliters of sediments and expressed as the solubility index (Thomas et al., 2004; Mistry et al., 1996). Methods used to determine solubility of milk powders were reviewed by Thomas et al. (2004).

Davenel et al. (2002) developed a nuclear magnetic resonance (NMR) method that gives information on the reconstitution process of milk powders in water. It allows the quantification of the water absorption by powder particles, their reconstitution rate, and the detection of the presence of insoluble materials. In this technique, a glass tube filled with water was put into the gap of the magnet of a Minispec Bruker NMR spectrometer operating at the resonance frequency of 10 MHz. The NMR measurements continued until the solution was completely reconstituted, except if insoluble materials were formed. The solubility index and NMR relaxometry gave compatible information on powder rehydration.

15.2.5 Microstructure

The microstructure of milk powders has been studied using polarized light microscopy confocal scanning laser microscopy (CSLM) and scanning electron microscopy (SEM).

Pictures of milk powder particles were obtained by Ilari and Mekkaoui (2005) using polarized light microscopy to check the homogeneity of a milk powder and the shape of milk particles. Electronic pictures provided complementary information on sieve and size measurements of powder particles.

The distribution of fat and protein in milk powder can be localized using CSLM by selecting an appropriate fluorescent probe and solvent. CSLM of spray-dried whole milk powder labeled with either Nile red-Nile blue or Nile red-fast green showed rounded particles consisting of a continuous amorphous lactose phase containing fat droplets, protein, and occluded air (Authy et al., 2001). Likewise, micrographs of whole milk powder obtained by Kim et al. (2002) showed that there are many fat globules around the boundary of particles, and larger and more coalesced fat droplets at the joining points of agglomerated milk powder particles.

Scanning electron microscopy has been widely used to study particle size, shape, and microscopic aspects like pores or wrinkles or the presence of lactose crystals in milk powders (Baechler et al., 2005; Hardy et al., 2002). Figures 15.5a and 15.5b are electron microscopy photographs showing the internal porosity of milk powder and the surface of milk powder particles after lactose crystallization (Hardy et al., 2002). Although the different equipment and process parameters highly influence the structure of milk powder particles, some general trends can be described. The surface of skimmed milk powder particles showed the typical characteristics with some deep and shallow dents on the surfaces, whereas whole milk particles were spherical and had a relatively smooth surface (Kim, 2002; Kim et al., 2002). After removal of free fat present on the surface with petroleum ether, the surface of skimmed milk powder particles seemed to be similar with that before fat extraction. In the case of whole milk powder after fat extraction, the particles were still spherical, but had many clear dents on the surfaces. This appears to suggest that the surface of whole milk powder is largely covered by free fat, sometimes in the form of irregular patches, or sometimes in the form of a fat layer. These results support the surface composition of whole milk powder measured by x-ray photoelectron spectroscopy, which indicated about 98% of fat coverage on the surface of spray-dried dairy powders.

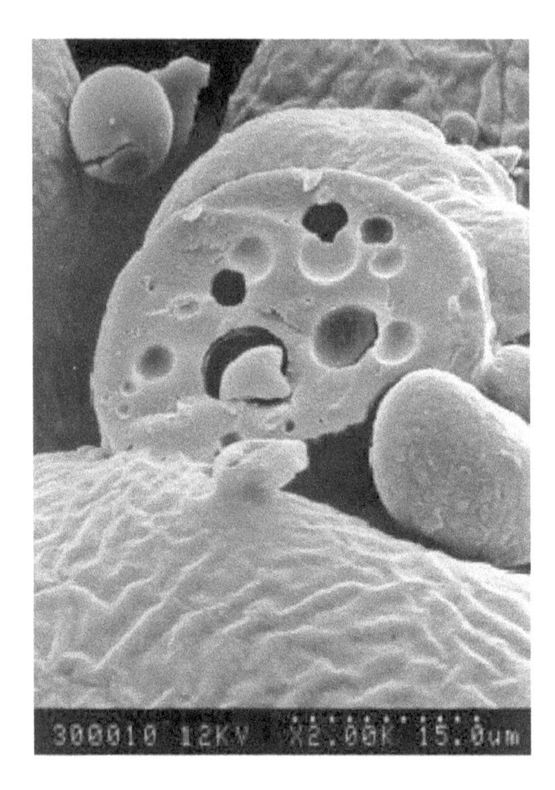

FIGURE 15.5(A) Electron microscopy photograph showing the internal porosity of milk powder particles. (From Hardy, J., J. Scher, and S. Banon. 2002. Water activity and hydration of dairy powders. *Lait* 82: 441–452. Reprinted with permission from EDP Sciences, dated March 24, 2011.)

The internal physical state of lactose in milk powders can be determined by x-ray diffraction. This technique is one of the most powerful techniques for analysis of crystalline compounds and has been widely used to investigate crystallinity in milk powders. In general, milk powders containing α-lactose monohydrates or β-lactose show x-ray diffraction patterns with characteristic peaks. However, a broad band at approximately 23°, without any characteristic peaks of lactose crystals, represents average distances between molecules in a liquid or amorphous state (Kim, 2009).

15.3 MEASUREMENT OF PHYSICAL PROPERTIES OF BUTTER

Butter is a water-in-oil emulsion that consists of a minimum of 80% milk fat, a maximum 16% water, and a maximum dry nonfat milk material content of 2%. The elaboration of butter requires a previous controlled cooling of the cream for 12–15 hours designed to give the fat the required crystalline structure. Afterward, the cream is violently agitated during the churning process to break down the fat globules (Avramis et al., 2003; Couvreur et al., 2006). Cream undergoes a phase inversion during the process as fat globule membranes are disrupted, globules coalesce, and oil leaks out to form the continuous phase. Therefore, butter is a multiphase

FIGURE 15.5(B) Electron microscopy photograph from the surface of milk powder particles after lactose crystallization. (From Hardy, J., J. Scher, and S. Banon. 2002. Water activity and hydration of dairy powders. *Lait* 82: 441–452. Reprinted with permission from EDP Sciences, dated March 24, 2011.)

emulsion consisting of damaged and intact fat globules, crystalline fat, and an aqueous phase, dispersed in a continuous oil phase (Wright et al., 2001).

Texture is the main physical quality factor that determines butter's acceptability, hardness, and spreadability, which are the most essential features perceived by consumers. Textural properties of butter are dependent on the chemical composition and the solid fat content (SFC) of the product, which also determine the thermal properties of butter. Furthermore, butter texture is influenced by the morphology of the fat crystal network and the different forms this might adopt at a given temperature (Glibowski et al., 2008; Krause et al., 2008; Vithanage et al., 2009).

15.3.1 Thermal Properties

The thermal properties of butter are usually determined by differential scanning calorimetry (DSC). Previously, butter was heated at 60–70°C to liquefy the sample, and thus to erase any crystals. Measurements are carried out in the temperature range of about −20°C to 70°C with heating and cooling rates to study melting behavior and crystallization. Three melting zones are usually found in the melting thermograms of butters. The three zones include a minor peak representing a low melting zone (between –30°C and 7°C), a major peak representing an intermediate melting zone (between 7°C and 26°C) and a broad shoulder representing a high melting zone (between 26°C and 45°C). The melting temperature can be determined from the peak maximum of the heating profiles. The temperature at the beginning of crystallization (T onset) can be calculated from the cooling profiles (Avramis et al., 2003; Couvreur et al., 2006; Jinjarak et al., 2006; Vithanage et al., 2009).

Recently, a low-field ^1H nuclear magnetic resonance (NMR) has been applied to determine phase transitions in cream (0–100% fat). The relaxation measurements were performed on a benchtop pulsed NMR analyzer with a resonance frequency of 23.2MHz for protons. Performing principal component analysis on the transverse relaxation data measured during cooling of the cream revealed a major change in the $T2$ relaxation characteristics at temperatures of 17°C and 22°C, which were almost identical to the crystallization temperatures determined by DSC analysis. Thus, it was suggested that low-field NMR could be used to measure on-line phase transitions in cream and butter (Bertram et al., 2005).

15.3.2 Solid Fat Content (SFC)

The extent of solidification, and hence the ratio of solid to liquid fat, is an important physical property that determines butter consistency (Wright et al., 2001). The determination of SFC in butter, at any temperature, has been performed using DSC by calculating the ratio of the partial area above the temperature of the peak maximum to the total area under the DSC melting curve to the baseline (Couvreur et al., 2006; Vithanage et al., 2009). However, this technique has been largely replaced by the pulsed NMR technique, which was adopted as an official method (Cd 16-81) by the American Oil Chemists' Society (AOCS, 1999). Crystallization curves can be constructed by plotting the SFC (%) as a function of time. The crystallization curves obtained by NMR can be fitted following the Avrami model (Avrami, 1941). Fitting the SFC data to this model allows for the determination of kinetic parameters, which provide information on the nature of the crystallization process (Campos et al., 2002). NMR instruments are capable of accurate, precise, rapid, and nondestructive measurement of solid fat content. This technique has been applied to study the effect of cooling rate on the structure and mechanical properties of anhydrous milk fat (Campos et al., 2002; Maranagoni and Narine, 2002), the physical and chemical properties of milk fat and phytosterol-ester blends (Rodrigues et al., 2007), and the influence on butter properties of feeding special diets to dairy cows (Avramis et al., 2003).

Singh et al. (2004) compared an ultrasonic velocimetry technique to direct pulsed nuclear magnetic resonance (pNMR) spectroscopy for the determination of the SFC of anhydrous milk fat (composed of 100% milk fat) and blends. In situ measurements of ultrasonic velocity were carried out during cooling and heating of the fat. Ultrasonic velocity was linearly correlated to solid fat content in anhydrous milk fat up to 40% solids during both crystallization and melting, although large deviations were observed above 20% solids. This ultrasonic spectrometry technique could be used to monitor and study the crystallization process of dairy fats on-line, so that their properties could be monitored during processing.

15.3.3 Textural Properties

The textural properties of butter have been extensively studied. Most of the tests used to characterize butter texture are empirical in nature and designed to imitate sensory perceptions for quality control operations. These are based mainly on the principles of extrusion, compression, sectility, and penetrometry, and involve large deformations that break down the material structure. Small deformation techniques have also been used to probe the response of butter to stress and have the advantage that strains are low enough to keep the structure intact (Wright et al., 2001).

In extrusion tests, the butter sample is extruded through an orifice at a constant speed and the force required to sustain motion is recorded (Prentice, 1972). The compression test involves placing the butter between two flat platens. Tests can be performed in two different ways: (1) a constant load is applied to the top platen the deformation of the butter is recorded (De Man et al., 1985) or (2) a constant speed is applied to the top platen and the load and the deformation are monitored (Dixon, 1974). Sectility tests involve the use of a stretched steel wire that is driven through the sample at a constant speed while the counteracting force is measured (Hayakawa et al., 1986).

Despite the fact that all these tests have been shown to be reliable in determining textural properties of butter, constant force penetrometry has been the most common empirical test used in the last few years. Brule (1893) first applied this technique using a steel rod to penetrate the butter sample and loading it with weights until it penetrated the fat. Penetrometry tests are usually performed using a texture analyzer (TA-TX2 machine) or on some occasions an universal testing machine (Couvreur et al., 2006) equipped with a cylinder (Glibowski et al., 2008; Vithanage et al., 2009) or a conical probe of known dimensions and weight (Bobe et al., 2003; Campos et al., 2002; Couvreur et al., 2006; Jinjarak et al., 2006; Rodrigues et al., 2007) which is loaded at a constant speed to a specific depth. Using this technique, the main parameter obtained is hardness, which is defined as the force measured at a specific depth penetration (usually between 5 and 18 mm) (Bobe et al., 2003; Couvreur et al., 2006; Glibowski et al., 2008; Jinjarak et al., 2006). The force-displacement measurement can be also recorded and the Young's modulus calculated (Vithanage et al., 2009).

Consistency of milk fat has also been determined using a penetration test equipped with a conical probe working at a constant speed to a certain depth, and consistency calculated as the yield value (Rodrigues et al., 2007). Adhesiveness can also be obtained using a penetrometry test. Adhesiveness is the work necessary to overcome the force of attraction between the area of butter and other solids coming into contact with it. Adhesiveness is defined as the negative force-time value generated during probe withdrawal (Bobe et al., 2003; Glibowski et al., 2008). Penetrometry tests offer a simple and economical method to evaluate textural properties of butter and the results obtained correlated well with sensory studies (Wright et al. 2001; Bobe et al., 2003).

Spreadability of butter has been determined by Glibowski et al. (2008) using a texture analyzer with a rig attachment. A female cone was filled with samples (90° angle). During the analysis, samples were displaced to within 0.5 mm of the base of the female cone using a corresponding male cone attachment for the texture analyzer. Force was measured for the duration of the test, and spreadability was equated to the area under the curve. Smaller forces required to spread samples reflects easier spreadability of butter.

Numerous attempts to correlate textural results obtained using large deformation techniques with sensory attributes of butter have been carried out. Despite the limitations of empirical testing, results obtained often correlate well with sensory evaluation (Bobe et al., 2003; Wright et al., 2001). Thus, because sensory tests are complicated, expensive, and time consuming, the possibility of replacing sensory evaluation with an instrumental method is very attractive.

Small deformation techniques have also been used to study textural properties of butter because it exhibits a viscoelastic behavior at small stresses. Viscoelasticity can be probed by evaluating the relationships among stress, strain, and time, using small deformations. This has to be carried out within the region where the relationship between stress and strain is linear, and when the structure of the sample remains intact (Wright et al., 2001). Butter viscoelasticity has been recently studied using rheometers with a number of configurations and geometries. Vithanage et al. (2009) used a rheometer with parallel plate fixtures to study the effect of temperature on the rheological properties of butter in order to correlate the magnitude of storage modulus and elastic modulus of the product at the linear viscoelastic region. They found a good correlation between the Young's modulus and shear elastic modulus of butter. These parameters also had a good relationship with those obtained using cone penetrometry in a texture analyzer at the temperatures studied. Glibowski et al. (2008) determined the flow curves and the apparent viscosity of butter using a rheometer, and observed that apparent viscosity was not very high correlated with hardness and spreadability determined with cone penetrometry.

Behkhelifa et al. (2008) developed a device to study the rheology of food products during heat processes in scraped surface heat exchangers. It consists of a cylindrical viscometer equipped with scraper blades and is called a *scraper-rheometer.* Measured viscosities were in good agreement with reported literature values or with measurements performed with a classical geometry. This technique could be used to follow the rheology of food products that are subjected to heat transfer and shear rates and that show Newtonian or shear thinning behavior, such as cream and butter.

15.3.4 Microstructure

Microstructural studies of butter have been carried out to determine the number and size of milk fat crystals as well as their morphology and polymorphism because they greatly influence the mechanical and thermal properties of the product. The fat crystal network of butter is affected by the processing conditions, cooling rate, and temperature to which butter is subjected. The microstructure of crystals present in milk fat obtained from butter has been studied using polarized light microscopy

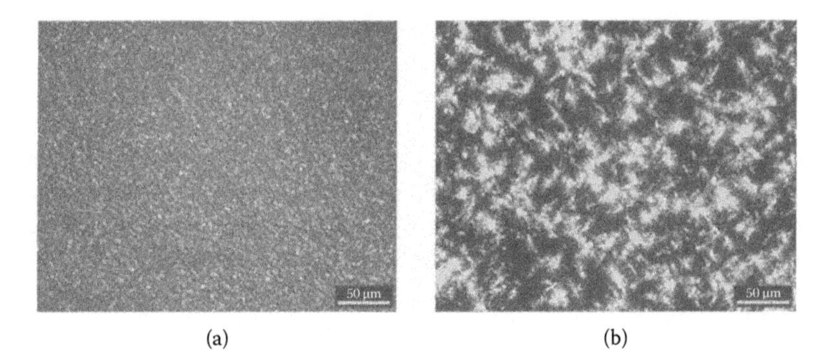

FIGURE 15.6 Polarized light micrographs of milk fat cooled rapidly (A) and slowly (B). (From Campos, R., S.S. Narine, and A.G. Marangoni. 2002. Effect of cooling rate on the structure and mechanical properties of milk fat and lard. *Food Research International* 35: 971–981. Reprinted with permission from Elsevier, License number 2637840505965, dated March 28, 2011.)

(PLM) (Rodrigues et al., 2007; Wright et al., 2000), CSLM (Van Lent et al., 2008), and electron microscopy (Precht and Peters, 1981).

Polarized light microscopy exploits the difference in the refractive index of a beam of incident light polarized in two perpendicular directions. Since fat crystals are birefringent, they will appear as sharp bright objects in a nonbirefringent, and therefore dark, background (Wright et al., 2000). Rodrigues et al. (2007) studied the different microstructures of milk fat from butter upon slow and rapid crystallization. As seen in Figure 15.6, milk fat that crystallized rapidly showed a granular morphology composed by crystals in great amounts and small size. This granular morphology arose from growth processes characterized by high crystallization rates. However, for slow crystallization processes, crystal aggregates are observed (Campos et al., 2002; Rodrigues et al., 2007).

The polymorphic forms present in milk fat can be determined by x-ray diffraction (XRD) techniques. The polymorphic forms are characterized by the d-spacings of the crystal lattices as observed in XRD patterns, which correspond to the distances associated with the lateral packing of the fatty acid hydrocarbon chains. Three different polymorphic crystal forms have been observed in milk fat: γ, α, and β'. The kind of polymorph formed during crystallization of milk fat from its melted state is dependent on the cooling rate and the final temperature. Moreover, transitions between the different polymorphic forms were shown to occur upon storing or heating the milk fat (Campos et al., 2002; ten Grotenhuis et al., 1999).

On the other hand, microstructural studies performed on butter have been also focused on the size distribution of the water droplets present in the product (Van Lent, 2008). The aqueous phase of butter is present in small droplets of less than 10 μm. The presence of this globular water may lead to an increase in the effective viscosity of the system and lends rigidity to the butter because of the high energy needed to overcome the surface tension of the globules (Wright et al., 2001). Therefore, the physical stability of food emulsions such as butter is determined in great part by the water droplet size distribution. It also determines the microbial stability as well as the rheological properties and flavor release of butter (Voda and Duynhoven, 2009; Van Dalen, 2002; Van Lent et al., 2008). Therefore, its experimental determination is of technological importance, and a fast and efficient method to determine the water droplet size distribution in butter is therefore valuable.

Some conventional techniques, such as laser diffraction, for the determination of the water droplet size distribution in butter, have been reported to be unsuitable because part of the continuous fat phase is in a solid state. Conventional wide-field light microscopy can also be used, but it requires thin sample slices that limit the detectable droplet size (Van Lent et al., 2008). The sample must be diluted and squeezed between glass slices, which may deform the large droplets (Van Dalen, 2002). The pulsed-field gradient low-resolution NMR (PFG-NMR), which operates at a low field, typically 20 MHz for ^{1}H, has gained popularity as a droplet-sizing tool in food emulsions. It works with undiluted samples, is easy, fast (15 min) and relatively cheap compared to high-resolution NMR (Van Duynhoven et al., 2002). On the other hand, the use of CSLM after staining the fat with Nile red has also been extended to study water droplet size distribution in food emulsions, including butter. The profiles of nonfluorescent water droplets in 2D images are identified and measured using image analysis. The water droplet size distribution is calculated from the distribution of the measured profile diameters using a transformation of log-normal distributions. This technique has the advantage of giving visual information and samples can be observed undiluted without the need for thin specimens (Van Dalen, 2002; Van lent et al., 2008).

Van lent et al. (2008) compared the use of CSLM and PFG-NMR to determine the water droplet size distribution in various commercial butter samples and significantly different results were obtained for samples analyzed by both techniques.

15.3.5 Color

Another physical property of butter that is important for consumer acceptability is color. The color of butter is usually determined using a colorimeter and the Hunter L values for lightness, and a and b values for chromaticities are usually determined (Jinjarak et al., 2006; Kim et al., 2006; Krause et al., 2008).

15.4 MEASUREMENT OF PHYSICAL PROPERTIES OF YOGURT

Yogurt is a fermented dairy product obtained by acid coagulation of milk due to the metabolic activity of lactic acid bacteria. There are three main types of yogurt: set yogurt, stirred yogurt, and drinking yogurt. *Set yogurt* is fermented in the same container and all the ingredients must be added to it before coagulation. The main ingredients of set yogurt are homogenized and pasteurized milk and two types of lactic acid bacteria, *Streptococcus thermophillus* and *Lactobacillus delbrueckii* subspecies *bulgaricus*. In addition, set yogurt usually contains milk derivates to increase dry matter, such as skimmed milk powder, and in some cases stabilizers or hydrocolloids, like gelatin or starch. These ingredients are used to increase firmness and water retention holding capacity, two important characteristics in set yogurt. In contrast, *stirred yogurt* is fermented in bulk and stirred after coagulation; consequently, other ingredients apart from milk and starters are added after stirring. Stirred yogurt is not usually enriched in dry matter as set yogurt, viscosity being more important than firmness. In stirred yogurt, viscosity is achieved by choosing certain strains of lactic acid bacteria that produce high levels of exopolysaccharides (Duboc and Mollet, 2001). Drinking yogurt is a stirred yogurt with a sufficiently low total solids content to achieve a liquid consistency and normally has undergone homogenization to further reduce the viscosity. In this section, we describe the methods used to measure the main physical properties of yogurt that are directly related to the organoleptic quality of this product.

15.4.1 Textural and Rheological Properties

Texture is one of the most appreciated features of set yogurt by consumers. The main sensory attributes related to yogurt texture are: thickness (or viscosity), smoothness, and sliminess (or ropiness) (Sodini et al. 2004). Although data obtained from instrumental analysis do not always correlate well with results obtained by sensory analysis, instrumental measurement gives some relevant information that can help to know how the milk base, lactic starters, and manufacturing process affect the final texture of yogurt (Sodini et al. 2004).

Yogurt is a nonNewtonian pseudoplastic material with shear-thinning, yield stress, viscoelasticity, and a highly time-dependent behavior (Basak and Ramaswamy, 1994; Benezech and Maingonnat, 1994). Measuring its rheological behavior is difficult because of its poor reproducibility, sensitivity to sample preparation, sensitivity to shear history, and wall slip (Yoon and McCarthy, 2002). These problems make rheological characterization a challenge in routine quality control, where it is necessary to analyze many samples in a short time.

Dynamic tests are used to measure viscoelastic properties of set and stirred yogurts. Viscosity is the property of a material to resist deformation. Two types of viscosity can be measured: extentional or shear-free viscosity and shear viscosity. In shear-free viscosity there is no velocity gradient among liquid molecules when liquid starts to flow (Mortazavian et al., 2009). Shear viscosity is normally measured in yogurt using rotational viscometers, which apply a given shear rate to the sample introduced between a support plate and a stirring cone, between two plates, or in a concentric cylindrical cup. In the review by Mortazavian et al. (2009) there is a good selection of publications on viscometric tests applied to yogurt reflecting the measuring system and recorded parameters used. In these tests, the torque measurements result from the resistance of the fluid to the movement, and allow the determination of apparent viscosity, which is often observed to decrease with increasing shear rates (Duboc and Mollet, 2001). Another important rheological property of yogurt is thixotrophy (or elasticity), which is the ability to recover the original structure after cessation of a shearing action. This property can be measured by applying an oscillatory stress to the sample. The dynamic response of a viscoelastic fluid to an oscillatory stimulation results from two contributions: the elastic or storage modulus G' and the viscous or loss modulus G''; corresponding to elasticity and viscosity, respectively, and the loss tangent (tan a), which is the ratio of the viscous and elastic properties (Duboc and Mollet, 2001). In the linear viscoelastic region, that is, when low strain

(<10%) or low stress (<5 Pa) are applied, the structure is maintained and the yogurt properties can be characterized (Gauche et al., 2009).

Rheological methods permit the calculation of true rheological behavior, yielding stress, and apparent viscosity. The test that is usually used is to apply to the sample different shear rates ranging from 0 to 1000 s^{-1} for a few minutes. Sometimes a linear down shear rate to 0 s^{-1} is applied afterward in order to study the thixotropic behavior of yogurt (Sodini et al., 2004).

Yield stress is a rheological property defined as the minimum shear stress that is required to start flow, which can be used to characterize the firmness of yogurt. This magnitude depends on the characteristic time of the sample, the time of the process, and on the history of structural alteration of the sample prior to measurement; therefore, it is not easily measured (Ares et al., 2006). Yield stress has been determined by the vane method in many dairy products including yogurt (Harte et al., 2003). According to the results of some studies, the vane yield stress of yogurt is highly correlated with the sensory firmness evaluated by trained panelists (Harte et al., 2007).

In some studies, rheometry has been combined with the addition of enzymes to mimic mouth melting. Alting et al. (2009) determined in vitro melting of set yogurt with different formulations by torque measurements using a rheometer equipped with vane geometry and adding alfa-amylase to reproduce human saliva activity.

To quantify the ropiness or sliminess of yogurt, which depends mainly on the production of exopolysaccharides by lactic acid bacteria, a simple test was developed by Hess et al. (1997) using the texture analyzer. This analyzer is equipped with a platen 3.8 cm in diameter and the sample of yogurt is placed on a flat block. A probe is lowered onto the sample to a height of 5 mm above the block and is moved upward 100 mm (rate of 10 mm/s). The time required to break the strand formed between the probe and the block is recorded, and extensibility is calculated from the distance to break.

Another rheological property important for yogurt manufacturing is its flow behavior, which can be characterized by empirical tests; these tests allow definition of specific parameters such as Brookfield viscosity, or flowing time for Posthumus funnel, or Cenco or Bostwick consistometers (Sodini et al., 2004).

It has been stated that the most appropriate technique for measuring the consistency of set yogurt is to use penetrometers, which constitute one of the simplest and most widely used types of texture-measuring instruments (Benezech and Maingonnot, 1994). The puncture test measures the force required to push a punch or a probe into a food product. A large variety of probes, penetration rates, penetration depths, and temperatures has been used, and consequently, it is hardly possible to compare results between different works. However, it is very useful to study the effect of different components and conditions of processing on yogurt texture. Derived from the penetrometic methods, texture-profile analysis is used to study set-type yogurts. The test consists of compressing a bite-sized piece of food twice in a reciprocating motion, which imitates the action of the jaw. A number of textural parameters were extracted from the resulting force–time curves, which are closely correlated to sensory evaluations (Benezech and Maingonnot, 1994).

The main parameters of the texture profile measured in yogurt by a General Food Texturometer are gumminess, cohesiveness, adhesiveness, and hardness from

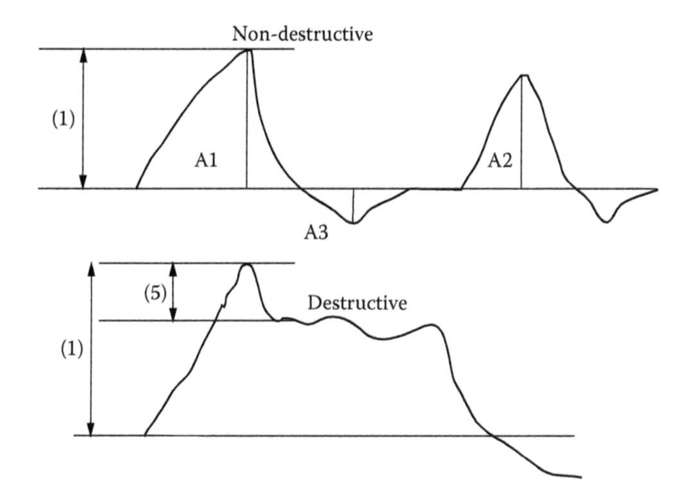

FIGURE 15.7 Texture profile analysis. (From Benezech, T., and J.F. Maingonnat. 1994. Characterization of the rheological properties of yoghurt: A review. *Journal of Food Engineering* 21: 447–472. Reprinted with permission from Elsevier, License number 2637831327589, dated March 28, 2011.)

the nondestructive curve and fracturability or brittleness from a destructive curve (Figure 15.7).

These parameters can give quite good information on the influence of different ingredients on milk texture, such as added solids or stabilizers or emulsifiers (Gauche et al., 2009; Fiszman et al., 1999) as is shown in Figure 15.8. Furthermore, the texture measurement of set yogurt may give valuable information on quality control of commercial production, storage stability, sensorial evaluation by consumers, and knowledge of the effects of mechanical processing on yogurt structure (Gauche et al., 2009). For texture analysis, as an example, Gauche et al. performed measured samples that were kept in plastic containers (45 mm diameter) of 80 mL (6 ± 1°C) for 3 days until the moment of analysis. The instrumental texture analysis was realized in a texturometer. The operation speed was 2.0 mm/s and the distance covered in the sample was 20 mm, using a cylindrical probe.

Empirical or imitative methods such as penetrometry tests, texture profile analysis, and Posthumus funnel have been preferred to characterize the textural properties of yogurts because they are inexpensive, can be generally correlated to sensory measurements, and do not require mathematical treatments (Ares et al., 2006). However, the main disadvantages of these empirical or imitative methods, when compared to fundamental measurement of rheological properties, are the use of relative scales and that results are for a given set of experimental conditions, making it hardly possible to compare results unless the same conditions are used (Ares et al., 2006).

The texture of stirred yogurt has been characterized by a back extrusion (pseudo-compression) method described by several authors (Pereira et al., 2003; Staffolo et al. 2004; Patrignani et al., 2007; Ciron et al., 2010). The textural parameters derived

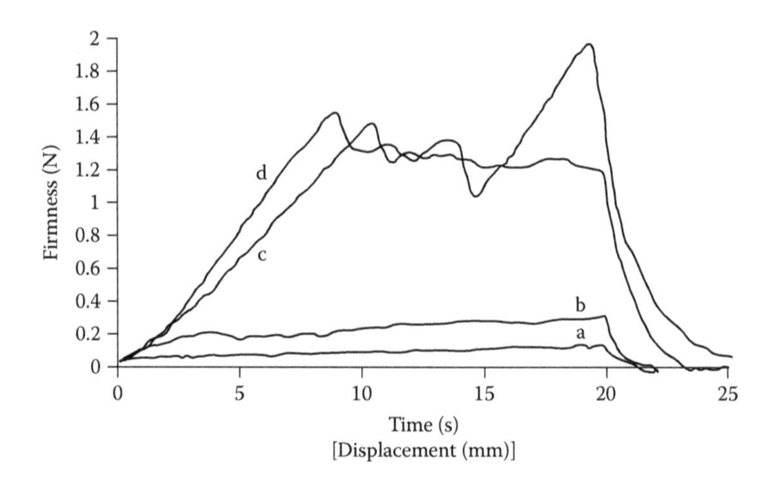

FIGURE 15.8 Penetrometry profiles of yogurts; plain yogurt (a), yogurt with 5% milk solids (b), yogurt with 1.5% gelatin (c), and yogurt with 5% milk solids, and 1.5% gelatin (d). (From Fiszman, S.M., M.A. Lluch, and A. Salvador. 1999. Effect of addition of gelatin on microstructure of acidic milk gels and yoghurt and on their rheological properties. *International Dairy Journal* 9: 895–901. Reprinted with permission from Elsevier, License number 2638210963549, dated March 28, 2011.)

from this measurement were maximum force in compression (firmness), positive area of the curve (consistency); maximum negative force, which indicates the resistance to withdrawal of the sample from the extrusion disc being lifted (cohesiveness); and negative area of the curve (viscosity index). Consistency and index of viscosity are related parameters; consistency indicates the thickness of the sample, while the index of viscosity gauges the resistance of the sample to flow off the disc during back extrusion (Ciron et al., 2010).

15.4.2 Particle Size

Particle size of stirred yogurt has been reported to be related to texture perception, in particular with creaminess (Cayot et al., 2008). The gel particle size distribution of yogurt after stirring conditions has been studied by laser light scattering techniques and it has been concluded that the smoothness perceived both visually and orally varies with the size of the particles of the stirred gels (Cayot et al., 2008).

The effect of high-pressure microfluidization or conventional homogenization on the particle size in milk to elaborate yoghurt, gel particle size, and textural quality of the product obtained by dynamic light scattering have been studied (Ciron et al., 2010). The gel particle size and microstructure of yogurt obtained from the two technologies were significantly different, with the microfluidized product having larger gel particles, apparently as a consequence of more fat globules connected and bound to the protein.

15.4.3 Microstructure

Microstructure of yogurt consists of a matrix of aggregated casein particles, with the fat globules embedded in this matrix. The cavities of the gels are filled with serum and bacterial cells; these are attached to the protein through polysaccharides. In yogurts with a good level of exopolysaccharides, the gel structure is more homogenous with randomly distributed small cavities, while yogurts obtained with nonropy strains show larger cavities (Duboc and Mollet, 2001). These interactions between components can be disrupted in stirred yogurt by shearing forces.

Microstructure studies make it possible to know about textural properties of yogurt and rheological behavior leading to improving, maintaining, and designing the texture (Mortazavian et al., 2009). The study and observation of gel microstructure and homogeneity of the network is essential to understand the molecular reasons for appearance changes of yogurt, and has been performed by scanning or transmission electronic microscopy (Sodini, 2004). Microscopic characterization by cryo-SEM has been used to evaluate the formation of gel using different additions of gelatin or milk solids (Fiszman, 1999). A relatively new technique called environmental scanning electron microscopy (ESEM) has also been applied to dairy product observation, in particular of yogurt. This technique has the advantage that the samples are viewed at any temperature in their natural, fully hydrated state; special preparation of the sample is not required for SEM (Mortazavian et al., 2009).

Light microscopy and CSLM have been also applied to study the microstructure of yogurt with interesting results when evaluating the impact of adding different ingredients in the gel structure. Thus, in the work by Alting et al. (2009), milk fat replacement by amylomaltase-treated starch was analyzed by light microscopy and CSLM. In that work, the microscopic analysis showed that the amylomaltase-treated starch is perfectly integrated in the protein phase and is not separated in the serum phase, and thus acts as a good fat replacement.

The application of CSLM by Girard et al. (2007) was directed at evaluating the effect of interaction of exopolysaccharides with the protein network on the resistance of yogurt to shearing and to observe the structure remaining after stirring. Other studies, like that carried out by Ciron et al. (2010), used confocal microscopy to analyze the effect on yogurt structure of previous treatments to milk, such as high-pressure homogenization in a microfluidizer compared with conventional homogenization. They used selective staining for fat and protein, and reported that the structure observed correlated well with the results obtained in texture analysis.

Unlike electron microscopy, CSLM has minimal sample preparation steps due to its optical sectioning capabilities, which can enable the microstructure of yogurt gels to be monitored without disturbing the gel structure or causing artifacts. Specific components of gels, such as protein, fat, and (live or dead) bacteria, can be identified using specific fluorescent probes.

15.4.4 Color

Although color is not as important an attribute of yogurt as texture, it has been measured for some specific purposes. For example, the influence of color has been determined to evaluate its influence on perception of other attributes, such as sweetness and flavor, in new products (Calvo et al., 2001). The effect of adding fiber on the final color and acceptability of yogurt has also been studied (Staffolo et al., 2004; Hashim et al., 2009). In many market studies, especially when the acceptance of a new product is evaluated, color is usually measured because it is a parameter considered very positively by the consumer. Color is measured by colorimetry, and the parameters determined are in accordance with the CIELAB system, the lightness, L*, redness, a*, yellowness, b* coefficients.

New information processing techniques like artificial neural network (ANN) modeling are being increasingly applied to food analysis, as they are nondestructive and permit the control of product continuously along shelf life. In fact, color is an influential attribute of visual information and a powerful descriptor of measurement in image analysis for food products (Sofu and Ekinci, 2007). The first application of ANN to evaluate yogurt quality was performed by digital image processing using a machine vision system to determine color changes during storage (Sofu and Ekinci, 2007). The data obtained were modeled with an ANN for prediction of yogurt shelf life with good results. The use of ANN may provide an inexpensive and easy technique for evaluation of yogurt quality parameters, and is proposed as an alternative method to control the expiration date, estimate the shelf life of yogurt, and ensure the safety of the product.

15.4.5 Water-Holding Capacity

Whey separation, which refers to the appearance of liquid (whey) on the surface of a milk gel, is a common defect in fermented milk products. Whey separation can occur if the gel network is damaged or if the gel undergoes substantial structural rearrangement. Spontaneous syneresis (the contraction of the gel without the application of any external force, e.g., centrifugation), is the usual cause of whey separation (Lucey and Singh, 1998). The amount of expulsed whey is inversely related to the water-holding capacity (WHC) of the gel. The percentage (w/w) of concentrated yogurt obtained after static or dynamic drainage is often used in literature to define the WHC. Yogurt WHC depends mainly on the total milk solids content and can be determined by permeability or drainage tests. Drainage is the most widely used technique and can be static or dynamic, by meshing or centrifugation (Staffolo et al., 2004). It is necessary to consider that centrifugation tests can give results that are not relevant to syneresis under normal storage conditions, though the mechanical stability of the yogurt protein network under g-forces is tested much more extensively than those under normal storage (Sodini et al., 2004).

15.5 PHYSICAL PROPERTIES OF CHEESE

Cheese is a milk product that is consumed around the world, with a great number of different varieties. In general, cheese is produced by coagulation of casein, drainage of whey, and in most cases, a final stage of maturation. This maturation occurs during a certain period of time and under specific conditions according to each type of cheese. Although there are many common features to all cheeses, the origin and characteristics of milk, the manufacturing process, and the maturation conditions, result in the great differences in appearance, texture, and flavor of the various cheese types. Control of some steps in cheese manufacturing and of certain characteristics of quality can be performed by measuring its physical properties.

15.5.1 Milk Coagulation: Determination of Curd Setting

Milk gels are produced in both cheese and yogurt manufacturing. Gel formation is produced by coagulation of milk that may be induced by enzymes, acidification, or heat treatment, or by a combination of them. The visual appearance, microstructure, and rheological properties of milk gels are important physical attributes that contribute to the sensorial and functional characteristics of the final product (Lucey, 2002). Control of milk coagulation on the production line has been attempted by different methods to determine the optimum time for gel cutting. This point is critical for cheese yield and quality of the final product at big-scale industrial manufacturing facilities (Bakkali et al, 2001). Methods assayed to determine the time for gel cutting are numerous and based on different principles; Lucey (2002) contains a very good review of them. Among the on-line techniques developed to determine the optimum firmness of milk coagulum are viscosity, turbidity, thromboelastography (known commercially as Formagraph), electrical conductivity, diffuse reflectance, dynamic light scattering, vibrational viscometry, oscillatory rheometry, near-infrared spectroscopy, thermal conductivity, refractometry, diffusing wave spectroscopy, microscopy, electroacoustics, fluorescence spectroscopy, dark field microscopy, and low- and high-frequency ultrasound (Lucey, 2002). Few of these techniques are industrially applied at this moment, and the factories still operate on standard time schedules or on the decision of experienced cheese makers assessed by empirical tests. The application of these techniques in the industry is difficult because of technical problems, such as the procedure to include sensors on the production line, the maintenance of clean and operative devices, and the requirement to frequently readjust the sensors to the different conditions of cheese processing and variation of milk composition.

Different approaches have been used to measure some parameters in order to follow cheese production. As an example, image texture analysis has been used to monitor syneresis with the final objective of introducing this technique as a rapid, consistent, and nondestructive method to control on-line cheese manufacturing (Fagan et al., 2008). In this study, images of the surface of the stirred curd–whey mixture were captured using a computer vision system during syneresis and subjected to image texture analysis. Multiscale analysis techniques of fractal dimension and wavelet transform were shown to be the most useful methods for predicting syneresis indices.

The monitoring of curd moisture content during syneresis has also been performed using an image capture system mounted at a sight glass on a wall of a cheese vat, in conjunction with four different image processing techniques (Mateo et al., 2010). These authors showed that the threshold technique, based on discrimination according to a reference intensity, was the simplest and fastest technique of those evaluated. Industrial application of this technique would still require the development of a special camera to be included on the production line.

15.5.2 Textural and Rheological Properties

The rheological properties of cheese are those that determine its response to stress or strain, as applied during compression, shearing, or cutting. These properties affect the handling of cheese, eating quality, use as an ingredient, ability to maintain its shape, and capacity to form eyes or cracks or to swell (O'Callaghan and Guinee, 2004).

Texture is one of the most important quality attributes of cheese for consumer acceptance of the product. Many instrumental techniques, such as small-strain methods (transient and dynamic oscillation) and large-strain tests (texture profile analysis, uniaxial compression, cone penetration, puncture, torsion, bending test, and wire cutting test), are used to measure the physical properties of cheese and are related to some extent to sensory attributes (O'Callaghan and Guinee, 2004; Truong et al, 2002). Large-strain (fracture) techniques break down the gel network; therefore, the properties measured are more related to the mastication process and sensory texture. Meanwhile, small-strain techniques allow investigation of the network without disrupting structure (Bowland and Foegeding, 2001).

Texture profile analysis (TPA) has been widely applied to study cheese texture (Bourne, 2002; Lobato-Calleros, 2007). Lobato-Calleros (2007) describes an example of conditions used to determine the texture profile of cheese in a texturometer as follows: cheese cylinders (1 cm in diameter by 1 cm in height) were taken from the central part of a cheese piece using a borer and a sharp knife. Each sample was compressed 50%, using two compression cycles at a constant crosshead velocity of 2 mm/s (after being at 20°C for 4 hours). Mechanical primary characteristics of hardness, springiness, adhesiveness, and cohesiveness were selected from the two successive uniaxial compression cycles of TPA. The secondary characteristic of chewiness (hardness × cohesiveness × springiness) was also estimated.

The main four rheological parameters (firmness, fracturability, elasticity, and cohesiveness) were defined for cheese by the International Dairy Federation (1991). *Firmness* is the force required to attain a given deformation; *fracturability* expresses the force at which the material fractures; *elasticity* or springiness is defined as the rate at which a deformed material returns to its undeformed condition after the deforming force is removed; and finally, *cohesiveness* is defined as the quantity simulating the strength of the internal bonds making up the body of the product.

Bowland and Foegeding (1999) investigated factors determining large-strain (fracture) rheological properties on a model processed cheese. The properties determined by torsional analysis were fracture or shear stress (force/area at fracture), fracture or shear strain (degree of deformation at fracture), fracture modulus (fracture stress/fracture strain), and slope ratio (a measure of the nonlinearity of the stress–strain curve). Fracture stress and strain have been shown to correlate with sensory textural properties of natural and processed cheese. The same authors studied processed cheese by small-strain techniques using oscillatory analysis (Bowland and Foegeding, 2001) and showed that the results obtained by these techniques provided indicators of the organization of the gel network responsible for fracture properties they had previously obtained.

The vane method has been investigated to measure the texture of Cheddar cheese in comparison with TPA and compression tests (Truong et al., 2002). The shear stress and deformation obtained using the vane method were lower than the shear stress and strain values obtained from other large-strain methods such as torsion. However, the texture maps constructed by plotting vane shear stress and apparent strain at failure described the texture of natural and processed cheeses in a similar manner to stress–strain plots from torsion analysis (Truong and Daubert, 2001). The vane method offers the following advantages over torsion and compression: ability to analyze cheeses too weak to withstand sample preparation for torsion and compression tests, minimal destruction of sample structure during loading, and simplicity. Furthermore, cheese firmness and cohesiveness evaluated by sensory panel were well predicted, respectively, by vane stress and apparent strain.

15.5.3 Textural Properties of Special Cheeses: Meltability

Meltability is an important characteristic of some cheeses, such as mozzarella, because of its application as a topping and ingredient of some food products. Meltability may be defined as the ease with cheese flows or spreads upon heating; it has been measured traditionally by two empirical methods: Arnott and Schreiber tests (Wang and Sun, 2002a). Both methods are performed by heating cylindrical samples of a certain diameter and thickness under predefined cooking conditions of temperature and time. Under those conditions, the increase in cheese area and decrease in thickness of cheese discs after cooking are used to determine the meltability index in each of the tests.

The Schreiber test was modified by Muthukumarappan et al. (1999) by changing some conditions and using the spread area measured by a computer imaging system, equipped with a camera, thus avoiding the problems derived from measuring the irregular shape of melted cheese.

Meltability of cheese has been also analyzed by dynamic rheological analysis in a rheometer with parallel plate geometry by a dynamic stress sweep test (Sutheerawattananonda and Bastian, 1998), and also by helical viscometry (Kindstedt and Kiely, 1992). In both studies, it has been shown that the viscoelasticity index based on viscoelastic parameters could be used for predicting cheese meltability.

Computer vision–based methods have been increasingly applied to agricultural and food industry analysis. Thus, a novel noncontact method using a computer vision system was developed to investigate the melting properties of Cheddar and mozzarella cheese, with results that prove it as an objective, efficient, and accurate method (Wang and Sun, 2001, 2002b).

15.5.4 Microstructure, Structure, and Appearance

The acceptance of a cheese product by consumers depends on its appearance, flavor, and texture, characteristics that are the result of a complex combination of microbiological, biochemical, and technological processes. These processes affect the microstructure in a direct or an indirect manner (Pereira et al., 2009).

Light microscopy has been applied to study the microstructure of processed cheese (Bowland and Foegeding, 2001). The samples were frozen with liquid nitrogen and, after being fixed, stained with protein- and lipid-specific stains. The color was then

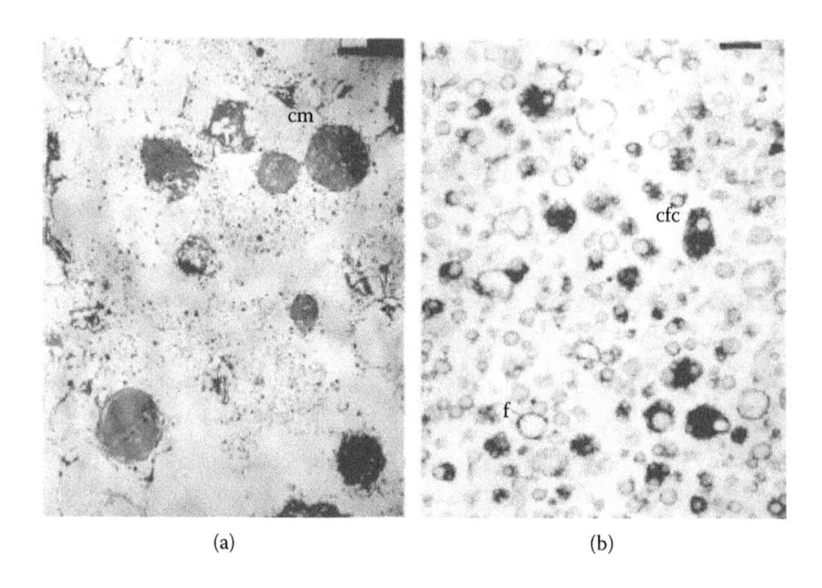

FIGURE 15.9 TM micrographs of pasteurized (a) and pressurized (b) whole milk. Casein micelle (cm), fat globule (f), casein-fat complex (cfc). Bars represent 200 nm. (From Kheadr, E.E., J.F. Vachon, P. Paquin, and I. Fliss. 2002. Effect of dynamic high pressure on microbiological, rheological and microstructural quality of Cheddar cheese. *International Dairy Journal* 12: 435–446. Reprinted with permission from Elsevier, License number 2637781114860, dated March 28, 2011.)

analyzed by image processing techniques to clearly separate the two phases. Color images were converted to L*a*b* images consisting of three channels: luminance, or lightness (L), the chromatic component a, which ranges from green to magenta, and the chromatic component b, which ranges from blue to yellow. Results showed that the distribution and size of lipid particles and their relation to the protein network determine some rheological parameters.

Transmission electron microscopy has also been used to study the effect of different preliminary treatments of milk on cheese microstructure. Thus, the effect of using high dynamic pressure-treated milk on cheese texture is revealed by observation of a casein matrix that is very compact and regular compared with the same cheese produced with pasteurized milk (Kheadr et al., 2002) (Figure 15.9).

Scanning electron microscopy has also been applied to study cheese and permits observation of the compactness of the structure, and the relationships among the different components of cheese: the protein network of casein fibers and the distribution of fat globules in this protein matrix. This is a very useful technique to analyze the quality of products obtained when elaborating cheese in which milk fat is partially or totally replaced by whey protein concentrates and vegetal oils. Differences in the microstructure of cheese are reflected in different textural characteristics (Lobato-Caballeros, 2007) (Figure 15.10).

Another optical method that has been used very extensively in recent years is CSLM. This technique has been applied to analyze the microstructure of cheese

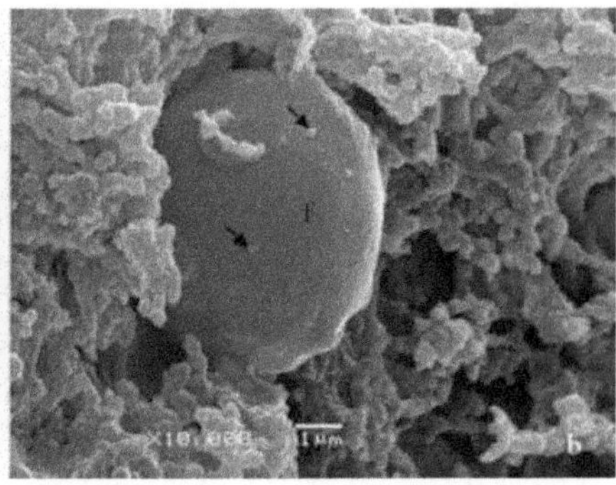

FIGURE 15.10 SME micrographs of full-fat control cheese with milk fat only ($MF_{1.0}$): (a) protein network, (b) milk fat globule, (f) integrated in the protein matrix in whose surface casein micelles (dark arrows) were adsorbed. (From Lobato-Calleros, C., J. Reyes-Hernández, C.I. Beristain, Y. Hornelas-Uribe, J.E. Sánchez-García, and E.J. Vernon-Carter. 2007. Microstructure and texture of white fresh cheese made with canola oil and whey protein concentrate in partial or total replacement of milk fat. *Food Research International* 40: 529–537. Reprinted with permission from Elsevier, License number 2637840162340, dated March 28, 2011.)

after staining with Nile blue A, observing alternately protein and fat phases by exciting at 568 nm and 488 nm, respectively (Capellas et al., 2001).

Food structure and physical state can be related to parameters determined by ultrasonic analysis. The three parameters that are measured most frequently in ultrasonic analysis applied to foods are ultrasonic velocity, attenuation coefficient, and acoustic impedance. The advantage of these methods is that they are nondestructive and can be applied on-line in some technological processes (McClements, 1995). Ultrasound measurement has been used to monitor the structural quality of Swiss cheese by using a single-transducer 2-MHz longitudinal mode pulse-echo setup (Eskelinen et al., 2007) (Figure 15.11). The volumetric ultrasonic image of a cheese sample allows characterization of gas holes (cheese eyes). These holes are the result of the biochemical activity of propionic bacteria, which involves lactate fermentation and production of $C0_2$, or due to some defects (cracks). Normally, the evaluation of hole formation in the ripening of Swiss cheese is made by measuring the cheese height and by analyzing a cheese sample. The units of cheese that have been subjected to sample extraction cannot reach complete maturation and have to be processed for other proposes, losing its whole economical value. Therefore, the application of nondestructive techniques to evaluate the quality of Swiss-type cheese would be very valuable to facilitate quality control and also for economic interests, as its quality, commercial value, and acceptability are greatly affected by eye structure and pattern.

The inner structure of Swiss-type cheeses has also been analyzed by magnetic resonance imaging (MRI) (Rosenberg et al., 1992) as a nondestructive method for the evaluation of eye formation and development during the ripening of cheese. Although it appears to be quite an expensive method, the resolution needed is not as high as in medical applications; consequently, the equipment is less sophisticated and less costly.

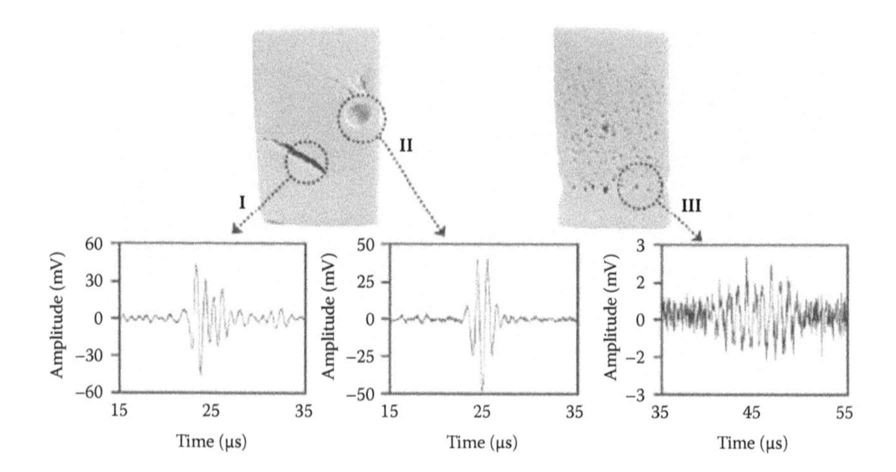

FIGURE 15.11 Ultrasonic reflections (1 MHz, 100 times averaged) from different structural elements found in two Swiss cheese samples (20 cm × 10 cm × 10 cm). Dotted circles and arrows indicate the measured structural elements and their corresponding radio-frequency signals. Characteristic signals from a crack (I) and an eye (II) in a 12-month-old sample (left). Whey nest structure (III) in a 4-month-old sample (right). (From Eskelinen, J.J., A.P. Alavuotunki, E. Hæggstro, and T. Alatossava. 2007. Preliminary study of ultrasonic structural quality control of Swiss-type cheese. *Journal of Dairy Science* 90: 4071–4077. Reprinted with permission from Elsevier, License number 2637790559364, dated March 28, 2011.)

Ultrasounds have also been used to estimate the maturity of Cheddar and Mahon cheeses by determining moisture content and it has been shown that ultrasound velocity increases with cheese maturation and is related to textural properties (Benedito et al., 2000a, 2000b).

Other specific problems in cheese, such as crystalline specks and haze on the surface of Cheddar have been studied by optical methods. This appearance is caused by calcium lactate crystals and is considered a quality defect because consumers mistake them for spoilage microorganisms. Calcium lactate crystals are believed to form when the serum phase of the cheese becomes supersaturated with calcium and lactate ions, followed by the nucleation of calcium lactate crystals and formation of larger aggregates (Rajbhandari and Kindstedt, 2005, 2008). Digital photography combined with image analysis has been used to develop an objective method to measure the area occupied by calcium lactate crystals and its growth on surfaces of naturally smoked Cheddar cheese. This analysis allows correlation of data obtained from processing and storage conditions, and the application of measures to avoid the problem.

15.5.5 Other Physical Properties

Water retention capacity is an important aspect of quality and yield of fresh cheese and is useful when changing certain conditions of cheese processing. Water retention capacity is estimated by measuring whey loss, mainly by two methods: gravity loss and centrifugation. Capellas et al. (2001) used these two methods to estimate the effect of high-pressure treatment of milk on the whey loss in fresh goat cheese. They found that although the gravity method gave very variable results, it was more similar to real conditions of storage. However, the centrifugation method showed lower variability, but could change the behavior of cheese, and results might be more difficult to interpret. Color was also measured in the work by Capellas et al. (2001) to evaluate the impact of high pressure on cheese color. They measured color by colorimetry using the CIELAB scale to calculate total color difference and found more changes on the surface than in the inner part of the cheese.

15.6 PHYSICAL PROPERTIES OF ICE CREAM

Ice cream and related aerated frozen desserts are complex colloidal systems consisting, in their frozen state, of ice crystals, air bubbles, partially coalesced fat globules, and aggregates, all in discrete phases surrounded by an unfrozen continuous matrix of sugars, proteins, salts, polysaccharides, and water. The relative amount of these four phases and the interactions among them determine the properties of ice cream, whether soft and whippy or hard (Goff, 2002).

From the physical point of view, ice cream is both a foam and an emulsion, made up of a dispersion of small particles (<0.5mm) of one phase into another. The air in the ice cream does not mix with the other substances and forms small bubbles in the bulk (defined as foam). Ice cream is also considered an emulsion because the milk cream is dispersed in the ice water. Colloidal aspects play an important role in this structure formation, such as interactions between proteins and emulsifiers at the fat interface, fat–fat partial coalescence, and interactions between proteins and partially coalesced fat at the air interface (Goff, 1997).

The quality of ice cream is characterized in part by a smooth, creamy texture. An icy defect caused by large ice crystals develops at constant and especially fluctuating temperatures through a process known as recrystallization. Depending on storage time, temperature, and the presence of other solutes in ice cream, small ice crystals disappear as larger ice crystals grow, which is an undesirable quality attribute. The function of using stabilizers in ice cream mix is to slow down ice crystal growth (Patmore et al., 2003).

15.6.1 Microstructure

The microstructure of ice crystals and air bubbles is critical to ice cream's quality and sensorial properties. Ice cream is inherently unstable, and temperature variations during transport, storage, or even at home can destroy the microstructure of small ice crystals, leading to recrystallization and to the development of a coarser structure (Patel et al., 2006).

Three types of microscopic techniques are currently used to study the structure of frozen foods: destructive methods, indirect methods, and direct methods. The first method is based on the optical observation of a sample mixed with a suitable medium for dispersing the observed phase and dissolving the other phases. The

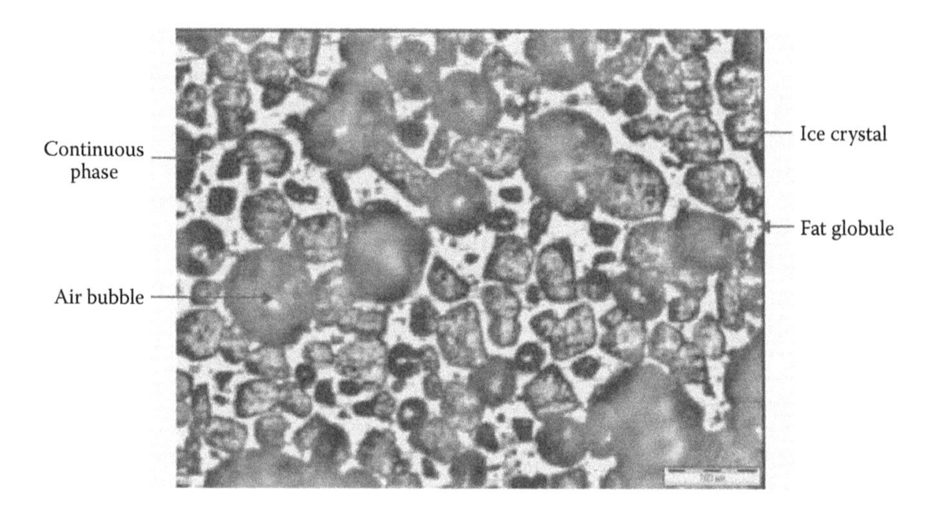

FIGURE 15.12 Structure of ice cream observed by direct microscopy method. $\varepsilon = 0.4$; $T_f = -5°C$; $T_h = -20°C$; $X = 100$. (From Caillet, A., C. Cogné, J. Andrieu, P. Laurent, and A. Rivoire, A. 2003. Characterization of ice cream structure by direct optical microscopy. Influence of freezing parameters. *Lebensmittel-Wissenschaft und-Technologie* 36: 743–749. Reprinted with permission from Elsevier, License number 2637861470927, dated, March 28, 2011.)

second method, performed by electron microscopy, only partially destroys the sample by cryo-substitution, cryo-fixation, freeze-etching, or freeze-drying. The third method is based on photonic microscopy with coaxial lighting and allows direct observation of the sample in its native state at cold temperature. Compared with electron microscopy, the third method has the advantage of preserving the original texture of the frozen sample and having a low running cost. Although it shows lower magnification, the results are comparable with the classical methods of electron microscopy, as shown in Figure 15.12 (Faydi et al., 2001; Caillet et al., 2003).

Light microscopy has been used to determine the mean ice crystal size. Thus, Patel et al. (2006) determined the ice crystal sizes at −18°C by measuring each crystal at its widest point by light microscopy. Patel et al. studied the influence on ice crystal formation, and consequently the texture of ice cream, of the addition (percentage and type) of proteins and processing conditions.

The whipping of ice cream in the manufacturing process is very important for the texture and quality perception of the final product. In a study by Chang and Hartel (2002), air cell size distribution in ice cream was measured quantitatively with two techniques, cryo-SEM and optical microscopy, and image analysis was applied to images obtained from both. Only slight differences were found between the two techniques, indicating that the optical microscope method gives sufficiently accurate results for air cell measurement, though obviously much greater detail of the fine structure in ice cream was observed using cryo-SEM (Figure 15.13). The effect of stabilizers that protect against recrystallization when ice cream is subjected to thermal fluctuations has been also studied by cryo-SEM and the measurement of crystal size performed by image analysis (Flores and Goff, 1999).

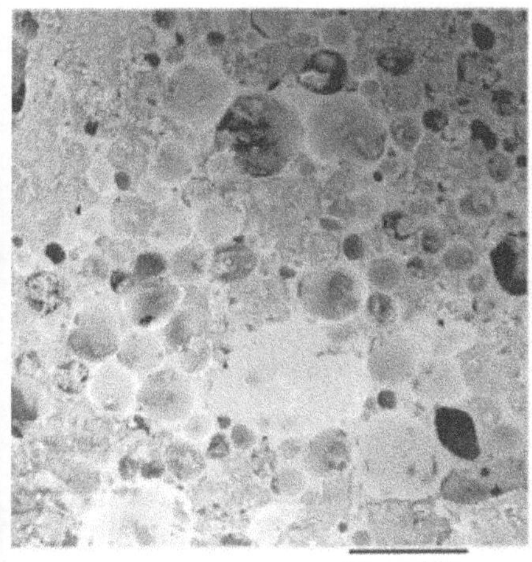

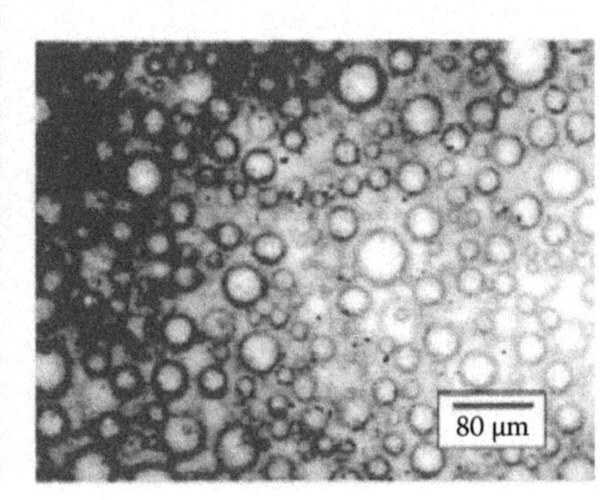

FIGURE 15.13 Comparison of air cells in ice cream taken at the draw of the continuous freezer, as obtained by SEM (left figure) and by optical light microscopy (right figure). (From Chang, Y., and R.W. Hartel. 2002. Measurements of air cell distributions in dairy foams. *International Dairy Journal* 12:463–472. Reprinted with permission from Elsevier, License number 2637790224660, dated, March 28, 2011.)

Regand and Goff (2003) studied fluorescence microscopy to analyze the effect of adding different hydrocolloid stabilizers labeled with rhodamine in recrystallization processes in ice cream after temperature fluctuations.

Some other parameters related to microstructure, such as particle size distribution of the mix samples of ice cream, have been studied by dynamic light scattering. This measurement is essential to evaluate the effect on the rheological parameters of ice cream of applying different homogenization pressures on the mix before and after freezing (Innocente et al. 2008; Sung and Goff, 2010).

15.6.2 Thermal Properties

The knowledge of the thermal properties of ice cream involved in its manufacturing process is essential for estimating freezing times and also for simulating temperature variations through the product during the freezing and storage periods. The most widely used methods to determine thermal properties of ice cream are calorimetry, conductivity measurement, thermal mechanical analysis, and meltdown (Clarke, 2004).

The heat capacity, freezing point curve, and the ice curve can be determined by calorimetry. As an example of a calorimetric analysis of ice cream, in the study by Alvarez et al. (2005), ice cream samples were subjected to cooling to –50°C, held isothermally for 5 minutes, and heated at a rate of 5°C per minute to 50°C. Data collected for each mix included the onset melting temperature, temperatures at initial and final deviations from baseline, peak area, and enthalpy (Figure 15.14).

Thermal conductivity is affected by two factors: temperature (as conductivity of ice is four times higher than that of water) and apparent density, which depends on

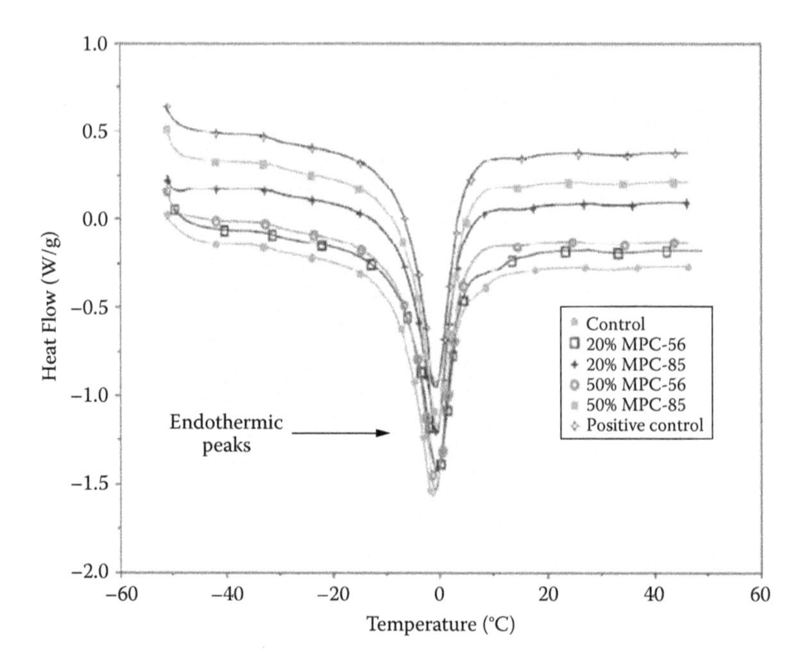

FIGURE 15.14 Differential scanning calorimetry melting curves for ice cream samples. (From Alvarez, V.B., C.L. Wolters, Y. Vodovotz, and T. Ji. 2005. Physical properties of ice cream containing milk protein concentrates. *Journal of Dairy Science* 88: 862–871. Reprinted with permission from Elsevier, License number 2637840813482, dated March 28, 2011.)

air insulation in the mix (Cogné et al., 2003b). In order to model the heat transfer phenomena that take place during ice cream freezing, predictive correlations of the ice cream's thermophysical properties as a function of temperature have been carried out, in particular, specific enthalpy and thermal conductivity (Cogné et al., 2003a, 2003b).

The glass transition temperature of ice cream depends on the composition of the matrix and is measured by thermal mechanical analysis (change of a dimension when subjected to a temperature regime) in an unaerated frozen product. The glass transition temperature is indicated by a change in the rate of expansion, which takes place at about −30°C for a typical ice cream.

The meltdown test measures the capacity of the ice cream to resist melting when exposed to warm temperatures during a certain period of time. The meltdown test of ice cream is an empirical method that reflects the influence of several factors, such as formulation and microstructure due to fat agglomeration and air incorporation (Bolliger et al., 2000; Sofjan and Hartel, 2004).

15.6.3 Textural and Rheological Properties

The first studies on the rheological behavior of frozen ice cream with varying fat content, temperature, and overrun were performed by Shama and Sherman (1966) using compression-type creep experiments and showing that ice cream behaved in a viscoelastic manner.

Traditionally, the firmness of frozen dairy products has been determined by a penetrometer test, which is a one-dimensional approach. However, measuring the complex nature of viscoelasticity, which reflects the existence of solid and fluid behaviors, as in the case of ice cream, has become easier with the use of rheometers. Goff et al. (1995) measured the rheological properties of ice cream mixes made with and without stabilizers using a controlled stress rheometer. They found that stabilized mixes exhibited significantly greater storage and loss moduli at temperatures of less than −8°C. They also reported that as overrun increased from 20% to 60% in the frozen product, tan δ decreased significantly, confirming that air contributes to the elasticity of the final product.

The ability of ice cream to be dipped or scooped is a direct consequence of yield stress, which is defined as the minimum stress required to produce flow. Various ice creams exhibited different degrees of scoopability, even when ice was held at the same temperature. A vane tester was designed and constructed by Briggs et al. (1996) and used to measure the yield stress of the ice cream at typical scooping temperatures of −16°C to −14°C. The vane method is advantageous for testing the physical characteristics of frozen ice cream because it does not destroy the product structure during sample loading.

Yield stress and frequency sweep (rotational) measurements have been performed in ice cream with a rheometer equipped with a cone to evaluate the addition of stabilizers and emulsifiers to ice cream and its effect on crystal growth. Samples were placed on a Peltier plate and the ram was raised slowly to prevent air bubbles under the plate. Solutions or emulsions were subjected on the Peltier plate to five temperature cycles from −1°C (held for 15 min) to −15°C (held for 15 min) to −1°C. There was no significant effect of emulsifier in either stabilizer system, as seen by similar G′ and G″ moduli with or without emulsifier (Patmore et al., 2003).

The properties of different formulations of ice creams were studied by sinusoidal oscillatory tests using parallel plate geometry (Adapa et al., 2000). The results showed that the amount of fat in ice creams and the degree of fat destabilization affected the elasticity of the frozen product. Furthermore, they found that the addition of protein-based and carbohydrate-based fat replacers increased the viscous properties of the mix, though did not enhance the elastic properties of the ice creams. All these data contribute to determine a good balance of milk fat, protein, and carbohydrate when producing lower-fat ice creams.

Oscillation thermorheometry has been used to determine the existence of the different microstructures in ice cream that appear when varying ice cream formulations and the impact of each ingredient function (i.e., fat, proteins, and lipid emulsifiers), and their interactions on the establishment of different networks (Granger, 2004). A stress rheometer fitted with streaked parallel plates (1 mm gap) was used in this study. Before being placed on the rheometer, the products were thermostatically controlled for 2 hours at −10°C. The results obtained demonstrated that the rheological characteristics of ice creams were related to the existence of different microstructures (ice, dispersed air, partially aggregated fat phase, and aqueous phase) that influenced the final texture either simultaneously or individually as a function of the temperature range considered (Figure 15.15).

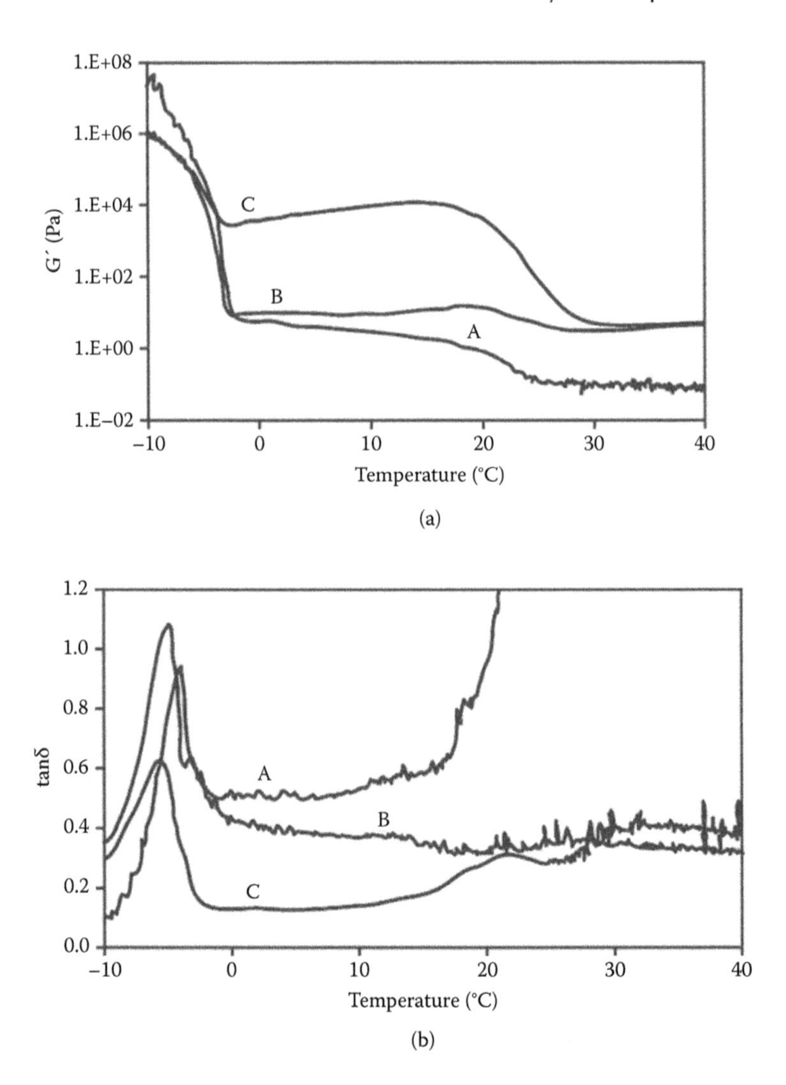

FIGURE 15.15 Rheological characteristics, storage modulus (G′) (a) and tand (b), of different ice cream formulations as a function of the temperature. Incomplete ice cream formulations without proteins (curve A), without lipid emulsifier (curve B), and complete ice cream formulation (curve C). The curves are typical of at least 2 different preparations. (From Granger, C., V. Langendorff, N. Renouf, P. Barey, and M. Cansell. 2004. Impact of formulation on ice cream microstructures: An oscillation thermo-rheometry study *Journal of Dairy Science* 87: 810–812. Reprinted with permission from Elsevier, License number 2637860992183, dated March 28, 2011.)

REFERENCES

Adapa S., Dingeldein H., Schmidt K. A., and T. J. Herald. 2000. Rheological properties of ice cream mixes and frozen ice creams containing fat and fat replacers. *Journal of Dairy Science* 83: 2224–2229.

Alting, A.C., F. van de Velde, M.W. Kanning, M. Burgering, L. Mulleners, A. Sein, and P. Buwalda. 2009. Improved creaminess of low-fat yoghurt: The impact of amylomaltase-treated starch domains. *Food Hydrocolloids* 23: 980–987.

Alvarez, V.B., C.L. Wolters, Y. Vodovotz, and T. Ji. 2005. Physical properties of ice cream containing milk protein concentrates. *Journal of Dairy Science* 88: 862–871.

AOCS (American Oil Chemists' Society) *Official Methods Cd16-81. Solid fat content (SFC) by low resolution magnetic resonance. The indirect method.* Urbana, IL: AOCS.

Ares, G., C. Paroli, and F. Harte. 2006. Measurement of firmness of stirred yogurt in routine quality control. *Journal of Food Quality* 29: 628–642.

Authy M.A.E., Twomey, M., Guinee T.P., and Mulvihill D.M. 2001. Development and application of confocal scanning laser microscopy methods for studying the distribution of fat and protein in selected dairy products. *Journal of Dairy Research* 68: 417-427

Avrami, M. (1941). Kinetics of phase change III. Granulation, phase change, and microstructure. *Journal of Chemical Physics*, 9, 177–184.

Avramis, C.A., H. Wang, B.W. McBride, T.C. Wright, and A.R. Hill. 2003. Physical and processing properties of milk, butter, and Cheddar cheese from cows fed supplemental fish meal. *Journal of Dairy Science* 86: 2568–2576.

Baechler, R., M.F. Clerc, S. Ulrich, and S. Benet, S. 2005. Physical changes in heat-treated whole milk powder. *Lait* 85: 305–314.

Bakkali, F., A. Moudden, B. Faiz, A. Amghar, G. Maze, F. Montero de Espinosa, and M. Akhnak. 2001. Ultrasonic measurement of milk coagulation time. *Measurement Science Technology* 12: 2154–2159.

Basak, S., and H.S. Ramaswamy. 1994. Simultaneous evaluation of shear rate and time dependency of stirred yoghurt rheology as influenced by added pectin and strawberry concentrate. *Journal of Food Engineering* 21: 385–393.

Belitz, H.D., W. Grosch, and P. Schieberle, P. 2009. *Food chemistry*, 4th ed.. Berlin: Springer-Verlag.

Benedito, J., J.A. Carcel, N. Sanjuan, and A. Mulet. 2000a. Use of ultrasound to assess Cheddar cheese characteristics. *Ultrasonics* 38: 727–730.

Benedito, J., J. Carcel, G. Clemente, and A. Mulet. 2000b. Cheese maturity assessment using ultrasonics. *Journal of Dairy Science* 83: 248–254.

Benezech, T., and J.F. Maingonnat. 1994. Characterization of the rheological properties of yoghurt: A review. *Journal of Food Engineering* 21: 447–472.

Benkhelifa, H., G. Alvarez, and D. Flick. 2008. Development of a scraper-rheometer for food applications: Rheological calibration. *Journal of Food Engineering* 85: 426–434.

Bertram, H.C., L. Wiking, J.H. Nielsen, and H.J. Andersen. 2005. Direct measurement of phase transitions in milk fat during cooling of cream: A low-field NMR approach. *International Dairy Journal* 15: 1056–1063.

Bobe, G., E.G. Hammond, A.E. Freeman, G.L. Lindberg, and D.C. Beitz. 2003. Texture of butter from cows with different milk fatty acid compositions. *Journal of Dairy Science* 86: 3122–3127.

Bolliger S., H.D. Goff, and B.W. Tharp. 2000. Correlation between colloidal properties of ice cream mix and ice cream. *International Dairy Journal* 10: 303–309.

Boonyai, P., B. Bhandari, and T. Howes. 2004. Stickiness measurement techniques for food powders: A review. *Powder Technology* 145: 34–46.

Boonyai, P., T. Howes, and B. Bhandari. 2007. Instrumentation and testing of a thermal mechanical compression test for glass–rubber transition analysis of food powders. *Journal of Food Engineering* 78: 1333–1342.

Bourne, M.C. 2002. Food texture and viscosity: concept and measurement, 2nd ed. New York: Academic Press.

Bowland, E.L., and E.A. Foegeding. 1999. Factors determining large-strain (fracture) rheological properties of model processed cheese. *Journal of Dairy Science* 82: 1851–1859.

Bowland, E.L., and E.A. Foegeding. 2001. Small strain oscillatory shear and microstructural analyses of a model processed cheese. *Journal of Dairy Science* 84: 2372–2380.

Briggs, J.L., J.F. Steffe, and Z. Ustunol. 1996. Vane method to evaluate the yield stress of frozen ice cream. *Journal of Dairy Science* 79: 527–531.

Brule, R. 1983. General methods for the analysis of butters. *Comptes Rendus Hebdomadaires des Seances de l´Academie des Sciences* 116: 1255.

Caillet, A., C. Cogné, J. Andrieu, P. Laurent, and A. Rivoire, A. 2003. Characterization of ice cream structure by direct optical microscopy. Influence of freezing parameters. *Lebensmittel-Wissenschaft und-Technologie* 36: 743–749.

Calvo, C., A. Salvador, and S.M. Fiszman. 2001. Influence of color intensity on the perception of color and sweetness in various fruit-flavoured yoghurts. *European Food Research and Technology* 213: 99–103.

Campos, R., S.S. Narine, and A.G. Marangoni. 2002. Effect of cooling rate on the structure and mechanical properties of milk fat and lard. *Food Research International* 35: 971–981.

Capellas, M., M. Mor-Mur, E. Sendra, and B. Guamis. 2001. Effect of high-pressure processing on physico-chemical characteristics of fresh goats' milk cheese (Mató). *International Dairy Journal* 11: 165–173.

Carr, R.G., Jr. 1965. Evaluating flow properties of solids. *Chemical Engineering* 72: 163–168.

Cayot, P., F. Schenker, G. Houzé, C. Sulmont-Rossé, and B. Colas. 2008. Creaminess in relation to consistency and particle size in stirred fat-free yogurt. *International Dairy Journal* 18: 303–311.

Chandan, R. C. 2006. Milk composition, physical and processing characteristics. In *Manufacturing yogurt and fermented milks*, ed. R.C. Chandan, C.H. White, A. Kilara, and Y.H. Hui, 17–40. Ames, IA: Blackwell Publishing.

Chang, Y., and R.W. Hartel. 2002. Measurements of air cell distributions in dairy foams. *International Dairy Journal* 12:463–472.

Chuy, L.E., and T.P. Labuza. 1994. Caking and stickiness of dairy-based food powders as related to glass transition. *Journal of Food Science* 59: 43–46.

Ciron, C.I.E., V.L. Gee, A.L. Kelly, and M.A.E. Auty. 2010. Comparison of the effects of high-pressure microfluidization and conventional homogenization of milk on particle size, water retention and texture of non-fat and low-fat yoghurts. *International Dairy Journal* 20: 314–320.

Clarke, C. 2004. Measuring ice cream. In *The science of ice cream*. Cambridge, UK: The Royal Society of Chemistry.

Cogné, C., J. Andrieu, P. Laurent, A. Besson, and J. Nocquet. 2003b. Experimental data and modelling of thermal properties of ice creams. *Journal of Food Engineering* 58: 331–341.

Cogné, C., P. Laurent, J. Andrieu, and J. Ferrand. 2003a. Experimental data and modelling of ice cream freezing. *Transactions of the Institution of Chemical Engineers* 81(Part A): 1129–1135.

Couvreur, S., C. Hurtaud, C. Lopez, L. Delaby, and J.L. Peyraud. 2006. The linear relationship between the proportion of fresh grass in the cow diet, milk fatty acid composition, and butter properties. *Journal of Dairy Science* 89: 1956–1969.

Davenel, A., P. Schuck, F. Mariette, and G. Brule. 2002. NMR relaxometry as a non-invasive tool to characterize milk powders. *Lait* 82: 465–473.

DeMan J.M., Gupta S., Kloek M., and Timbers G.E. 1985. Viscoelastic properties of plastic fat products. *Journal of American Oil Chemists' Society* 62: 1672–1675.

Dixon, B.D. 1974. Spreadability of butter: Determination .1. Description and comparison of five methods of testing. *Australian Journal of Dairy Technology* 29: 15–22.

Duboc, P., and B. Mollet. 2001. Applications of exopolysaccharides in the dairy industry. *International Dairy Journal* 11: 759–768.

Eskelinen, J.J., A.P. Alavuotunki, E. Hæggstro, and T. Alatossava. 2007. Preliminary study of ultrasonic structural quality control of Swiss-type cheese. *Journal of Dairy Science* 90: 4071–4077.

Fagan, C.C., C.-J. Du, C.P. O'Donnell, M. Castillo, C.D. Everard, D.J. O'Callaghan, and F.A. Payne. 2008. Application of image texture analysis for online determination of curd moisture and whey solids in a laboratory-scale stirred cheese vat. *Journal of Food Science* 73: E250–E258.

Faydi, E., J. Andrieu, and P. Laurent. 2001. Experimental study and modelling of the ice crystal morphology of model standard ice cream. Part I: Direct characterization method and experimental data. *Journal of Food Engineering* 48: 283–291.

Fiszman, S.M., M.A. Lluch, and A. Salvador. 1999. Effect of addition of gelatin on microstructure of acidic milk gels and yoghurt and on their rheological properties. *International Dairy Journal* 9: 895–901.

Fitzpatrick, J.J., S.A. Barringer, and T. Iqbal. 2004 b. Flow property measurement of food powders and sensitivity of Jenike's hopper design methodology to the measured values. *Journal of Food Engineering* 61: 399–405.

Fitzpatrick, J.J., M. Hodnett, M. Twomey, P.S.M. Cerqueira, J. O'Flynn, and Y.H. Roos. 2007. Glass transition and the flowability and caking of powders containing amorphous lactose. *Powder Technology* 178: 119–128.

Fitzpatrick, J.J., T. Iqbal, C. Delaney, T.Twomey, and M.K. Keogh. 2004a. Effect of powder properties and storage conditions on the flowability of milk powders with different fat contents. *Journal of Food Engineering* 64:435–444.

Fitzpatrick, J.J., E. O'Callaghan, and J. O'Flynn. 2008. Application of a novel cake strength tester for investigating caking of skim milk powder. *Food and Bioproducts Processing* 86: 198–203.

Flores, A.A., and H.D. Goff. 1999. Recrystallization in ice cream after constant and cycling temperature storage conditions as affected by stabilizers. *Journal of Dairy Science* 82: 1408–1415.

Freudig, B., S. Hogekamp, and S. Schubert. 1999. Dispersion of powders in liquids in a stirred vessel. *Chemical Engineering and Processing* 38: 525–532.

Gaiani, C., J.J. Ehrhard, J. Scher, J. Hardy, S. Desobry, and S. Banon. 2006. Surface composition of dairy powders observed by x-ray photoelectron spectroscopy and effects on their rehydration properties. *Colloids and Surfaces B: Biointerfaces* 49: 71–78.

Gauche, C., T. Tomazi, P.L.M. Barreto, P.J. Ogliari, and M.T. Bordignon-Luiz. 2009. Physical properties of yoghurt manufactured with milk whey and transglutaminase. *LWT – Food Science and Technology* 42: 239–243.

Girard, M., and C. Schaffer-Lequart. 2007. Gelation and resistance to shearing of fermented milk: Role of exopolysaccharides. *International Dairy Journal* 17: 666–673.

Glibowski, P., P. Zarzycki, and M. Krzepkowska. 2008. The rheological and instrumental properties of selected table fats. *International Journal of Food* Properties 11: 678–686.

Goff, H.D. 1997. Colloidal aspects of ice cream: A review. *International Dairy Journal* 7: 363–373.

Goff, H.D. 1997. Colloidal aspects of ice cream: A review. *International Dairy Journal* 7: 363–373.

Goff, H.D. 2002. Formation and stabilisation of structure in ice-cream and related products. *Current Opinion in Colloid and Interface Science* 7: 432–437.

Goff, H.D., B. Freslon, M.E. Sahagian, T.D. Hauber, A.P. Stone, and D.W. Stanley. 1995. Structural development in ice cream B dynamic rheological measurements. *Journal of Texture Studies* 26: 517–536.

Granger, C., V. Langendorff, N. Renouf, P. Barey, and M. Cansell. 2004. Impact of formulation on ice cream microstructures: An oscillation thermo-rheometry study *Journal of Dairy Science* 87: 810–812.

Hamann, J., and Zecconi, A. 1998. Evaluation of the electrical conductivity of milk as a mastitis indicator. *International Dairy Federation Bulletin* n° 334, Brussels, Belgium, pp. 5–26.

Hardy, J., J. Scher, and S. Banon. 2002. Water activity and hydration of dairy powders. *Lait* 82: 441–452.

Harte, F., S. Clark, and G.V. Barbosa-Cánovas. 2007. Yield stress for initial firmness determination on yogurt. *Journal of Food Engineering* 80: 990–995.

Harte, F., L. Luedecke, B. Swanson, and G.V. Barbosa-Cánovas. 2003. Low-fat set yogurt made from milk subjected to combinations of high hydrostatic pressure and thermal processing. *Journal of Dairy Science* 86: 1074–1082.

Hashim, I.B., A.H. Khalil, and H.S. Afifi. 2009. Quality characteristics and consumer acceptance of yogurt fortified with date fiber. *Journal of Dairy Science* 92: 5403–5407.

Hayakawa, M., S. Hayakawa, and R. Nakamura. 1986. Studies on the consistency of butter: A review. *Japanese Journal Dairy and Food Science* 35: A81–A92.

Hennigs, C., T.K. Kockel, and T.A.G. Langrish. (2001). New measurements of the sticky behavior of skim milk powder. *Drying Technology* 19: 471–484.

Hess, S.J., R.F. Roberts, and G.R. Ziegler. 1997. Rheological properties of nonfat yogurt stabilized using *Lactobacillus delbrueckii ssp. bulgaricus* producing exopolysaccharide or using commercial stabilizer systems. *Journal of Dairy Science* 80: 252–263.

Hogan, S.A., M.H. Famelart, D.J. O'Callaghan, and P. Schuck. 2010. A novel technique for determining glass–rubber transition in dairy powders. *Journal of Food Engineering* 99: 76–82.

IDF Standard 87. 1979. Determination of the dispersibility and wettability of instant dried milk. *International Dairy Federation.*

Ilari, J.L. 2002. Flow properties of industrial dairy powders. *Lait* 82: 383–399.

Ilari, J.L., and L. Mekkaui. 2005. Physical properties of constitutive size classes of spray-dried skim milk powder and their mixtures. *Lait* 85: 279–294.

Innocente, N., M. Biasutti, E. Venir, M. Spaziani, and G. Marchesini. 2008. Effect of high-pressure homogenization on droplet size distribution and rheological properties of ice cream mixes. *Journal of Dairy Science* 92:1864–1875.

International Dairy Federation (1991). Rheological and fracture properties of cheese. *International Dairy Federation Bulletin* n° 268. Brussels, Belgium.

Intipunya, P., A. Shrestha, T. Howes, and B. Bhandari. 2009. A modified cyclone stickiness test for characterizing food powders *Journal of Food Engineering* 94: 300–306.

Jenike, A.W. 1964. *Storage and flow of solids.* Bulletin 123. Engineering Experiment Station, University of Utah.

Jinjarak, S.,A. Olabi, R. Jiménez-Flores, and J.H. Walker. 2006. Sensory, functional, and analytical comparisons of whey butter with other butters. *Journal of Dairy Science* 89: 2428–2440.

Keogh, M.K., C.A. Murray, and B.T. O'Kennedy. 2003. Effects of ultrafiltration of whole milk on some properties of spray-dried milk powders. *International Dairy Journal* 13: 995–1002.

Kheadr, E.E., J.F. Vachon, P. Paquin, and I. Fliss. 2002. Effect of dynamic high pressure on microbiological, rheological and microstructural quality of Cheddar cheese. *International Dairy Journal* 12: 435–446.

Kim, E.H.J., X.D. Chen, and D. Pearce. 2001. Surface characterization of four industrial spray-dried dairy powders in relation to chemical composition, structure and wetting property. *Colloids and Surfaces B: Biointerfaces* 26: 197–212.

Kim E.H.J., Chen X.D., Pearce D. 2002. Surface characterization of four industrial spray-dried dairy powders in relation to chemical composition, structure and wetting property. *Colloids and Surfaces B: Biointerfaces* 26:197–212.

Kim, E.H.J., X.D. Chen, and D. Pearce. 2005. Effect of surface composition on the flowability of industrial spray-dried dairy powders. *Colloids and Surfaces B: Biointerfaces* 46: 182–187.

Kim, E.H.J., X.D. Chen, and D. Pearce. 2009. Surface composition of industrial spray-dried milk powders. 3. Changes in the surface composition during long-term storage. *Journal of Food Engineering* 94: 182–191.

Kim, J.J., T.H. Jung, J. Ahn, and H.S. Kwak. 2006. Properties of cholesterol-reduced butter made with β-cyclodextrin and added evening primrose oil and phytosterols. *Journal of Dairy Science* 89: 4503–4510.

Kindstedt, P.S., and L.J. Kiely. 1992. Revised protocol for the analysis of melting properties of mozzarella cheese by helical viscometry. *Journal of Dairy Science* 75: 676–682.

Kindstedt, P.S., J.K. Rippe, and C.M. Duthie. 1989. Measurement of mozzarella cheese melting properties by helical viscometry. *Journal of Dairy Science* 72: 3117.

Krause, A.J., R.E. Miracle, T.H. Sanders, L.L. Dean, and M.A. Drake. 2008. The effect of refrigerated and frozen storage on butter flavor and texture. *Journal of Dairy Science* 91: 455–465.

Lee, W.J., and J.A. Lucey. 2010. Formation and physical properties of yogurt. *Asian-Australian Journal of Animal Science* 23: 1127–1136.

Lobato-Calleros, C., J. Reyes-Hernández, C.I. Beristain, Y. Hornelas-Uribe, J.E. Sánchez-García, and E.J. Vernon-Carter. 2007. Microstructure and texture of white fresh cheese made with canola oil and whey protein concentrate in partial or total replacement of milk fat. *Food Research International* 40: 529–537.

Lucey, J.A. 2002. ADSA Foundation Scholar Award. Formation and physical properties of milk protein gels *Journal of Dairy Science* 85: 281–294.

Lucey, J.A., and H. Singh. 1998. Formation and physical properties of acid milk gels: A review. *Food Research International* 30: 529–542.

Marangoni A.G., and Narine S.S. 2002. Identifying key structural indicators of mechanical strength in networks of fat crystals. *Food Research International* 35: 957–969.

Mateo, M.J., D.J. O'Callaghan, C.D. Everard, C.C. Fagan, M. Castillo, F.A. Payne, and C.P. O'Donnell. 2009. Influence of curd cutting programme and stirring speed on the prediction of syneresis indices in cheese-making using NIR light backscatter. *LWT – Food Science and Technology* 42: 950–955.

Mateo, M.J., D.J. O'Callaghan, A.A. Gowen, and C.P. O'Donnell. 2010. Evaluation of a vat wall-mounted image capture system using image processing techniques to monitor curd moisture during syneresis with temperature treatments. *Journal of Food Engineering* 99: 257–262.

McClements, D. J. 1995. Advances in the application of ultrasound in food analysis and processing. *Trends in Food Science & Technology* 6: 293–299.

Mistry, V.V., and J.B. Pulgar.1996. Physical and storage properties of high milk protein powder. *International Dairy Journal* 6: 195–203.

Mortazavian, A.M., K. Rezaei, and S. Sohrabvandi. 2009. Application of advanced instrumental methods for yogurt analysis. *Critical Reviews in Food Science and Nutrition* 49: 153–163.

Murti, R.A., A.H.J. Paterson, D.L. Pearce, and J.E. Bronlund. 2009. Stickiness of skim milk powder using the particle gun technique. *International Dairy Journal* 19: 137–141.

Murti, R.A., A.H.J. Paterson, D.L. Pearce, and J.E. Bronlund. 2010. The influence of the particle velocity on the stickiness of milk powder. *International Dairy Journal* 20: 121–127.
Muthukumarappan, K., Y.C. Wang, and S. Gunasekaran. 1999. Short communication: Modified Schreiber test for evaluation of mozzarella cheese meltability. *Journal of Dairy Science* 82: 1068–1071.
O'Callaghan, D.J., and T.P. Guinee. 2004. Rheology and texture of cheese. In *Cheese: Chemistry, physics and microbiology*, Volume 1, 3rd ed., ed. P.F. Fox, P.L.H. McSweeney, T.M. Cogan, and T.P. Guinee. Amsterdam: Elsevier.
Ozkan, N., N. Walisinghe, and X.D. Chen. 2002. Characterization of stickiness and cake formation in whole and skim milk powders. *Journal of Food Engineering* 55: 293–303.
Patel, M.R., R.J. Baer, and M.R. Acharyaa. 2006. Increasing the protein content of ice cream. *Journal of Dairy Science* 89: 1400–1406.
Paterson, A.H., J.E. Bronlund, J.Y. Zuo, and R. Chatterjee. 2007. Analysis of particle-gun-derived dairy powder stickiness curves. *International Dairy Journal* 17: 860–865.
Patmore, J.V., H.D. Goff, and S. Fernandes. 2003. Cryo-gelation of galactomannans in ice cream model systems. *Food Hydrocolloids* 17: 161–169.
Patrignani, F., L. Iucci, R. Lanciotti, M. Vallicelli, J.M. Mathara, W.H. Holzapfel, and M.E. Guerzoni. 2007. Effect of high-pressure homogenization, nonfat milk solids, and milkfat on the technological performance of a functional strain for the production of probiotic fermented milks. *Journal of Dairy Science* 90: 4513–4523.
Pereira, C.I., A.M.P. Gomes, and F.X. Malcata. 2009. Microstructure of cheese: Processing, technological and microbiological considerations. *Trends in Food Science & Technology* 20: 213–219.
Pereira, R.B., H. Singha, P.A. Munro, and M.S. Luckman. 2003. Sensory and instrumental textural characteristics of acid milk gels. *International Dairy Journal* 13: 655–667.
Precht, D., and K.H. Peters. 1981. The consistency of butter II. Relationships between the submicroscopic structures of cream fat globules as well as butter and the consistency in dependence on special physical methods of cream ripening. *Milchwissenschaft* 36: 673–676.
Prentice, J.H. 1972. Rheology and texture of dairy products. *Journal of Texture Studies* 3: 415–458.
Rajbhandari, P., and P.S. Kindstedt. 2005. Development and application of image analysis to quantify calcium lactate crystals on the surface of smoked cheddar cheese *Journal of Dairy Science* 88: 4157–4164.
Rajbhandari, P., and P.S. Kindstedt. 2008. Characterization of calcium lactate crystals on Cheddar cheese by image analysis. *Journal of Dairy Science* 91: 2190–2195.
Regand, A., and H.D. Goff. 2003. Structure and ice recrystallization in frozen stabilized ice cream model systems. *Food Hydrocolloids* 17: 95–102.
Rennie, P.R., X.D. Chen, C. Hargreaves, and A.R. Mackereth. 1999. A study of the cohesion of dairy powders. *Journal of Food Engineering* 39: 277–284.
Rodrigues, J.N., R.P. Torres, J. Mancini-Filho, and L.A. Gioielli. 2007. Physical and chemical properties of milkfat and phytosterol esters blends. *Food Research International* 40: 748–755.
Roos. Y.H. 2002. Importance of glass transition and water activity to spray drying and stability of dairy powders. *Lait* 82: 475–484.
Rosenberg, M., M. McCarthy, and R. Kauten. 1992. Evaluation of eye formation and structural quality of Swiss-type cheese by magnetic resonance imaging. *Journal of Dairy Science* 75: 2083–2091.
Schuck P., Mejean S., Dolivet A., Jeantet R, and Bhandari B. 2007. Keeping quality of dairy ingredients. *Le Lait* 87: 481–488.
Shama, F., and P. Sherman. (1966). The texture of ice cream. 2. Rheological properties of frozen ice cream. *Journal of Food Science* 31: 699–706.

Sherbon, J.W. 1999. *Physical properties of milk in fundamentals of dairy chemistry*, 3rd ed., ed. R. Jennes, N.P. Wong, and E.H. Marth, 409–460. Gaithersburg, MD: Aspen Publishers.

Shrestha A.K,. Howes T., Adhikari B.P., Wood B.J., and B. R. Bhandari. 2007. Effect of protein concentration on the surface composition, water sorption and glass transition temperature of spray-dried skim milk powders. *Food Chemistry* 104: 1436–1444.

Singh, A.P., D.J. McClements, and A.G. Marangoni. 2004. Solid fat content determination by ultrasonic velocimetry. *Food Research International* 37: 545–555.

Sodini, I., F. Remeuf, S. Haddad, and G. Corrieu. 2004. The relative effect of milk base, starter, and process on yogurt texture: A review. *Critical Reviews in Food Science and Nutrition* 44: 113–137.

Sofjana, R.P., and R.W. Hartelb. 2004. Effects of overrun on structural and physical characteristics of ice cream. *International Dairy Journal* 14: 255–262.

Sofu, A., and F.Y. Ekinci. 2007. Estimation of storage time of yogurt with artificial neural network modeling. *Journal of Dairy Science* 90: 3118–3125.

Sofu, A., and F.Y. Ekinci. 2007. Estimation of storage time of yogurt with artificial neural network monitoring. *Journal of Dairy Science* 90(7): 3118–3125.

Staffolo, M.D., N. Bertola, M. Martino, and A. Bevilacqua. 2004. Influence of dietary fiber addition on sensory and rheological properties of yogurt. *International Dairy Journal* 14: 263–268.

Sung, K., and D. Goff. (2010). Effect of solid fat content on structure in ice creams containing palm kernel oil and high-oleic sunflower oil. *Journal of Food Science*, 75: C274–C279.

Sutheerawattananonda, M., and E.D. Bastian. 1998. Monitoring process cheese meltability using dynamic stress rheometry. *Journal of Texture Studies* 29: 169–183.

Tamine, A. Y. and R.K. Robinson. *Yogurt science and technology*. Pergamon Press Ltd, Oxford, 1985. pp-365–373.

Ten Grotenhuis, E., G.A. van Aken, K.F. van Malssen, and H. Schenk. 1999. Polymorphism of milk fat studied by differential scanning calorimetry and real-time X-ray powder diffraction. *Journal of the American Oil Chemists' Society* 76: 1031–1039.

Teunou, E., J.J. Fitzpatrick, and E.C. Synnott. 1999. Characterisation of food powder flowability. *Journal of Food Engineering* 39: 31–37.

Thomas, M.E.C., J.L. Scher, S. Desobry-Banon, and S. Desobry. 2004. Milk powders ageing: Effect on physical and functional properties. *Critical Reviews in Food Science and Nutrition*, 44: 297–322.

Truong, V.D., and C.R. Daubert. 2001. Textural characterization of cheeses using vane rheometry and torsion analysis. *Journal of Food Science*, 66: 716–721.

Truong, V.D., C.R. Daubert, M.A. Drake, and S.R. Baxter. 2002. Vane rheometry for textural characterization of Cheddar cheeses: Correlation with other instrumental and sensory measurements. *Lebensmittel-Wissenschaft. und-Technology* 35: 305–314.

Tsenkova, R., S. Atanassova, K. Toyoda, Y. Ozaki, K. Itoh, and T. Fearn. 1999. Near infrared spectroscopy for dairy measurement: Measurement of unhomogenized milk composition. *Journal of Dairy Science*, 82: 2344–2352.

Tsenkova, R., S. Atanassova, K. Itoh, Y. Ozaki, and K. Toyoda. 2000. Near infrared spectroscopy for biomonitoring: Cow milk composition measurement in a spectral region from 1,100 to 2,400 nanometers. *Journal of Animal Science*, 78: 515–522.

Van Dalen, G. 2002. Determination of the water droplet size distribution of fat spreads using confocal scanning laser microscopy. *Journal of Microscopy* 208: 116–133.

Van Duynhoven, J.P., G.J.W. MGoudappel, G. van Dalen, P.C. van Bruggen, J.C.G. Blonk, and A.P.A.M. Eijkelenboom. 2002. Scope of droplet size measurements in food emulsions by pulsed field gradient NMR at low field. *Magnetic Resonance Chemistry* 40: 551–559.

Van Lent, K., B. Vanlerberghe, P.Van Oostveldt, O. Thas, and P. Paul Van der Meeren. 2008. Determination of water droplet size distribution in butter: Pulsed field gradient NMR in comparison with confocal scanning laser microscopy. *International Dairy Journal* 18: 12–22.

Vega, C., and Y.H. Roos. 2006. Spray-dried dairy and dairy-like emulsions. Compositional considerations. *Journal of Dairy Science* 89: 383–401.

Vithanage, C.R., M.J. Grimson, and B.G. Smith. 2009. The effect of temperature on the rheology of butter, a spreadable blend and spreads. *Journal of Texture Studies* 40: 346–369.

Voda, M.A., and J.V. Duynhoven. 2009. Characterization of food emulsions by PFG NMR. *Trends in Food Science and Technology* 20: 533–543.

Wang, H.H., and D.W. Sun. 2001. Evaluation of the functional properties of Cheddar cheese using a computer vision method. *Journal of Food Engineering* 49: 49–53.

Wang, H.H., and D.W. Sun. 2002a. Correlation between cheese meltability determined with a computer vision method and with Arnott and Schreiber tests. *Journal of Food Science* 67: 745–749.

Wang, H.H., and D.W. Sun. 2002b. Melting characteristics of cheese: Analysis of effect of cheese dimensions using computer vision techniques *Journal of Food Engineering* 52: 279–284.

Wright, A.J., S.S. Narine, and A.G. Marangoni. 2000. Comparison of experimental techniques used in lipid crystallization studies. *Journal of the Association Official Chemists Society* 77: 1239–1242.

Wright, A.J., M.G. Scanlon, R.W. Hartel, and A.G. Marangoni. 2001. Rheological properties of milkfat and butter. *Journal of Food Science* 66: 1056–1071.

Yoon, B., and K.L. McCarthy. 2002. Rheology of yogurt during pipe flow as characterized by magnetic resonance imaging. *Journal of Texture Studies* 33: 431–444.

Index